APPLIED
CRYSTALLOGRAPHY

Proceedings of the XVII Conference

APPLIED CRYSTALLOGRAPHY

Wisla, Poland 31 August – 4 September 1997

Editors

Henryk Morawiec & Danuta Stróż

University of Silesia

Katowice, Poland

World Scientific

Singapore • New Jersey • London • Hong Kong

Published by

World Scientific Publishing Co. Pte. Ltd.
P O Box 128, Farrer Road, Singapore 912805
USA office: Suite 1B, 1060 Main Street, River Edge, NJ 07661
UK office: 57 Shelton Street, Covent Garden, London WC2H 9HE

British Library Cataloguing-in-Publication Data
A catalogue record for this book is available from the British Library.

APPLIED CRYSTALLOGRAPHY
Proceedings of the XVII International Conference

ISBN 981-02-3283-7

Printed in Singapore.

Foreword

This volume contains most of the papers presented at the XVII Conference on Applied Crystallography held in Wisła - Poland from 1 to 7 September 1997.
These conferences started in 1962 when Prof. Zbigniew Bojarski from the University of Silesia organised a meeting dealing exclusively with X-ray powder diffraction and its application for phase identification in different materials. A short history of these conferences written by Prof. Bojarski is presented just below this text.
Since the first one the range of subjects dealt with at the conferences has significantly broadened. The main topics of interest of the present XVII Conference on Applied Crystallography include:
- determination of crystal structure from powder diffraction by the Rietveld method,
- application of the Rietveld method for quantitative phase analysis,
- crystallography computing - software and data bases,
- development of methods and techniques in X-ray, electron and neutron diffraction,
- electron crystallography - crystal structure determination by HREM and electron diffraction,
- crystallography of phase transformations,
- texture studies and microstructure characterisation,
- material structure.

Participants of these conferences do not treat the field of crystallography narrowly i.e. associated only with crystal structure determination. Some of the presented papers represent an appreciable achievement in propagating and developing methods of structure studies applied to materials utilised in industrial applications. Studies of the real structure of materials have become a special mark of these conferences.
The conference was attended by many distinguished specialists who presented their papers on plenary and oral sessions. Also about 100 papers were presented at the poster sessions.
The conference was attended by about 140 participants coming from 23 countries. In my opinion these conferences held in Poland still play an important role in promoting contacts between scientists from West and Eastern European countries, reinforce the existing links and promote new contacts and friendships.
For the third time the conference was accompanied by the Summer School devoted to the Rietveld Method in X-ray diffraction. This school was organised under the patronage of the Commission on Powder Diffraction of the International Union of Crystallography. The International Union of Crystallography was also a sponsor of the school. Additionally the school was sponsored by the Polish Committee for Scientific Research. The organiser and director of the school was Prof. R. Young from Georgia University in Atlanta.
For the first time this conference was also accompanied by the Summer School on Electron Crystallography. The organiser and director of the school was Prof. S. Hovmöller from Stockholm University.
I would like to point out here that it would not have been possible to organise the conference without financial support of different institutions and societies to whom I would like to express my thanks.
Finally, I would also like to thank all the members of the Organising Committee and the Advisory Board who where unsparing in their efforts to prepare the XVII Conference on Applied Crystallography.

Henryk Morawiec

Acknowledgements

The organisers of the conference would like to thank Hotel Stok and Huta Bankowa Ltd. for their generous sponsorship.

BRIEF HISTORY OF THE CONFERENCES

(introductory talk given at the conference)

Z. Bojarski, Honorary Chairman of the XVII Conference on Applied Crystallography
University of Silesia, Katowice

Ladies and Gentelmen,
It's hard to believe that 35 years have already passed since our first Conference in 1962. It was held at the Institute of Ferrous Metallurgy - my place of work at the time. That first Conference that we organised, was a very modest undertaking and I never even anticipated or dreamed of it becoming such a widely known meeting place for crystallographers all over the world. The present one, already the XVII - the Conference on Applied Crystallography is organised jointly by the University of Silesia and the Polish Academy of Sciences. It is my belief that they owe their success to the determination of the organisers and to the need of such meetings in this part of Europe in those days.

The main aim of our first Conference, 35 years ago, was the exchange of experience on the use of diffraction methods for material studies, their application and propagation. At the time we felt a strong need for such an exchange of ideas. Our Conferences made scientific contacts possible - the lack of which we were so painfully aware of. The long period of time, that has elapsed since our first conference and the fundamental changes which have taken place in this part of Europe have not weakened their primary aims. However, the present ranking of our Conferences would not have been possible without the goodwill and support of the International Union of Crystallography and especially its presidents Prof. T. Hahn and Prof. Andre Authier who frequently took part in them. I also wish to emphasise the role of the ICDD. This institution represented by Prof. D.Smith and Dr J. Visser has organised a number of exhibitions and sessions, on the studies carried out by ICDD. These have undoubtedly enriched the programme of our meetings. Thanks to our conferences we have made friends with many distinguished crystallographers and material specialists who, in turn, gave us the necessary stimulus to continue our work. It is their support that made it possible to organise such accompanying events as the summer schools on the development of new methods for material structure studies.
During the last three Conferences we have also run summer schools on the Rietveld Method. These started thanks to Prof. Young's great personal involvement, Prof. Young is now a Foreign Member of the Polish Academy of Sciences.

The history of our Conferences with its traditions is so abundant and stimulating that, in my opinion, it is worth a separate study, apart from my brief speech here.

It is my sincere hope, that the tradition of organising these Conferences will be kept up in the future and that they will not only contribute substantially to the development of crystallography but also help to unite Europe, so deeply divided until recently.

I, personally, wish you all a pleasant and fruitful stay in Wisła and hope that you and your colleagues will participate in our future meetings.

Advisory Board

Organising Committee

Conference sponsors:

Polish Ministry of Education
Polish Committee of Research
Polish Academy of Sciences
Stalexport SA
Steel plant „Bankowa"
JEOL Ltd.
Philips Analytical X-ray

CONTENTS

Structure Determination by X-ray and Electron Diffraction

Applications of Rietveld Method in X-ray Diffraction Analysis

Paracrystallinity

Amorphous Materials

Small Angle Scattering

Texture Analysis

Phase Transformation

Surface and Layers

Materials Structure

Defect Structure

Structure Determination by X-ray and Electron Diffraction

LABORATORY X-RAYS AND STRUCTURE DETERMINATION FROM POWDER DIFFRACTION DATA

D. LOUËR

Laboratoire de Chimie du Solide et Inorganique Moléculaire (UMR CNRS 6511), Groupe de Cristallochimie, Université de Rennes, Avenue du Général Leclerc, 35042 Rennes cedex, France

The use of conventional X-ray sources to collect powder diffraction data for crystal structure determination is discussed. Two diffraction optics, Bragg-Brentano and Debye-Scherrer, with monochromatic X-rays are used for data collection. The successive stages involved in an *ab initio* structure determination are described and the examples studied in the Rennes Group are listed. A few examples are discussed and, finally, the impact of this modern development of the powder method is illustrated by the study of members of chemical families and materials obtained in the course of the decomposition of solids with low thermal stability.

1 Introduction

X-ray diffraction by single crystals has been responsible for important structural results in varied areas of materials science. Nevertheless, for crystalline solids obtained in a polycrystalline form only powder diffraction data are available. Though all information on the structural and microstructural properties of the solid is contained in its powder diffraction pattern, it is only in the last ten years that powder crystallography is appeared as an important technique for structural characterisations. This modern development is now popular as shown by the large number of crystal structures solved *ab initio* from powder data reported to date. The factors having contributed to this advance have been recently reviewed by Langford and Louër [1]. Similarly to the widely and routinely use of single-crystal diffraction data to determine structural properties, powder diffraction data are now employed for the study of powder materials, leading to fundamental and new results on hitherto unexplored compounds. Conventional X-ray sources have played an essential rôle in this progress [2] and, with the advent of integrated software for desk computers [3], the technique is almost used routinely in X-ray laboratories, provided that the structure is moderately complex and that diffraction line broadening arising from structural imperfections is small compared to the instrumental contribution. The aim of this article is to describe the power of conventional X-ray sources for solving *ab initio* crystal structures from X-ray powder data and to illustrate, from a number of examples studied in our Group, the profound implications of these structural research on such diverse scientific disciplines as solid state chemistry and pharmaceutical science.

2 One-dimensional powder data

The major difference between single-crystal and powder diffraction is the loss of information with powder data that results from the rotational projection of the three-dimension location of the reciprocal lattice points on to the one-dimension of the pattern. An immediate consequence is the overlap of the diffraction lines, whose number rises as

d^{*3} ($d^* = 2\sin\theta/\lambda$). This overlap can be exact or accidental. An additional factor contributing to the loss of resolution is the presence of line broadening arising from structural imperfections. The powder pattern is then the result of two mathematical operations. One is the *summation* of $h_j(x)$ components, associated to each projected reciprocal node, contributing to the superposition of the diffraction lines. Hence, the discrete intensity $y(x_i)$ [$= \Sigma_j h_j(x_i)$] at point x_i is given by summing the contributions from neighbouring Bragg reflections at this point. The recovery of the individual components $h_j(x)$ is a *dessummation* operation. The second operation is a *convolution* of the instrumental broadening $g(x)$, including wavelength dispersion, with a profile $f_j(x)$ which contains the information on the microstructure of the sample. The breadth and shape of the individual components $h_j(x)$ [$= f_j(x) * g(x)$] observed in the pattern are then influenced by the microstructure of the sample. Three major advances have contributed to overcome the loss of resolution of the powder data:

- the advent of powerful methods for the geometrical reconstruction of the three-dimensional reciprocal lattice. This is carried out by pattern indexing procedures. Methods and experimental requirements have been reviewed elsewhere [4].
- the introduction of pattern modelling techniques, i.e. pattern decomposition without structure model, used for extracting profile parameters (peak position, breadth, line shape, integrated intensity), and the Rietveld method used for the refinement of the atomic coordinates of an approximate structure model.
- the improvement of instrument resolution, ultra high resolution with X-ray synchrotron sources and high resolution with conventional X-ray sources when monochromatic radiation is employed [5].

3 Instrumentation and experimental considerations

The use of high quality data (high resolution, precision on peak positions and good counting statistics) is essential in structural analyses to minimise the line overlap problem. The relative merits of diffraction optics used for collecting powder data with conventional X-rays have been recently reviewed [1,2,6,7]. Some aspects of the instrumentation commonly used in X-rays laboratories for solving structures from powder data are shortly discussed.

Guinier focusing camera have often been used for structure determination in the early period of *ab initio* structure determination from powder data, due to their high resolution, the accuracy on measured diffraction angles and the current use of monochromatic radiation [8]. Representative examples were the complex structure determination of two lithium tungstates prepared by high-pressure synthesis [9,10]. Powder diffractometers have nowadays largely replaced film cameras in most laboratories. There are different X-ray optics for diffractometers, operating in reflection or transmission modes. Although the former, using the parafocusing Bragg-Brentano geometry, is the most popular, the transmission mode with thin film or capillary presents some definitive advantages which must be considered for collecting powder data prior to start a structural analysis. Indeed, a small amount of sample can be used in the transmission mode and the technique is suitable for materials containing light atoms only. Additionally, with capillaries the preferred orientation effect is usually minimised. Representative examples of structure solved from data collected with the two diffraction modes are listed in Table 1. On the other hand, the widely used reflection geometry is ideal for absorbing materials since there is no absorption correction for a sample with 'infinite' thickness. Although the use of monochromatic wavelength is required in diffraction experiments operating in the angle-dispersive mode, the most widely used radiation in conventional powder diffractometry is still the $K\alpha_{1,2}$ doublet. Two diffraction patterns are thus present simultaneously. They contribute to a scrambling of the pattern as 2θ increases, particularly at moderate and high angles. For an improved resolution, the unwanted radiation can be completely removed at source by means of a focusing monochromator [5]. There is some reduction in the intensity of the $K\alpha_1$ line, but this is partly compensated by using high-power X-ray tubes. Typically, powder data until 150°(2θ) can be collected with good counting statistics within 48 hours. The advantages of using a monochromatic radiation have been listed by Louër and Langford [5]:

- the number of peaks is halved with respect to the $K\alpha_{1,2}$ radiation.
- instrumental line profiles can be readily modelled with a single analytical function.
- the shape of instrumental $g(x)$ profiles varies with angle in a straightforward manner.
- the background is low and is essentially constant over a wide angular range, while it becomes ill-defined at lower angles when the doublet is used.

The instrument most commonly used with conventional divergent-beam X-ray sources is based on Bragg-Brentano parafocusing geometry, whereby the source, sample and receiving slit lie on the 'focusing circle', which has a radius dependent on θ. Coherently scattered X-rays from a flat sample then converge on a receiving slit located in front of the detector. The detector normally rotates about the goniometer axis through twice the angular rotation of the sample (θ/2θ scans). Both the breadth and shape of instrumental line profiles vary continuously with 2θ. As shown by Louër and Langford [5], the instrumental resolution function (IRF), *FWHM versus* 2θ, for a system using monochromatic X-rays, shows a minimum of 0.06°(2θ) at intermediate angles, increasing to twice this value at 130°(2θ). The angular dependence of *FWHM* is usually correctly modelled by the usual quadratic form in tanθ incorporated in Rietveld method software.

In addition to a high instrument resolution, peak positions must also be precise, both for indexing and Rietveld refinement. The quality of the diffractometer alignment can be evaluated from standard reference materials. Differences between observed and calculated peak positions lower than ~0.005°(2θ) can be reasonably obtained with a conventional powder diffractometer properly adjusted [4]. Such a precision is excellent for indexing routinely most powder diffraction patterns for materials with moderate unit-cell volumes (< ~3000 Å^3).

4 Steps in structure determination

The steps involved in the determination *ab initio* of the structure from powder diffraction data have been described in recent reviews [1,11,12]. They are given in outline form below. Though it is recognised that X-ray diffraction is, in general, desirable for solving the structure due to the contrast between atomic scattering factors, the combination with neutron diffraction can be useful at the refinement stage to improve the overall accuracy of the structure.

(a) Data collection

High quality diffraction data from carefully prepared samples are required. If the Bragg-Brentano mode is used a top-loaded sample holder is preferable in order to reduce preferred orientation effects. Tests with varied smoothing effects of the front surface of the sample can give an indication on the sensitivity of the material to a preferred orientation of crystallites. The use of monochromatic radiation is recommended and the alignment of the diffractometer must be checked with standard reference materials (see, for example, ref. 13).

(b) Search for an isostructural phase

Search/match programs, e.g. the powerful approach from Socabim [14-15], to interrogate the PDF database from the ICDD [16] can be used to identify an isostructural material with a known structure. Consequently, a starting model is immediately derived for a subsequent refinement by the Rietveld method (step *h*).

(c) Measurement of peak positions

Precise angular positions are required to index correctly a powder pattern [13]. Pattern fitting techniques without any crystallographic information (see, for example, ref. 17), must be used to extract the Bragg angles. An additional benefit of the analysis is to extract simultaneously the breadth of the diffraction lines. These data can be used to detect the presence of anisotropic line broadening along the pattern. The origin of line broadening (strain, size, stacking faults) can thus be ascertained and used for modelling line breadths in the Rietveld refinement [18].

(d) Unit cell determination

Unit cell parameters and crystal symmetry are found from pattern indexing methods (see ref. 4). Figures of merit are used to check the reliability of the solution. Additionally, all data available in the pattern must be indexed by the solution. A consequence of this data reviewing is the detection of systematic absences and, then, of possible space groups.

(e) Search for an isostructural phase

From the crystallographic data obtained in step *d*, databases can again be interrogated to find a possible structure model from chemically related materials. NIST CDF [16] and ICSD [19] are useful databases for this search. In case of success, the next step is a Rietveld refinement (step *h*).

(f) Extraction of structure factor magnitudes

Individual reflection intensities are extracted from pattern decomposition techniques, using already known crystallographic data, i.e. lattice parameters and selected space group, from which only the positions of symmetry allowed reflections are calculated. Algorithms commonly used have been proposed by Pawley [20] and Le Bail [21]. The principles used in the last method have been employed in some recent programs for extracting integrated intensities [22,23]. The method is also implemented in most Rietveld codes. In these pattern-decomposition methods, the intensity of overlapping lines is divided evenly among contributing reflections. This is a crude approximation, e.g. certain non-equivalent planes for a cubic powder give lines that are exactly superposed on one another, it is then unclear how much each set of planes contributes to the observed lines. This is an important source of information scrambling for subsequent structure analysis. In some cases, the percentage of symmetry-independent reflections is provided [23]. This is a useful indicator prior to continue the structure analysis, since this number can be changed by acting on the angular limit of the pattern used in the fitting procedure. Result of this stage of the analysis is a list of *hkl*'s and their respective intensities, which are used in the next step as far as the structure solution is derived from the traditional Patterson and direct methods. Indeed, an advantage of modern direct-space methods to generate structure models is to avoid the problem of partitioning intensities.

(*g*) *Phase determination*

Using the extracted $|F_{obs}|$ the phase problem is usually solved by means of Patterson or direct methods, from which fragments of the structure are found. The partial model is therefore used to assign probable phases to the structure factors in calculating an electron-density map, from which the remaining atoms are localised. Although the programs used for solving the structure are often based on single-crystal structure determination software, there is an intense activity on structure solution determination from powder data. Harris and Tremayne have recently reviewed the methods used for solving a crystal structure from powder data [12].

(*h*) *Refinement of atomic coordinates*

The approximate coordinates of the structure model are refined by means of the Rietveld method. In this treatment, the instrumental factors as well as anisotropic line broadening must be considered to model correctly the whole powder diffraction pattern. Additionally, in case of high contrasts between atoms, neutron diffraction can also be used. From X-ray and neutron data two different images of the same object are obtained. The propagation of errors is, then, different with the two data sets and a substantial improvement of the precision on the atomic coordinates of light atoms can result from neutron data.

5 The impact of conventional X-rays in materials science

The methodology reported above has been applied in the Rennes Group since 10 years. Most materials were selected from our research program in solid state chemistry, in various fields where single crystals are rarely prepared. The knowledge of the crystal structure always constitutes a great advance to understand the properties of families of powder compounds. In Table 1 are listed the materials for which the structure was solved *ab initio* from conventional X-ray powder diffraction data. Different crystal symmetries (from cubic to triclinic), unit cell volumes (up to 7471 $Å^3$) and complexity (up to 29 atoms in the asymmetric unit) are reported. Bragg-Brentano geometry with monochromatic X-rays were generally used, except for $LiB_2O_3(OH).H_2O$, $C_6H_{10}N_2O_2$ and the low temperature stable form of 2-bromonaphtalene, $C_{10}H_7Br$, for which the capillary technique with monochromatic X-rays was employed. For all these examples the Patterson method or direct methods were used, except for the case of piracetam ($C_6H_{10}N_2O_2$) for which the atom-atom potential method was employed. In the following some noteworthy examples are shortly described as well as the impact of this modern technique to explore families of compounds obtained in a powder form.

Table 1: Summary of crystal structures solved *ab initio* by the Rennes Group from powder diffraction data collected with a conventional X-ray source.

S.G.	Formula	No. Atoms	V (Å^3)	R_F	References*
*C*2/*m*	$Nd(OH)_2NO_3.H_2O$	9	468	0.07	JSSC 68, 292, 1987
*C*2/*m*	$KCaPO_4.H_2O$	7	461	0.05	AC B44, 463, 1988
Pmmn	$Cd_3(OH)_5NO_3$	9	378	0.05	JAC 22, 470, 1989
*P*2$_1$/*c*	$Zn(OH)NO_3.H_2O$	14	757	0.07	JSSC 81, 9, 1989
*C*2/*m*	$Cd_5(OH)_8(NO_3)_2.2H_2O$	10	767	0.05	PD 6, 10, 1990
$P\bar{1}$	$Zr(OH)_2(NO_3)_2.4.7H_2O$	16	531	0.04	JSSC 94, 27, 1991
Pnna	$LiB_2O_3(OH).H_2O$	12	780	0.05	JAC 25, 617, 1992
$P\bar{1}$	$C_9H_5O_5NSCu.2H_2O$	18	518	0.07	NJC 17, 187, 1993
*P*2$_1$/*c*	$C_{18}H_{12}O_2N_2Cu$	12	696	0.05	CM 6, 116, 1994
*P*2$_1$/*m*	$MoO_3.0.5H_2O$	9	248	0.02	JSSC 108, 170, 1994
*P*2$_1$/*c*	$Zr(OH)_2SO_4.3H_2O$	11	672	0.03	JSSC 27, 399, 1994
*Pa*3	$CeMg(NO_3)_6.6H_2O$	7	1986	0.06	EJSSIC 22, 35, 1995
*P*4/*nmm*	$YONO_3$	5	145	0.04	JSSC 112, 182, 1994
$P\bar{4}3n$	$Ge_4S_{10}(N(CH_3)_4)$	15	7472	0.07	CM 6, 827, 1994
$P\bar{1}$	$U(UO_2)(PO_4)_2$	14	417	0.04	CM 6, 1049, 1994
*C*2/*m*	$Zr(OH)_3NO_3$	7	458	0.03	JPCS 56, 1345, 1995
*P*2$_1$/*n*	$C_6H_{10}N_2O_2$	10	723	0.04	AC B51, 182, 1995
$I\bar{4}$	$[(CH_3)_4N]_2Ge_4MnS_{10}$	9	1282	0.05	JAC 219, 111, 1995
*C*2/*c*	$CuTh_2(PO_4)_3$	10	988	0.03	AC B51, 908, 1995
$P\bar{1}$	$C_{10}H_7Br$	r-b LS*	843	0.08	JPCS 56, 777, 1995
Cmca	$U_2O(PO_4)_2$	6	815	0.08	AQ 92, 79, 1996
*P*2$_1$/*m*	α-$La(NO_3)_3.4H_2O$	11	510	0.05	JSSC 126, 127, 1996
Pbca	β-$La(NO_3)_3.4H_2O$	17	2077	0.06	JSSC 126, 127, 1996
*I*4/*m*	$UXPO_4.2H_2O$	8	1426	0.03	JSSC, in press, 1997
$P\bar{1}$	α-$Zr(OH)_2(NO_3)_2.1.65H_2O$	25	743	0.04	JSSC 128, 295, 1997
*P*2$_1$/*c*	$Bi(H_2O)_4(OSO_2CF_3)_3$	29	1837	0.05	CM, in press, 1997
Pbcm	γ-$Zn_2P_2O_7$	14	1089	0.05	in preparation, 1997
Pc	$Fe_2O(OH)_3(NO_3).H_2O$	11	155	0.04	in preparation, 1997

* JSSC: *J. Solid State Chem.*; AC: *Acta Cryst.*; JAC: *J. Appl. Cryst.*; PD: *Powder Diffr.*; NJC: *New J. Chem.*; CM: *Chem. Mater.*; EJSSIC: *Eur. J. Solid State Inorg. Chem.*; JPCS: *J. Phys. Chem. Solids*; AQ: *Anales Quim. Int. Ed.*; r-b LS: Rietveld rigid-body least-squares refinement.

5.1 Some noteworthy examples

The diffraction pattern of lithium diborate hydrate [24], $LiB_2O_3(OH).H_2O$, was indexed with an orthorhombic unit cell [a = 9.798(1) Å, b = 8.2759(7) Å, c = 9.6138(8) Å, M_{20} = 56, F_{30} = 109(0.0081,34), S.G. *Pnna*]. Structure-factor moduli were extracted for $\sin\theta/\lambda < 0.48$ Å^{-1} with the Le Bail pattern-decomposition procedure and the structure solution was obtained with the direct methods. The solution was obtained from both the program MULTAN and SIRPOW.92 [25]. The number of statistically independent observations in the data set was calculated to be 173, which means an independent reflections percentage of 47% in the complete data set. All nonhydrogen atoms were found in one calculation. The final Rietveld refinement converged to the structure model indicator R_F = 0.05. Recently, the crystal structure of $LiB_2O_3(OH).H_2O$ has been solved again from single-crystal data collected up to $\sin\theta/\lambda$ = 0.807 Å^{-1} [26], which is considerably more than the limit of the powder data. The results were essentially the same at each of the successive stages of the structural analysis. However, due to the extended single-crystal data set, a higher precision on atom coordinates was obtained and, on average, the results agreed within 1 combined e.s.d. for x and z and 2 combined e.s.d.'s for y.

The organic compound (2-Oxo-1-pyrrolidinyl) acetamide (piracetam) is a pharmaceutical compound used in human therapeutics. Three polymorphs are known. The structure of the triclinic and monoclinic stable phases have been determined from single-crystal data, showing that the conformation of the piracetam molecule are nearly identical in both phases. The third phase is formed as a result of phase transformation upon heating at 135°C of one of the stable phases. At room temperature, the phase is stable for approximately two hours. Powder diffraction data could, however, be collected with the Debye-Scherrer technique combined with the curved PSD [27]. For organic phases diffraction details usually vanish at moderate angles (~60°2θ). However, pattern indexing and space group assignment, based on Bragg-Brentano data obtained from a thin layer of the material were successful [a = 6.747(2) Å, b = 13.418(3) Å, c = 8.090(2) Å, M_{20} = 25, F_{30} = 49(0.0145,42), S. G. $P2_1/n$]. Due to the short angular limit of the data available the structure solution was obtained from a direct space method, from which structure models are generated, independently of the powder diffraction data. The atom-atom potential method was employed in the structure determination of the metastable piracetam polymorph [27]. The method assumes that the molecule conformation is known and it involves the search for most favourable crystal packing constrained by the known unit cell dimensions and space group symmetry. The method has been applied first to the two piracetam polymorphs with known structures, showing that the optimised solutions were in good agreement with the observed one [28]. It was, then, applied to predict the structure of the metastable phase. As a result of packing calculations, two distinct energy minima were found. However, only one energy structure was found to be correct with respect to the powder data, as shown by the fast convergence of the Rietveld refinement (R_B = 0.04).

5.2 Zirconium hydroxide nitrates

Basic zirconium nitrates are used in the synthesis of gels and of zirconium oxide from thermal decomposition. A number of phases were reported in literature, but no crystal structures was described, due to some phase instability and, also, to the fact that only powders were available. From powder diffraction data collected with conventional monochromatic X-rays, the structure of three phases has been solved *ab initio*. Structure solutions were found from Patterson maps and subsequent Fourier analyses They revealed the interesting crystal chemistry of this class of compounds, based on one- and two-dimensional structure types:

- $Zr(OH)_2(NO_3)_2.4.7\ H_2O$ [$a = 9.541(1)$ Å, $b = 9.590(1)$ Å, $c = 6.753(1)$ Å, $\alpha = 98.64(1)°$, $\beta = 92.89(1)°$, $\gamma = 118.60(1)°$, $M_{20} = 54$, $F_{30} = 55(0.0109,50)$, S. G. $P\bar{1}$] [29]. The structure is based on isolated infinite linear macrocations $[Zr(OH)_2(NO_3)(H_2O)_2^+]_n$, which are balanced by an equivalent amount of nitrate ions. Additional water molecules are located between the chains, which explains the non-stoichiometry in water molecules.

- α-$Zr(OH)_2(NO_3)_2.1.65H_2O$ [$a = 10.051(3)$ Å, $b = 10.357(3)$ Å, $c = 8.131(2)$ Å, $\alpha = 108.30(2)°$, $\beta = 112.02(2)°$, $\gamma = 79.75(3)°$, $M_{20} = 22$, $F_{30} = 40(0.0111,67)$, S. G. $P\bar{1}$] and β-$Zr(OH)_2(NO_3)_2.H_2O$ [$a = 9.180(3)$ Å, $b = 10.465(4)$ Å, $c = 8.076(2)$ Å, $\alpha = 109.58(3)°$, $\beta = 98.96(3)°$, $\gamma = 80.43(3)°$, $M_{20} = 30$, $F_{30} = 64(0.0112,42)$, S. G. $P\bar{1}$] [30]. The two structures are built from infinite neutral zigzag chains $[Zr(OH)_4(NO_3)_2]_n$ with 'free' water molecules located between the chains, with again the possibility of non-stoichiometry.

- $Zr(OH)_3NO_3$ [$a = 9.1146(9)$ Å, $b = 6.7670(7)$ Å, $c = 7.6137(9)$ Å, $\beta = 102.61(1)°$, $M_{20} = 66$, $F_{30} = 99(0.0087,35)$, S. G. *C2/m*] [31]. The structure consists of neutral $[Zr(OH)_6(NO_3)]_n$ layers. Each layer results from the condensation of zigzag chains similar to that observed in the first compound.

5.3 Uranium phosphates

Uranium phosphates were extensively studied in the past, however significant clarifications on the chemistry of these phases were reported only recently (see ref. 32). This is the consequence of the structure determination of a few phases available in a powder form. Indeed, modern powder diffraction has shed new light on this chemistry in revealing the crystal structure and the chemical formula of three major phases:

- $U(UO_2)(PO_4)_2$, a mixed valence compound [$a = 8.8212(4)$ Å, $b = 9.2173(5)$ Å, $c = 5.4772(3)$ Å, $\alpha = 102.622(4)°$, $\beta = 97.748(4)°$, $\gamma = 102.459(5)°$, $M_{20} = 71$, $F_{30} = 120(0.0047,54)$, S. G. $P\bar{1}$] [33]. The structure solution revealed that the previous formula reported in the literature, $U_3(PO_4)_4$, and initially expected in the synthesis was wrong.

- $U_2O(PO_4)_2$ [$a = 7.0878(5)$ Å, $b = 9.0362(5)$ Å, $c = 12.702(1)$ Å, $M_{20} = 57$, $F_{30} = 83(0.0113,32)$, S. G. *Cmca*] [34]. Again, the structure solution showed that the chemical formula previously reported, an uranium oxide diphosphate, was incorrect.

- $UXPO_4.2H_2O$ (X = Cl, Br) [Cl/Br: $a = 14.631(2)/14.7480(7)$ Å, $c = 6.662(1)/6.6810(4)$ Å, $M_{20} = 50/122$, $F_{30} = 74(0.0126,32)/151(0.0052,38)$, S. G. *I4/m*] [35]. The structure is characterised by two kinds of tunnels, in which are located the halogen atoms and half the water molecules, respectively.

5.4 Materials obtained from thermal decomposition

This is the *real world* of powders since the synthesis of single crystals in the course of thermal decomposition is questionable. Powders are then obtained and there is again a need to know the crystal structures to understand the crystal chemistry governing the successive phases obtained until the formation of oxides. On the other hand, it is a favourable situation to generate materials with small crystallite sizes and strain and to observe broadened diffraction lines. Then, there is often a loss of resolution due to structural imperfections. Nevertheless, in a few cases we have been able to solve and refine the structure of decomposition products from powder diffraction data:

- α-$La(NO_3)_3.4H_2O$ [a = 6.7778(9) Å, b = 11.3673(1) Å, c = 6.5843(1) Å, β = 90.644(5)°, M_{20} = 31, F_{30} = 64(0.011,42), S. G. $P2_1/m$] and β-$La(NO_3)_3.4H_2O$ [a = 11.834(1) Å, b = 12.973(1) Å, c = 13.531(1) Å, M_{20} = 60, F_{30} = 85(0.0048,74), S. G. *Pbca*] [36]. The two polymorphs were obtained from thermal decomposition of the hexahydrated phase by acting on the water vapour pressure. The two structures were solved *ab initio* from powder data. The structures consist of chains formed from La polyhedra connected by bridging nitrate groups. An important difference between the two phases is the coordination number of La atom, which is 11 and 10 in the α and β phases, respectively. This result offered the opportunity to discuss the relationships with the structure of the precursor, in which 11-fold coordinated La are also found.

- $Bi(H_2O)_4(OSO_2CF_3)_3$ [a = 18.977(2) Å, b = 8.0034(9) Å, c = 21.918(3) Å, β = 146.503(6)°, M_{20} = 41, F_{30} = 99(0.0076,40), S. G. $P2_1/c$] [37]. This phase has been obtained from the decomposition of the related nonahydrate. The structure was partially solved from the direct methods [25] and completed from somewhat laborious subsequent Fourier analysis. This example is quite remarkable since it is a complex structure containing 29 atoms in the asymmetric unit. Obviously, the number of structural parameters to be refined (87) limited, to a certain extent, the precision on the coordinates of the light atoms. Nevertheless, the structure solution revealed the presence of discrete dimers in which the Bi atoms are 8-fold coordinated by water molecules and triflate groups. This arrangement contrasts with that found in the structure of the precursor.

6 Conclusion

The different examples of *ab initio* structure determination from conventional powder data listed in Table 1 demonstrates the power of the method and its usefulness in solid state chemistry. Clearly, a particular attention must paid to the quality of data and, then, to the quality and resolution of the diffractometer. However, the last studies on decomposition products, where intrinsic diffraction line broadening is observed in the pattern, have shown that, sometimes, satisfactory results can still be obtained in these less favourable cases. On the other hand, some fields of materials science can be totally transformed by the progress occurred in powder diffraction. There are still a few domains of solids to explore, e.g. pharmaceutical compounds. Conventional X-rays can provide a useful tool in these investigations and, in all cases, must be used for a preliminary study of complex structures, prior to use ultra-high resolution facilities offered at the synchrotron sources.

Acknowledgements

Much of this work originated from a collaboration over a period of many years with several collaborators. The author expresses grateful thanks to Prof. M. Louër, Prof. J.P. Auffrédic, Dr P. Bénard-Rocherullé, Miss A.-E. Gobichon and Mr G. Marsolier.

References

1. J.I. Langford and D. Louër, *Rep. Prog. Phys.* **59**, 131 (1996).
2. D. Louër, in *Structure Determination from Powder Diffraction Data*, ed. W.I.F. David, Oxford Univ. Press, in press (1998).
3. M. Louër and D. Louër, *Adv. X-ray Anal.* **27**, 21 (1994).
4. D. Louër, *Accuracy in Powder Diffraction II*, ed. E. Prince & J.K Stalick, NIST Spec. Pub. No. 846 (Gaithersburg MA, USA) pp. 92-104 (1992).
5. D. Louër and J.I. Langford, *J. Appl. Cryst.* **21**, 430 (1988).
6. P.-E. Werner, *Accuracy in Powder Diffraction II*, ed. E. Prince & J.K Stalick, NIST Spec. Pub. No. 846 (Gaithersburg MA, USA), pp. 503-509 (1992).
7. C. Baerlocher and L.B. McCusker, *Stud. Surf. Sci. Catal.* **85**, 391 (1994).
8. P.-E. Werner, *Chem. Scripta* **26A**, 57 (1986).
9. K. Waltersson, P.-E. Werner and K.-A. Wilhelmi, *Cryst. Struct. Comm.* **6**, 227 (1977).
10. K. Waltersson, P.-E. Werner and K.-A. Wilhelmi, *Cryst. Struct. Comm.* **6**, 231 (1977).
11. A.K. Cheetham, *The Rietveld Method*, ed. R.A. Young, Oxford Univ. Press, pp. 276-292 (1995).
12. K.D.M. Harris and M. Tremayne, *Chem. Mater.* **8**, 2554 (1996).
13. D. Louër, *Difrakcia na Polykryštalických Látkach*, ed. L. Smrcok, Vydalo Vydavatelstvo R & D Print, Bratislava, pp. 203-229 (1994).
14. P. Caussin, J. Nusinovici and D.W. Beard, *Adv. X-ray Anal.* **31**, 423 (1988).
15. P. Caussin, J. Nusinovici and D.W. Beard, *Adv. X-ray Anal.* **32**, 531 (1989).
16. International Centre for Diffraction Data, Newtown Square, PA (USA).
17. D. Louër, *Adv. X-ray Anal.* **37**, 27 (1994).
18. R. Delhez, T.H. de Keijser, J.I. Langford, D. Louër, E.J. Mittemeijer and E.J. Sonneveld, *The Rietveld Method*, ed. R.A. Young, Oxford Univ. Press, pp. 132-166 (1995).
19. Inorganic Crystal Structure Database, Karlsruhe (Germany).
20. G.S. Pawley, *J. Appl. Cryst.* **14**, 357 (1981).
21. A. Le Bail, H. Duroy and J.L. Fourquet, *Mater. Res. Bull.* **23**, 447 (1988).
22. J. Rius, J. Sañé, C. Miravitlles, J.M. Amigó, M.M. Reventós and D. Louër, *An. Quím. Int. Ed.* **92**, 223 (1996).
23. A. Altomare, M.C. Burla, G. Cascarano, C. Giacovazzo, A. Guagliardi, A.G.G. Moliterni and G. Polidori, *J. Appl. Cryst.* **28**, 842 (1995).
24. D. Louër, M Louër and M. Touboul, *J. Appl. Cryst.* **25**, 617 (1992).
25. A. Altomare, G. Cascarano, C. Giacovazzo, M.C. Burla, G. Polidori and M. Camalli, *J. Appl. Cryst.* **27**, 435 (1994).
26. M. Louër, D. Louër, E. Bétourné and M. Touboul, *Adv. X-ray Anal.* **40**, in press.
27. D. Louër, M. Louër, V.A. Dzyabchenko, V. Agafonov and R. Céolin, *Acta Cryst.* B**51**, 182 (1995).
28. V.A. Dzyabchenko and V. Agafonov, in *Proc. 28th Int. Conf. on System Sciences*, Vol. 5: Biotechnology Computing, ed. L. Hunter & B.D. Shriver, IEEE Computer Soc. Press, Los Alamitos, CA (1995).
29. P. Bénard, M. Louër and D. Louër, *J. Solid State Chem.* **94**, 27 (1991).
30. P. Bénard-Rocherullé, J. Rius and D. Louër, *J. Solid State Chem.* **128**, 295 (1997).

31. P. Bénard and D. Louër, *J. Phys Chem. Solids* **56**, 1345 (1995).
32. V. Brandel, N. Dacheux and M. Genet, *J. Solid State Chem.* **121**, 467 (1996).
33. P. Bénard, D. Louër, N. Dacheux, V. Brandel and M. Genet, *Chem. Mater.* **6**, 1049 (1994).
34. P. Bénard, D. Louër, N. Dacheux, V. Brandel and M. Genet, *An. Quím. Int. Ed.* **92**, 79 (1996).
35. P. Bénard-Rocherullé, M. Louër, D. Louër, N. Dacheux, V. Brandel and M. Genet, *J. Solid State Chem.*, in press (1997).
36. A.-E. Gobichon, M. Louër, J.P. Auffrédic and D. Louër, *J. Solid State Chem.* **126**, 127 (1996).
37. M. Louër, C. Le Roux and J. Dubac, *Mater. Chem.*, in press (1998).

A TEMPERATURE-DEPENDENT POWDER DIFFRACTION STUDY OF LANTHANUM (POTASSIUM) NITRATES

A.-E. GOBICHON, J.-P. AUFFRÉDIC and D. LOUËR

Laboratoire de Chimie du Solide et Inorganique Moléculaire (UMR CNRS 6511), Groupe de Cristallochimie, Université de Rennes, Avenue du Général Leclerc, 35042 Rennes cedex, France

The power of Temperature-Dependent X-ray Diffraction (TDXD) is illustrated through the study of the thermal decomposition of $La(NO_3)_3.6H_2O$ and $LaK_2(NO_3)_5.2H_2O$. The atmosphere-dependent decomposition processes are described and intermediate phases are identified and characterized. The structures of two varieties of $La(NO_3)_3.4H_2O$, $LaONO_3$, $La_2K_3(NO_3)_9$, cubic La_2O_3 and a new oxide nitrate stabilized by potassium are discussed.

1 Introduction

Temperature-dependent X-ray powder diffraction (TDXD), used with conventional X-ray sources, is a powerful technique which reflects the structural changes taking place in solid phases subjected to external pertubations, e.g. temperature and atmosphere. It is particularly suitable for the identification and characterization of the intermediate phases which arise from the thermal decomposition of oxide precursors. The present study deals with the rôle of TDXD in the elucidation of the successive stages of the thermal decomposition of lanthanum nitrates, i.e. $La(NO_3)_3.6H_2O$ and $LaK_2(NO_3)_5.2H_2O$.

2 Experimental

$LaK_2(NO_3)_5.2H_2O$ was synthesized from an aqueous solution of commercial $La(NO_3)_3.6H_2O$ and KNO_3 [1]. TDXD experiments were carried out with an INEL curved position sensitive detector combined with a high-temperature attachment [2]. The detector was used in a semi-focusing geometry by reflection ($CuK\alpha_1$). Structural studies were carried out from data collected with a Bragg-Brentano diffractometer.

3 Thermal dehydration and decomposition of $La(NO_3)_3.6H_2O$

The thermal behaviour of $La(NO_3)_3.6H_2O$ was extensively studied in the past, but the reported results were contradictory [3]. By using the TDXD method, the decomposition of this precursor was found atmosphere dependent. The complete mechanism of the successive transformations leading to hexagonal La_2O_3 is described in the following.

3.1 Dehydration

The dehydration of $La(NO_3)_3.6H_2O$ into a hemihydrate was carried out using selected water-vapour pressures. The TDXD plots in Fig. 1 show that different mechanisms occur according to the magnitude of P_{H2O}.

The results of powder pattern indexing have been reported elsewhere [4]. The structure of the two polymorphic forms of $La(NO_3)_3.4H_2O$ have been solved from high resolution X-ray powder diffraction data [5]. The two structures consist of infinite chains running along [001], in which lanthanum polyhedra are connected by means of nitrate groups. They are built from 11-fold coordinated (α phase) and 10-fold coordinated (β phase) La atoms. The relationship between the structures of the precursor and these two varieties can be explained by a simple re-arrangement of the framework found in the precursor after the departure of the two water molecules. The formation of chains results from a condensation of individual polyhedra of the precursor through bridging nitrate groups. Furthermore, the structures of the α and the hexahydrate phases are directly related, due to the presence of 11-fold coordinated lanthanum in both compounds.

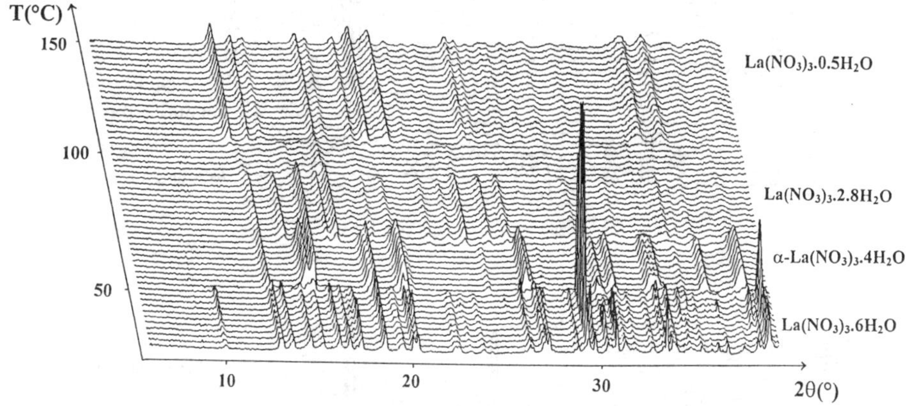

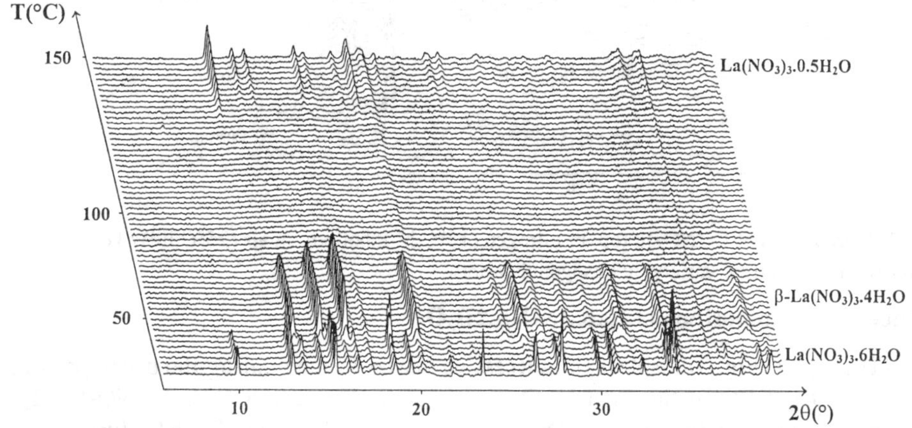

Fig. 1: TDXD plots of $La(NO_3)_3.6H_2O$ in P_{H2O} greater (top) and lower (bottom) than 4.5 Torr (heating rate: 3°C.h^{-1}; 3100 s per pattern).

3.2 Thermal decomposition in nitrogen and in air

The TDXD plots (Fig. 2) show that the decomposition of $La(NO_3)_3$, formed above 200°C, into La_2O_3 proceeds in a different way according to the atmosphere. In nitrogen, $La(NO_3)_3$ yields $LaONO_3$ and $La_3O_4NO_3$, successively. In air, the formation of monoclinic $La_2O_2CO_3$ is observed at 410°C, due to the reactivity of $LaONO_3$ with carbon dioxide from air.

The structure data of $YONO_3$ was used as starting model for the refinement of the structure of $LaONO_3$ by the Rietveld method [4]. Due to the low resolution of data, only an average structure could be described in S.G. *P4/nmm*. It is a layer structure, closely related to the PbFCl-type structure.

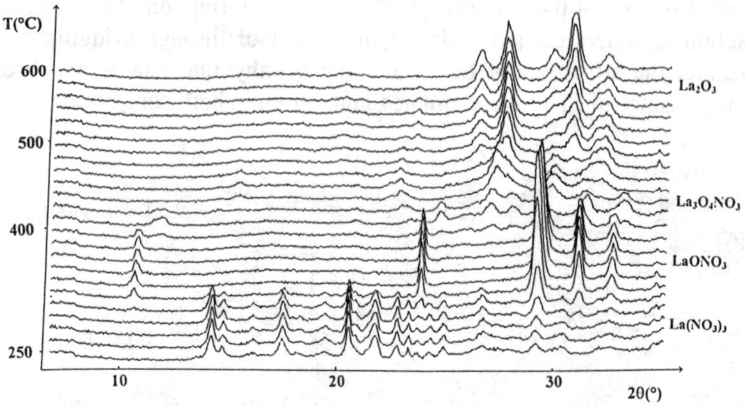

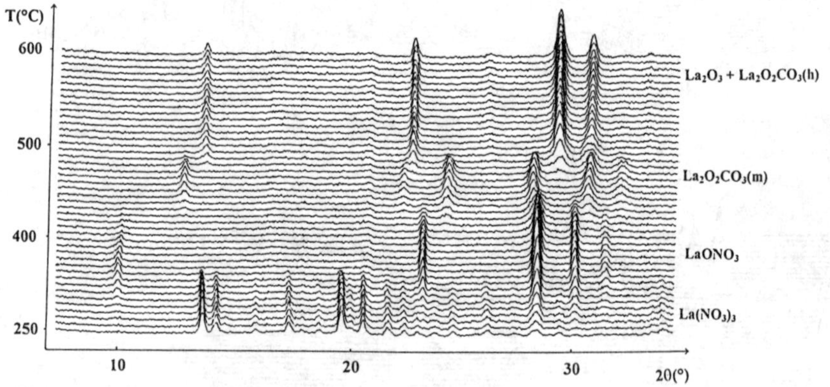

Fig. 2: TDXD plots of $La(NO_3)_3$; top: in nitrogen (18°C.h^{-1}), bottom: in air (25°C.h^{-1}) (3100 s per pattern), m: monoclinic variety, h: hexagonal variety.

4 Thermal decomposition of $LaK_2(NO_3)_5.2H_2O$

The TDXD plot (Fig. 3) displays the successive phases formed during the decomposition of $LaK_2(NO_3)_5.2H_2O$ to the final La_2O_3, in vacuum. Above 115°C, $LaK_2(NO_3)_5.2H_2O$ decomposes into a mixture of $La_2K_3(NO_3)_9$ [a = 13.6607(5) Å], and β-KNO_3, instead of a $LaK_2(NO_3)_5$ phase suggested from TG analysis [3]. The phase obtained above 260°C has an orthorhombic symmetry [a = 4.1818(5) Å, b = 4.2188(7) Å, c = 24.223(2) Å, V = 427.34(7) Å^3]. It undergoes a phase transformation at 357°C, with a slight modification of the cell parameters [a = 4.1819(7) Å, b = 4.2365(7) Å, c = 24.467(3) Å, V = 433.47(9) Å^3]. Subsequently, it decomposes at 440°C into the unexpected cubic La_2O_3 [a = 11.414(3) Å], which, in turn, gives the hexagonal oxide at higher temperature. Structural investigations gave the following results:

i) $La_2K_3(NO_3)_9$ is isotructural with $Ce_2K_3(NO_3)_9$ [6]. The structure is built from a 3-D network of $[La(NO_3)_6]^{3-}$ icosahedra linked through bidentate nitrate groups.

ii) The orthorhombic phase presents some similarities with tetragonal $LaONO_3$. Indeed, the cell parameters *a* and *b* compare well with the parameter *a* of $LaONO_3$ (4.127 Å). On the contrary, parameter *c* (24.223 Å) is much greater than *c* in $LaONO_3$ (9.239 Å). This result suggests that the structure is derived from that of $LaONO_3$ and is described from $(LaO)_n^{n+}$ layers, to which nitrate groups are linked. The insertion of potassium nitrate between the layers could explain the greater interlayer spacing.

iii) The structure of the cubic La_2O_3 was refined by the Rietveld method using the isostructural properties with the known cubic rare-earth sesquioxides. It is built from an arrangement of OLa_4 tetrahedra forming $(LaO)_n^{n+}$ layers.

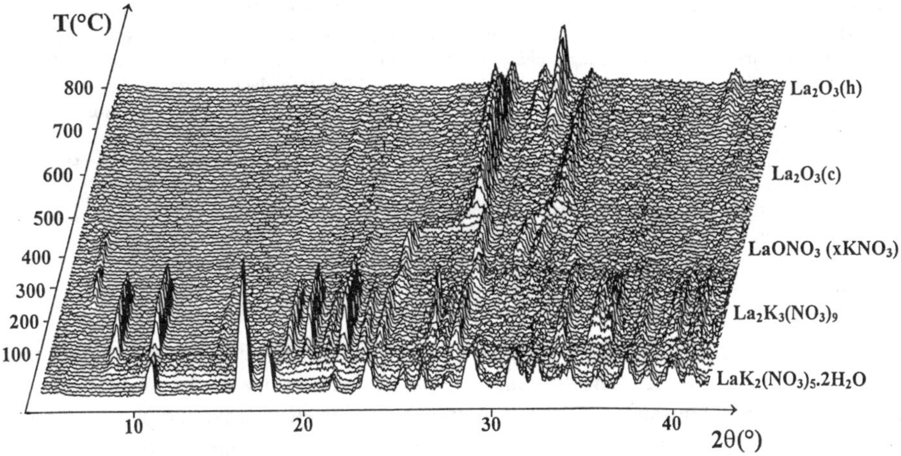

Fig. 3: TDXD plot of $LaK_2(NO_3)_5.2H_2O$, in vacuum (heating rate 20°C.h^{-1}, 1800 s per pattern), c: cubic variety, h: hexagonal variety.

5 Conclusion

The results described above demonstrate that TDXD is a complementary technique to conventional thermal analysis. The thermal decomposition schemes of lanthanum (potassium) nitrates has been completely determined. The structure of five intermediate phases have been described: α- and β-$La(NO_3)_3.4H_2O$, $LaONO_3$, a new phase derived from $LaONO_3$ by incorporation of K atoms between the layers, and cubic La_2O_3. Finally, TDXD has revealed the reactivity of $LaONO_3$ towards carbon dioxide.

Acknowledgments

Grateful thanks are expressed to the *Conseil Régional de Bretagne* for an economical support to one of the authors (A.-E. Gobichon).

References

1. B. Eriksson, *Acta Chem. Scand.* **A34**, 567 (1980).
2. J. Plévert, J.P. Auffrédic, M. Louër and D. Louër, *J. Mater. Sci.* **24**,1913 (1989).
3. M. Karppinen, P. Kyläkoski, and L. Niinistö, *J. Therm. Anal.* **35**, 347 (1989).
4. A.-E. Gobichon, J.P. Auffrédic and D. Louër, *Solid State Ionics* **93**, 51 (1997).
5. A.-E. Gobichon, M. Louër, J.P. Auffrédic and D. Louër, *J. Solid State Chem.* **126**, 127 (1996).
6. N. Guillou, J.P. Auffrédic and D. Louër, *Acta Cryst.* **C51**, 1032 (1995).

SINGLE CRYSTAL STRUCTURE DETERMINATION ON POWDERS BY ELECTRON CRYSTALLOGRAPHY

Sven Hovmöller, Xiaodong Zou and Thomas Weirich
Structural Chemistry, Stockholm University, S-106 91 Stockholm, Sweden

Electron crystallography has great potentials for becoming the method of choice for accurate structure determination of extremely small crystals. The cumbersome step of solving an unknown crystal structure is greatly facilitated by the fact that electrons, unlike X-rays, can be focused into an image. This solves the phase problem of X-ray crystallography. With modern electron microscopes having a resolution of below 2 Å, individual atoms can be determined to an accuracy of about 0.2 Å, which is sufficient for solving an unknown structure. Many crystal structures, ranging from metal oxides over organic molecules to membrane proteins, have now been solved by electron microscopy. For a refinement, phase information is not necessary, but on the other hand very many high quality amplitudes are needed. Electron diffraction data extend to the same resolution as X-ray diffraction. Most inorganic crystals diffract well beyond 1 Å resolution. A complete 3D data set can be obtained by tilting the crystal and taking diffraction patterns along different zone axes. The two most severe problems of powder diffraction are overcome by (selected area) electron diffraction. There is no problem of overlapping reflections, and the indexing of reflections is straight-forward. The first complete structure determination of an inorganic crystal using electrons, including both solving and refining atomic positions was recently accomplished.

1. The phase problem in crystallography

The phase problem is one of the most intriguing phenomena in X-ray crystallography. Among all mathematical problems, the phase problem may be one of the most studied, in terms of man-years. Thousands of crystallographers have tried to solve the phase problem. None has succeeded but many have been successful. Still today there is no general and complete solution to the phase problem in X-ray crystallography, but partial solutions, or solutions of special cases, have been essential parts of at least 8 Nobel prizes in chemistry and physics over the years. The two latest of these were awarded in chemistry in 1982 and 1985. Aaron Klug at the MRC Laboratory of molecular biology in Cambridge received the Prize in 1982 "for his development of crystallographic electron microscopy and his structural elucidation of biologically important nuclei acid-protein complexes"[1]. In 1985 it was awarded to Herbert Hauptman and Jerome Karle "for their outstanding achievements in the development of direct methods for the determination of crystal structures"[1]. These direct methods revolutionised the field of structure determination of organic molecules with up to and even above 100 atoms.

It is easy to get into the idea that the phases are mere mathematical abstractions, since the crystallographic structure factor phases cannot be directly determined experimentally in an X-ray diffraction pattern. The use of words such as "real" and "imaginary" for phases contribute to this feeling. Actually, the phases are very concrete physical entities, as shown in Figure 1. It is quite the other way with the unit cell - it is a concept which is easily understood although it is an abstraction - there are no small boxes in the crystal where the molecules are put inside! One remarkable advantage of electron crystallography is that the enigmatic phases can actually be observed and measured in the electron microscope. This paper will discuss the phase problem and relations between the scattering of X-rays and electrons, real and reciprocal space, images and diffraction. The two questions that need to be answered positively in order to justify the claim that ***single crystal structure determination on powders is possible by electron crystallography*** are:

* Can unknown crystal structures be ***solved*** only from electron microscopy images?
* Can crystal structures be ***refined*** against electron diffraction amplitudes?

2. Electrons and X-rays

If we want to determine the positions of atoms, we need a probe which is able to detect distances of the order of 1 Ångström (10^{-10}m). The three kinds of radiation used in crystallography are X-rays, electrons and neutrons. It is possible to tune the wavelength of each of them, such that an optimal wavelength is used for the experiment. By far most crystallographic studies are carried out using X-rays, with neutrons on second place, at least if we talk about determination of atomic co-ordinates. Neutrons are especially useful because they, unlike X-rays, interact relatively strongly with hydrogen, which is an extremely important element in biology. However, here we will concentrate on electrons and X-rays and their interrelations.

X-rays are created by bombarding a metal plate with electrons accelerated to some 40 kV. If the metal is copper, then the main wavelength (λ) of the X-rays is CuK_α at 1.54 Å, incidentally the same as the length of the C-C single bond, the most important bond for life. When X-rays pass through matter, they interact with the electrons surrounding the atomic nuclei, but not with the nuclei themselves. Yet, since most electrons are located very close to the atomic nuclei, X-rays give information mainly about the positions of atoms, rather than about the more diffuse shapes of valence electrons. The interaction is rather weak - the mean free path length is in the order of a millimetre. If the wavelength is shorter (for example MoK_α with $\lambda = 0.71$ Å), the interaction becomes even weaker.

If the accelerated electrons are produced in an electron microscope, they can themselves be used to study the structure of matter. The most common accelerating voltages for electron microscopes today are 100 kV to 400 kV. The electrons reach a velocity of about half the speed of light and have a wavelength of about 0.04 to 0.02 Å. The electrons interact with the electrostatic potential in matter, *i.e.* they sense the positive and negative charges. All atoms are made up of positive nuclei surrounded by negative electrons. Since electrons are themselves negatively charged, they interact very strongly with all these charges when passing through an object. The electrons are attracted by the positive nuclei and repelled by the electrons. The distribution of charges in solids is such that the positive charges are concentrated in the extremely small nuclei, with most of the negatively charged electrons also located close to these nuclei, partially shielding the positive charges. The electrostatic potential is thus a set of rather sharp positive peaks on a low and relatively smooth background. The mean free path length is very short for electrons, in the order of nanometers. Because of the strong interaction of electrons with matter, extremely small samples can be studied. The optimal crystal thickness is less than 100 Å. A crystal structure determination can be carried out on a crystal of less than 100 unit cells, as compared to 10^{12} to 10^{15} needed for single crystal X-ray diffraction. In fact, even single atoms can be studied by electron microscopy, opening up the world of crystal defects and even amorphous materials for study at the atomic level. As with X-rays, if the wavelength is shorter (higher accelerating voltage), the energy becomes higher but the interaction becomes weaker.

In spite of the facts that X-rays and electrons are so different and interact with different features in matter, in both cases it all boils down to sensing the positions of the centers of atoms. We may therefore expect similar results from structure investigations carried out with X-rays and electrons, and a close interaction between the scientific fields of X-ray crystallography and electron microscopy. For various practical and historical reasons, some of which will be discussed in the following, this is not really the case. But there are strong reasons why science in general and crystallography in particular would gain from a stronger interaction between these two fields.

3. Images, diffraction and phases

The mathematical definition of an image involves a one-to-one relation between the points in an object and its image. One fascinating aspect of the Fourier transform (FT) of an object is that it is not at all an image, yet contains the same information as the image. Each point in an object contributes to every point in its FT and *vice versa*. The FT of a periodic object is especially useful, in that the continuous (analogue) function of the object is condensed into a discrete (digital) function; the Fourier coefficients.

Only the first Fourier coefficient is non-zero on average - it gives the average value of the function. In the case of an X-ray diffraction study it is the F(000) term which is the average electron density within the unit cell. For proteins this term is 0.38 electrons/$Å^3$. For inorganic compounds it may be up to 5 electrons/$Å^3$. All the other Fourier coefficients are cosine waves, only modulating the average electron density, so as to concentrate the density into sharp peaks (where the atomic nuclei are) on a rather smooth background of some 0.01 electrons/$Å^3$. Each Fourier coefficient is characterised by three numbers; the index (h,k,l) giving the direction and periodicity of the cosine wave, an amplitude and a phase. The phase is the position of the maximum of the wave relative to the origin of the co-ordinate system.

If a function is known, the calculation of its FT is straight-forward, although very time-consuming. The FT of a 2D image consisting of 256 x 256 points (pixels) also consists of 256 x 256 points. Each point in the FT is the sum of every point in the image after having been multiplied by a 2D cosine function. Even at the rate of making one such calculation per second, it would take more than a life-time to calculate a 256 x 256 FT by hand. Thus image processing involving Fourier transforms did not emerge as a science until the computers became fast enough. In the late 1960ies, when Klug and co-workers started image processing, a 256 x 256 FT was an over-night job at the single main-frame computer of the university. Today such FTs can be calculated in a fraction of a second on a personal computer. The inverse FT gives back the original function.

From the positions of the reflections we can immediately get their indices. The intensities of the bright spots seen in a single crystal diffraction pattern are proportional to the squares of the Fourier coefficients. However, we can not calculate the inverse Fourier transform, since there is no information about where the maxima of the cosine waves are positioned relative to each other or to the origin of the unit cell. The phase information is lost in a diffraction pattern. If the atomic co-ordinates are known, the FT (with phases) can easily be calculated. The crux is that we do not know the atomic co-ordinates when we start a crystal structure determination. In a powder diffraction pattern the situation is even worse, since the 3D diffraction pattern is projected to a 1D set of diffraction maxima. This adds the problem of ambiguous indices in powder diffraction, which becomes especially severe at high resolution.

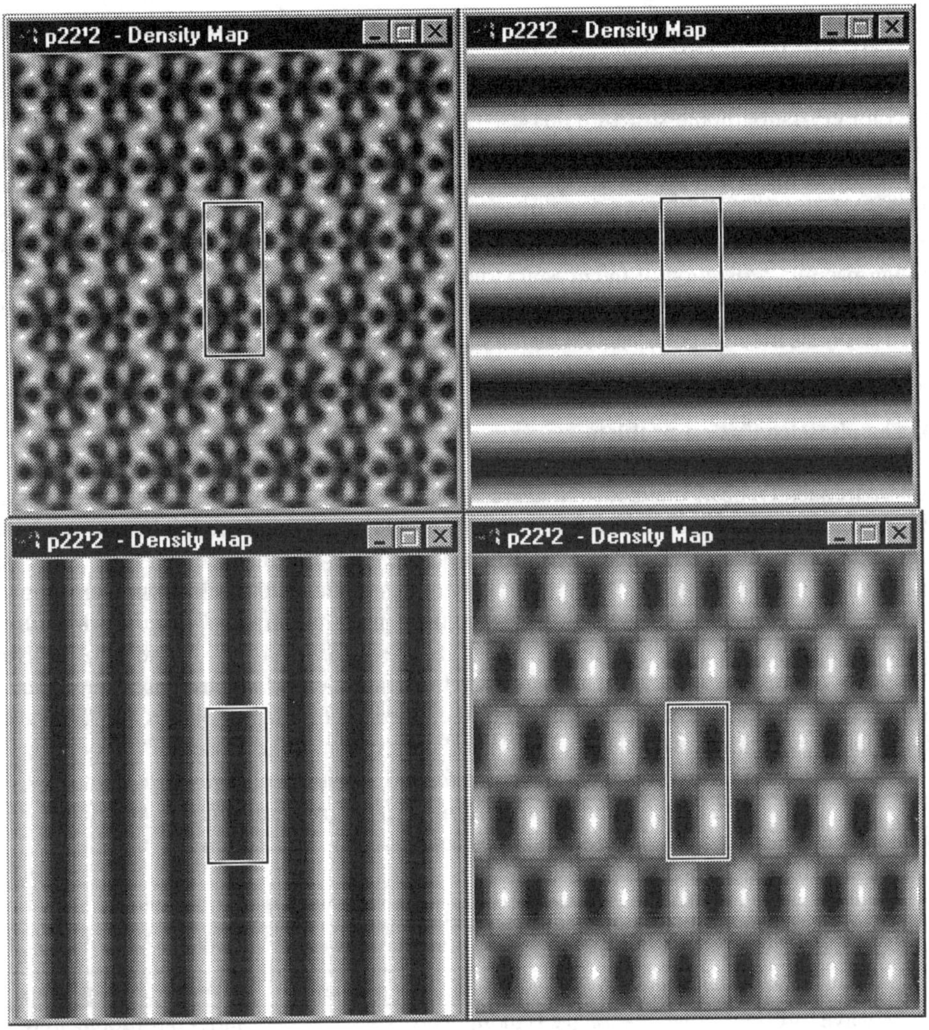

Figure 1. a) An EM image of a metal oxide crystal. Each 5-fold star ★ is the projection of five NbO_6 octahedra surrounding a central NbO_7 pentagonal bipyramid. At this resolution (2.5 Å) each octahedron is seen as a black blob - the oxygen atoms at only 1.9 Å from the metal atoms are not resolved. The crystal is orthorhombic with a projected centro-symmetric symmetry pmg. One unit cell is indicated; a = 9Å, b = 22Å. Atoms show up as black features representing projected potential, which is nearly equivalent to the electron density.

b) The (0 2) reflection has a phase of 180° since its minimum passes through the origin.

c) The phase of (1 0) is similarly 180°.

d) The two symmetry-related reflections (1 1), going \\ and (1 -1) going //, are shown together. Their phases are 0° and 180° respectively. The image in a) is composed of 24 unique reflections. If their phases were not known from an EM image, each of these 24 reflections could have any of the two allowed phase values for centro-symmetric crystals; 0° or 180°, resulting in as many as $2^{24} = 16 \cdot 10^6$ possible combinations. ***A single EM image gives directly the correct phase values of all tnese 24 reflections.***

An image, for example an electron micrograph, is a two-dimensional function. Its FT is easily calculated on modern computers and ***does contain the phases***. That this is really the case should be clear from Figure 1. It is worth emphasising that here the word *phase* is used for crystallographic structure phases. The reason why the crystallographic phases are present in EM images, but not in electron diffraction or X-ray diffraction patterns, is that the different diffracted beams are focused into an image by the lenses in the EM. Each diffracted beam carries information about the relative atomic positions on Bragg planes through the crystal. In the image all the contributing reflections interact, with phases, giving rise to a projection of the crystal structure. Since there are no lenses for X-rays, an X-ray microscope with atomic resolution has not yet been constructed and thus we only have X-ray diffraction patterns, in which the phase information is lost.

4. The phase confusion problem

It has frequently been claimed that there should be a phase problem also in electron microscopy. [2,3]. However, there is no phase problem, only a phase confusion problem in electron microscopy. The phase confusion problem has arisen since the word phase is used for (at least) two very different physical entities in the field of electron microscopy. When crystallographers speak about phases, they mean the ***crystallographic structure factor phases***, while physicists usually mean the ***wave front phases***. In order to resolve the phase confusion problem, it is necessary to define clearly which phases are meant. Some of the most prominent differences between the structure factor phases and the wave front phases are shown in Table 1 below. A full description of the process

crystal structure → structure factors → electron wave →
→ lenses → image formation → structure restoration

with mathematical and physical formulas has been published by Zou [4].

TABLE 1. A comparison of some central properties of crystallographic structure factor phases and wave front phases.

Entity	Crystallographic structure factor phases	Wave front phases
The waves are	stationary standing waves	moving - propagating
Wave length	related to unit cell size, ≥1 Å	≈ 0.03 Å
Physical entity	electron density or potential	electromagnetic wave
Phases relative to	symmetry elements in crystal	incident electron wave front
Phase information	present in EM images	lost in EM images
Method of retrieval	Fourier transform of EM image	holography, focal series EM
Interrelations	modulates the wave front	records the structure factor

5. Small and large objects at high and low resolution

It is illuminating to compare the sizes of objects studied by electron microscopy and X-ray crystallography during different periods of the 20th century. In spite of the much shorter wavelength of electrons (0.02 - 0.04 Å) compared to that of X-rays (0.5 - 2 Å), electrons are actually best suited for studies of large objects at low resolution, while X-rays are ideal for structure determination of smaller structures at atomic resolution.

X-ray crystallography started in 1912 with the determination of exact atomic co-ordinates for inorganic crystals with small unit cells and then gradually developed to handle larger and larger molecules and unit cells. Electron microscopy, on the other hand, started off 20 years later, with a resolution comparable to that of a light microscope, which is useful for studying large systems such as whole cells, bacteria and organelles. As the electron optical lenses were improved, the resolution of electron microscopes improved. It was only in the early 1970-ies that the EM resolution reached values around 5 Å, allowing the visualisation of details within a unit cell of an inorganic crystal. The historical development of X-ray crystallography and electron microscopy can be considered as running in opposite directions, as shown in Table 2.

TABLE 2. Main new categories of structures studied by X-rays and electrons during different periods of the 20th century

Period	X-ray crystallography	Unit cells (in Å)	Electron microscopy	Resolution (in Å)
1910-30	Inorganic crystals	3-10	-	-
1930-50	Organic molecules	5-20	Whole cells and organelles	50
1950-70	Proteins	30-100	Viruses and ribosomes	15
1970-90	Viruses	200-400	Proteins, inorganic crystals	5-2
1990-	Ribosomes	600-1000	Atomic resolution	1-2

6. Electron crystallographic studies of organic and inorganic compounds

Although the scattering of X-rays and electrons are so different, in effect they both probe positions of atomic nuclei. It is now nearly half a century since Vainshtein and Pinsker in 1950 localised hydrogen atoms in paraffin by electron diffraction data [5]. The Moscow group succeeded in the structure determination of several inorganic compounds from electron diffraction data already in the 1950ies. One advantage of electrons pointed out by Vainshtein was that the relative scattering power of lighter elements, including hydrogen, is higher for electrons than for X-rays. The heavy elements are not so heavy for electrons.

There was an unfortunate and long period of almost no structure determinations by electron crystallography outside the Soviet Union, mainly because of exaggerated fears for the effects of multiple scattering of electrons within the crystals. One of the very few who pursued such investigations was Douglas Dorset whose book [6] is the only modern comprehensive treatise on electron crystallography on organic molecules. Dorset has solved numerous lipids and other organic molecules from electron diffraction data and shown that such data can be solved by the direct methods from X-ray crystallography.

In 1975, Unwin and Henderson applied electron crystallography to membrane proteins and were able to solve the 3D structure of bacteriorhodopsin to 7Å resolution [7]. This was 10 years before a membrane protein was solved by X-ray crystallography. These spectacular results inspired a restart of electron crystallography studies also of inorganic compounds, but now using the full power of electron microscopy and image processing rather than only electron diffraction as had been done by the Moscow group. Several metal oxides were solved by electron microscopy and image processing, giving atomic co-ordinates of the metal atoms accurate to within about 0.1 Å [8]. The first refinement of an inorganic structure against selected area electron diffraction data to better than 1 Å resolution, reaching an accuracy of 0.02 Å was recently completed by Weirich *et al.* [9].

7. Comparison of phases obtained by electron microscopy and X-ray diffraction

As shown above, electron microscopy can give structure factor phases. If these phases are similar to the X-ray structure factor phases, they may be used as a starting set for phase extension, for example by direct methods or maximum entropy [10]. Even if only a few low-resolution reflections can be found, this can be very helpful. If reliable phases can be determined to a resolution of about 2 Å or so, then the complete solution of a crystal structure should be straight-forward.

We must ask: how similar are the amplitudes and phases obtained from the two techniques X-ray and electron crystallography? Can phases obtained by electron microscopy be used for phasing X-ray diffraction data sets? Strangely enough, it turns out that very little work has been carried out trying to answer these questions.

For inorganic structures, it has been shown that indeed the phases obtained from electron microscopy images, are nearly identical to the ones obtained after refinement of X-ray crystallography data on the same crystal, at least to 2.5 Å resolution [11].

8. Crystal structure refinement against electron diffraction data

The first unknown inorganic crystal structure to be solved and refined by electron crystallography was the metal cluster compound $Ti_{11}Se_4$ [9]. Attempts to grow single crystals suitable for X-ray diffraction and to solve the structure from X-ray powder diffraction had both failed.

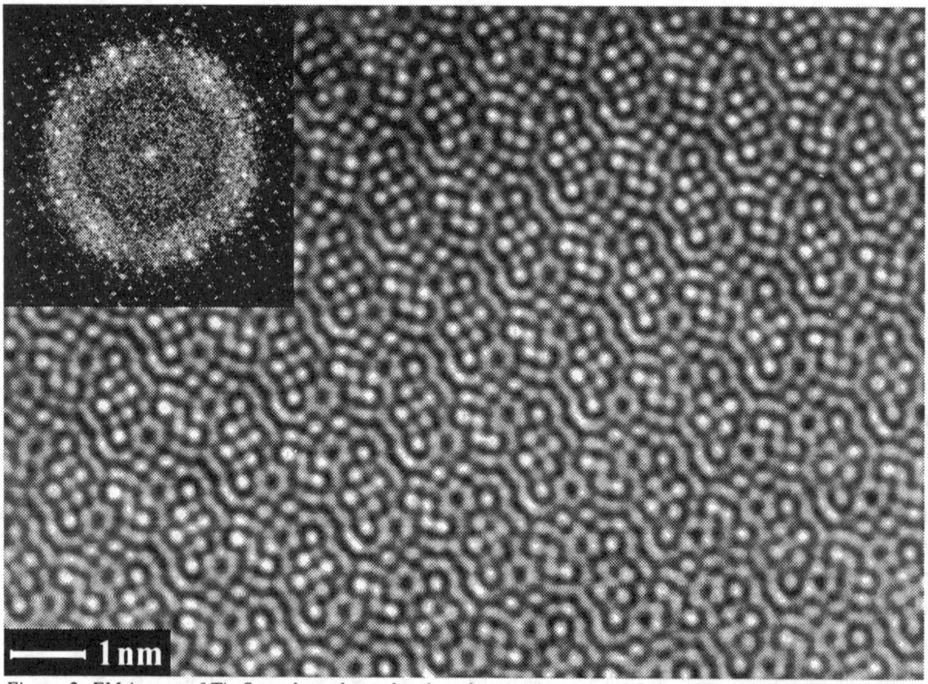

Figure 2. EM image of $Ti_{11}Se_4$ taken along the short b-axis. An area of 512 x 512 pixels (corresponding to 190Å x 190Å) at the edge of the crystal was selected and its Fourier transform (FT) calculated (inset). The dark ring in the FT marks a cross-over in the contrast transfer function (CTF). All phases are reversed, i.e. shifted 180 degrees, outside such a cross-over. The lattice of diffraction spots was refined and amplitudes and phases extracted from this calculated FT.

The crystal symmetry was found to be *C2/m*. The unit cell dimensions were a = 25.516, b = 3.4481, c = 19.201Å, β = 117.84° (refined from X-ray powder diffraction). HRTEM images (Fig. 2) and SAED patterns were taken, along the short b axis, on a 300 kV Philips CM30 electron microscope. The photographic negatives were digitised by a video-rate CCD camera. The 23 unique atoms were all found from one electron microscopy image (1.75 Å resolution) after applying crystallographic image processing (Fig. 3), using the computer program system CRISP [12]. The defocus value was determined to be -930 Å from the radius of dark ring in the calculated Fourier transform of the image.

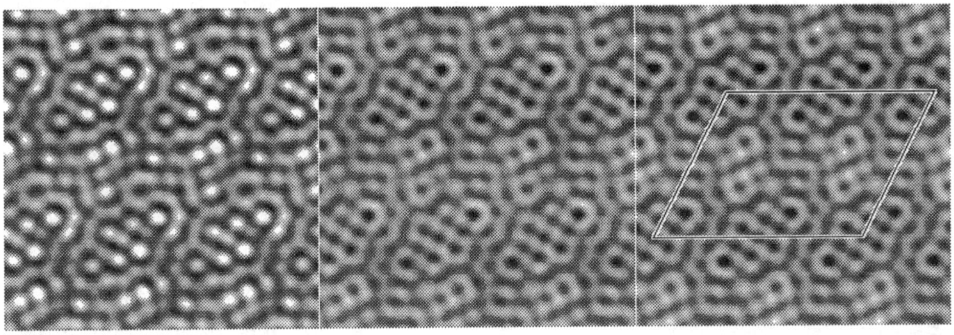

FIG 3. Projection maps of $Ti_{11}Se_4$ at various stages of image processing. (a) Only lattice averaging. Here white dots are atoms (reversed contrast). (b) After compensating for the phase cross-over in the CTF, the contrast becomes correct with black atoms. (c) A further improvement is obtained by imposing the crystallographic symmetry in projection, p2. All the 90 atoms in the unit cell (marked) were localised from this map.

Projected potential maps (Fig. 3) were calculated using all 113 diffraction spots out to 1.75 Å resolution, by inverse Fourier transformation of the amplitudes and phases extracted from the FT [9].

The atomic co-ordinates were refined to e.s.d. 0.02Å using selected area electron diffraction (SAED) data to 0.75 Å resolution, recorded on photographic film (Fig. 4). A range of exposure times from 0.5 to 7.5 seconds were used, and the data was quantified by the ELD program [13]. Diffraction data of similar quality was obtained using a slow-scan CCD camera. There were 113 unique reflections to 1.75 Å resolution in the Fourier transform of the HRTEM images, whereas 408 reflections out to 0.75 Å could be quantified from the SAED patterns. The 23 atomic co-ordinates (x and z) and temperature factors were refined against all 408 reflections from SAED. SHELXL-93 [14].the standard program for least-squares refinement in X-ray crystallography was used, but with electron scattering factors for Ti and Se, giving a crystallographic R-value of 14.7%. The estimated standard deviations were close to 0.02 Å for all atoms. As can be seen in Fig. 5, the structure of $Ti_{11}Se_4$ has great resemblances with that of Ti_8Se_3 which has been solved by single crystal X-ray diffraction [15]. This renders further evidence of the reliability of crystal structure determination using electrons.

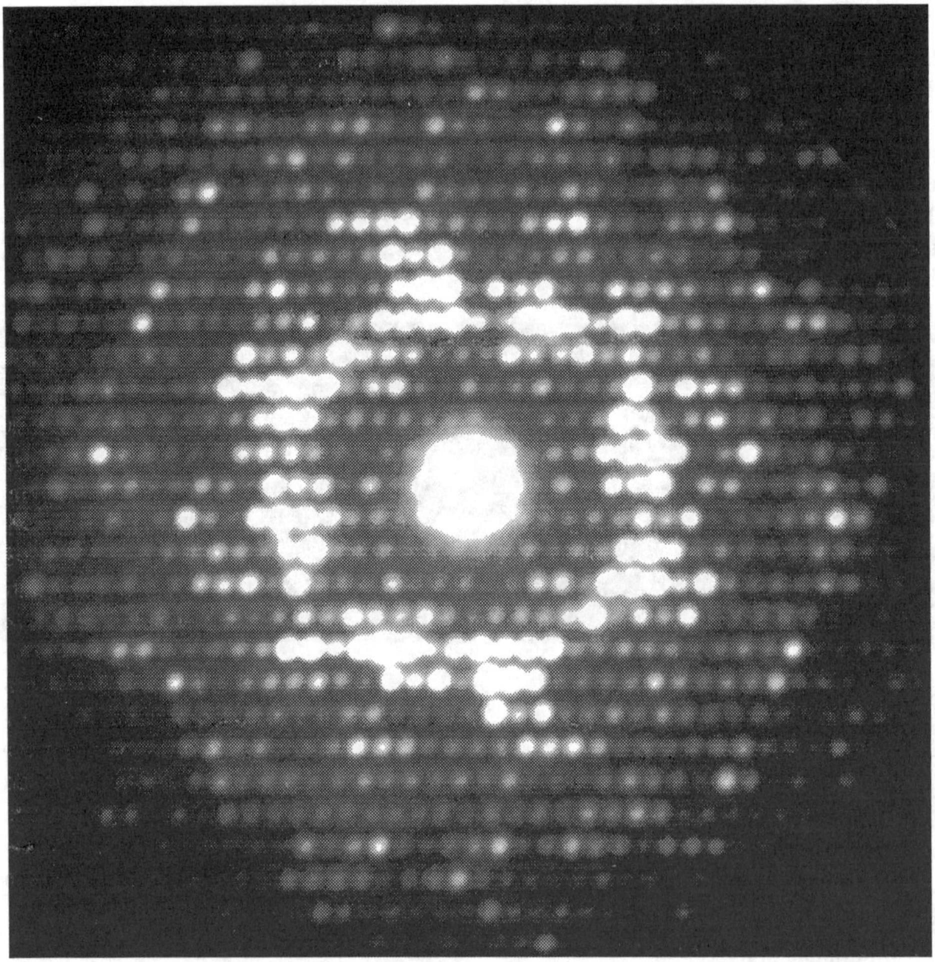

FIG 4. Electron diffraction pattern of $Ti_{11}Se_4$ along the b-axis. 408 unique reflections to 0.75 Å resolution were collected from this projection.

In order to verify that structures obtained from HRTEM images after crystallographic image processing can be trusted, even for inorganic compounds with heavy elements, another seven related structures which had already been solved by X-ray crystallography were studied [16].These were Ti_9Se_2, Ti_2Se, Ti_2S, Zr_2Se, Ta_2P, Ti_8S_3 and Ti_8Se_3. In all cases every atom was found, with an average error in atomic co-ordinates of 0.2Å.

One may wonder if the low e.s.d.s obtained for $Ti_{11}Se_4$ (0.02 Å) are just numbers coming out of the refinement program. How can we be sure these atomic co-ordinates are within 0.02 Å from the true positions? In order to answer this question of accuracy of the method, we are now refining some structures that have been refined to high accuracy by X-ray diffraction. Preliminary results on Ta_2P show that the heavy Ta atoms (atomic # 73) refine to within 0.02 Å of the X-ray co-ordinates. The lighter P atoms (atomic # 15) are more difficult to refine in the presence of the heavier Ta atoms.

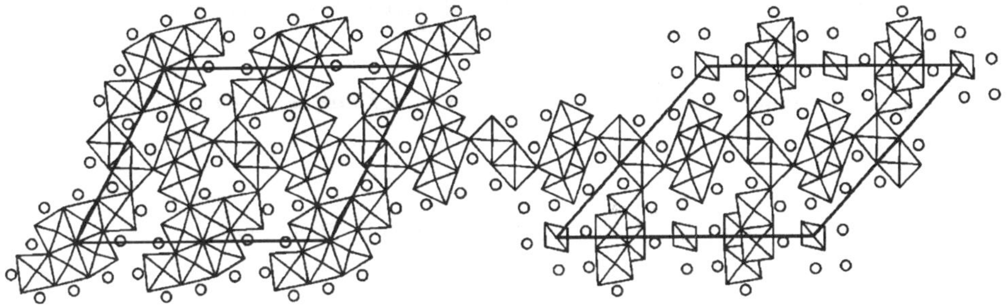

FIG. 5. Structure model of (left) $Ti_{11}Se_4$ obtained by electron crystallography and (right) the related structure Ti_8Se_3 solved by single crystal X-ray diffraction ⊠ = Ti_6 octahedral clusters and ● = Se. The central layer of alternating strings of two and four Ti octahedra parallel to the ab plane is identical in the two structures.

9. Concluding remarks

After nearly one hundred years, the circle is closed; both electrons and X-rays can now be used for crystal structure determination in all fields of chemistry, from inorganic solids over organic molecules to proteins, with comparable accuracy. Both electron crystallography and X-ray crystallography may gain from a combined approach, for example in the study of complicated crystal structures which form only very small crystals, powders, which are difficult to solve by X-ray diffraction and refine by the Rietveld method.

10. Acknowledgements

This research was carried out with financial support from the Swedish Natural Science Research Council (NFR). TW thanks Wenner-Gren Foundation for a post doctoral grant.

11. References

1. http://www.nobel.se/ Home page of the Nobel Foundation, Sweden
2. D. Van Dyck, and M. de Beeck, *Scanning Microscopy Supplement* **6**, 115 (1992).
3. BRITE-EURAM Project No. BE-3322 *Ultramicroscopy* **64** Entire volume (1996).
4. X.D. Zou, *Electron Crystallography of inorganic structures - Theory and practice* Chem. Commun. 5. Stockholm University. Doctoral thesis (1995).
5. B.K. Vainshtein and Z.G Pinsker, *Dokl. Akad. Nauk SSSR* **72**, 53 (1950).
6. D. Dorset, *Structural Electron Crystallography,* Plenum Press, New York (1995).
7. R. Henderson and P.N.T. Unwin, *Nature* **257**, 28 (1975).
8. S. Hovmöller In: Electron Crystallography, Trans. of the American Crystallographic Association. Ed. D.Dorset. **28**, 95 (1992).
9. T. Weirich, R.Ramlau, A.Simon, X.D. Zou and S. Hovmöller, *Nature* **382**, 144 (1996).
10. W. Dong, T. Baird,, J.R. Fryer., C.J. Gilmore, D.D. MacNicol, G. Bricogne, D.J.Smith, M.A. O'Keefe, and S. Hovmöller, *Nature* **355**, 605 (1992).
11. X.D.Zou, M. Sundberg, M.Larine, and S. Hovmöller. *Ultramicroscopy* **62**,103 (1996).
12. S. Hovmöller *Ultramicroscopy* **41**, 121 (1992).
13. X.D. Zou. Y. Sukharev and S. Hovmöller, Ultramicroscopy **52,** 436 (1993).
14. G.M. Sheldrick, *Acta Cryst.* **A46,** 467 (1990).
15. T. Weirich, R. Pöttgen, and A. Simon, *Z. Kristallogr.* **212,** 929 (1996).
16. T. Weirich, Ph.D. thesis, Univ. Osnabrück (1996).

X-RAY DIFFRACTION STUDY ON K_2ZnCl_4 DOWN TO10 K

J. KUSZ
Institute of Physics , University of Silesia, Katowice,. Poland

H.BÖHM

Institut für Geowissenschaften, Universität Mainz, D-55099 Mainz, Germany

1 Introduction

K_2ZnCl_4 belongs to the A_2BX_4 structure family exhibiting a sequence of transitions from a normal to an incommensurate and to a commensurate phase. This sequence is repeated at lower temperatures a second time with the modulation direction perpendicular to the high temperature modulation. The idealised, basic structure at high temperatures is shown in Fig. 1a (a_0 = 7.30 Å; b_0= 12.54 Å; c_0=8.96 ÅK SG=Pmcn R=8.7% [5]). This model does not account for the dynamics in the structure. In fact, a split model had to be introduced in order to obtain a satisfactory agreement with the observed data (R= 4.1% [5]). Fig 1b exhibits the two equilibrium positions, uncorrelated thermal oscillations occur between them. The realistic model for the N-phase is a disordered one.

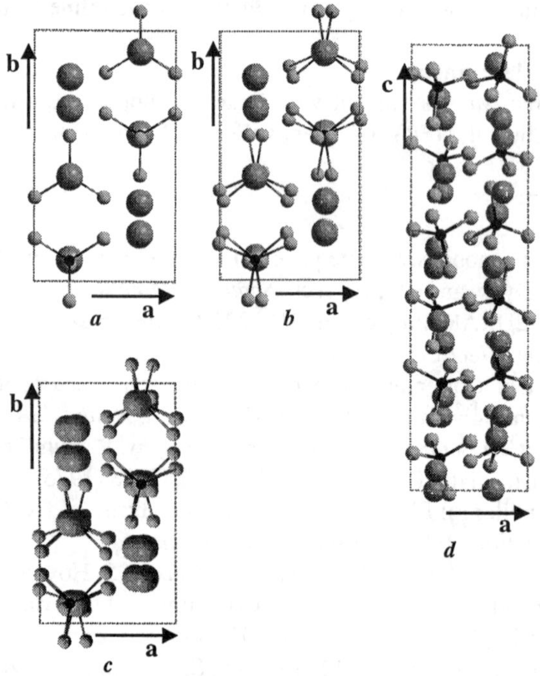

Fig. 1 The structures of K_2ZnCl_4 : the idealised, ordered N-phase (*a*); the realistic, disordered N-phase (*b*); the commensurate C-phase, projection along the **c**-(*c*) and the **b**-axis (*d*).

In Fig. 2 a schematic of the sequence of phases is shown. Beside the phases in an tentative p-T phase diagram also the characteristic unit cells in the reciprocal space are depicted. The phase transition from the high temperature N-phase to the IC-phase with a wave vector $q_1 = (1/3-\delta_1)c^*$ occurs at 553 K. It is an order/disorder transition with the onset of a correlated motion along the **c**-direction in the IC-phase. The phase transition from the IC- into the C-phase is found at 403 K [1] (wave vector $q_1 = 1/3c^*$). The 3-fold superstructure at 155 K [10] is shown in Fig. 1c-d.

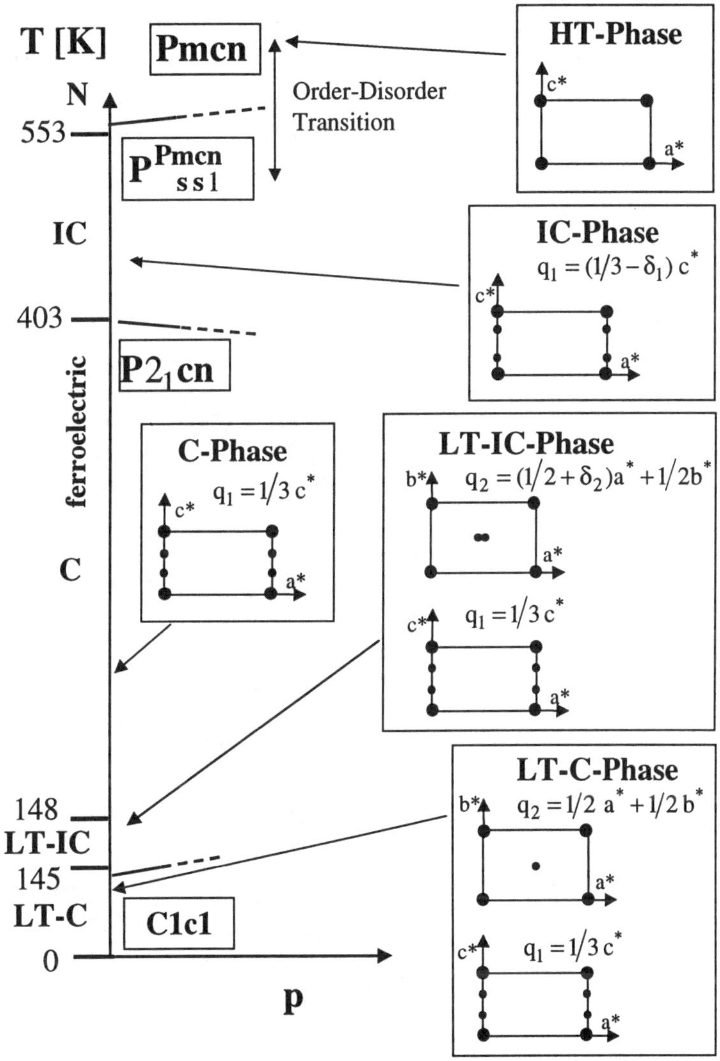

Fig.2 The sequence of phases of K_2ZnCl_4 and schematics of the corresponding patterns in the reciprocal space

Another commensurate to incommensurate transition was found by neutron scattering [6]. Apparently this LT-IC-phase exists in the range between 148 K and 144.5 K with a wave vector of $q_2 = (1/2+\delta_2)a^*+1/2b^*$. Below 144.5 K δ_2 vanishes and the commensurate phase with doubled a- and b-lattice constants as compared with the basic structure is formed. The transitions and the structures have been studied by several authors [1-6]. A few open questions remained concerning the transitions into the LT-phases. The LT-IC-phase manifested by the occurrence of the split satellite reflections has only been verified by neutron scattering [4,5], however, diffuse q_2-reflections have been observed using synchrotron radiation [8]. The phase transition to the LT-C-phase is of first order but the space group of the phase has been proposed as **C1c1** assuming that the criteria of the Landau-theory for second order phase transititons can be applied [2]. The LT-C-structure has been refined at 140 K in the monoclinic SG **C1c1** [7], however, the monoclinic angle ß was found to be 90°. The present paper is focussed at the question whether the LT-C-phase is monoclinic or not.

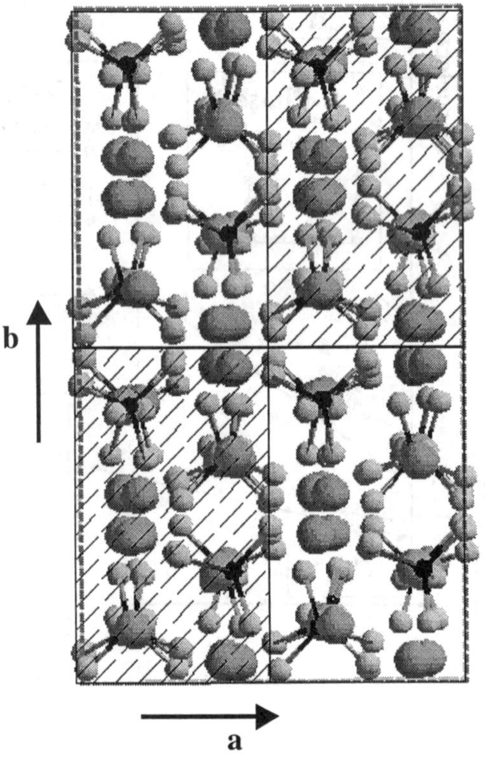

Fig. 3 The LT-C-structure , a projection along the **c**-axis.

2 Experimental

The crystals were grown from an aqueous solution as described in [10]. We suspect that the resolution of the diffractometer in a previous experiment [10] was not sufficient. Therefore in the present study a four-circle HUBER goniometer (graphite monochromator, CuKα radiation) with an 250mm χ-circle and a distance crystal-detector equal to 425 mm was used. The X-ray source was an ENRAF-NONIUS FR571 rotating anode; for cooling a two-stage closed-cycle helium cryostat (CTI-CRYOGENICS) was used in order to perform experiments between 10 K and room temperature. The STOE driving software (STADI4) is well adapted to allow angle scans but also Q-scans and more dimensional Q-scans.

3 Results and discussion

In a previous experiment [10] the attempt was made to prove the existence of monoclinic domains with a monoclinic angle very close to 90°. If such domains should be twinned with respect to a mirror plane the superimposed reflections would cause a broadening in the **a***- and **c***-direction but not along **b***. Another cause for broadening is the formation of microdomains which cause a broadening because of the small particle size. Such a broadening had been found, in deed, for the isostructural K_2CoCl_4 [9]. Using a conventional diffractometer, however, a similar result could not be obtained for the K_2ZnCl_4 [10].

Also in the present experiment no splitting of the q_2-reflection was observed on cooling. However, when performing ω-scans as a function of temperature a significant broadening of the peak width is observed as shown, for example, for the (3 3 0) reflection (Fig. 4). This is an indication that microdomains are formed at the phase transitions into the LT-C-phase. If, in addition, also Q-scans are carried out as a function of temperature along the three basis vectors in the reciprocal space a significant difference between the FWHM in the **b***-direction and perpendicular to it is observed. Perpendicular to **b*** the FWHM is greater by a factor of about 4. This can be explained by the formation of microdomains which have very small dimensions in the monoclinic (a-c)-plane.

The LT-C-structure is characterised by tilts of the chlorine-tetrahedra which are correlated along the **a**- and **c**-direction which result in a twofold superstructure in both directions as compared with the basic structure. This superstructure is achieved from the C-phase in two steps. In a first step the correlations only lock to the lattice along the **b**-direction, along the a-direction they have a long incommensurate period (LT-IC-phase). In a second step (first order phase transition) the correlations along **a** also lock to the lattice yielding a twofold superstructure in the **a**- and **b**-direction (Fig. 3). It is found that this superstructure is C-centred, because the superstructure reflection only appears at the C-centering position. In Fig. 3. the C-centering is indicated by differentiating between hatched and non-hatched areas. Thus, identical subcells form a chess board pattern.

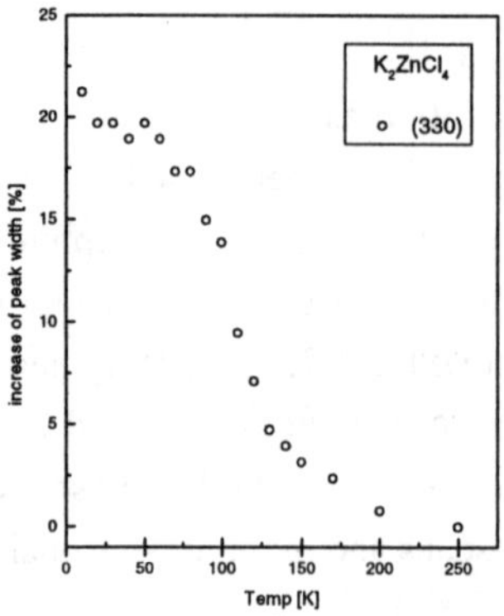

a

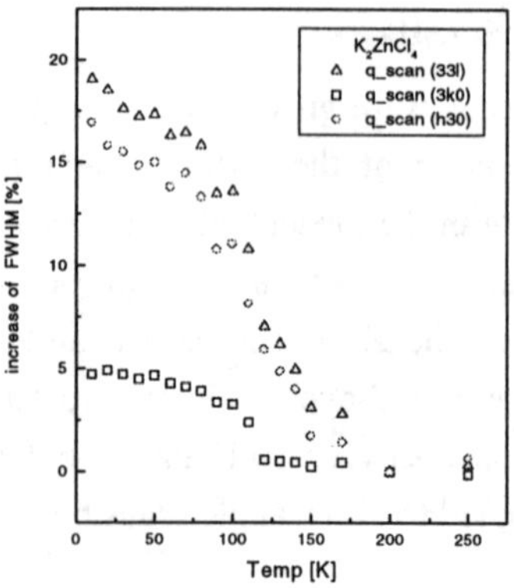

b

Fig. 4 The peakwidth as a function of temperature : an ω-scan (a) and Q-scans in the three main directions (b) of the (3 3 0) reflection

Acknowledgment

The financial support of the *Materialwissenschaftliches Forschungszentrum der Universität Mainz* is gratefully acknowledged

References

1. K. Gesi, *J. Phys. Soc. Jpn*, **45**, 1431 (1978)
2. V. Dvorak et al. , *Phys. Status Solidi*, **107**, K109 (1981)
3. I. Mikhail et al., *Acta Cryst.* **B35**, 1200 (1983)
4. J. Kusz, *Thesis*, University of Silesia (1988)
5. M. Quilichini et. al., *J. Phys. Condes. Matter*, **2**, 4543 (1990).
   J. Phys. I France, **1**, 1321-1323 (1991).
6. K. Gesi, *J. Phys. Soc. Jpn*, **59**, 416-419 (1990)
7. H. Mashiyama, *Acta Cryst.*, **C49**, 9 (1993).
8. Hasebe et al., *J. Phys. Soc. Jpn.*, **63**,3340-3343 (1994)
9. J. Kusz et al., *Phase Transitions*, **56**, 147-152 (1996)
10. J. Kusz et al. *Phase Transitions*, **60**, 1-13 (1997)

THE PHASE DIAGRAM OF THE K_2MoO_4 - K_2WO_4 SYSTEM

B. BZOWSKI, J. KUSZ, H. DUDA AND J. WARCZEWSKI
University of Silesia, Institute of Physics, ul. Uniwersytecka 4, 40007 KATOWICE, POLAND

Single crystals of K_2MoO_4 and K_2WO_4 were grown at room temperature from the filtered solutions of H_2MoO_4 resp. H_2WO_4 in the concentrated solution of KOH. These crystals are isomorphous and belong to the monoclinic system. From the measurements with the aid of the 4-circle diffractometer KM-4 the crystal structure and lattice parameters of the compounds $K_2Mo_xW_{1-x}O_4$ (for x=0.0, 0.5, 1.0) were determined. To study the structure modulation in the intermediate compounds the phase diagram of the K_2MoO_4 - K_2WO_4 system was obtained.

1. INTRODUCTION

The compounds K_2MoO_4 and K_2WO_4 are isomorphous and monoclinic at room temperature [1,2], while their high-temperature phase is shown to have modulated structure which transforms in higher temperature in the hexagonal phase without modulation [3]. Till now the room-temperature structure for both compounds [3,4] and high-temperature structure of the K_2MoO_4 [5] have been determined. The aim of the present work is to determine the structure of the intermediate compounds as well as the phase diagram.

2. EXPERIMENTAL

The $K_2Mo_xW_{1-x}O_4$ compounds, where $0 \leq x \leq 1$, were obtained by dissolution of stoichiometric mixtures composed of x MoO_3 and (1-x) WO_3 oxides in the concentrated KOH solution used in excess. This exothermic reaction runs according to the following formula:

$$MeO_3 + 2KOH = K_2MoO_4 + H_2O, \text{ where Me=Mo,W.}$$

The resulting solutions have strongly alkaline reaction and readily react with CO_2 from the air, therefore the synthesis was carried out in the protective and desiccative atmosphere of nitrogen. In this manner we have obtained the single crystals of the compounds with the following concentrations: x=1.0, 0.75, 0.5, 0.25, 0.0. Although most of the crystals were twinned, it was possible to select a few crystals of each composition, which were sealed in the Lindemann capillaries and some of them measured with the aid of the four-circle KM-4 diffractometer.

The powder samples were obtained from the single crystals using the procedure including the filtering, washing up in ethanol and drying in the inert atmosphere for a few days. Powder diagrams taken from these samples with the HZG-4 diffractometer exhibited only the peaks generated by potassium molybdate and potassium tungstate in agreement with the data base PDF-1. These samples were measured with the aid of the DSC-7 differential scanning calorimeter.

3. RESULTS

For each composition the phase transition temperature in the range from 50 to 500°C was measured in two cycles, i.e. heating up and cooling down. The results for some compositions are shown in Figures 1, 2. One can see, that the temperature of the first phase transition increases with the change of the composition (from 335.9°C for K_2MoO_4 to 384.6°C for K_2WO_4) in the contrast to the second phase transition, where such a correlation does not take place (Fig. 1). Both cycles confirmed the hysteresis of the phase transitions [4]. The following phase sequence in the first cycle mentioned above, is possible: monoclinic- modulated orthorhombic- nonmodulated hexagonal. For the second cycle the following two sequences of phases are possible (Fig. 2):
-as described in [4]: nonmodulated hexagonal-modulated orthorhombic which transforms after several days into a monoclinic one;
-as described in [6]: nonmodulated hexagonal-modulated orthorhombic which transforms into a monoclinic one at about 200°C[6].
The phase diagram (Fig. 3) confirms hysteresis of the phase transitions also for the intermediate compounds. For the concentrations x=1.0, 0.5, 0.0 every crystal structure was refined in two space groups: C2 and C2/m. The results are:

x= 0.0	C2 R= 0.0438	C2/m R= 0.0373		
	a= 12.346(3) Ä	b=6.093(2) Ä	c=7.536(2) Ä	and β=115.75(3)°
x= 0.5	C2 R= 0.0536	C2/m R= 0.0499 (see Tables 1, 2)		
	a= 12.349(2) Ä	b=6.088(2) Ä	c=7.540(2) Ä	and β=115.77(2)°
x= 1.0	C2 R= 0.0747	C2/m R= 0.0731		
	a= 12.373(1) Ä	b=6.116(1) Ä	c=7.550(1) Ä	and β=115.94(1)°

One can then conclude, that symmetry of the intermediate compound is the same as that of the boundary compounds.

4. LITERATURE

[1] Gatehouse B.M., Leverett P.,(1969) J.Chem. Soc.(A) 849.
[2] Koster A.S., Cools F.K.N.M., Rieck E.D.,(1969) Acta Cryst.B.25,1704.
[3] Berg A.J.van den, Tuinstra F., Warczewski J.(1973) Acta Cryst.B.29,586.
[4] Warczewski J.(1979), Phase Transitions 1(2), 131.
[5] Berg A.J.van den, Overeijnder H., Tuinstra F. (1983), Acta Cryst.C.39,678-680.
[6] Hahn T.(1997), private information.

Tab. 1. Structure refinement of K_4MoWO_8 for C2

ATOM	x	y	z	sof	U11	U22	U33	U23	U13	U12	Ueq
Mo	0.17578	0.00069	0.22697	0.25000	0.01185	0.01499	0.01318	0.00022	0.00458	0.00131	0.01365
	0.00004	0.00024	0.00006	0.00000	0.00019	0.00021	0.00020	0.00048	0.00013	0.00047	0.00015
W	0.17578	0.00069	0.22697	0.25000	0.01185	0.01499	0.01318	0.00022	0.00458	0.00131	0.01365
	0.00004	0.00024	0.00006	0.00000	0.00019	0.00021	0.00020	0.00048	0.00013	0.00047	0.00015
K1	0.50870	-0.00704	0.23465	0.50000	0.01437	0.01716	0.01225	-0.00176	0.00521	0.00138	0.01480
	0.00016	0.00193	0.00024	0.00000	0.00060	0.00107	0.00061	0.00181	0.00046	0.00184	0.00038
K2	0.84460	-0.00646	0.25462	0.50000	0.01245	0.02184	0.01354	-0.00398	0.00546	-0.00093	0.01601
	0.00016	0.00188	0.00028	0.00000	0.00057	0.00107	0.00060	0.00284	0.00045	0.00204	0.00039
O1	0.33086	-0.00727	0.38234	0.50000	0.00574	0.03939	0.01504	-0.00408	-0.00206	0.00391	0.02241
	0.00053	0.00541	0.00101	0.00000	0.00188	0.00469	0.00241	0.00867	0.00168	0.00798	0.00164
O2	0.08715	-0.00921	0.36003	0.50000	0.01664	0.03483	0.01287	0.00233	0.00961	0.00462	0.02030
	0.00065	0.00517	0.00101	0.00000	0.00246	0.00462	0.00230	0.00829	0.00188	0.00846	0.00151
O3	0.14250	0.23368	0.07690	1.00000	0.01822	0.01394	0.01895	0.00666	0.00485	0.00319	0.01819
	0.00039	0.00090	0.00064	0.00000	0.00156	0.00154	0.00163	0.00132	0.00127	0.00127	0.00070

Tab. 2. Structure refinement of K_4MoWO_8 for C2/m

ATOM	x	y	z	sof	U11	U22	U33	U23	U13	U12	Ueq
Mo	0.17577	0.00000	0.22702	0.25000	0.01234	0.01533	0.01368	0.00000	0.00477	0.00000	0.01410
	0.00003	0.00000	0.00006	0.00000	0.00018	0.00018	0.00019	0.00000	0.00012	0.00000	0.00014
W	0.17577	0.00000	0.22702	0.25000	0.01234	0.01533	0.01368	0.00000	0.00477	0.00000	0.01410
	0.00003	0.00000	0.00006	0.00000	0.00018	0.00018	0.00019	0.00000	0.00012	0.00000	0.00014
K1	0.50867	0.00000	0.23446	0.50000	0.01444	0.01823	0.01250	0.00000	0.00521	0.00000	0.01529
	0.00014	0.00000	0.00021	0.00000	0.00055	0.00060	0.00055	0.00000	0.00042	0.00000	0.00025
K2	0.84452	0.00000	0.25438	0.50000	0.01289	0.02249	0.01384	0.00000	0.00557	0.00000	0.01649
	0.00014	0.00000	0.00024	0.00000	0.00052	0.00066	0.00053	0.00000	0.00041	0.00000	0.00025
O1	0.33128	0.00000	0.38344	0.50000	0.00617	0.03846	0.01481	0.00000	-0.00260	0.00000	0.02237
	0.00048	0.00000	0.00090	0.00000	0.00171	0.00352	0.00216	0.00000	0.00152	0.00000	0.00120
O2	0.08673	0.00000	0.35972	0.50000	0.01617	0.03570	0.01387	0.00000	0.00967	0.00000	0.02079
	0.00056	0.00000	0.00091	0.00000	0.00220	0.00333	0.00214	0.00000	0.00172	0.00000	0.00110
O3	0.14246	0.23720	0.07666	1.00000	0.01787	0.01186	0.01786	0.00502	0.00556	0.00248	0.01665
	0.00036	0.00067	0.00059	0.00000	0.00144	0.00135	0.00148	0.00113	0.00117	0.00112	0.00062

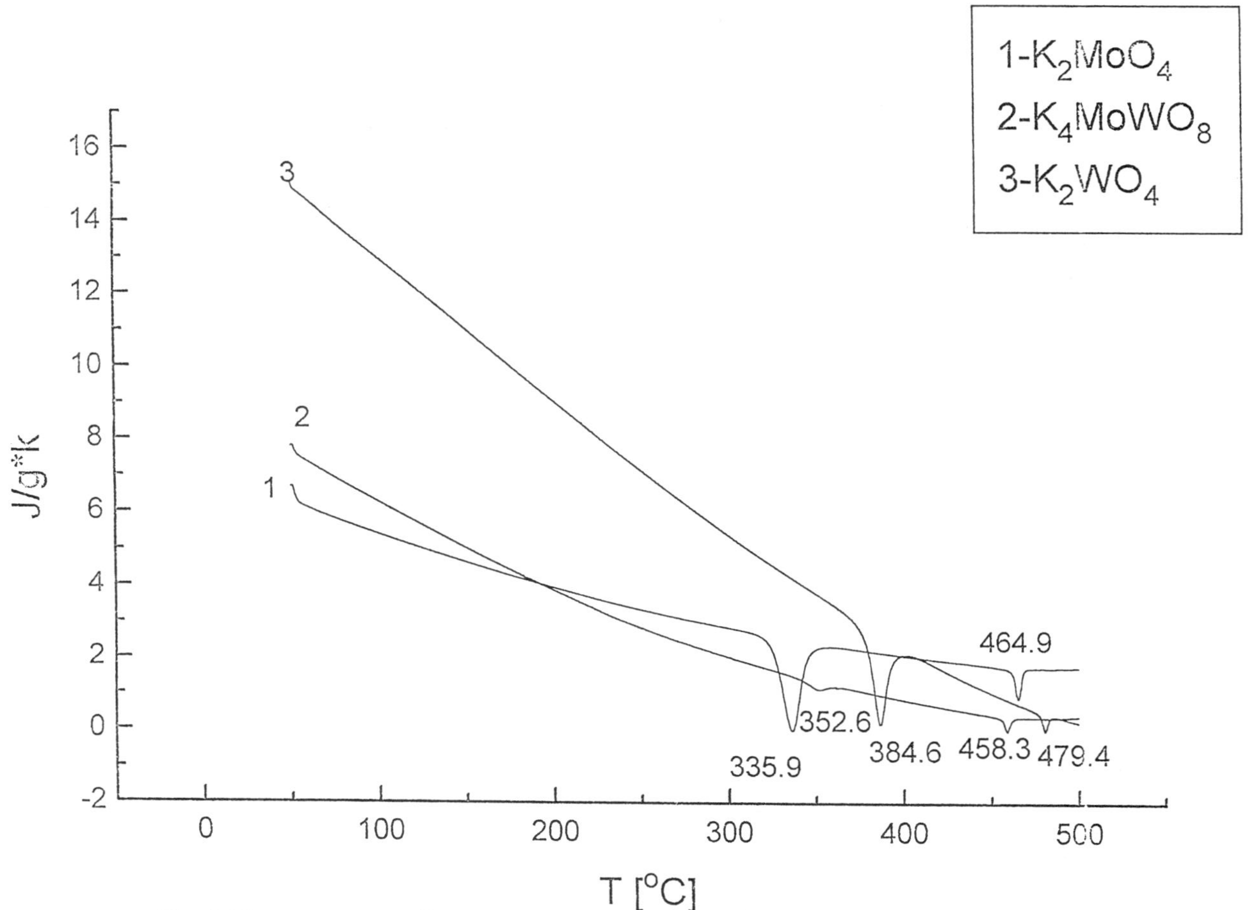

Fig.1 The curves of the second heating for the investigated compositions.

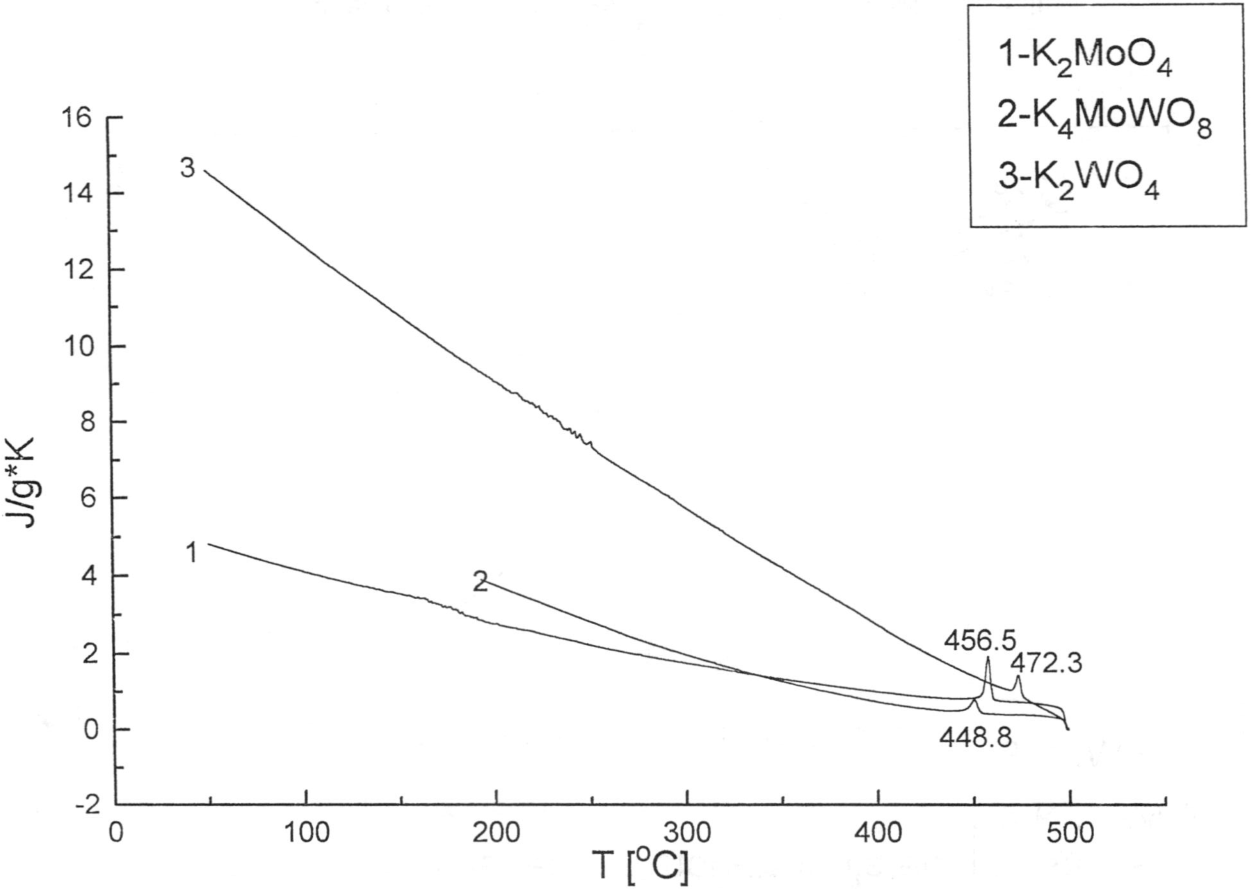

Fig.2 The curves of the second cooling for the investigated compositions.

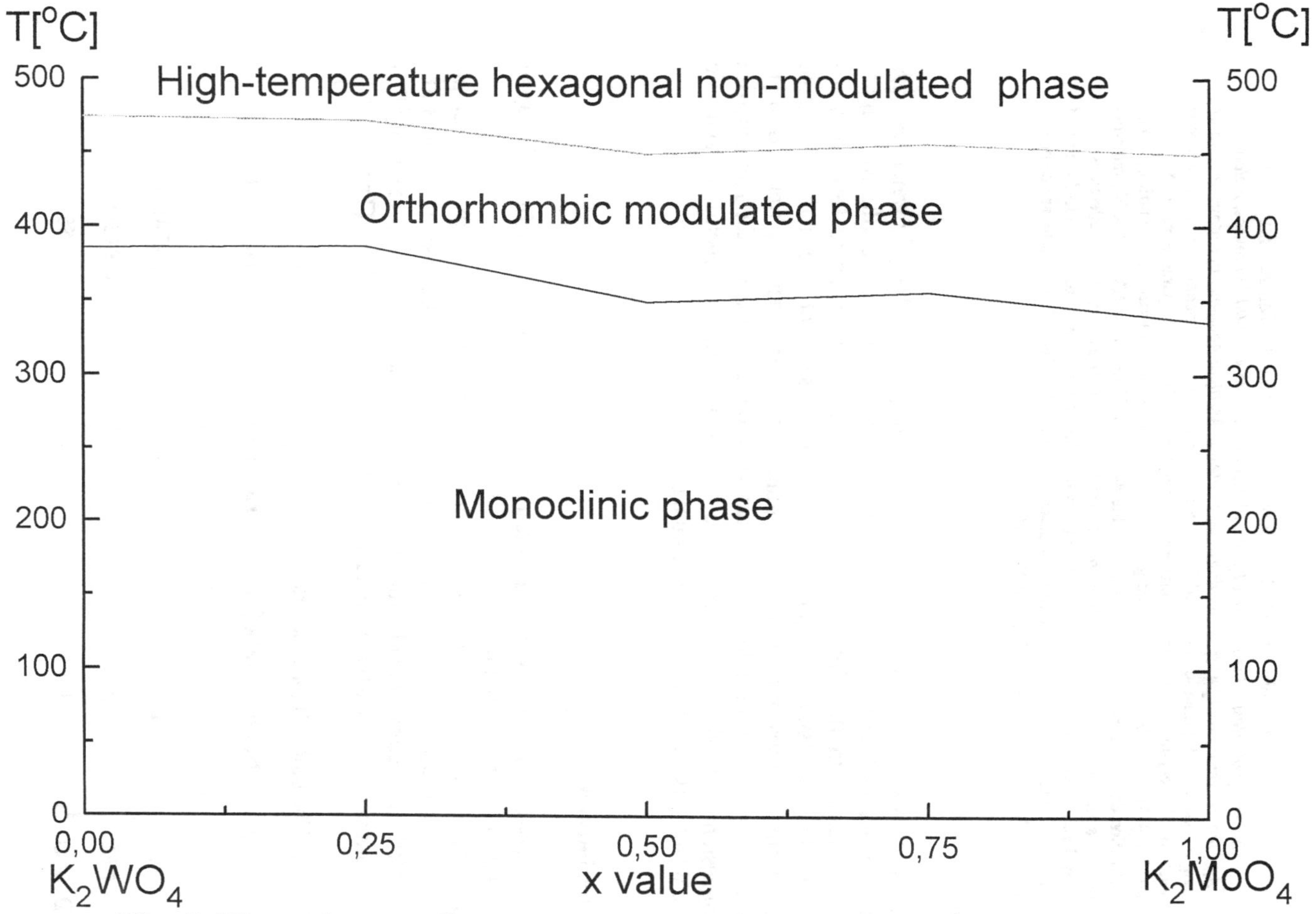

Fig.3 The phase diagram of the K_2MoO_4-K_2WO_4 system

STRUCTURE REREFINEMENT OF L-GLUTAMIC ACID WITH THE AID OF THE X-RAY 4-CIRCLE DIFFRACTOMETER

W. Marcoin*, H. Duda, J. Kusz, B. Bzowski, J. Warczewski
University of Silesia, Institute of Physics, ul. Uniwersytecka 4, 40-007 Katowice, Poland
**Applied Pharmacy and Drug Technology Dept. Silesian Medical Academy in Katowice, Poland*
Single crystals of l-glutamic acid ($C_5H_9NO_4$) were grown at room temperature from the aqueous solution. The crystals belong to the orthorhombic system with the cell parameters: a=5.154(1), b=17.274(3), c=6.942(1) Å. The space group is $P2_12_12_1$ with four molecules in the unit cell. The diffraction data were collected with the aid of the 4-circle diffractometer KM-4. The K_α radiation of Mo anode was used. 5075 unique reflections have been measured. 1713 reflections were considered in the structure analysis and refinement. The structure was solved with the use of SHELXTL/PC programs and the agreement index R is equal to 0.060. These results are compared with the X-ray [1] and neutron [2] data obtained earlier.

Introduction

Although most of inorganic magnesium salts show good solubility in water, they are not accepted by human organism, because they make non-soluble salts with higher aliphatic acids in the alimentary canal. One of the direction to look for well-acceptable magnesium compounds is to study salts of the natural organic aminoacids. The structures of different isomers of glutamic acid are well known [1-3]. It is very important to know the structure of the acid used in the reaction with magnesium hydroxide. In this paper we try to determine the structure of glutamic acid which was used to obtain magnesium salts .

Experimental

Single crystals of glutamic acid were obtained by crystallisation from the water solution at room temperature.

Laue photographs were performed for crystals to check their quality. The sample used for the measurements had the following approximate dimensions: 0.2×0.3×0.4 mm. Cell parameters were obtained by the measurement of 29 reflections using four circle diffractometer KM-4 (Kuma Diffraction) with MoK_α radiation (Table 1).

Table 1. The unit cell parameters

	Neutron data [2]	X-ray data [1]	X-ray data
a (Å)	5.159(5)	5.17(1)	5.154(1)
b (Å)	17.30(2)	17.34(3)	17.274(3)
c (Å)	6.948(7)	6.95(1)	6.942(1)
D_c (g cm^{-3})	1.59	1.57	1.55

Intensities of 5075 reflections were measured up to $2\theta=60^\circ$ by the ω scan technique witch the graphite monochromator ($-2<h<7$, $-2<k<10$, $-24<l<24$). During the integrated intensity measurements the stability of the intensity was checked by the measurements of the intensities of three reference reflections after every 60 reflections. The intensities of these reflections did not change very significantly throughout the experiment. The intensities were corrected for Lorentz and polarization factor and for absorption. 1713 reflections with $I>3\sigma$ were used for analysis. Full matrix least-square method, anisotropic temperature factors of all the non-H atoms as well as isotropic temperature factors of the H atoms were used for the refinement. The final value of the agreement factor R after refinement was R=0.060. All calculations were performed with the aid of SHELXTL/PC programs. The final atom coordinates are given in Table 2.

Results and discussion

The obtained cell parameters and the atom coordinates are in good agreement with other investigations [2] for l-glutamic acid. The agreement factor is much better than R from the earlier X-ray investigations [1], and close to the agreement factor from the neutron investigations [2]. The obtained results suggest that the investigations with the aid of an X-ray four-cyrcle diffractometer can give results on the same accuracy level as the neutron investigations. Therefore, this kind of investigations of the organic compounds became very attractive comparing to the a neutron method.

Acknowledgments

The authors are grateful to Professor Z.Gałdecki (Technical University of Łódź, Poland) for allowing us to use the Siemens SHELXTL/PC program system in his laboratory.

References

[1] S. Hirokawa, *Acta Cryst.* (1955). **8**, 637
[2] M. S. Lehmann, T. F. Koetzle, W. C. Hamilton, *J. Cryst. Mol. Struct.* (1972). **2**, 225
[3] N.Boyd, T.Morisson, Organic chemistry, Pergamon Press (1974) p. 333.

Table 2. Atomic coordinates (R=0.060)

Atom	X/A	Y/B	Z/C	U11	U22	U33	U23	U13	U12
O1	0.66707(52)	0.57832(46)	-0.23591(15)	0.01904	0.04490	0.03272	-0.00720	0.00043	0.00454
O2	0.27745(53)	0.58791(47)	-0.29233(13)	0.02492	0.05665	0.01764	-0.00724	-0.00139	-0.00043
O3	0.36068(60)	0.56931(50)	0.02509(15)	0.03722	0.05534	0.02554	-0.00301	-0.00391	-0.01322
O4	0.00327(62)	0.44553(48)	0.07811(14)	0.03880	0.05194	0.01869	0.00870	0.00294	-0.00708
N1	0.09032 (63)	0.79220(47)	-0.16866(17)	0.02278	0.02765	0.01952	0.00158	0.00085	0.00283
C1	0.42762(74)	0.60221(55)	-0.23539(19)	0.02644	0.02374	0.02013	-0.00393	0.00414	0.00034
C2	0.29604(71)	0.64460(50)	-0.15759(19)	0.01701	0.02182	0.01899	-0.00227	-0.00040	0.00094
C3	0.18608(85)	0.45394(54)	-0.12779(19)	0.04060	0.02183	0.01842	0.00220	0.00022	-0.00180
C4	0.01008(86)	0.46244(60)	-0.05682(20)	0.03368	0.03587	0.01895	0.00574	-0.00151	-0.00961
C5	0.14533(78)	0.49839(52)	0.01895(20)	0.02923	0.02081	0.02191	0.00166	-0.00102	0.00406
H4	0.05200	0.41810	0.13370						
H41	-0.12980	0.55350	-0.06220						
H21	0.42620	0.69410	-0.12650						
H31	0.32550	0.35480	-0.12000						
H11	0.15360	0.88080	-0.19650						
H12	0.03360	0.83520	-0.12570						
H32	0.06920	0.41380	-0.16970						
H42	-0.10930	0.34500	-0.04330						
H13	0.02190	0.75650	-0.19910						

GRAZING INCIDENCE X-RAY DIFFRACTION STUDIES OF TWO-DIMENSIONAL PACKING PROPERTIES OF AMPHIPHILIC ALCOHOLS AT AIR-WATER INTERFACE

J. MAJEWSKI*, M. LAHAV, L. LEISEROWITZ
Department of Materials & Interfaces, Weizmann Institute of Science, 76100, Rehovot, Israel
**Present address: MLNSC, Los Alamos National Laboratory, Los Alamos, NM 87545, USA.*

K. KJAER
Risø National Laboratory, 4000 Roskilde, Denmark

J. ALS-NIELSEN
Copenhagen University, DK2100, Denmark

A systematic analysis of grazing incidence synchrotron x-ray diffraction data of uncompressed amphiphilic alcohols $C_nH_{2n+1}OH$ (n=13-31) on water subphase at 5°C is presented. For n=31-18 the molecules are arranged in a rectangular cell with plane symmetry $p1g1$. For n<18 a tilted free-rotator phase is adopted. The various structural parameters of the monolayer series, such as degree of crystallinity, in-plane coherence length, chain orientation and molecular motion were established.

1 Two-Dimensional Crystal Structure Determination of $C_nH_{2n+1}OH$ (n=13-31)

In general, the grazing incidence x-ray diffraction (GID) using synchrotron radiation[1-5] measurements of the monolayers $C_nH_{2n+1}OH$ (n=13-31) yielded two low-order reflections which are indexed as {1,1} and {0,2} describing a rectangular cell. As n decreases from 31, the resolved {1,1} and {0,2} reflections tend to converge until n=16 at which point they coincide and so the corresponding $d_{\{1,1\}}$ and $d_{\{0,2\}}$ spacings are equal (see Fig. 1). Below n=16 the trend continues and so the Bragg positions of the two reflections cross over and once again tend to be resolved as is evident for n=14 and 13 from the two-dimensional intensity plots $I(q_{xy}, q_z)$, (Fig. 1). The 2-D packing arrangements in the $C_nH_{2n+1}OH$ series were determined by fitting a molecular model to the observed Bragg rod data. This analysis yielded good fits to the Bragg rod profiles for all monolayers. In the refinement of each crystal structure we assumed the plane group symmetry $p1g1$ (namely a primitive cell containing two molecules related by a glide plane perpendicular to the a axis, see Fig. 2). In order to improve the fit to the observed Bragg rod profiles the plane group symmetry was relaxed by allowing the glide plane to lean from the vertical plane by an angle δ about an horizontal axis parallel to b. Refinement of the Bragg rod data yielded δ values ranging from 1.5° for n=31 to 3° for n=13 (Table 1). The resulting plane symmetry group is thus no longer $p1g1$, but $p1$ with two independent molecules per unit cell. Such a "pseudoglide" packing arrangement yields only a minor distortion from the pure glide motif and still preserves the herring-bone motif of the hydrocarbon chains[6].

2 Comparison of the 2-D Crystalline Properties of the $C_nH_{2n+1}OH$ Series.

The change in crystal structure as a function of chain length, in terms of cell dimensions, molecular tilt and cross-sectional area per molecule are shown in Table 1.

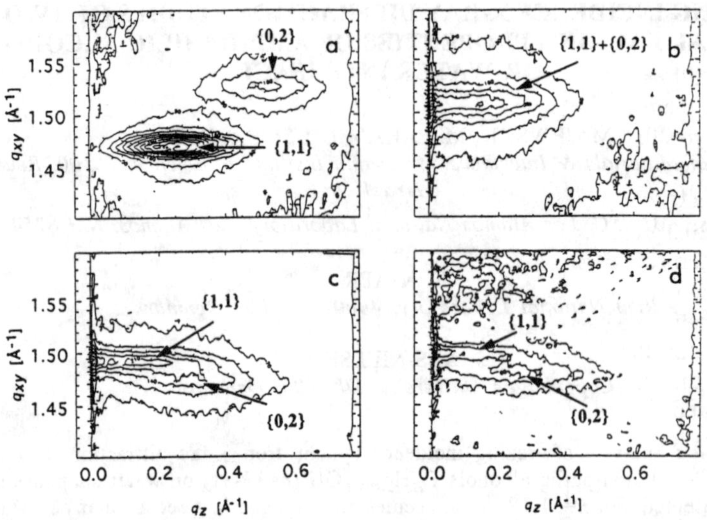

Figure 1. Grazing incidence X-ray diffraction measurements from four different, uncompressed monolayers of (a) $C_{18}H_{37}OH$, (b) $C_{16}H_{33}OH$, (c) $C_{14}H_{29}OH$, (d) $C_{13}H_{27}OH$ alcohols at 5°C on the water surface. Shown are two-dimensional contour plots $I(q_{xy},q_z)$ of the intensity distribution along the horizontal (q_{xy}) and the vertical (q_z) scattering vectors. The (q_{xy},q_z) positions of intensity maxima of the {1,1} and the {0,2} reflections of the four alcohols are not the same, indicating different unit cell dimensions and angles of molecular tilt.

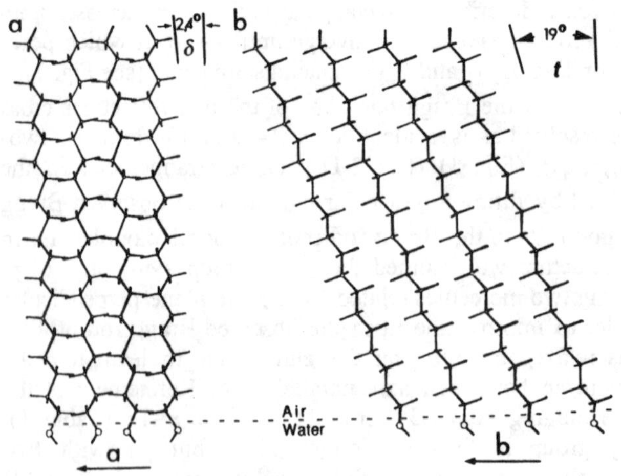

Figure 2. Molecular packing arrangement of $C_{20}H_{41}OH$ alcohol in the 2-D crystal. All molecules are tilted by t=19° from the vertical in the direction of the b axis and by δ=2.4° in the direction of the a axis. (a) View along the b axis. (b) View along the a axis. (c) View along the molecular axis. Because of the ambiguity in fixing the absolute azimuthal orientation of the hydrocarbon chain, we have arbitrarily chosen one of the two possible orientations.

The a axis shows only a minor decrease in length, by about 0.15Å, from n=18 to 13. There is a shallow increase in the length of the b axis from n=31 to 23 followed by a steeper increase till n=13. The molecular tilt angle t follows an irregular profile. Starting with 7° for n=31 there is a minor increase of 2° for n=23, followed by a jump of about 10° on reaching n=20, 19, 18. Surprisingly, the chain tilt decreases to 12° at the lower limit n=13, despite a continuous increase in the length of the b axis. The cross-sectional molecular area A_x (=$A\cos t$, where A is the surface molecular area $ab/2$) rises continuously from 18.4Å^2 to 20.3Å^2 across the whole range of n[7]. This increase extends only in the b direction. We interpret this result in terms of a continuous increase in molecular motion parallel to the water surface with decreasing chain length[8]. As mentioned previously, the full width at half maximum of the Bragg peak, corrected by resolution of the spectrometer, yields the 2-D crystalline coherence length L, associated with the hk crystallographic direction. The coherence lengths are anisotropic, particularly for the longer chains, and become smaller with decreasing chain length. The crystalline coherence lengths for $C_{31}H_{63}OH$ were found to be anisotropic with values of ~1000Å and ~300Å along the {1,1} and {0,2} crystallographic directions. For $C_{13}H_{27}OH$ the coherence length became almost isotropic and as small as 300 Å.
We have been able to extract the relative amount of 2-D crystalline material from a comparison of the factor which scales the measured and calculated Bragg rod intensities of the monolayers $C_nH_{2n+1}OH$ (n=31, 20, 16) measured under the same experimental conditions. The relative amount of crystalline material for n=31, 20, 16 is equal to 1:0.33:0.25.

3 Molecular Motion in the $C_nH_{2n+1}OH$ Series. Testing the Free-Rotator Phase.

The increase in molecular cross-sectional area A_x as a function of shortening in chain length may be associated with an increase in molecular motion. A direct estimate of the average atomic displacement (or Debye-Waller) parameter[9] parallel to the water surface u_{xy} may be derived from the GID data for the longer chain alcohols n=31, 30, 23, 20, 19[10]. This estimate was found by varying u_{xy} to best fit the Bragg rod profiles, particularly for the higher order reflections. A value of ~0.1Å^2 had been computed for n=31 and 30 and 23 by making use of the {1,1}, {0,2} and {2,0} reflections[5]. The presence, albeit weak, of the higher order {2,0} reflection for n=20 and 19 yielded values of u_{xy}=0.13Å^2 and 0.25Å^2 respectively. Below n=19 the {2,0} reflection was unobserved, from which we could deduce by structure factor calculations that $u_{xy} \geq 0.3$Å^2. For all the monolayer systems we have assumed a herring-bone structure of the hydrocarbon chains. Assuming the a, b unit cell as being c-centered, the dimensions of such a cell projected along the molecular axis a_p , b_p ($a_p = a\cos\delta \approx a$, $b_p = b\cos t$), but expressed in terms of the pseudo-hexagonal representation ($a_H = b_H = 0.5(a_p^2+b_p^2)^{1/2}$, $\gamma_H = 2\arccos(0.5a_p/a_H)$), are given in Table 1. We observe from Tab. 1 that γ_H tends to an angle of 120° as n approaches 13, indicative of a hexagonal cell in projection. The projected molecular area A_x reaches a value of 20.3Å^2 for n=13, which is very close to that corresponding to a presumably free rotator phase in the alcohol $C_{21}H_{43}OH$ at 33°C in the uncompressed state[11] and the rotator I-II phases of $C_{20}H_{41}COOH$ in the compressed state[12]. We therefore tested by structure factor calculations the effect of a *tilted* free-rotator phase on the Bragg rod intensity profiles for n=13. For practical purposes we constructed a model composed of 12 azimuthal orientations of the chain, each with 1/12th molecular occupancy, covering the full 360° angle.

Table 1. Structural data of the uncompressed alcohol monolayer series $C_nH_{2n+1}OH$.

n	a axis [Å]	b axis [Å]	b axis proj. [Å]	area/mol. [Å^2]	projected area/mol. [Å^2]	tilt in b t [deg.]	tilt in a δ [deg.]	a_H [Å]	γ_H [deg.]	η [deg.]
31	4.98	7.45	7.39	18.6	18.4	7.0 *	1.5	4.46	112.1	90
30	4.99	7.49	7.42	18.7	18.5	7.7	1.7	4.47	112.1	86
23	5.00	7.56	7.46	18.9	18.6	9.5	1.2	4.49	112.3	82
20	5.05	8.00	7.56	20.2	19.1	19.0	2.4	4.55	112.6	130
19	5.03	8.13	7.60	20.4	19.1	20.8	2.0	4.56	113.1	130
18	4.99	8.19	7.73	20.4	19.2	19.4	2.9	4.60	114.5	120
17	4.94	8.34	7.91	20.6	19.5	18.5	2.5	4.66	116.0	130
16	4.86	8.41	8.09	20.4	19.6	15.8 *	2.8*	4.72	118.2	120
14	4.84	8.47	8.19	20.5	19.8	14.8*	3.0*	4.75	118.9	120
13	4.86	8.55	8.36	20.8	20.3	12.0	3.0	4.83	119.7	130

* tilt derived from observed Bragg rod fitted with overlapped {1,1} and {0,2}; b axis projected=$b\cos t$; a axis projected=$a\cos\delta \approx a$; area/molecule=$ab/2$; projected area/molecule=$(ab\cos t)/2$; t-tilt angle, measured from the normal to the water surface in the direction of b axis; δ-tilt angle, measured from the normal to the water surface in the direction of the a axis; a_H, γ_H -pseudo-hexagonal representation of the rectangular cell treated as c-centered and projected along the molecular axis; η -dihedral angle between carbon planes of two molecules related by pseudo-glide symmetry, obtained from Bragg rod fits.

The fit of such a model to the scattering data is no worse than for the herring-bone arrangement. A similar analytical procedure was carried out for n=18, which had a projected cross-sectional area A_x=19.4Å^2 and γ_H=114.5°. A model which simulated the free rotator, yielded a poorer fit to the Bragg rod data, particularly for the {0,2} reflection. Indeed assuming pronounced librational motion about the molecular axis corresponding to a libration angle greater than ± 20° from the mean azimuthal angle in the herring-bone arrangement already gave a noticeably poorer fit to the {0,2} reflection.

4 Effect of Water Subphase on Molecular Packing

We are confronted with the question why the continuous rise in cross-sectional molecular area, over the range n=31 to 13 requires a concomitant increase in tilt angle along the b axis from n=31 to 19 and then a fall in tilt angle from n=19 to 13. At first sight, it might seem, that an increased tilt angle leads to a lower stabilization energy in the 2-D crystal structures because of possible end-of-chain effect and relative offsetting between neighboring chains. We may explain the preference for tilt along the b axis rather than the a axis in terms of hydrogen bonding to the subphase water molecules. Our basic premise is that a water molecule shall form acceptable hydrogen bonds with two neighboring alcohol molecules provided the distance between the two alcohols lies in the range of 4.5 to maximally 5.0Å, based on the 3-D crystal structures of ice, methanol, and other

alcohols[13]. Thus, a tilt along the *a* axis could be detrimental to hydrogen bonding since the length of the *a* axis with the molecular chains vertically aligned is 5Å, the maximum threshold distance. On the other hand, a rise in molecular tilt from 7° to 21° along the *b* axis, results in an increase in axial length from ~7.4Å to ~8.1Å, which is satisfactory for hydrogen bonding since the distance between glide-related molecules along the diagonal $1/2|(\boldsymbol{a}+\boldsymbol{b})|$ would then increase from 4.5 to 4.8Å respectively. We propose that the loss in lattice energy associated with the rise in molecular tilt angle from 7° to 21° with decreasing chain length (n=31 to 19) may be partially compensated by an improved orientation of the C-OH group for hydrogen bonding to water.
A tilt angle of 21° allows either the two oxygen lone pair electron lobes and the O-H bond to be exposed to water or say, one lone pair electron lobe and the O-H bond to form hydrogen bonds with water depending on the absolute orientation of the hydrocarbon chain. The decrease in tilt angle from 21° for $C_{19}H_{39}OH$ to 12° for $C_{13}H_{27}OH$, but with the concomitant increase in molecular cross-sectional area, may be rationalized in terms of hydrogen bonding. Had the molecular chains retained the tilt angle of 21° but with the observed increase in cross-sectional area A_x, the *b* axis would have increased from ~8.1Å to ~8.9Å, resulting in a separation between glide-related chains of 5.1Å which is longer than the threshold hydrogen bonding distance.

5 Conclusions

Various structural properties of uncompressed crystalline monolayers of the $C_nH_{2n+1}OH$ (n=13-31) alcohols on water at 5°C have been determined by grazing incidence X-ray diffraction. These properties encompass relative amount of crystalline material formed, crystalline coherence length, molecular packing arrangements, and molecular motion. Pronounced structural changes were observed on reduction of chain length from n=31 to 13. The two glide-related molecules in the unit cell form a herring-bone arrangement in which the chain axes are parallel and separated by $|(\boldsymbol{a}+\boldsymbol{b})/2|$. The molecular chains are tilted from the vertical in the *b* direction; the tilt angle increases from ~7° to ~21° over the range n=31 to 19, and then drops to 12° when n has reached a value of 13. There is a continuous increase in molecular cross-sectional area from 18.4Å^2 to 20.3Å^2 and in the atomic displacement parameter parallel to the water surface; it increases from 0.1Å^2 for n=31 to 0.25Å^2 for n=19. We explain the preference for chain tilt along the *b* axis, rather than *a*, in terms of hydrogen bonding to the water subphase.

References

1. J. Majewski, R. Popovitz-Biro, K. Kjaer, J. Als-Nielsen, M. Lahav, L. Leiserowitz, *Chem. Eur. J.*, **1**. no. 5, 302-309 (1995).
2. J. Majewski, L. Margulis, D. Jacquemain, F. Leveiller, C. Böhm, T. Arad, Y. Talmon, M. Lahav, L. Leiserowitz, *Science*, **261**, 899 (1993).
3. J. Majewski, R. Popovitz-Biro, K. Kjaer, J. Als-Nielsen, M. Lahav, L. Leiserowitz, *J. Phys. Chem.*, **98**, 4087 (1994).
4. D. Jacquemain, F. Leveiller, S. P. Weinbach, M. Lahav, L. Leiserowitz, K. Kjaer, J. Als-Nielsen, *J. Am. Chem. Soc.*, **113**, 7684 (1991).
5. J. L. Wang, F. Leveiller, D. Jacquemain, K. Kjaer, J. Als-Nielsen, M. Lahav, L. Leiserowitz, *J. Am. Chem. Soc.*, **116**, 1192 (1994).
6. F. Leveiller, D. Jacquemain, K. Kjaer, J. Als-Nielsen, L. Leiserowitz, *J. Phys. Chem.*, **96**, 10380 (1992).

7. J. F. Legrand, A. Renault, O. Konovalov, E. Chevigny, J. Als-Nielsen, G. Grübel, B. Berge, *Thin Solid Films*, **248**, 95 (1994).
8. This motion is probably anisotropic, extending primarily in the *b* direction.
9. J. D. Dunitz, *X-ray Analysis of the Structure of Molecules*; Cornell University Press: Ithaca, NY, p. 43-49 (1979).
10. Q. Du, R. Superfine, E. Freysz, Y. R. Shen, *Phys. Rev. Lett.,* **70**, 2313 (1993).
11. M. C. Shih, T. M. Bohanon, J. M. Mikrut, P. Zschack, P. Dutta, *J. Chem. Phys.*, **97**, 4485 (1992).
12. M. C. Shih, T. M. Bohanon, J. M. Mikrut, P. Zschack, P. Dutta, *Phys. Rev. A* **1992**, *45* 5734.
13. D. W. Peterson, H. A. Levy, *Acta Cryst.,* **10**, 70 (1957). K. Shimaoko, *J. Phys. Soc. Japan*, **15**, 106 (1960). K. J. Tauer, W. N. Lipscomb, *Acta Cryst.* **5**. 606 (1952).

PHASE IDENTIFICATION IN Al-Fe-Mn-Si ALLOYS

H. MORAWIEC, M. GIGLA, T. GORYCZKA
Institute of Physics and Chemistry of Metals, Silesian University,
Bankowa 12, 40-007 Katowice, Poland

M. SUKIENNIK
Department of Non-Ferrous Metals, University of Mining and Metallurgy,
Mickiewicza 30, Kraków, Poland

The formation, growth and sedimentation of intermetalic phase is one of the methods used to produce from scrap - commercial purity aluminium alloys.

The formula of the sedimented phase was established basing on the chemical composition determined by the x-ray microprobe analysis of the sedimented particles. Until now the quaternary Al-Fe-Mn-Si equilibrium phase diagram is not known. The structure of this phase was studied using the x-ray powder diffraction technique and electron diffraction on thin foils using TEM.

Tentatively it was assumed that the structure has a cubic symmetry and belongs to one of the space group *Pm3* or *Im3*. Basing on this assumptions the x-ray diffraction pattern was calculated using the Rietveld method. In a similar way the electron diffraction patterns were simulated. Both methods confirmed that the structure of the $Al_7(Fe,Mn)_2Si$ phase has a cubic lattice with a parameter a_0=12.56 A and 144 atoms per unit cell and belongs to the *Im3* space group. The formation of this structure was confirmed in alloys with a chemical composition similar to the studied intermetallic phase.

1 Introduction

It is well known that a controlled precipitation of some iron-rich intermetallic phases can be used to remove this element from aluminium - silicon scrap [1]. It is difficult to identify these phases because there is still lack of a quaternary Al-Fe-Mn-Si equilibrium phase diagram [2]. But according to the authors of [2] their studies have shown that some of the phases in the quaternary system are similar to those found in the binary and ternary system. Papers concerning the structure of α-AlMnSi [3] and α-AlFeSi [4] were especially interesting for our purpose. The occurrence of equilibrium phase in some quaternary Al-Fe-Mn-Si alloys were also identified in [5]. The aim of this paper was to identify the phase precipitated during solidification from Al-Fe-Mn-Si alloys.

2 Experimental

To study the segregation of the intermetallic phase in the Al-Fe-Mn-Si system a dilute alloy was chosen of a composition given in Table 1.

Table 1: Chemical composition of the dilute alloy.

Alloy	Al [at.%]	Fe [at.%]	Mn [at.%]	Si [at.%]
A	89.02	0.77	1.08	9.13
B	66.70	14.40	14.3	16.60

After melting the alloy in crucible a cooling procedure was applied (Table 2) which lead to segregation and sedimentation of the particle in the crucible bottom.

Table 2: Cooling procedure applied to the melt alloy in order to obtain precipitation of the intermetallic phases.

Temperatures of cooling start and finish T_s [°C] →	T_f [°C]	Cooling rate v [°C/min]	Time of holding at T_f t [min]
900	796	0.58	60
796	640	0.65	60
640	625	0.25	-
625	617	0.07	-

3 Results and discussion

Distribution of the sedimented phase particles in the bottom of the ingot as well as morphology of these particles are shown on Fig. 1

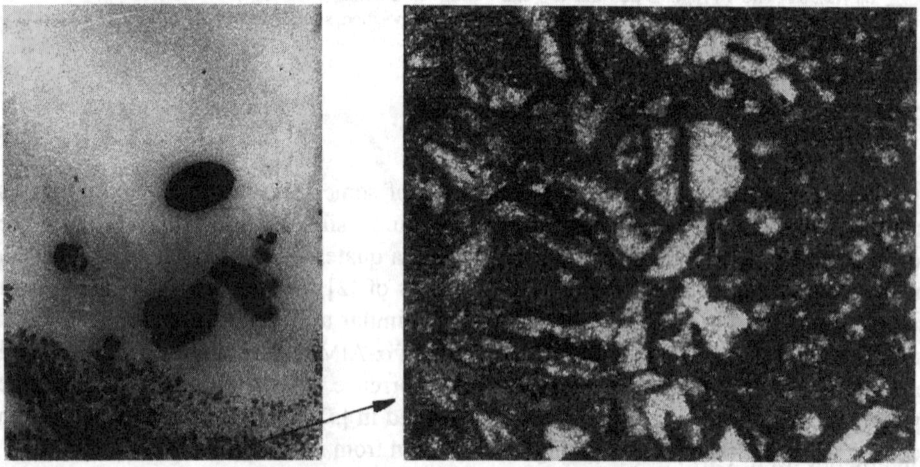

Fig.1. Sedimented particles on the bottom of the ingot and their morphology

Table 3: Chemical composition of the precipitated intermetallic phase.

Description of analysed particles	Al at.%	Fe at. %	Mn at. %	Si at.%
Particles analysed in situ in the matrix (A)	70.56	7.28	10.62	11.42
Particles extracted from the matrix (A)	71.66	6.80	10.53	10.99
Phase regions analysed (B)	70.50 ±0.34	8.41±0.21	10.12±0.22	10.30±0.31

The chemical analysis of these particles was carried out by a microprobe analyser in situ in the matrix and on particles extracted by preferential chemical dissolution of the matrix. Results of these analyses are shown in Table 3.

The chemical composition of the particles allows to define the formula of this phase as: $Al_7(Fe,Mn)_2Si$. This formula was confirmed for particles obtained during cooling different dilute alloys according to different cooling procedures.

In order to prove the occurrence of this phase, alloys of a composition close to the studied particles were prepared and the phase analysis was carried out. Example of such alloy composition is shown in Table 1.

The microanalysis carried out on these alloys confirmed presence of the $Al_7(Fe,Mn)_2Si$ phase. However, it should be stated that it was not possible to obtain a single-phase alloy, relatively large regions of this phase were embedded in an eutectic Al+Si matrix. The results of these studies are also shown in Table 3.

Attempts to grow a single crystal by the Bridgman method were unsuccessful. Therefore, determination of the studied phase structure was based on polycrystalline samples. The X-ray powder diffraction patterns were taken for two specimens: a bulk piece cut from the bottom of the ingot containing the sedimented particles and powders obtained from the selective chemical extraction.

The diffraction pattern is presented in Table 4. It is easy to notice that this diffraction pattern fulfils the rule: $h^2+k^2+l^2 = 2n$ and the determined lattice parameter was a_0=12.56 A. A structure of this type was established by Cooper and Robinson [3] for the α-AlMnSi phase and by Cooper [4] for the α-AlFeSi. The space groups for the phases were determined by the author as Pm3 and Im3, respectively. Basing on the atomic positions for phases a theoretical diffraction pattern was calculated. Both, the experimental and theoretical patterns are presented in Table 5.

The calculated intensities of the diffraction pattern certainly bear an error because it was impossible to choose proper position for 144 atoms in the unit cell. Thibballs, Davis and Parker [6] showed that in the α-AlMnSi phase silicon atoms occupy approximately 50% of three crystallographic aluminium sublattices. However, the X-ray measurement show that the volume of the unite cell decreases with the increasing silicon content faster than it could be expected for one-to-one Al-Si substitution.

To eliminate diffraction from other phases also studies with the use of electron microscopy were carried out on thin foils. The studies of the AlFeMnSi alloy were conducted on thin foils which were cut and mechanically thinned to a thickens of 0.1 mm and then polished in an ion mill. It was found that a phase crystallising in a cubic system with a lattice parameter a_0=12.56 A and a space group Im3 was a dominating phase in the studied samples. The literature data concerning structure of unit cells for α-Al-Mn-Si and α-Al-Fe-Si phases are not complete [3, 4, 7]. The studied phase structure was initially compared to the data of these phases. The diffraction pattern simulations for basic zone orientations based on positions of 138 atoms in a unit cell showed significant differences in intensities of experimental and theoretical spots. However, taking into account possibilities of position modifications and assuming higher number of atoms in a unit cell equal to 144 (which follows from a possibility of description of a unit cell with clusters of 54 atoms around vacancies positioned in an origin and centring lattice points [8]) - a good agreement of the simulated and

experimental diffraction patterns was obtained. The data for 3 basic orientations are presented in Fig. 2.

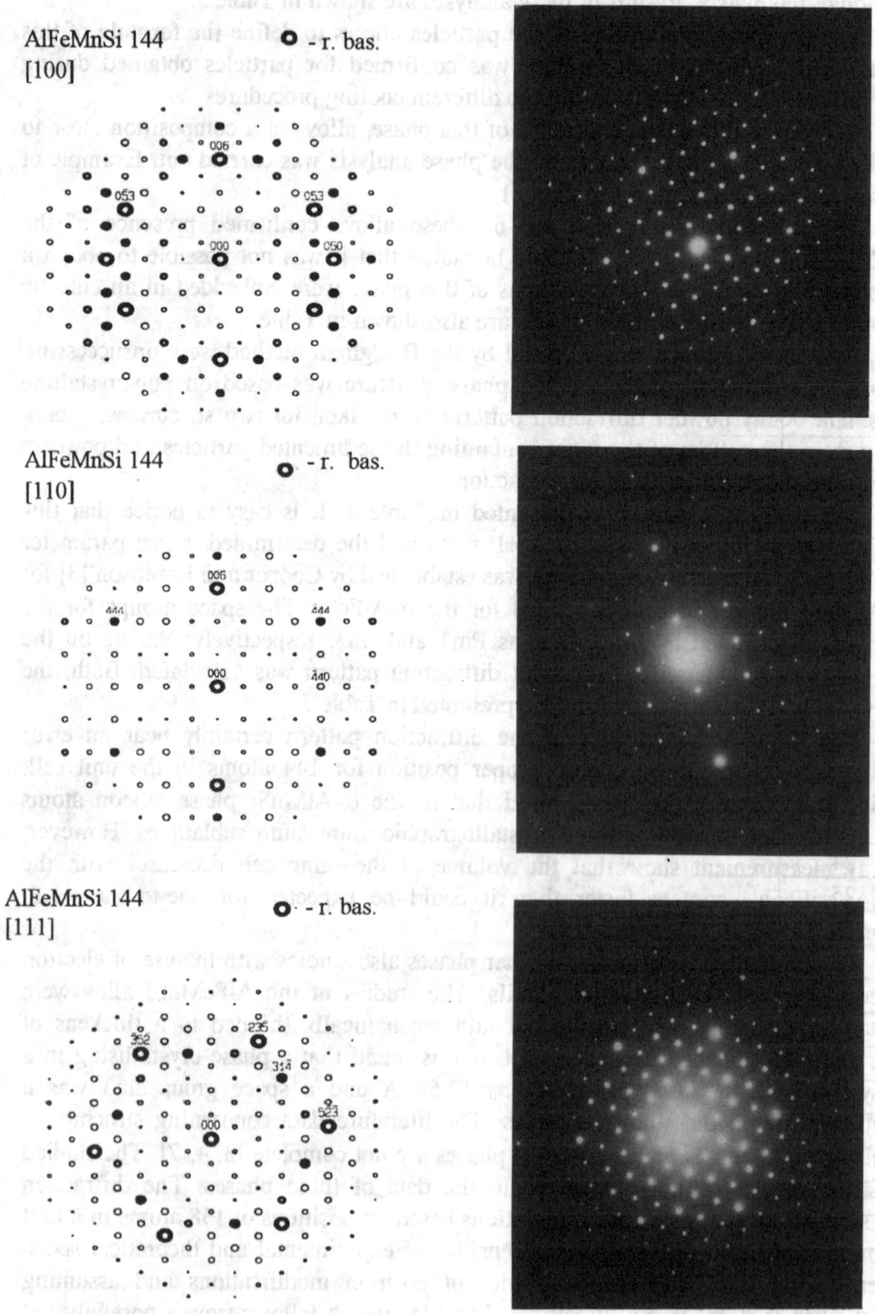

Fig. 2. The simulated and experimental diffraction patterns for the three basic zone axes.

Table 4: X-ray powder diffraction pattern of the $Al_7(Fe,Mn)_2Si$.

No.	Experimental results		Calculated data		
	d_{hkl}	I_r	h k l	d_{hkl}	I_{calc}
1	8.5	7	0 1 1	8.8819	26.6
2	6.24	27	0 0 2	6.2802	28.4
3	4.44	8	0 2 2	4.4409	17.3
4	3.97	18	0 1 3	3.9720	35.5
5	3.62	8	2 2 2	3.6259	61.4
6	3.36	9	1 2 3	3.3570	16.4
7	3.14	8	0 0 4	3.1401	14.0
8	2.97	15	0 3 3	2.9605	35.4
9	2.80	9	0 2 4	2.8086	55.4
10	2.68	3	2 3 3	2.6779	0.8
11	2.56	3	2 2 4	2.5639	3.6
12	2.46	22	1 4 3	2.4633	22.2
13	2.29	13	1 2 5	2.2932	16.8
14	2.22	6	0 4 4	2.2204	5.8
15	2.15	83	0 3 5	2.1541	79.3
16	2.08	100	0 0 6	2.0934	42.6
17	2.04	100	2 3 5	2.0376	100.0
18	1.99	11	0 6 2	1.9860	8.8
19	1.93	6	1 4 5	1.9381	31.1
20	1.88	21	2 2 6	1.8936	2.0
21	1.85	7	1 3 6	1.8519	22.5
22			4 4 4	1.8130	2.2
23	1.78	3	3 5 4	1.7763	9.2
24	1.74	7	0 6 4	1.7418	6.5
25	1.72	3	1 2 7	1.7093	2.2
26	1.66	4	2 6 4	1.6785	8.6
27	1.59	3	0 0 8	1.5701	8.3
28	1.55	4	4 5 5	1.5461	10.3
29	1.51	7	4 4 6	1.5232	2.3
30	1.50	10	3 5 6	1.5013	21.5
31	1.48	18	0 6 6	1.4803	13.0
32	1.46	10	3 7 4	1.4601	5.2
33	1.39	8	3 3 8	1.3871	5.1
34	1.37	1	2 4 8	1.3705	2.9
35			5 5 6	1.3544	6.8
36	1.34	6	4 6 6	1.3390	4.4
37	1.32	6	0 3 9	1.3240	7.2
38	1.29	10	3 7 6	1.2955	9.2
39	1.27	45	3 8 5	1.2688	66.3
40	1.25	20	0 0 10	1.2561	21.1
41	1.23	10	2 8 6	1.2317	6.5
42	1.22	10	0 5 9	1.2200	3.8

continuation of Table 4.					
43	1.20	4	6 6 6	1.2086	2.0
44	1.19	10	1 3 10	1.1976	8.8
45	1.17	7	5 5 8	1.1764	2.1
46			4 8 6	1.1662	3.3
47	1.15	4	1 6 9	1.1563	2.9
48			2 10 4	1.1466	3.6
49			4 9 5	1.1372	3.4
50	1.11	9	0 8 8	1.1102	1.1
51	1.10	4	0 11 3	1.1016	9.6
52	1.09	6	2 8 8	1.0933	11.7
53	1.08	11	6 7 7	1.0851	18.0
54			0 6 10	1.0771	12.4

Conclusions

- The carried out studies showed that the phase precipitated in Al-Fe-Mn-Si alloy during its solidification can be described by a formula $Al_7(Fe,Mn)_2Si$. This phase has a cubic lattice cell with a_0=12.56 A and belongs to the Im3 space group.
- The phase of the characterised structure may occur as an intermediate phase with a large solid solubility.

Acknowledgements

This work was sponsored by the National Committee of Research through grant No. 7T081300509.

References

1. A. Flores, J. Escobedo, J. Mendez, M. Mendez - Proc. TMS 1992 Annual Meeting, Light Metals (1992) 729, California
2. G. Davignon, A. Serneels, B. Verlinden, L, Delaey, *Metall. Mater. Trans.* **27**A 3357 (1996).
3. M. Cooper, K. Robinson, *Acta Cryst.* **20** 614 (1966).
4. M. Cooper, *Acta Cryst.* **23** 1106 (1967).
5. A.M. Zacharov, J.T. Guldin, A.A. Arnold, Ju.A. Macenko, *Cvetnaja Metall.* **4** 89 (1988).
6. J.E. Tibballs, R.L. Davis, B.A. Parker - *J. Mater. Sci.* **24** 2177 (1989).
7. P. Villars, L.D. Calvent, Pearson's Handbook of Crystallographic Data for Intermetalic Phases, vol. 1, Amer. Society for Metals (1985) 746.
8. L.A. Bendersky, J.W. Cahn, *Phil. Mag.* B **60** (1989) 837

STRUCTURE OF Al-Cu-Fe ALLOY WITH A QUASICRYSTALLINE PHASE

M. GIGLA, J. LELĄTKO, M. SUROWIEC
Institute of Physics and Chemistry of Metals, University of Silesia,
Bankowa 12, 40-007 Katowice, Poland

The TEM was applied for the investigations of the phase composition of an Al-Cu-Fe alloy containing a quasicrystalline phase. It was found that during conventional casting the quasicrystalline ψ phase occurred together with other crystalline phases (λ, ω, and β_2). Multitwinned regions with artificial five-fold axis were also found.

1. Introduction

Quasicrystals constituting a new class of materials are characterised by non-crystallographic symmetry elements such as five-fold, eight-fold, ten-fold and twelve-fold symmetry axes [1]. Moreover, they exhibit interesting physical properties significantly different from properties of crystalline materials.

Many quasicrystals were obtained by rapid solidification of a liquid. Quality and stability of quasicrystals obtained in such way are not very good. Better properties can be achieved using conventional slow casting [2]. The ternary Al-Cu-Fe system, with a stable icosahedral ψ - Al_6Cu_2Fe phase [3] seems to be very interesting from this point of view. The aim of this work was to determine the structure of slowly cooled Al-Cu-Fe alloy of a composition providing the quasicrystalline phase.

2. Experimental procedure

The $Al_{65}Cu_{20}Fe_{15}$ alloy was prepared from elements of 99.9% purity. The components were melted in an induction furnace, then mechanically mixed and cooled down with the furnace. Specimens for studies were cut from the areas near both ends of the ingot after directional crystallisation.

Structural investigations were carried out in a JEM 3010 electron microscope attached with a CCD Slow Scan Camera. Chemical compositions were determined using ISIS EDS spectrometer.

3. Results and discussion

It was found that the obtained alloy is chemically inhomogeneous. The aluminium content increased with the vertical direction of the ingot. With the change in chemical composition the change in phase composition was also observed.

In the lower part of the ingot three phases denoted β_2, λ and ω were observed (notation introduced by [4]). On the electron micrographs these phases exhibit different morphology. The low-defected β_2 phase (Al(Fe,Cu) composition, space group Pm3m, a_o=0.29076 nm) includes small density of dislocations and few low-angle boundaries only (Fig. 1a). The particles of the ω phase (Al_7Cu_2Fe composition, P4/mnc, a_o= 0.6336 nm, c_o= 1.482 nm) observed inside of the λ-phase grains (Fig. 1b) phase contain also relatively small amount of defects comparing to the λ-phase. The λ-phase ($Al_{13}(Fe,Cu)_4$ composition, C2/m, a_o=1.5849 nm, b_o=0.8083 nm, c_0=1.2476 nm, β=107.7°) exhibits very strong contrast connected with internal defects (Fig. 1). This phase contains many twins and nano-twins (Fig. 2). The twinning plane was determined as the (001) plane. The thickness of the nano-twins showed on HREM image is about 3 nm.

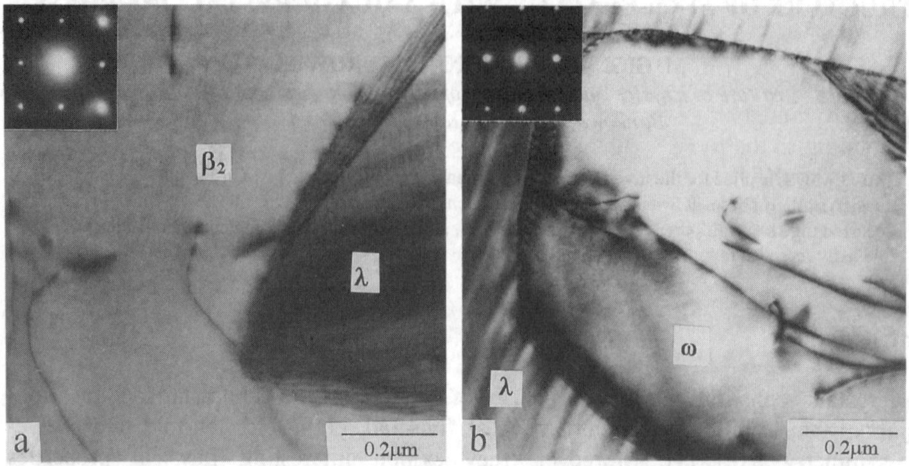

Fig. 1. The λ phase insided the grain of the β_2 phase (a) and the ω phase particle inside the λ phase (b)

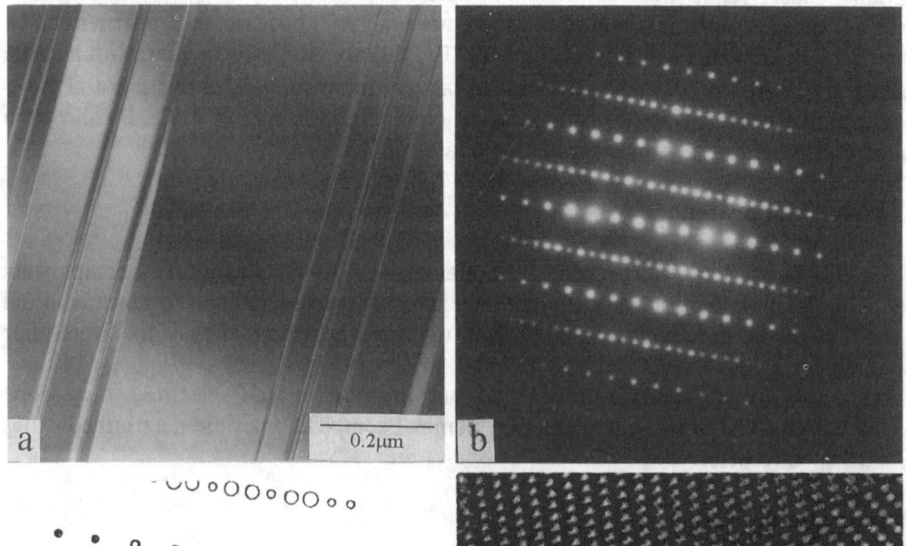

Fig. 2. The microtwins observed in the λ phase (a), corresponding electron diffraction pattern (b), theoretical diagram showing (001) twin plane (c), HREM image showing a nano-twin in the λ phase (d)

In the upper area of the ingot three phases were also observed. Apart from the λ - phase and a small amount of the β_2 phase the quasicrystalline ψ phase was present. The ψ phase exhibits characteristic „tweed" structure on the electron micrographs and the five - fold symmetry axis on electron diffraction patterns (Fig. 3). The HREM image corresponding to the five - fold zone axis shows pentagonal arrangement of atoms (Fig. 4). Contrary to the structure visible on Fig. 2c no periodicity can be defined in this case. The structure showed on Fig. 4 may be regarded as a perfect icosahedral structure.

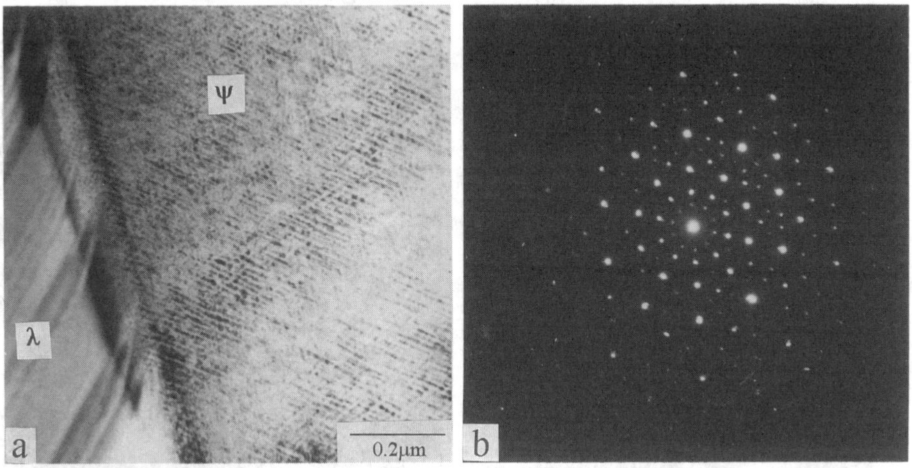

Fig. 3. The ψ phase with characteristic „tweed" contrast (a) and corresponding electron diffraction pattern along five-fold axis (b)

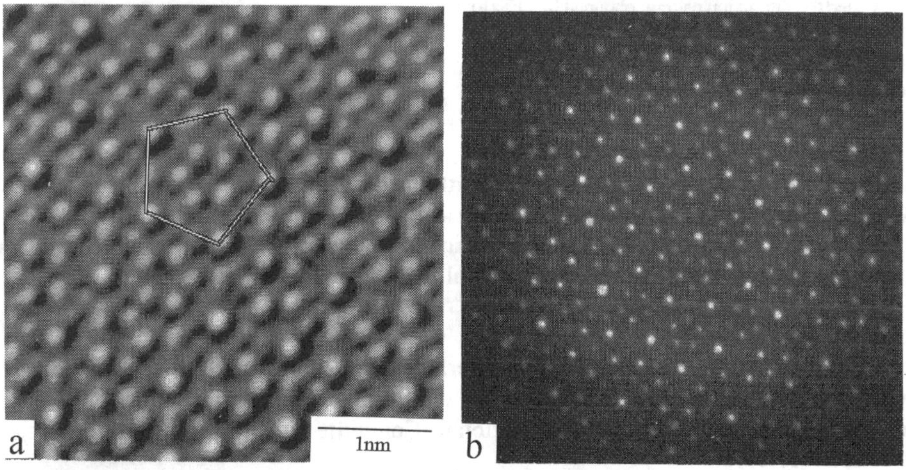

Fig. 4. The HREM image obtained from the perfect icosahedral ψ phase (a) and corresponding electron diffraction pattern recorded by SSC (b)

Other HREM images taken from the same sample (Fig. 5) show a phase which is not in the icosahedral state, but instead is a multitwinned microcrystalline structure. Within small domains (with sizes about 5nm) periodic fringes are clearly visible. These domains are not oriented randomly, but follow an icosahedral symmetry. The diffraction pattern obtained in this case is a superposition of five diffraction patterns corresponding to five domains connected by artificial five-fold axis. Such superposition leads to a splitting of the diffraction spots (Fig. 5b).

A similar structure reported in [5] was considered as a rhombohedral structure with parameters a_o=1.886 nm and α=63.43°. It was suggested that the differences in the structures showed on Fig. 4 and 5 may be connected with differences in the chemical composition. In the $Al_{62}Cu_{25.5}Fe_{12.5}$ alloy the icosahedral ψ phase occurs, but for slightly different composition $Al_{63.5}Cu_{24}Fe_{12.5}$ the structure is in multitwinned microcrystaline state. Because of chemical inhomogeneous of the ingot in our case both phases were observed.

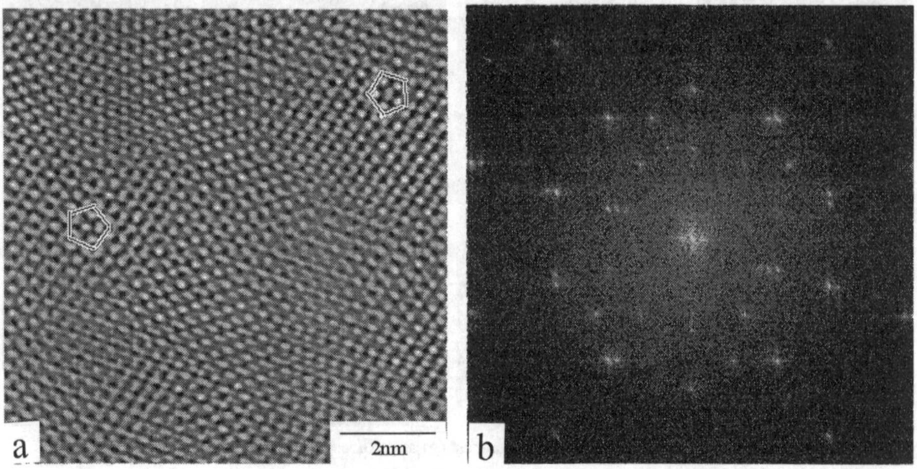

Fig. 5. The HREM image from microcrystalline domains with icosahedral symmetry (a) and corresponding electron diffraction pattern obtained by FFT (b)

4. Summary

It was found that during conventional casting of the $Al_{65}Cu_{20}Fe_{15}$ alloy the quasicrystalline ψ phase coexists together with the β_2, λ and ω crystalline phases. The ψ phase exhibits tweed structure on classical TEM micrographs and pentagonal arrangement of atoms on HREM images which is characteristic for a perfect icosahedral structure. Multitwinned regions with artificial five-fold axis were also found.

Acknowledgement

The JEM 3010 transmission electron microscope was presented to the Institute of Physics and Chemistry of Metals, University of Silesia by Polish Foundation for Science. This work was sponsored by the National Committee of Research through project No 7T08B 017 09.

References

1. C. Janot, *Quasicrystals: a primer,* Clarendon Press, Oxford, 1994
2. A.P. Tsai, A. Inoue, T. Masumoto, *Jap. J. Appl. Phys.* **26**, 1505 (1987)
3. A.P. Tsai, A. Inoue, T. Masumoto, *J. Mater. Sci. Letters* **6**, 1403 (1987)
4. A.J. Bradley, H.J. Goldschmidt, *J. Inst. Met.* **65**, 389 (1939)
5. M. Audier, P. Guyot, in *Quasicrystals*, ed. M.V. Jaric and S. Lundquist, p.74 (World Scientific, Singapore, 1990)

ON UNIQUE DESCRIPTION OF SPACE GROUPS

K. STRÓŻ
Institute of Physics and Chemistry of Metals, University of Silesia,
40-007 Katowice, Bankowa 12, Poland

In the third edition of International Tables for Crystallography the space groups are specified together with the used generators. These generators together with the generating method and the rule that one set of generators is used for each space group standardise description of a space group. This work presents additional rules for standardisation of geometrical description of symmetry elements and compares the generated description with items contained in the Tables.

Among different groups the three-dimensional space groups are fundamental for crystallography. The volume A of International Tables fulfils their complete description. This volume edited by Theo Hahn[1] is an extension and complete revision of the two previous editions from 1935 and 1952. An important step in standardisation of the space-group description is specification of generators which are used to generate all symmetrically equivalent points of the general position and other information related to this ordered set.

Basing on the generators and on a method of generation in the form of composition series it is possible to obtain full compatibility of a generated description and items contained in the Tables. Such feature is very important for didactic purposes. The personal computer program SPACER was developed by Stróż[2] to help obtaining different information on space groups. While the generating of symmetry elements in the form of coordinate triplets is well established, obtaining their geometrical forms is frequently not unique without special conventions. In addition to the rule for coordinate system transformation, three standardisation rules were deduced from the symbols of symmetry operations contained in the Tables. This work presents the rules together with comparison of the items generated by the program SPACER with that contained in the Tables.

Coordinate system transformation

The idea of characterising each space group by one set of generators was described by Hahn and Vos[1] and is adopted here as a standardisation rule.

Rule 1. *Any group, even such which is several times presented in the Tables, can be characterised by one set of generators. The transformation relating the two coordinate systems transforms also the generators from one description into those of the other.*

Later the authors point out that there is one exception to this rule: for five groups R32, R3m, R3c, R-3m. and R-3c one generator is treated specially to obtain the sequence of coordinate triplets compatible with that description present in the previous version of Tables (1952).

It was found that the exception stated earlier can be reinterpreted as a regular transformation. There are three right-hand obverse cells a,b,c which correspond to primitive cell a_p, b_p, c_p: $R_1\{ a = a_p - b_p,\ b = b_p - c_p,\ c = a_p + b_p + c_p \}$, $R_2\{ a = b_p - c_p,\ b = c_p - a_p, c = a_p + b_p + c_p\}$ and $R_3\ \{ a = c_p - a_p,\ b = a_p - b_p,\ c = a_p + b_p + c_p\}$. Assuming R_1 relation as is done in the Tables, the two-fold symmetry axis described in the hexagonal system as 2 x,x,0 (fig. 1) ought to be transformed in the symbol 2 -x,0,x (in the rhombohedral system) but is changed into the 2 x,-x,0 generator. But if we select the R_2 relation then designation 2 x,-x,0 is properly transformed from the 2 x,x,0. Thus, for regular transformation of generators between hexagonal and rhombohedral axes system the R_2 relation should be assumed.

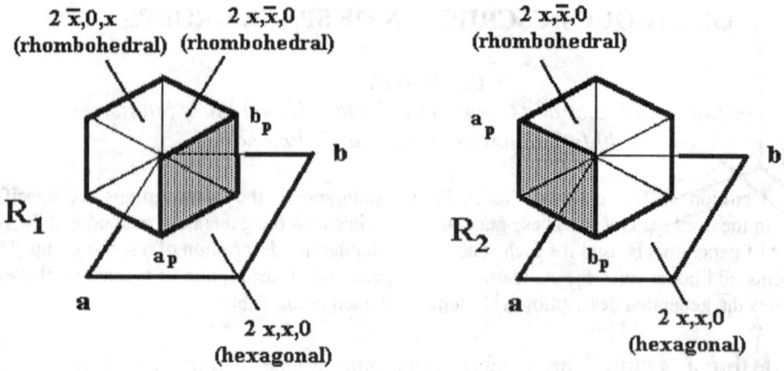

Figure 1. Overcoming the exception from the Rule 1 in the rhombohedral group description by selecting the R_2 cell.

The not stated inconsistency was revealed in description of the group Fd-3 (that is 203 group). It is well seen when we compare the generators for that group and the group Fddd (70). For Origin choice 1 the generators are the same for both groups except to the the presence z,x,y symmetry operation corresponding to 3-fold axis. After shifting the origin to the inversion point (in both cases the shift vector is same) the two generators differ in translation part specification. These differences (1/2,1/2,0) and (1/2,0,1/2) are equal to centring vectors in the F-cell. For origin choice 2 in the group 203 the data themselves are correct in every case but their sequence is interchanged and dependent on a selected set of generators. In order to obey rule 1 the generators in group 203 for origin choice 2 should be changed according to the group 70.

Sense of a symmetry axis

According to the rule 1, to the selected generators and generating method a unique, ordered set of coordinate triplets can be obtained. More problems occur if one tries to convert this algebraic description of a symmetry operation to its geometric description. One important problem is that the axis can be interpreted as a line of rotation, not as a vector which has a sense. Three fold axis parallel to a space diagonal of unit cell can be equivalently described as 3 [111] or 3[-1-1-1]. But it is not true for symmetry operation: rotation by the angle of 120 degrees can be interpreted as rotation by the angle of -180 if the point of view is opposite. This ambiguity was avoided by connecting the sense of the axis with increasing values of the parameter describing that axis. The sense of axis direction is arbitrary, but analysis of items placed in „Symmetry operations" showed that in the Tables the following convention is chosen:

***Rule** 2. A symmetry axis parallel to a system axis is described in such a way, that these two axes point to the same directions (fig. 2a). For an axis lying in a basal plane the positive sense is chosen according to figures 2b-2d. For the other axes the product of the three coordinates of a point must be positive (x y z > 0) if the parameter describing that point on the axis is positive (fig. 2e,2f).*

According to this rule the proper designation of an axis contains following descriptions: x,0,0; 0,y,0; 0,0,z; x,x,0; x,-x,0; 0,y,0; 0,y,-y; x,0,x; -x,0,x; x,x,x; -x,-x,x; x,-x,-x; -x,x,-x. In general case in front of x,y or z one can place a number. The same conventions is used for the description of plane traces.

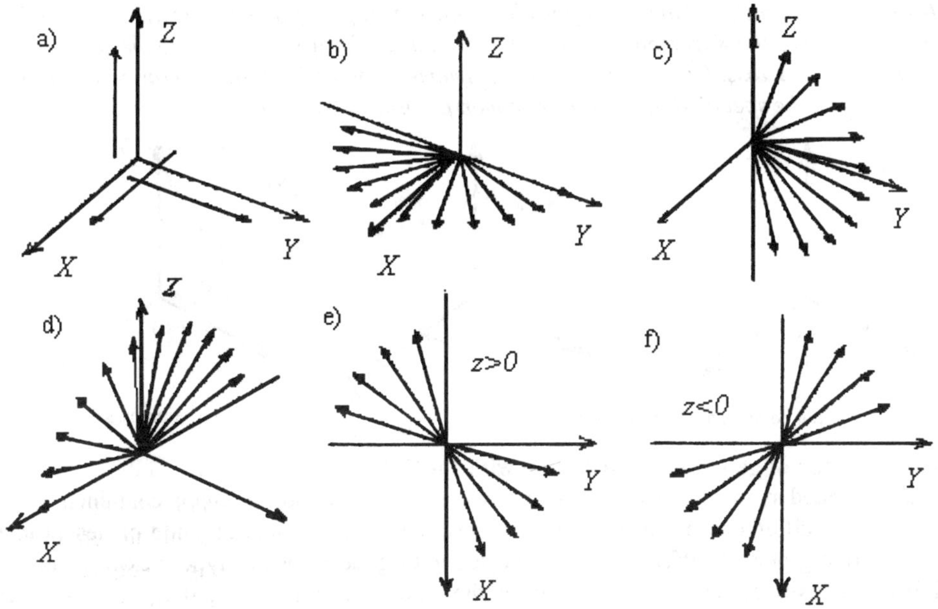

Figure 2. Sense of symmetry axis

Special point on a symmerty axis

A location symbol of a symmetry axis is composed with description of an oriented line and specification of one point on that symmetry axis. For a unique specification of the axis location this point must be conventionally stated. It is natural to choose the point closest to origin of the system (shift vector). But for simplicity of graphical presentations of an axis on the diagram the special rule is applied in the Tables.

Rule 3. *If a symmetry axis is not parallel to the basal plane XY (fig. 3a) then in the location description the coordinates of the common point of that axis and the plane XY are specified, otherwise the common point with the basal plane YZ is described (fig. 3b). Only in the case of an axis parallel to the axis system Y, the specified point lies on the basal plane XZ.*

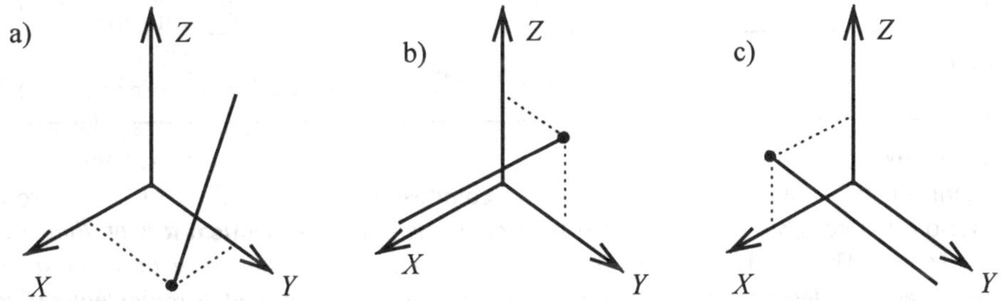

Figure 3. Special point on symmetry axis.

Special point on a symmetry plane

Geometrical description of a plane in the Tables can be treated as an analytical description of two lines: two traces of that plane on basal plane. The special point on the plane is a crossing point of three lines: two traces and the system axis. Thus, there are at most two points on a symmetry plane and for proper selection a next rule was applied.

***Rule 4.** If a symmetry plane is not parallel to the axis (fig. 4a) then its location is given by the point where the X axis passes through that plane, otherwise (fig. 4b) by the common point with the Y axis. Only in the case when symmetry plane is parallel to the basal plane XY (fig. 4c), the specified point is the common point with the Z axis.*

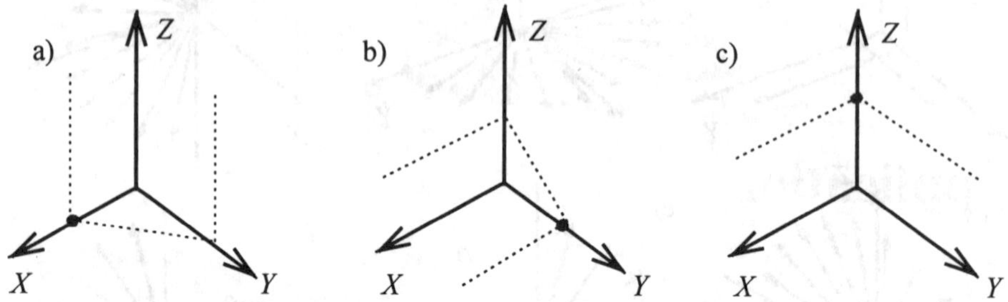

Figure 4. Special point on a symmetry plane.

Comparision of generated description with the Tables

The generated symmetry operation names were compared with the items contained in the Tables.The difference occurred only in description of two kinds of glide planes (Table 1).The first group of differences is connected with glide planes of type x,2x,z present in hexagonal axes systems; exception from the rule 3. The second group of differences containes the glide planes of the type -x,y,x; exception from the rule 4.

Table 1. List of differences in symmetry operation names.

Group symbol	Origin choice	Lattice point	Tables	SPACER
R3m R-3m		2/3,1/3,1/3 2/3,1/3,1/3	g(1/6,1/3,1/3) x,2x-1/2,z	g(1/6,1/3,1/3) x+1/4,2x,z
R3c R-3c		2/3,1/3,1/3 2/3,1/3,1/3	g(1/6,1/3,5/6) x,2x-1/2,z	g(1/6,1/3,5/6) x+1/4,2x,z
Fm-3 m Fd-3m	 1	0,1/2,1/2 0,0,0	g(-1/4,1/2,1/4) x+1/4,y,-x	g(-1/4,1/2,1/4) -x+1/4,y,x
Fm-3m Fd-3m	 1	1/2,1/2,0 1/2,0,1/2	g(1/4,1/2,-1/4) x+1/4,y,-x	g(1/4,1/2,-1/4) -x+1/4,y,x
Fd-3 m Fd-3m	2 2	0,0,0 1/2,0,1/2	g(1/4,1/2,-1/4) x+1/2,y,-x	g(1/4,1/2,-1/4) -x+1/2,y,x

Conclusions

Basing on the selected generators and the composition series method of space group generation it is possible to obtain the lists of coordinate triplets identical to that containes in the Tables. This is also true for the groups presented in the tables several times; the inconsequence in description of the group Fd-3 can be easily removed. Simple geometrical rules allow to transform coordination triplets to conventional geometrical descriptions. The space group description is well standardised in the third edition of the Tables and this can be easily implemented into the programs dealing with the space groups.

References

1. *International Tables for For Crystallography, vol.A*, ed. Th. Hahn (D. Reidel, Dordrecht, 1983).
2. Stróż, *J. Appl. Cryst.* 30, 178 (1997).

Applications of Rietveld Method in X-ray Diffraction Analysis

QUANTITATIVE PHASE ANALYSIS WITH THE RIETVELD METHOD

R.J. HILL

CSIRO Minerals, Box 312, Clayton South, Victoria 3169, Australia

Use of the full powder diffraction pattern for the determination of phase abundance in multi-phase mixtures can provide more accurate and precise results that those obtained from the employment of pre-selected pairs (or small sets) of integrated peak intensities. These advantages accrue from the fact that (i) all reflections in the pattern are explicitly included, irrespective of overlap, (ii) the background is better defined through the fitting of a continuous function to the whole pattern, and (iii) the effects of systematic errors like preferred orientation and extinction are reduced since all reflection types are considered in the analysis. The most commonly used full-pattern technique evolved as a by-product of the Rietveld method of crystal structure refinement, after it was recognised that a simple relationship pertains between the Rietveld scale factors, the unit cell contents of the component phases and their relative phase abundances[1,2]. In another established full-pattern method, the phase abundance is obtained by comparison of the full-pattern digital data from the sample with an extensive database of observed patterns compiled from standard samples of all possible contributing phases[3]. Others have derived the phase abundances by solution of simultaneous equations using the calculated mass absorption coefficients of the component phases and the scale factors derived by pattern decomposition of the mixture and its component pure phases[4]. More recent developments involve the application of powerful chemometric or statistical-mathematical methods that allow the inclusion of a wide range of property-related parameters[5,6]. This paper describes the development of full-profile phase analysis methods, with an emphasis on the Rietveld technique, including a discussion of recent advances, residual challenges and examples of their application in industry and mineralogy.

1 Traditional methods of quantitative phase analysis

1.1 Non-diffraction methods

Traditional non-diffraction methods of quantitative phase analysis (QPA) involve point counting with an optical microscope, calculation of phase areas from digital images collected by scanning electron microscopy (SEM) or optical grey scales, or manipulation of chemical analyses according to a prescribed recipe (i.e., normative calculations such as CIPW, Niggli, Bogue, etc). Some of these methods may utilize additional observations from thermal analysis or cation exchange to assist in resolving ambiguities in the analysis.

Difficulties with optical methods can arise because the samples are often too fine grained to resolve phases (*e.g.* sediments), or involve potential sampling problems when the grains are large. In addition, phases with similar optics may be misidentified due to the overlapping of transparent and coloured phases[7]. For normative methods, the phases are assumed to form in certain sequences and to have ideal compositions. While the methods are reasonably successful for certain systems, such as granites and cements, the 'recipes' are not applicable in the general case. Furthermore, accessory phases are often ignored and polymorphs cannot be treated effectively.

Image analysis using digitised SEM data can provide excellent results, although extreme care must be taken in sample preparation, in the accumulation of data of sufficiently reliable precision, and in the conversion of point phase identifications to appropriate modal analyses. Several instruments have been developed with this capability inbuilt, including the so-called QEM*SEM system[8]. Table 1 shows the results of an elemental reconciliation undertaken on a QEM*SEM phase analysis of a copper ore concentrate. The comparison provides a very good check on the phase analysis since Cu, Fe and S are each present in many phases.

Table 1: Reconciliation of the results (wt%) of QPA on a copper core concentrate using QEM*SEM (D. Sutherland, pers. comm. 1997).

Element	QEM*SEM QPA	Chemical analysis
Cu	30	32.2
Fe	25.5	27.1
Mo	1.7	2.0
S	33.9	33.2

1.2 Diffraction methods

All of the traditional diffraction methods of QPA rely on the fundamental relationship described by Klug and Alexander[9]:

$$I_{ri} = C_{ri} W_r \,/\, \rho_r \mu_m^m \tag{1}$$

where I is the intensity of reflection(s) r in phase i, C is a constant for the particular diffraction setup, W is the phase abundance, ρ is the density of the phase, and μ is the mass absorption coefficient of the mixture. Commonly used methods include the well-known internal standard technique, which allows matrix absorption effects to be eliminated (*i.e.*, calibrated out) by the addition of a known weight of a standard material. The so-called reference intensity ratio (*RIR*) or Chung[10] method is equivalent to the internal standard method, but the function of the internal standard is performed by a phase already present in the sample. In some cases the diffraction data is supplemented by chemical analysis information and the combined observations treated by linear programming techniques[11]. In other cases, Monte Carlo, principal component and other chemometric or statistical methods are applied to multiple data sets to extract the phase analyses without the need for standards or calibration[12,13]. In addition, the diffraction information may be supplemented by additional phase-specific data inferred from the presence of anomalous scattering, extinction and preferred orientation.

Difficulties with, and the limitations of the traditional diffraction methods arise primarily from pattern complexity (*i.e.*, the need for free-standing reflections), which is related to the number and symmetry of the component phases and to the instrumental resolution. Other problems can arise from (i) inhomogeneity of the sample, leading to a non-representative analysis, (ii) solid solution effects and other compositional variabilities of the components, (iii) particle size effects, leading to inadequate particle statistics; extinction and absorption contrast (microabsorption), (iv) unknown or variable crystallinity arising from surface or bulk amorphous components, (v) preferred orientation, and (vi) the difficulty in determining the intrinsic accuracy and precision of the measurements. In many case, the achievement of an accurate analysis may require very laborious customised calibration.

2 Phase analysis using the whole diffraction pattern

The solution to many of these problems lies in the use of the full diffraction pattern. This has the advantages that (i) the peak overlap problem is substantially reduced - there is no need to extract a free-standing set of integrated intensities, (ii) extinction and preferred orientation are minimised due to 'averaging' over all reflection types in the full pattern, (iii) better determination of background by fitting a continuous function over the whole pattern, (iv) the detection limit often can be reduced to <1 wt%, and (v) more reflections are used, resulting in a more accurate and precise analysis.

Several methods have emerged for quantitative phase analysis utilizing the whole diffraction pattern. These are:

(i) analysis of the pattern with chemometric or mathematical-statistical methods involving the collection of a database or 'learning' set of digitised patterns representing the range of phases likely to be encountered and with which the real sample patterns are compared and contrasted[5,6,14];

(ii) solution of simultaneous equations using the calculated mass absorption coefficients of the component phases and the corresponding scale factors derived by pattern decomposition of the mixture[4];

(iii) use of the scale factors derived in a Rietveld analysis for which the calibration constants are derived from the mass and volume of the unit cell[1,2].

2.1 Chemometric and principal component methods

Chemometrics can loosely be defined as the application of mathematical and statistical methods and logic to improve chemical measurement processes and to extract chemical information from chemical and physical data[6,15]. It is not a single technique, but rather a range of methods including calibration, statistics, method optimization, factor analysis, factorial design, pattern recognition, fuzzy logic, signal processing, library searching and neural networks. The goal of many chemometrics techniques is the production of an empirical model that allows one to estimate the properties of a system from measurements. Recent advances in this discipline have led to a new breed of analytical tools, including microprocessor-controlled 'intelligent' instrumentation and data analysis systems. The chemometric process commonly involves application of principal component analysis, discriminant analysis, covariance analysis and partial least squares.

Various early manifestations of the method to X-ray powder diffraction data have involved the analysis of selected subsets of peaks, including linear regression of peaks from multiple sets of patterns spanning the composition range of interest[12,13], combination of bulk chemical and X-ray data[16,17], application of Monte Carlo methods[18], cluster analysis[19], and fuzzy set theory[5].

One long-standing application of these methods to the whole powder diffraction pattern has been pioneered by Smith and coworkers for more than a decade[20,21,22]. It involves the simple least squares comparison of an unknown pattern with a customized database of appropriate observed patterns; this method is described in detail in Section 2.2. In more recent work, fuzzy set theory has been applied to the analysis of inorganic mixtures[5], while partial least squares methods have been applied to the quantitative phase analysis of complex bauxite ores using both X-ray and NIR data[14] and to the grain boundary phases of magnesia refractories[23], but these more sophisticated chemometrics methods have not yet received wide application in the X-ray field.

In these latter chemometrics examples, the systems were calibrated using a 'learning' set of diffraction patterns that spanned the full range of potential or actual phase compositions in the unknown samples. Mathematical factors were extracted during the calibration process using partial least squares or fuzzy set theory that permit the subsequent derivation of the phase identity/composition or, indeed, any other learned property of the unknowns in a related analysis. A typical set of results obtained for small quantities of magnesium silicate phases present in the grain boundaries of magnesite refractory bricks is shown in Figure 1[23]. This system is particularly difficult for standard X-ray QPA because of the complexity of many of the phases, their extremely low abundance, and the dominant effect of the MgO pattern.

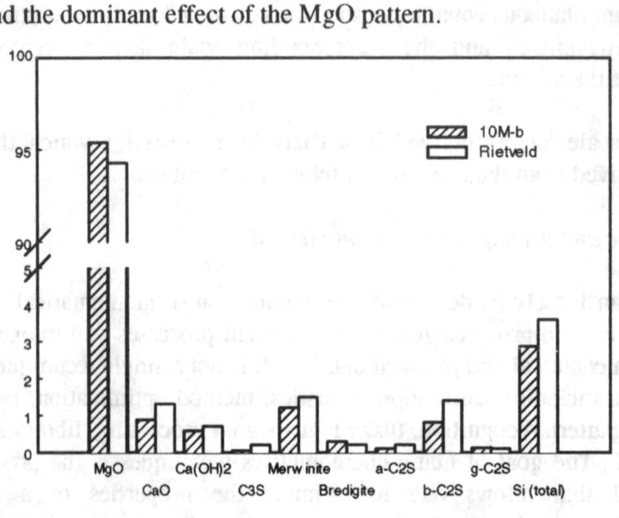

Figure 1: Comparison of QPA results (wt%) obtained using the chemometrics method (10M-b) with an analysis calculated using the Rietveld method (see below)

The limitations of chemometrics analyses of whole diffraction patterns are (i) the requirement for pure or near-pure samples, alone or in mixtures, and/or structural models to generate patterns for the database/learning set, (ii) the need to determine calibration constants or correlation 'factors' from these pure phases or mixtures, (iii) the lack of generality and representability of the benchmark database (*i.e.*, each new system requires a new learning set and multiple patterns are required for solid solution series and/or variable crystallinities), (iv) the need for reproducibility of the sample preparation methods, and (v) the incorporation within the database patterns of the characteristics of the instrument used for data collection, not just the sample.

2.2 Database of observed and/or calculated patterns

This is a subset of the generalized chemometrics method. The phase abundances of the component phases are calculated from a simple least-squares comparison of the sample pattern with the weighted sum of selected reference patterns contained in a purpose-built whole-pattern database (WPD) constructed from 'standard' samples of all possible contributing phases. The reference patterns in the database can be observed or calculated, and each has a calibration constant previously determined from synthetic mixtures of the reference phases.

This technique was first proposed by Karlak and Burnett[24] but has been substantially overhauled and enhanced by Smith and colleagues over the past 10 years. In particular, the developments have included storage of the full pattern and use of point-by-point matching of oil shales[20], use of calculated patterns and consolidation of the software into an integrated suite of programs for clay minerals[21], modelling of peak broadening effects[22,25], and inclusion of chemical constraints over a nominated compositional range for sedimentary rocks[26].

As for the generalised chemometrics techniques, execution of the Smith method can be divided into two stages: construction and calibration of the database, and analysis of the unknown mixture using this database. The first stage involves (i) collection of standard data on pure materials of likely component phases using fixed instrumental conditions, (ii) removal of background and the effects of minor impurity phases, and (iii) determination of *RIR* values from 50:50 mixtures of the pure phases. In a novel development and enhancement of the method, if suitable pure forms of the phases cannot be found, the observed patterns and *RIR*s may be replaced by calculated versions obtained from the crystal structure parameters or from d and I lists in the ICDD database.

In the second stage (i) data are collected on the unknown mixtures using identical conditions to those used for the standards, (ii) background and artifacts from the sample are removed, (iii) each database pattern is compared with the sample pattern to determine the probability of its presence in the sample, and (iv) if required, the mixture and database patterns are aligned on the 2θ scale and peak broadening effects are applied. The component phase abundances W_i are calculated from the best fit between selected database patterns and the mixture by minimizing the expression

$$\delta I(\Delta 2\theta) = I_{unk}(\Delta 2\theta) - \sum_i W_i (RIR)_i I_i (\Delta 2\theta) \quad (2)$$

where $\Delta 2\theta$ is the step interval in the diffraction angle (*e.g.*, 0.02°), I_{unk} is the diffraction intensity for the mixture over this interval, I_i is the intensity for the reference phase i, and RIR_i is the calibration constant for this phase.

It has been demonstrated for mixtures of inorganic phases that the source of the reference patterns used in the database has little effect on the resulting phase abundances (Table 2)[27].

Table 2: Effect of source of reference patterns on phase analysis results[27].

Phase	As Weighed	Experimental patterns	Calculated patterns	From ICDD database
CaF_2	30	30.8(7)	32.6(4)	31.0(3)
$CaCO_3$	10	12.8(3)	9.0(4)	10.4(4)
Al_2O_3	20	19.2(7)	16.7(7)	20.4(9)
ZnO	40	37.3(2)	41.9(3)	38.3(2)

Compositions are in wt %; figures in parenthesis show the estimated standard deviation of the determination, in terms of the least significant figure to the left

2.3 Whole pattern decomposition

Powder diffraction patterns can be decomposed into their component integrated reflection intensities for the refinement of crystal structure parameters using conventional 'single crystal' methods and, with increasing frequency over recent years, for the determination/solution of crystal structures using Patterson or direct methods[28]. By combining pattern deconvolution with the principles of traditional single-line methods, Toraya and Tsusaka[4] have derived a two-step method for phase abundance determination in which the concentrations in a mixture are obtained by solution of simultaneous equations using the calculated mass absorption coefficients of the component phases and the corresponding scale factors derived by pattern decomposition of the mixture and its component pure phases.

In the first step of this process, the patterns of single-phase samples of each of the phases present in the mixtures are decomposed separately, retaining the derived parameters of integrated intensity, unit cell and peak profile shape. In the second stage, the whole pattern for the mixture is decomposed using the pattern parameters derived for each phase in the first stage, keeping the integrated intensities fixed and allowing the fitting to proceed through individual phase scale factors. The weight fractions, W_i, are obtained by solving simultaneous equations, the coefficients of which include the scale factors and the mass absorption coefficients calculated from the chemical formulae of the phases, as follows[4]:

$$W_i = S_i \prod_{j \neq i}^{n} \mu_j^* / \sum_{j=1}^{n} (S_j \prod_{j' \neq j}^{n} \mu_{j'}^*) \qquad (3)$$

This method has all of the flexibility and other advantages of whole pattern fitting but, like the chemometrics techniques, it is a two-step process that requires pure samples (with known mass absorption) of the individual component phases to be available for analysis prior to the study of the unknown mixtures.

In a mixed full-profile, integrated-intensity variant of the Smith and Toraya methods applied to rock cuttings, Emmermann and Lauterjung[29] used full pattern profile analysis of 280 pure mineral phases held in a databank to reconstruct the complete observed diffraction pattern by fitting background functions and modeling peak profile functions. A system of linear equations was then solved to produce a set of individual phase scale factors which were related to abundance through the mass absorption coefficients and mineral densities. This method was used both to identify the constituent minerals in the sample and to quantify their abundance with a detection limit of 1-3 wt%..

2.4 Rietveld analysis

The Rietveld method of whole-pattern analysis was initially developed for the refinement of crystal structure parameters from neutron powder diffraction data[30,31] but has been expanded over the intervening years to treat multi-phase, multi-wavelength and multi-pattern diffraction data collected with neutrons, X-rays or electrons[32]. The method involves a least squares fit between step-scanned points in an observed diffraction profile and an overlaid simulated profile calculated using (for a mixture of phases):

- The crystal structure and unit cell parameters of each component phase
- Parameters describing the peak width and shape (and their change with 2θ)
- An observed or calculated background model
- A scale factor for each component phase to adjust their relative intensities
- Models for preferred orientation, size-strain, extinction, instrumental aberrations, etc

The key feature of Rietveld analysis that distinguishes it from pattern decomposition is the feedback, *during refinement*, of improved knowledge about the structure and thus improved allocation of observed intensity between partially overlapping individual Bragg reflections.

The least squares process minimizes the differences between the observed and calculated intensity at each step in the pattern, where

$$y_{ic} = y_{ib} + \sum_{k=k1}^{k2} G_{ik} I_k \qquad (4)$$

and y_{ic} is the calculated intensity at step i in the pattern, y_{ib} is the background intensity at step i, G_{ik} is a normalized peak profile function, I_k is the Bragg intensity calculated from the crystal structure of phase k, and $k_1 \dots k_2$ are all of the reflections contributing intensity to step i (from *all* component phases).

There have been many developments to the Rietveld method since its first exposure by Rietveld at a conference in 1966. These include (non-exhaustively):

- Extension to time-of-flight neutron data[33];
- Extension to X-ray data[34];
- Extension to handle two-phases in the model for neutrons[35] and for two-wavelength X-rays[36];
- Generalized peak shape functions[36];
- Introduction of bond length / bond angle restraints[37];
- Generalization for more than two phases[38];
- Introduction of the March model for preferred orientation[39];
- Simultaneous analysis of multiple data sets[40];
- Direct convolution of specimen and instrument profiles[41];
- Extension to the analysis of incommensurate structures and superstructures[42];
- Inclusion of an observed file of peak intensities when the crystal structure of a component is unknown[43];
- Application of variable step counting time[44] and variable step width[45];
- Coupled structure and energy minimization[46].

This level of accelerating activity and interest in Rietveld analysis arises from the wide variety of information and data, actual and potential, that can be obtained from a Rietveld analysis, including:

- Crystal structure parameters (this was the original goal for method development), including the determination of phase chemistry from atomic site occupancies, and atomic displacements and/or disorder from 'thermal' parameters;
- Very precise, although not necessarily accurate, unit cell dimensions[47];
- Crystallite size and strain parameters from peak shape/width parameters, provided the instrumental contribution to the peak profile is known[48,49];
- Sample texture from preferred orientation parameters, sometimes involving spherical harmonic functions[50];
- Characteristics of amorphous / disordered material by radial distribution function analysis of the background[51];
- Phase abundance of the components of multiphase mixtures from the relative scales of the individual component patterns in mixtures[52,1,2,53].

It is the last outcome, namely QPA, that is of most interest in the current context. In summary, the phase abundance W_i of a component i in a mixture of j phases is obtained from:

$$W_i = S_i(ZMV)_i / \sum_j S_j(ZMV)_j \tag{5}$$

where S_j is the refined scale factor for component phase i in a mixture of j phases, Z is the number of formula units in the unit cell, M is the mass of the formula unit in atomic mass units and V is the volume of the unit cell in Å^3 [1,2]. In essence, the product ZMV is the Rietveld analysis equivalent of the traditional *RIR* for the phase. In other manifestations of this methodology[53], the constant ZMV is replaced by the product ρV^2, where ρ is the phase density.

The major advantage of the Rietveld analysis method in phase analysis is that the component pattern intensities are calculated on an absolute scale during the refinement so that, under ideal conditions, the 'calibration' constant ZMV can be simply calculated from the mass and volume of the unit cell, rather than from a laborious series of experiments undertaken on a series of 'standard' samples.

3 Practical application of Rietveld analysis for QPA

3.1 *History of development*

The application of the Rietveld method for quantitative phase analysis can be traced through a series of developments from Werner *et alia*[52] who used the Rietveld atomic occupancy factors to determine the amount of quartz in binary mixtures:

- First use of the Rietveld phase scale factors, embodied in the ZMV relationship[1];
- Generalized derivation of the ZMV relationship from diffraction optics[2,53];
- Use of mass absorption coefficients in place of ZMV constants[54];
- Construction of an inbuilt database of crystal structure parameters and default refinement options as an aid to novice users of the method[55,56];
- Post-refinement correction for absorption contrast, or microabsorption, according to the Brindley[57] model[58];
- Experimental calibration of the ZMV 'constants' in order to account for the effects of microabsorption and other effects not able to be modeled in the Rietveld refinement[59];
- Inclusion of a file of experimental/observed peak intensities for those components in the sample for which no adequate crystal structure data is available for use in the Rietveld refinement model[43];
- Inclusion of double peakwidth variation functions to account for systematic, Miller-index-related variations in peak width, such as that arising from disordered clay materials[60].

3.2 *General requirements and protocol for Rietveld analysis*

The general requirements for *ab-initio* (*i.e.*, no calibration) quantitative analysis with the Rietveld method are that: (i) good quality step-scan diffraction data are available over a wide range of diffraction angles[61,44]; (ii) all components in the sample are identified and have known crystal structures (but note recent developments by Taylor and Zhu Rui[43], listed above); (iii) the sample displays no significant microabsorption, extinction, preferred orientation or surface roughness effects; (iv) the variations in peak width and shape are all able to be modeled effectively; (v) the user has a reasonable knowledge of crystallography and experience with the Rietveld method in structure analysis.

A practical protocol for data collection as a precursor to Rietveld refinement has been provided in the IUCr CPD Round Robin on Rietveld Refinement[62,47]. The more important of the recommendations from these studies are:

- Collect, and include in the refinement, the maximum effective range of *d*-spacings. This requires that the data include *d*-spacings down to at least 1Å and yield an observations to parameter ratio of at least 5.
- Maximize the instrumental resolution and peak-to-background ratio. This latter ratio should be at least 50.
- Ensure adequate counting statistics, especially at the small *d*-spacing (high-angle) end of the pattern. This usually means that more than 5,000 counts should be collected at the top of the major peaks and that a step width should be used that is around 1/5 of the peak full width at half maximum (FWHM) of the narrowest peak.

3.3 Calibration of the Rietveld method

In the event that the sample displays significant microabsorption, extinction, multiple preferred orientation, surface roughness, structural uncertainty, partial amorphicity or variations in peak width and shape that are not able to be modeled effectively, the *ZMV* values calculated *ab initio* will not provide an accurate estimate of phase abundance. In these cases, it is possible to determine experimental values A_i for the *ZMV* 'constants' for each phase *i* from a one-off Rietveld analysis of a synthetic mixture of all phases with a standard phase *s*[59].

$$A_r = W_r S_s (ZMV)_s / W_s S_r (ZMV)_r \tag{6}$$

An extreme example of the need for calibration of the *ZMV* values in a Rietveld analysis is demonstrated in Table 3 using the results of X-ray and neutron studies of a five-phase simulated lead-acid battery positive plate consisting of lead oxides and lead sulphate with KCl added as an internal standard, all in equal abundance. The neutron data results show good agreement with the as-weighed abundances when normalized to the content of the internal standard, and the phase abundances sum to 96%, showing that there is little if any surface amorphous material present. These results are typical of the accuracy that can be achieved with the *ab-initio* Rietveld method when the data is relatively unaffected by microabsorption and other systematic errors.

On the other hand, the uncalibrated X-ray results are in very poor agreement with the as-weighed compositions due to the presence of substantial size differences and severe absorption contrast between the different lead phases in the sample. The final row of the table gives the calibration factors which must be applied to the *ab initio ZMV* values in order to reproduce the known phase composition. The corrected *ZMV* values can be used in subsequent analyses of 'real' samples when these form a related series of mixtures.

Table 3: Calibration of X-ray Rietveld *ZMV* constants for a synthetic lead-acid battery positive plate.

	α-PbO_2	β-PbO_2	$PbSO_4$	α-PbO	β-PbO	KCl	Σ
Weighed	16.7	16.7	16.7	16.7	16.7	16.7	100
Neutrons	15.5(14)	14.9(10)	15.5(13)	15.1(10)	18.3(13)	16.7	96.0
X-rays	9.2(10)	16.7(9)	27.3(17)	20.0(11)	7.1(7)	16.7	97.0
Calibration factor	1.82	1.00	0.612	0.835	2.35	-	-

Compositions are in wt %; figures in parenthesis show the estimated standard deviation of the determination, in terms of the least significant figure to the left

3.4 What if the crystal structure of one or more components is unknown?

The Rietveld method relies on the fact that the crystal structure parameters of the component phases are known and can therefore be used to provide a calculated pattern against which the observed one is compared. In the event that this structural information is not available, Taylor and Zhu Rui[43] have proposed a novel adaptation and enhancement of the Rietveld method in which the observed reflection intensities are incorporated into the analysis. In this method, a file of reflection intensities F_{obs}(hkl) is measured from an observed pattern of the phase whose structure is unknown and these intensities are substituted for the calculated intensities F_{calc}(hkl) normally derived in the Rietveld analysis from crystal structure parameters. The Rietveld analysis then proceeds as usual. The F_{obs}(hkl) can be extracted from the patterns of pure samples or from the ICDD database, as appropriate. Their intensities are scaled analytically by assuming ΣF_{calc}(hkl) = ΣF_{obs}(hkl), or by using *RIR*s derived from standard mixtures.

Of course, in the case where several of the component phases in the mixture have unknown, partially-known, or improperly-modeled structures, the advantages of using Rietveld analysis are substantially diminished. It is then more appropriate to use one of the structure-independent methods, such as the pattern decomposition technique[4] or one of the chemometrics techniques that treats the pattern purely as a spectrum, rather than a diffraction pattern[3,5]. These alternative methods have been discussed above.

3.5 Comparison between Rietveld and whole pattern database methods

Figure 2 provides a comparison of the results of phase analyses undertaken on a seven-phase mixture of minerals using the Rietveld and whole pattern database (WPD) methods[20]. The results show that, for this system, the accuracies of the two techniques are very similar over a wide composition range.

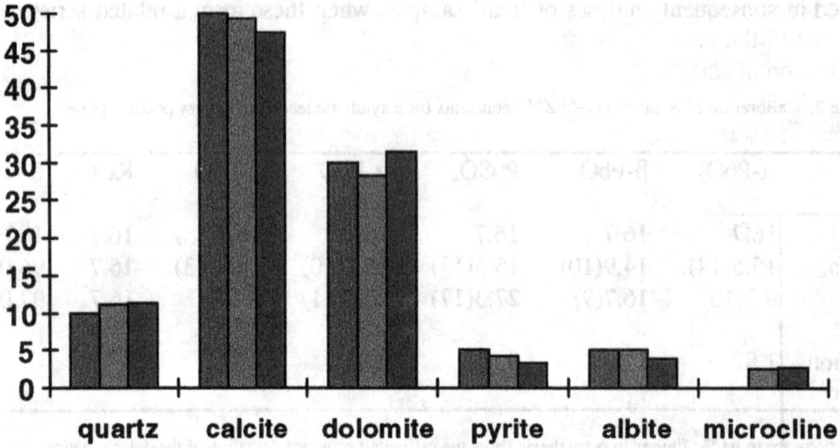

Figure 2: Comparison of QPA results obtained with the Rietveld and WPD methods[20].
From left to right within each block, the histograms represent the real, WPD and Rietveld abundances.

4 Applications of Rietveld analysis for QPA

The Rietveld method has emerged in recent years as the method of choice for QPA in the chemistry, mineralogy and materials science literature. The following section provides a small selection of recent QPA applications of the method, mostly from within the author's laboratory.

4.1 Rocks and ores

One of the most stringent tests of quantitative X-ray powder diffraction phase analysis is in the modal analysis of rocks and ores. These systems are particularly challenging because of the often large number of phases in the samples, the complexity (i.e., low symmetry) and hence high degree of overlap within and between their powder diffraction patterns, and the presence of multiple preferred orientation, microabsorption and chemical variability of the component phases. A number of author's[53,63,7,64,65] have demonstrated the utility of Rietveld analysis for the modal analysis of these systems by undertaking systematic studies of a wide range of rock and ores suites and comparing the results so obtained with those from other methods.

A comparison of the results obtained for an adamellite (fine-grained igneous rock) using optical, normative and X-ray Rietveld methods of modal analysis[7] is shown in Table 4. In this example, X-ray Rietveld analysis was the only method (*c.f.*, optical point counting and normative calculations) able to discriminate between all three of the feldspar minerals in the sample. Where direct comparisons between the methods are possible, as in the case of the measured abundances of quartz and orthoclase, the agreement between the methods is very good. However, the amount of biotite is overestimated in the optical, relative to the Rietveld and normative methods, due to the misidentification of some colourless phases which overlap with the biotite.

Table 4: Comparison of optical, normative and X-ray Rietveld determinations of phase abundance (wt%) in adamellite[7].

Phase	Optical point counting	Normative	X-ray Rietveld
Quartz	34.3(23)	35.0	34.4(11)
Orthoclase	} 53.2(8)	19.1	19.2(9)
Albite		} 39.4	9.6(10)
Plagioclase			30.0(16)
Biotite	12.6(10)	6.5	6.9(3)

Figure 3 shows the results of X-ray and neutron Rietveld refinement of six phases in a Pb-Zn-Ag ore[65]. In this case, comparison of (i) the X-ray and neutron results and (ii) the measured bulk chemical composition of the sample and that calculated from the X-ray modes shows that the analysis is corrupted by the presence of absorption contrast (*viz,* microabsorption) between the sulphide and silicate components. The chemical, X-ray and neutron analyses are brought into closer alignment by applying a correction for microabsorption according to the Brindley model[58], but a substantial discrepancy remains. In this case, the assumption of a uniform particle size for all phases of 5 microns (and perhaps the Brindley model itself[57]) is clearly inadequate and separate calibration of the *ZMV* values using a synthetic mixture of the component phases is required for an accurate analysis.

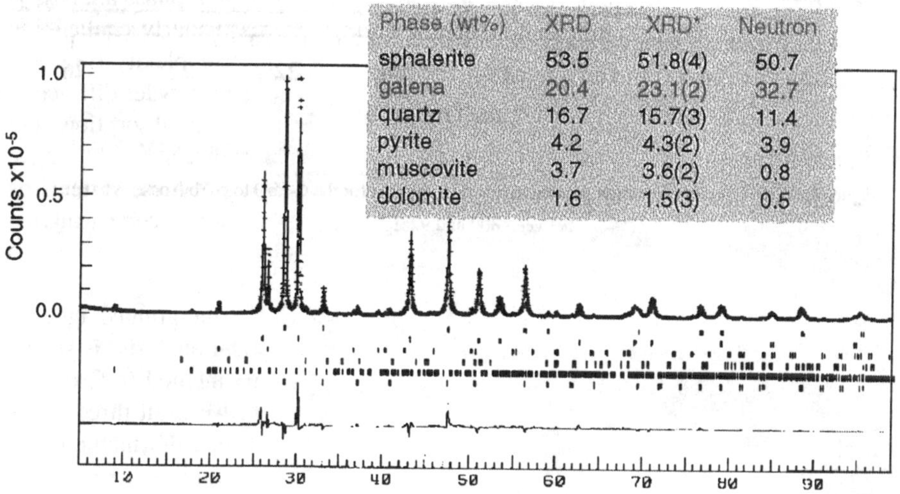

Figure 3: Rietveld refinement plot for a Pb-Zn-Ag sulphide ore[65].

4.2 α-PbO to β-PbO phase transformation

Figure 4 is a 'waterfall' plot of a small combined section of the neutron powder diffraction patterns collected from α-PbO and β-PbO at temperatures between 295 and 925K on the D2B diffractometer at ILL[66]. The figure shows the gradual appearance of β-PbO at temperatures above 850K and the simultaneous disappearance of α-PbO up to about 910K. Rietveld analyses of these patterns, collected every 60 minutes, provide a quantitative measure of the phase transformation kinetics (Figure 5), while at the same time revealing the details of the atomic structural changes experienced by both phases during the transformation (*i.e.*, atomic coordinate and displacement parameter shifts).

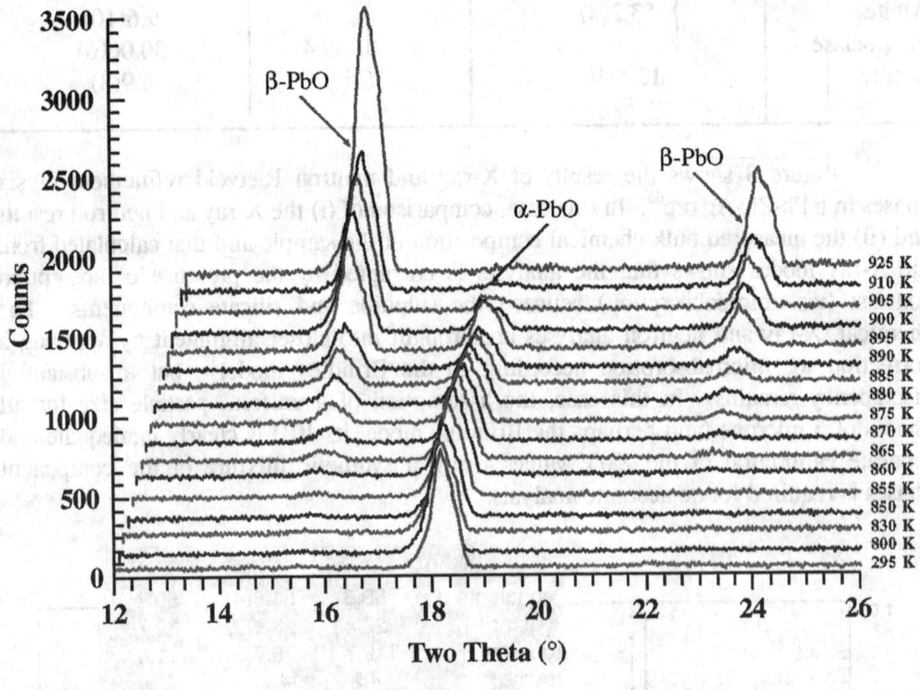

Figure 4: Waterfall plot of neutron powder diffraction patterns for the α-PbO to β-PbO phase change between 800 and 925K[66].

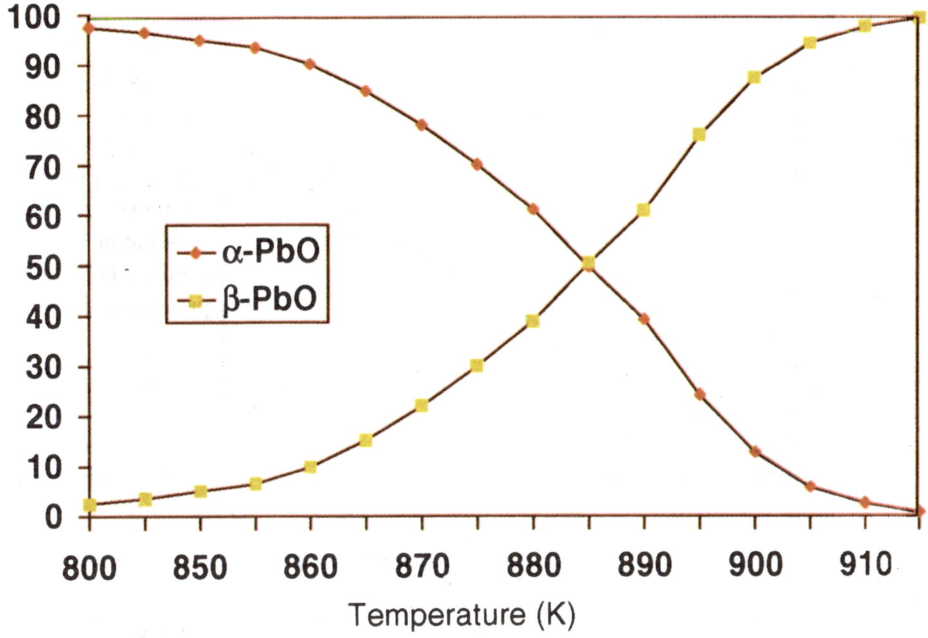

Figure 5: Rietveld QPA results for the α-PbO to β-PbO phase change between 800 and 915K[66]

4.3 Bauxites

More than 160 minerals have been identified in bauxite ores, and 20 of these can be present in rock-forming quantities[11]. Bauxites are of substantial industrial importance since they provide feedstock for the production of alumina by the Bayer process for subsequent smelting into aluminium metal. Hence, there have been many X-ray diffraction studies of this complex mineralogical system[67,68].

The Rietveld-based studies of Aylmore and Walker[68] have provided perhaps the most comprehensive and accurate X-ray diffraction QPA results to date on bauxites. An example of the results from their work is shown in Figure 6 which provides a comparison of the component mineral weight percentages determined for a range of 'real' bauxites by Rietveld QPA with calibration modes calculated by a 'standard' method akin to a Bogue analysis. The relatively wide spread of points around the 1:1 relationship is a demonstration of the extreme complexity of this system and of the resultant difficulty in adequately accounting for all of the chemical and structural variabilities of the phases within and between samples. The accuracy and precision of these determinations was improved by the ability of the Rietveld analysis to successfully model several goethites with different levels of isomorphous substitution. Furthermore, from analysis of the peak shapes and widths, information was obtained on phase crystallinity and crystal size that influence their reactivity in the Bayer process.

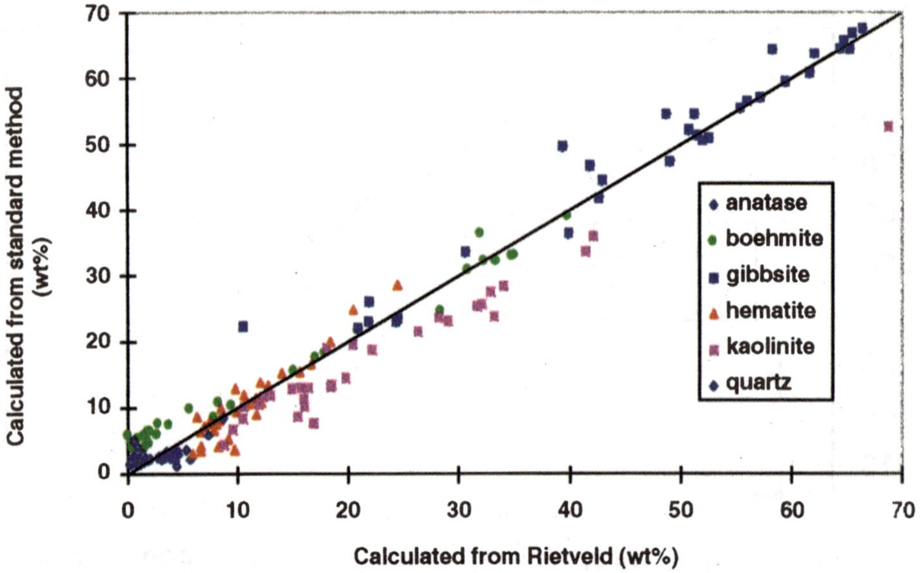

Figure 6: Rietveld QPA results versus 'standard' modes calculated by a Bogue-like method[68].

Walker has recently extended the bauxite X-ray studies through the use of chemometrics methods[14]. These have not only improved the QPA accuracy and precision, but have the potential to quantify the determination of reactivity and other 'soft' but industrially important parameters such as 'available' alumina and 'reactive' silica. In these chemometric studies, as for Rietveld analysis, the quality of the results is heavily dependent on the time invested in sample preparation, data collection and calibration of the learning set of standard patterns.

4.4 Portland cement

QPA of Portland cement by X-ray methods is a particularly difficult task due to the combined effects of (i) complex and relatively weak diffraction patterns, with the major peaks from the main phases all concentrated (and overlapped) in the region 30-40°2θ (for Co Kα radiation), (ii) significant variability in composition, crystal structure and crystallinity, (iii) the presence of many polymorphs with closely similar crystal structures and powder patterns, and (iv) the lack of adequate reference materials. These and other problems have been discussed by several authors[69,70,71].

The application of Rietveld analysis to this system is hindered by the fact that the crystal structures of some of the key phases are complex and poorly characterised and therefore cannot be easily incorporated into the model for the calculated diffraction patterns; this is especially true for alite, a substituted impure form of tricalcium silicate (Ca_3SiO_5 or C_3S). In a recent advance, Taylor and Aldridge[71] have derived a set of observed integrated intensities (the '*hkl*' file) from experimental alite diffraction profiles and included these into the Rietveld method, as though they had been calculated from the crystal structure parameters, as normally occurs. This technique not only produced the best fits to the observed cement profiles, but achieved this in computing times that were shorter by up to a factor of five over conventional Rietveld refinements. The final QPA results are compared in Table 5 with average results obtained in a round robin study of a standard cement sample[72].

Table 5: Rietveld and standard QPA results for Portland cement (wt%)[71].

Phase	Rietveld with '*hkl*' file[43]	Known composition[72]
Alite, C_3S	69.7(8)	68.9
Belite, β-C_2S	13.6(6)	13.4
C_3A	6.2(5)	6.7
C_4AF	7.6(7)	8.0
Gypsum	3.0(2)	-

C = CaO; S = SiO_2; A = Al_2O_3; F = Fe_2O_3

5. Summary and Conclusions

5.1 Comparison of the Chemometrics and Rietveld methods for QPA

The characteristics and relative advantages of the chemometrics full-pattern database / learning set method, and the Rietveld method of phase analysis are summarized, for simplicity, in Tables 6 and 7.

5.2 The future

The chemometrics and Rietveld methods of QPA have already begun to merge into intermediate ground through (i) in the Rietveld method, the incorporation of observed diffraction patterns when calculated data cannot be generated, and (ii) in the chemometrics method, through the increasing use of calculated patterns and/or factors that model the peculiarities of the data from unknown samples when 'standard' patterns are not available in the database or learning set for all components in the mixture. This merging of the two methods is summarized schematically in Figure 7.

Table 6: Characteristics of chemometrics and Rietveld methods of QPA.

	Chemometrics / database method	Rietveld analysis method
Based on	Customized database of observed patterns of mixtures or pure phases	Calculated patterns from crystal structures
Fitting process	Least-squares sum of data-base patterns or more complex statistical-mathematical methods	Full matrix least-squares refinement of structure and pattern profile parameters
Calibration	Measured *RIR*s or factor analysis using a learning set	Calculated '*ZMV*' constants or analysis of a standard mixture
Phase analysis	From *RIR*s or correlation factors	From scale factor and *ZMV*
Utilization	Limited, partly due to need for calibration through an extensive learning set of observed patterns	Extensive, and growing rapidly as experience and software availability increases

Table 7: Relative advantages of chemometrics and Rietveld methods of QPA.

Chemometrics / database method	Rietveld analysis method
Knowledge of crystal structures required	No pattern database or learning set required for calibration
'Difficult' features of sample and instrument calibrated out by use of observed data in the learning set	Updated cell and compositional information (i.e., *ZMV* values) obtained for component phases interactively during the analysis
-	Single pass operation

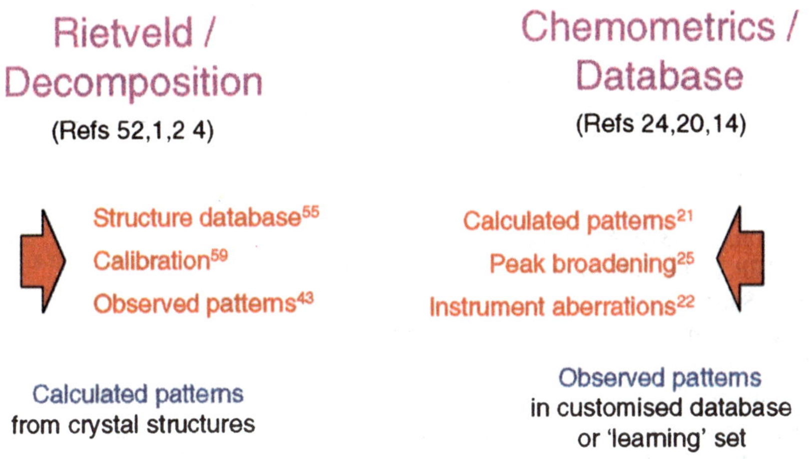

Figure 7: Relative developments of the Rietveld and chemometrics methods of QPA.

For Rietveld analysis the future of quantitative phase analysis applications is likely to involve the following developments:

- Better models for the effect of the sample and the instrument - increasingly necessary as the resolution of instruments improves and the complexity of the phases under study increases;
- Widespread use of spherical harmonics for the modelling of preferred orientation;
- More effective models for extinction and microabsorption (especially in the case of X-ray data);
- Generalised methods to treat anisotropic variations of the cell dimensions and peak widths and shapes - the best resolution instruments are already showing up anisotropies to some degree in almost all materials;
- Widespread incorporation of constraints and restraints such as rigid-body, distance least squares, bulk chemistry and lattice energy minimisation - required as the crystal structure complexity increases and the robustness of the Rietveld refinement suffers accordingly.

In parallel with the development and enhancement of Rietveld analysis for structure and phase abundance determinations, in chemometrics nothing short of a quiet revolution is taking place. The method is already used across a very broad range of applications in chemistry, materials science and industry[6] and it is inevitable that it will be applied increasingly to the more intractable problems in QPA..

Thus, despite many potential challenges, whole pattern fitting methods of one form or another are here to stay and independently, or in combination, they will continue to provide among the best quantitative phase analysis results reported in the literature to date.

6 Acknowledgements

The author is grateful to Messrs Graham Walker and Stephen Tassios and Dr David Sutherland for kindly making available figures and data prior to publication.

7 References

1. R.J. Hill, (1983) *J. Power Sources*, **9**, 55-71.
2. R.J. Hill, and C.J. Howard, *J. Appl. Crystallogr.* **20**, 467-474 (1987)
3. D.K. Smith, G.G. Johnson, A. Scheible, A.M. Wims, J.L. Johnson, and G. Ullmann, *Powder Diffraction* **2**, 73-77 (1987).
4. H. Toraya, and S. Tsusaka, *J. Appl. Cryst.* **28**, 392-399 (1995)
5. B. Adler, P. Schütze, and J. Will, *J. Anal. Chim. Acta*, **271**, 287 (1993).
6. S.D. Brown, S.T. Sum, and F. Despagne, *Anal. Chem.* **68**, 21R-61R (1996)
7. R.J. Hill, G. Tsambourakis and I.C. Madsen, *J. Petrology* **34**, 867-900 (1993)
8. D.N. Sutherland and P. Gottlieb, *Minerals Engineering* **4**, Nos 7-11, pp 753-762, (1991)
9. H.P. Klug, and L.E. Alexander, *X-ray Diffraction Procedures for Polycrystalliune and Amorphous Materials*, 2nd Ed., New York, Wiley, 1974
10. F.H. Chung, *J. Appl. Cryst.* **7**, 519-525 (1974)
11. G. Bardossy, L. Bottyau, P. Gado, A. Griger and J. Sasvari, *Amer. Mineral.* **65**, 135 (1980)
12. J. Fiala, *Anal. Chem.* **52**, 1300-1304 (1980)
13. L.S. Zevin, and I.M. Zevin, *Powder Diffraction*, **4**, 196-200 (1989)
14. G.S. Walker, Personal communication, 1997
15. J.J. Workman, P.R. Mobley, B.R. Kowalski, and R. Bro, *Appl. Spectr. Rev.* **31**, 73-124 (1996)
16. G.E. Braun, *Clays and Clay Minerals*, **34**, 330 (1986)
17. K.P. Zangalis, *J. Appl. Crystallogr.* **24**, 197 (1991)
18. C.R. Gonzalez and R. Roque, *Acta Cryst.* **A43**, 622 (1987)
19. B. Liao and J. Chen, *J. Appl. Crystallogr.* **25**, 336 (1992)
20. D.K. Smith, D.M. Kerrick, S.M. Sterner and W.S. Meddaugh, *Adv. X-ray Anal.* **28**, 169-174 (1985)
21. D.K. Smith, G.G. Johnson and C.O. Ruud, *Adv. X-ray Anal.* **29**, 217-224 (1986)
22. D.K. Smith, G.G. Johnson and S.Q. Hoyle, *Adv. X-ray Anal.* **37**, 67-77 (1994)
23. S. Tassios, Personal communication, 1997
24. R.F. Karlak and D.S. Burnett, *Anal. Chem.* **38**, 1741-1745 (1966)
25. D.K. Smith, M.C. Nichols and M.E. Zolensky, *POWDIO. A FORTRAN IV program for calculating X-ray powder diffraction patterns - version 10.* The Pennsylvania State University, University Park, PA. (1983)
26. D.K. Smith, G.G. Johnson, M.J. Kelton and C.A. Anderson, *Adv. X-ray Anal.* **32**, 489-496 (1989)
27. D.K. Smith, G.G. Johnson and A.M. Wims, *Aust. J. Physics* **41**, 311-321 (1988)
28. J.I. Langford and D. Louër, *Rep. Prog. Phys.* **59**, 131-234 (1996)
29. R. Emmermann and J. Lauterjung, J. *Scientific Drilling*, 269-282 (1990)
30. H.M. Rietveld, *Acta Cryst.* **22**, 151-152 (1967)

31. H.M. Rietveld, *J. Appl. Crystallogr.* **2**, 65-71 (1969)
32. R.A. Young, (ed.) *The Rietveld Method*, IUCr, Oxford Univ. Press, 1993
33. D.L. Decker, R.A. Beyerlein, G. Roult and T.G. Worlton, *Phys. Rev.* **B10**, 3584 (1974).
34. R.A. Young, P.E. Mackie and R.B. von Dreele, *J. Appl. Crystr.* **10**, 262-269 (1977)
35. T.G. Worlton, J.D. Jorgensen, R.A. Beyerlein and D.L. Decker, *Nucl. Instrum. Methods*, **137**, 331-337 (1976)
36. D.B. Wiles and R.A. Young, *J. Appl. Cryst.* **14**, 149-151 (1981)
37. C. Baerlocher, *XRS-82*. The X-ray Rietveld System. Institut für Kristallographie, ETH, Zurich, 1982
38. R.J. Hill, *SR1*: A Program for the Refinement of Crystal Strctures with the Rietveld Method (unpublished), 1985
39. R.J. Hill and C.J. Howard, *LHPM, A Computer Program for Rietveld Analysis of Fixed-Wavelength X-ray and Neutron Powder Diffraction Patterns*. Report M112, AAEC (now Ansto) 1986
40. J.K. Maichle, J. Ihringer and W. Prandl, *J. Appl. Crystallogr.* **21**, 22-27 (1988)
41. S.J. Howard and R.L. Snyder, *J. Appl. Cryst.* **22**, 238-243 (1989)
42. A. Yamamoto, M. Onoda, E. Takayama-Muromachi, F. Izumi, T. Ishigaki and H. Asano, *Phys. Rev.* **B42**, 4228-4239 (1990).
43. J.C. Taylor and Zhu Rui, *Powder Diffraction*, **7**, 152-161 (1992)
44. I.C. Madsen and R.J. Hill, *J. Appl. Crystallogr.* **27**, 385-392 (1994)
45. A.C. Larson and R.B. von Dreele, Los Alamos National Laboratory Report No. LA-UR-86-748 (1987)
46. J.J. Liang, F.C. Hawthorne, Novák, M. and Cerny, P. *Can. Mineral.* **33**, 859-865 (1995)
47. R.J. Hill and L.M.D. Cranswick, *J. Appl. Crystallogr.* **27**, 802-844 (1994)
48. J.P. Cline and R.L. Snyder, *Adv. X-ray Anal.* **30**, 447-456 (1987)
49. P. Thompson, D.E. Cox and J.B. Hastings, *J. Appl. Cryst.* **15**, 581-589 (1987)
50. R.B. von Dreele, Personal communication, 1996
51. J.W. Richardson and J. Faber jr, *Adv. X-ray Anal.* **29**, 143-152 (1986)
52. P.-E. Werner, S. Salomé, G. Malmros and J.O. Thomas, *J. Appl. Cryst.* **12**, 107-109 (1979)
53. D.L. Bish and S.A. Howard, *J. Appl. Cryst.* **21**, 86-91 (1988)
54. B.H. O'Connor and M.D. Raven, *Powder Diffraction*, **3**, 2-6 (1988)
55. I.C. Madsen and R.J. Hill, *Powder Diffraction*, **5**, 195-199 (1990)
56. J.C. Taylor, *Powder Diffraction*, **6**, 2-9 (1991)
57. G.W. Brindley, *Phil. Mag.* **36**, 347-369 (1945)
58. J.C. Taylor and C.E. Matulis, *J. Appl. Crystallogr.* **24**, 14-17 (1991)
59. R.J. Hill, *Powder Diffraction*, **6**, 74-77 (1991)
60. J.C. Taylor and C.E. Matulis, *Powder Diffraction*, **9**, 119-123 (1994)
61. R.J. Hill and I.C. Madsen, *Powder Diffraction*, **2**, 146-162 (1987)
62. R.J. Hill, *J. Appl. Crystallogr.* **25**, 589-610 (1992)
63. D.L. Bish and J.E. Post, *Amer. Mineral.* **78**, 932-940 (1993)
64. W.G. Mumme, G. Tsambourakis, I.C. Madsen and R.J. Hill, *J. Sedimentary Res.* **66**, 132-138 (1996)

65. W.G. Mumme, G. Tsambourakis and L.M.D. Cranswick, *Neues Jahrb. Mineral.-Abhand.* **170**, 231-255 (1996)
66. R.J. Hill, L.M.D. Cranswick and T. Vogt, Unpublished ILL data (1989)
67. D.K. Smith, *Adv X-ray Anal.* 35, 1-15 (1992)
68. M.G. Aylmore and G.S. Walker, *Powder Diffraction*, (in press, 1997)
69. H.F.W. Taylor, *Cement Chemistry*, Academic Press Inc., San Diego, CA, 1990
70. L.J. Struble, *Cement, Concrete and Aggregates*, **13**, 97-102 (1991)
71. J.C. Taylor and L.P. Aldridge, *Adv X-ray Anal.* **36**, 309-314 (1993)
72. L.P. Aldridge, *Cem. Concr. Res.* **12**, 381-398 (1982)

THE NEW SEIFERT RIETVELD PROGRAM AND ITS APPLICATION TO QUANTITATIVE PHASE ANALYSIS

T. TAUT

Rich. Seifert & Co. Freiberger Präzisionsmechanik GmbH, Am St. Niclas Schacht 13, 09599 Freiberg, Germany

R. KLEEBERG

Freiberg University of Mining and Technology, Mineralogical Institute, Brennhausgasse 14, 09596 Freiberg, Germany

J. BERGMANN

Ludwig-Renn-Allee 14, 01217 Dresden, Germany

The newly developed BGMN Rietveld program is presented. Its unique features which ensure an excellent stability are described. An application of fully automatic refinement of samples with wide composition ranges and strong preferred orientation is included. Complex real structure models are introduced by BGMN into Rietveld refinement. As an example the phase analysis for different chlorite minerals with a common real structure model is described.

1 Introduction

The Rietveld method has become a powerful tool for quantitative phase analysis since many years. Nevertheless some inconveniences have limited its use in routine analysis. Such problems are the joint refinement of profile and structural parameters which causes parameter correlation between the two and can lead to physically meaningless values. Often the program crashes if parameters reach forbidden ranges. Phases with strong preferred orientation (PO), e.g. phyllosilicates, and/or multiple cleavage, e.g. feldspars, cannot be refined with the commonly used March model. Although this list is not complete, it illustrates the possible frustration of the novice Rietveld user. The only solution to get reliable results is a carefully planned refinement strategy, in which - step by step - the number of free parameters increases. Of course this requires long computer sessions if the problem is not simple.

The goal of our development was the complete elimination of numerical instability and bad convergence to make a fully automatic refinement possible. In addition a better model for correction of PO had to be implemented. The description of disordered structures like layer silicates should be introduced as a complete new feature into routine Rietveld refinement.

2 Features of the Program BGMN

2.1 *Numerical Basis: Advanced Peak Model and New Optimization Algorithm*

First, the strong correlation between profile and other parameters had to be eliminated. The only solution was a development of a complete new profile model (Bergmann et al. 1997). This new profile model strictly separates the wavelength distribution of the X-ray source, the instrumental function and the sample function (individual sum of a variable number of squared Lorentzian functions at each angular position) and models the sample function taking into account peak broadening due to crystallite size and microstrain.

The wavelength distributions which are emitted by various X-ray tubes are predetermined. Before starting the refinement the user has only to check if the instrumental function is already computed. If not, one has to compute this function only once for a given geometrical setup. During the refinement only the parameters of crystallite size and microstrain are varied. The profile portion which describes the instrument remains constant.

Next, a new numerical algorithm was introduced. This algorithm has in common an optimum convergence behavior for the refinement of X-ray diffraction data. But it has a valuable extension: In the new algorithm physical meaningful parameter ranges with lower and upper limits can be defined without lengthening the computation time. This is extremely useful to avoid parameter values which could be causes for a program crash, for example negative peaks widths.

In sum the above mentioned developments lead to a very important feature of the program: **There is no need for intermediate models and the refinement is carried out in only one step.** Any user interaction during the refinement is not necessary nor possible. The success of the refinement depends only on the chosen starting model.

2.2 Correction of Preferred Orientation

On of the most important rules in X-ray powder diffractometry is: 'The sample has to be prepared carefully to avoid PO'. But often it is not possible to prepare such an ideal sample. Many minerals have multiple cleavage and/or crystallite shapes which force PO. As an example needle-shaped crystallites of synthetic diaspore for use in ceramics were investigated.

Analyzing such a mineral requires a suitable model for correction of PO. The well known March model does not work because of the multiple preferred orientation combined with large correction factors.

Because of that a model using spherical harmonics was implemented. This model can correct not only strong PO in one direction but also multiple ones. The program uses spherical harmonics up to the 8th order. This leads in maximum to 45 parameters in the triclinic crystal system. Because of restrictions due to crystal symmetry the number of free parameters reduces if the symmetry raises. As a rule one has to choose a higher order model if the crystal symmetry is higher.

The example diaspore (15 parameters for PO) exhibits correction factors between 0.4 and 3 in different crystallographic directions. A satisfactory agreement between measured and calculated data is the result.

In addition to the implementation of this extended correction model the program includes an algorithm which automatically reduces the order of the model if the intensities of the peaks are too low. So you can always start with the maximum model even if the sample contains only low abundance of the phase.

2.3 Complex disordering models

There are a number of minerals which crystalline order is noticeable disturbed. Important examples for this behavior are layer silicates.

They are distorted by stacking faults. The layers are shifted against each other. These shifts may be well defined and have discrete values or can be poorly defined and yield to two-dimensional diffraction effects. Natural samples may contain domains of different degree of disorder. In comparison to an ideal structure, the resulting diffraction pattern shows depending on hkl variable peak broadening, shifted peak positions and asymmetric, tailed reflections.

Conventional Rietveld programs are of limited use in this case because they cannot handle these effects. That's why a new approach to introduce real structure models into Rietveld refinement was implemented in the BGMN software. A hard coded model was not practicable because the nature of real structure varies between different mineral groups or structure types. An interpreter for a structure description language was the solution. It can be used to introduce additional parameters for describing disordering effects. Unlimited coupling of parameters is possible. Shifting of peak positions for reflections of different classes can be introduced to approximate the two-dimensional diffraction effects. To approximate the presence of domains having different degree of disorder sub-phases are definable. They share global parameters, for instance lattice constants, and have individual parameters, for instance broadening model for selected Bragg peaks.

Chlorite minerals are an example for an already implemented real structure model. They are characterized by a well defined disordering (semi-random stacking). The layers are shifted from one to another by exactly b/3 along the three pseudo-hexagonal Y axes. The resulting X-ray diagram shows sharp and intensive 00l reflections, relatively sharp and less intensive reflections with $k=3n$ and broadened or erased reflections with $k\neq 3n$.

The first refinement using the ideal structure of Steinfink (1958) without disordering model showed a correct fit of the $k=3n$ peaks and the problem of disappeared lines with $k\neq 3n$. The tailed band 02, 11 was not fitted at all by the peaks in this region. The reliability factor R_{wp} was about 21 %.

Finally, a far more complex model had to be formulated (Kleeberg et al. 1997). A rod-like intensity distribution in reciprocal space was assumed for the peaks with $k\neq 3n$. These rods are projected by the direction cosine on the diffraction pattern. The length of the rods are a new parameter introduced in the structure description. As an approximation, the resulting band was described by individually broadened hkl reflections. The angular shift of 0k0 peaks with $k\neq 3n$ was refined by a new parameter depending on the dimension of the rods. The width of 0k0 was also parametrized. The relatively sharp reflections with $k=3n$ were described by an anisotropic broadening model due to microstrain. We used two sub-phases to match differing degrees of disorder in the sample. The two sub-phases were refined by using identical lattice parameters and identical PO correction. Their mixing ratio was also calculated. The dimension of reciprocal lattice rods and the normal peak broadening of the sub-phases were refined in a constrained ratio.

The result (Fig. 1) is a much better difference plot and a dramatic decrease of the reliability factor (R_{wp} = 13 %). The separately broadened $k\neq 3n$ reflections merge into a sufficiently described bank-like region. The other reflections are better fitted because of the distinct shape parameters for different classes of peaks.

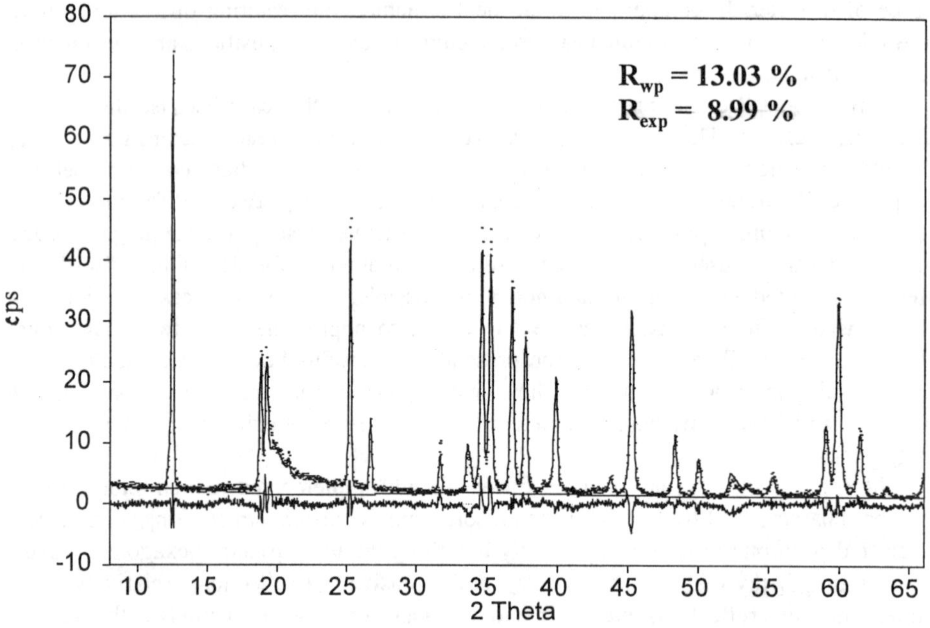

Fig. 1: Refinement result for ripidolite using the disordering model. CuKα radiation, transmission geometry.

3 Application in Phase Analysis

3.1 Minor Phases in Aluminum Oxide-Hydroxide System

A number of samples containing diaspore, bayerite, boehmite, corundum and gibbsite with strongly varying compositions had to be analyzed. This sample system is a good example for the program use in principle:

- First, the device function has to be chosen.
- Next, all phases have to be included in the starting model with their full parameter set.
- Next, the calculation has to be carried out in fully automatic mode.
- Finally, the results have to be verified.

The measurements were performed on an URD-6 diffractometer (SEIFERT-FPM GmbH, Freiberg/Germany) in reflection mode with an angular range from 10 ° to 80 ° in 2Θ. The step width was 0.05 ° (except samples 10 and 44 with 0.02 °) and the measuring time per step was 5 s. An appropriate instrumental function was calculated. The following starting model was set up:

- diaspore: correction of PO with 8th order spherical harmonics; isotropic peak broadening due to crystallite size with an upper limit (20 parameters)
- corundum: no PO; isotropic peak broadening due to crystallite size with an upper limit (4 parameters)
- bayerite, boehmite, gibbsite: correction of PO with 6th order spherical harmonics; isotropic peak broadening due to crystallite size with an upper limit (each 22 parameters except boehmite with 15 parameters)

All lattice parameters had a variation range of about 2 %. In sum the number of refined parameters varied between 26 and 58 due to the complex modeling of PO.

All calculations were carried out in one step without user interaction. The actual choice of the model for preferred orientation was done automatically by the program.

Sample	Diaspore		Corundum		Bayerite		Boehmite		Gibbsite		Para-	R_{wp}/
	wt.-%	PO	wt.-%	PO	wt.-%	PO	wt.-%	PO	wt.-%	PO	meters	%
05	98.4	8th	0.1	-	0.3*	-	1.0	-	0.2*	-	41	14.4
10	98.5	8th	0.1*	-	0.1*	-	1.0	-	0.3*	-	41	14.2
19	16.5	6th	0.5	-	76.3	6th	0.6*#	-	6.1*#	2nd	52	16.1
21	0.9*	-	13.9	-	78.6	6th	0.5*	-	6*#	2nd	43	15.9
23	13.1	6th	0.6	-	77.2	6th	0.4*#	-	8.7#	2nd	52	16.1
29	0.1	-	99.8	-	0.1*	-	0	-	0	-	26	13.0
30	59.2	4th	0	-	3.5*	-	33.3#	-	4.0*	-	31	27.7
30 2nd	20.9	4th	0.2*	-	1.2	-	77.4	6th	0.3*	-	31	9.4
31	29.8	8th	1.0*	-	0.1	-	69.1	6th	0	-	58	13.5
44	98.5	8th	1.0	-	0.3*	-	0.2	-	0	-	41	13.6

Table 1: Refinement results of samples containing phases with strong PO and varying phase content;
***: at least one lattice constant moved to one of its limits,**
#: the peak width moved to its upper limit

For all samples except number 30 a good result for the phase content was calculated. Some minor phases with contents below 1 % had parameter values which were shifted to their limits. Such phases can be excluded in a final calculation if one needs a more exact numeric result. But in general this is not necessary because of their small contents.

Samples 19, 20 and 23 showed a noticeable amount of gibbsite. Because of the strong overlapping with peaks of bayerite some gibbsite parameters moved to their limits.

Only sample 30 stepped out of line. It had a very large reliability value and the large content of boehmite was refined only with an isotropic model for PO. A short view on the diffraction pattern showed the reason. Boehmite had an extreme peak broadening due to very small crystallites. A second calculation with a 20 times higher upper limit for the peak broadening gave a good fit also for boehmite.

The presented set of samples demonstrates that a wide range of phase content can be refined without user interaction with only one common starting model. That works because the program reaches many decisions automatically.

3.2 Chlorite Minerals

To check the behavior of the above mentioned disordering model for chlorite in quantitative phase analysis, mixtures of different chlorites with quartz, muscovite and oligoclase were prepared. Muscovite was refined using an anisotropic peak broadening model. All refinements were done in a single run without any interaction of the user, starting from the same structures. Typical calculation times for such mixtures are one to four hours.

Chlorite mineral	Refined Fe	Chlorite wt%		Quartz wt%		Muscovite wt%		Oligoclase wt%		R_{wp}/%
		nom.	ref.	Nom.	ref.	nom.	ref.	nom.	ref.	
penninite	0.7	50	50.9	50	49.1	-	-	-	-	15.5
penninite	0.8	33.3	36.2	33.3	31.1	33.3	32.7	-	-	14.2
penninite	0.7	30	31.4	30	28.7	30	31.2	10	8.7	13.9
ripidolite	2.7	48	46.6	52	53.4	-	-	-	-	21.3
ripidolite	2.1	32	32.9	34.7	32.9	33.3	34.2	-	-	17.6
thuringite	3.9	47	47.4	50	50.4	-	-	-	-	25.8

Table 2: Refinement results of reference mixtures containing chlorites; transmission measurements

The quantitative results are satisfactory (Table 2). The refined total iron content of chlorite differs between the measurements, but there is a general correspondence to the mineralogical classification of the samples.

The modeling of disorder in chlorites is a first step of implementing disordering models for layer silicates which will be followed by more complicated real structures in the future.

4 Summary

The presented BGMN program has a set of unique features which are very useful for quantitative phase analysis.

Because of the underlying numerical algorithms the program can handle samples with varying phase content and strong preferred orientation in a fully automatic way. If a suitable starting model based on the qualitative analysis was established no user interactions during the calculation are required. So, the fully automatic refinement of samples with varying compositions and phase contents using the same starting model becomes a routine task.

Furthermore, the structure-based modeling of broadening and shifting of individual Bragg peaks is an useful tool to approximate the diffraction patterns of disordered layer silicates. The problem of refining such a complicated set of parameters is soluble by means of the program. If the formulation of the dependencies was carried out once, a convenient refinement of multiphase samples including disordered layer silicates is possible.

References

1. Bergmann, J., Kleeberg, R., Taut, T. and Haase, A., 'Quantitative Phase Analysis Using a New Rietveld Algorithm - Assisted by Improved Stability and Convergence Behavior'. *Adv. X-ray Anal.* 40 (1997), in press
2. Kleeberg, R and Bergmann, J., 'Rietveld Analysis of Disordered Layer Silicates' EPDIC-V, Parma (1996), in press
3. Steinfink, H., Acta Cryst. 11 (1958), 191-195.

FROM PRECURSORS TO CERAMICS : HETEROGENEOUS STRUCTURES IN THE ANHYDROUS LANTHANIDE PERCHLORATO COMPLEXES

F. FAVIER, J.L. PASCAL
LAMMI, UPRESA 5072, Université Montpellier II, Place E. Bataillon, C.C.015, F-34095 Montpellier cedex 5

A. FITCH, G. VAUGHAN
ESRF, BP 220, F-38043 Grenoble Cedex

The structures of a heterogeneous series of lanthanide perchlorato complexes, $Ln(ClO_4)_3$ with Ln = La, Ce, Pr, Sm, Eu, Ho, Er, Tm, Yb and Lu, were determined by Rietveld analysis of high resolution powder X-ray diffraction data. From lanthanum to erbium, the space group is $P6_3/m$ while lutetium crystallizes in a trigonal R3c unit cell. Depending on the temperature, thulium and ytterbium exist with either structure. We have focussed our work on the structural relationships existing at the different stages of the thermolysis of $Yb(ClO_4)_3.H_2O$ giving the two polymorphic forms of anhydrous $Yb(ClO_4)_3$ and finally, Yb_3O_4Cl as submicronic spherical particles.

1 Introduction

In the development of new preparative methods for ceramics as dispersed powders, the thermolysis at moderate temperature of a fully inorganic precursor with weak oxy-ligands such as nitrates or perhallogenates, is an alternative way of synthesis which can promote attractive properties.[1,2] Perchlorates generally decompose bellow 250-350°C giving fine powders of oxides in which the metal is in its highest oxidation state. However, when we recently tried to prepare oxides of lanthanides by this method, we found that, with the exception of $Ce(ClO_4)_3$, the resulting powders were oxychlorides rather than oxides. In order to control the structure and morphology of the resulting material, it nevertheless remains of importance to know the structure of the precursor as well as those of the successive intermediate solids. We have recently had the opportunity to be among the first users of the new beam line 16 at the European Synchrotron Radiation Facility (ESRF), devoted to X-ray powder diffraction. Recording data on this line has allowed us to determine the various crystalline and molecular structures for a series of $Ln(ClO_4)_3$ compounds with Ln = La, Ce, Pr, Sm, Eu, Ho, Er, Tm, Yb and Lu.

2 Experimental

The general methods of synthesis, purification and handling of reagents have been previously described.[3] X-ray patterns recorded in-house showed that $Ln(ClO_4)_3$ complexes can be separated into two different groups of isostructural compounds. The first group includes complexes of La, Ce, Pr, Sm, Eu, Ho, Er, and the low temperature forms (LT) of Tm and Yb, and the second involves the high temperature forms (HT) of Tm and Yb, and Lu perchlorates. High resolution X-ray diffraction patterns of europium, ytterbium (HT and LT), thulium (HT) and lutetium

Table 1: Details of Rietveld refinements for $Yb(ClO_4)_3$ LT and HT forms.

Formula	$Yb(ClO_4)_3$ (LT)	$Yb(ClO_4)_3$ (HT)
T / sampling	25°C / 0.5 mm glass capillaries	
Space group	$P6_3/m$ Z = 2	R3c Z = 6
Lattice parameters (Å)	a = 9.21970(7)	a = 8.12406(3)
	c = 5.53518(3)	c = 24.0818(1)
Wavelength (Å)	0.54994(1)	0.54994(1)
2 θ range (°) / step incr (2 θ)	6.50 - 35.00 / 0.0025	6.50 - 45.00 / 0.005
Peak shape / asym. correct.	pseudo-Voight / Berar-Baldinozzi	
Background	linear interpolation	
No. of reflections	251	446
No. of parameters	23	31 + 10 const. dist.
No. of atoms in asym. unit	5	6
R_F and R_B	0.0366 and 0.0397	0.0414 and 0.0472
R_p and R_{wp}	0.0508 and 0.0728	0.0437 and 0.0584

complexes were collected at the ESRF. A brief description of the BM16 line and of the characteristics of the diffractometer have been reported elsewhere.[4] The indexing of the powder patterns was carried out using the TREOR90 program.[5] For $Yb(ClO_4)_3$ (LT), the best indexing solution with M(25) = 122 and F_{25} = 213 (.001951, 58) indicated an hexagonal unit cell (a = 9.216(3)Å and c = 5.534(2)Å). Systematic absences were consistent with the space group $P6_3/m$ indicating the existence of an isostructural relationship with the anhydrous perrhenate related compound, $Yb(ReO_4)_3$.[6] For $Yb(ClO_4)_3$ (HT) the patterns were indexed in a hexagonal unit cell (a = 8.124(4)Å and c = 24.08(2)Å) with figures of merit of M(25) = 60 and F_{25} = 135 (.001363, 142). This compound appears to be isostructural with $Sc(H_2PO_4)_3$ which crystallizes in the R3c space group.[7]

3 Results and Discussion

Rietveld refinements and profile matchings were performed using the FULLPROF program on a PC computer.[8] As an example, details of the Rietveld refinements for $Yb(ClO_4)_3$ LT and HT forms are given in Table 1, while plots of the experimental and calculated X-ray diffraction patterns, and difference data after the final Rietveld refinements, are depicted in Figure 1. The profile matching option of FULLPROF allows us to calculate the refined lattice parameters for other synthezised $Ln(ClO_4)_3$ compounds.[9]

Since $Yb(ReO_4)_3$ appears to be isostructural with LT ytterbium perchlorate, the atomic coordinates of Yb, Re and O atoms were used as starting positions for Yb, Cl and O atoms respectively. $Sc(H_2PO_4)_3$ was used as model for the HT form. Thus, scandium and phosphorus atomic positions were used as starting coordinates for the heaviest atoms, Yb and Cl respectively, but at this stage, it was not possible to correctly fit the oxygen positions. In $Sc(H_2PO_4)_3$, bidendate phosphato groups build a distorted octahedral environment around Sc atoms, while for lanthanide perchlorato complexes, a higher coordination might be expected. On the basis of ninefold coordinated Yb atoms, the initial positions of oxygen atoms were then calculated from a relationship d(M-O) = f(R) [a] established elsewhere [10] while Cl-O distances and angles were constrained to values consistent with coordinated perchlorato groups.

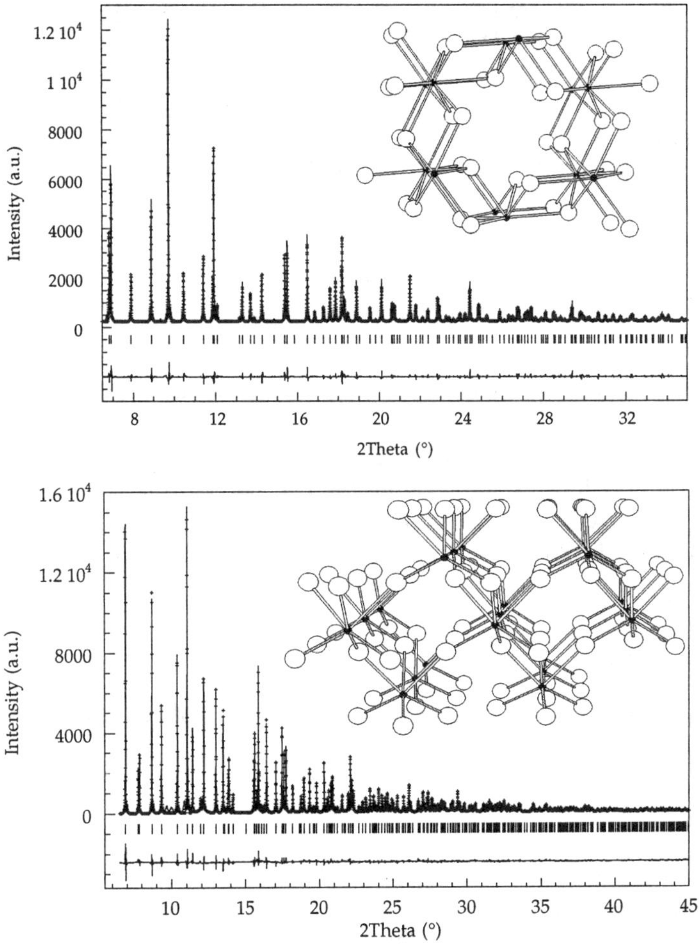

Figure 1: Observed (line), calculated (crosses) and difference plots for $Yb(ClO_4)_3$ LT (top) and HT (bottom). Corresponding schematic packings are also depicted; black circles for Yb atoms and large open circles represent $[ClO_4]$ groups.

As depicted in Figure 1, the LT form has a channeled arrangement with bridging tridendate perchlorate groups bonded to three different Ln atoms. The HT form is a layered structure with unusual tridendate ClO_4 ligands, simultaneously bridging and chelating. Final atomic positions, and distances and angles, are given in ref. 9.

[a] d(M-O) = metal-oxygen distance and R = Brown and Shanon effective ionic radii of the metallic center

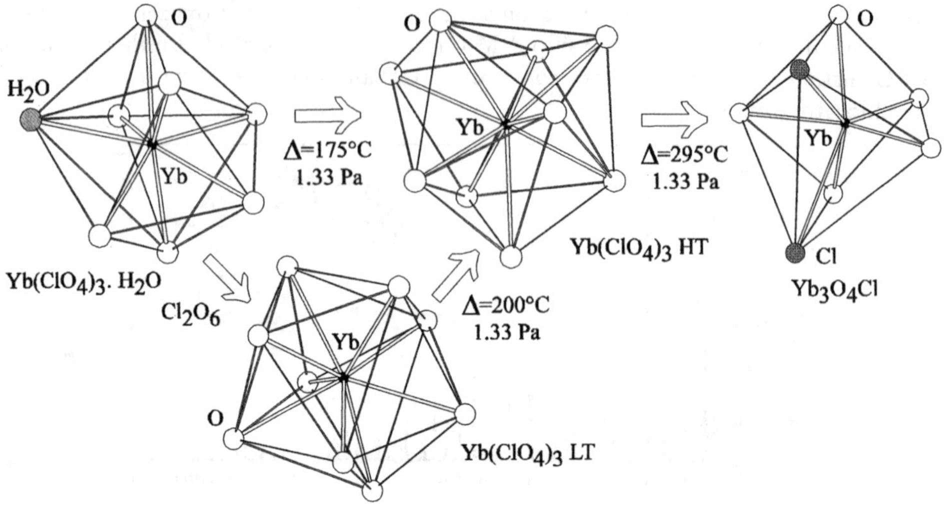

Figure 2: Ytterbium coordination polyhedra at different stages of the thermolysis.

Figure 2 shows the various ytterbium surroundings at different stages of the thermolysis of $Yb(ClO_4)_3.H_2O$ [11] giving the two polymorphic forms of $Yb(ClO_4)_3$ and, finally, Yb_3O_4Cl. This material is obtained as fine dispersed powder. The size of the spherical particles was estimated by electronic microscopy in the range 0.2-0.5 μm.

Acknowledgments

Funding from the European Synchrotron Radiation Facility is gratefully acknowledged.

References

1. E. Matijevic, Chem. Mater. **5**, 412(1993).
2. N. Guillou, J.P. Auffrédic and D. Louër, J. Solid State Chem. **112**, 45(1994).
3. F. Favier and J.L. Pascal, J. Chem. Soc. Dalton Trans. , 1997(1992).
4. A.N Fitch, Materials Science Forum **219**, 228(1996).
5. P.E. Werner, L. Eriksson and M. Westdahl, J. Appl. Crystallogr. **18**, 367(1985).
6. V.N. Krustalev, M.B. Varfolomeev, N.B. Shamrai, U.T. Struchkov and A.P. Pisarevskii, Koord. Khim. **19**, 871(1993).
7. Y.I. Smolin, Y.F. Shepelev and A. Domanskii, Sov. Phys. Crystallogr. **27**, 146(1982).
8. J. Rodriguez Carjaval in *Collected Abstracts of Powder Diffraction Meeting*, (Toulouse, France, 1990), p. 127.
9. F. Favier, J.L. Pascal, F. Cunin, A. Fitch and G. Vaughan, J. Chem. Soc. Dalton Trans.*(in press)*.
10. J.E. Huheey in *Inorganic Chemistry. Principles of Structures and Reactivity*, Second Edition (Harper and Row, New York, 1978), p. 71.
11. C. Belin, F. Favier, J.L. Pascal and M.Tillard-Charbonnel, Acta Cryst. **C52**, 1872(1996).

ESTIMATION OF INVERSION PARAMETER OF MAGNESIOFERRITE BY RIETVELD REFINEMENT METHOD.

J. Podwórny
Institute of Refractory Materials, Department of Structural Research,
ul. Toszecka 99, 44-100 - Gliwice

Abstract

The Rietveld refinement method was applied for estimation of inversion parameter of magnesioferrites. The proposed method require the knowledge of accurate stoichiometry of the tested magnesioferrite.

The stoichiometric samples of $MgFe_2O_4$ were prepared by firing $Mg(OH)_2$ and α-Fe_2O_3 pure mixture at 1400 °C / 8h. The differences of inversion degree were obtained by annealing in different temperatures and quenching or cooling the magnesioferrite samples.

The results showed good agreement with literature data obtained by traditional XRD method and SM (saturation magnetic) method.

Introduction

Magnesioferrite is the base ferrite which have cubic closest packing of oxygen anions. The structure (Fd3m)accommodates 8 chemical entities in the conventional cubic unit cell, the 64 tetrahedral sites and 8 Mg^{2+} + 8 Fe^{3+} in octahedral sites. In fact only 9 of 10 octahedral sites are occupied by Mg^{2+} and 1 of 10 tetrahedral sites is occupied by Mg^{2+}. Such a structure have inversion parameter $\lambda^{\dagger}$ equal 0.45 which correspond to 90% degree of inversion [1]. Changes in the coordination numbers, or cation disordering, are usually induced by heating or cooling with different rates [2,3].

Several techniques have been used to study properties related to cation order-disorder and structure of $MgFe_2O_4$ and other spinels [4]. In the case of $MgFe_2O_4$ X-ray diffraction technique can be applied because both cations Mg^{2+} and Fe^{3+} differ sufficiently in atomic scattering factors, which influence relative intensity of diffraction lines, as shown in table 1.

Table 1. Relative intensities of some diffraction lines ordered and disordered magnesioferrites.

Magnesioferrite ordered (17-646 ICDD)		Magnesioferrite disordered (36-398 ICDD)	
Miler indices	relative intensity	Miler indices	relative intensity
(111)	4	(111)	12
(222)	2	(222)	3
(400)	25	(400)	20
(422)	14	(422)	10
(531)	-	(531)	1

Sample preparation

The samples were prepared as stoichiometric magnesioferrite $MgFe_2O_4$ obtained by sintering mixtures of pure $Mg(OH)_2$ and α-Fe_2O_3 powders. The mixed powders were dry pressed in form of cylinders under pressure 120 MPa, calcined at 450

---

† $y=2\lambda\cdot100\%$, λ- percent of 3+ cations in tetrahedral sites (inversion parameter), y- degree of inversion

°C and fired at 1650 °C for 12 h. The obtained magnesioferrite was divided into 3 samples which were subsequently differently thermally treated, as shown in table 2.

Table 2. The way of thermal treatment of samples.

Number of sample	temperature treatment	way of cooling
1	as received	————
2	350 °C / 8 h	slow cooling per 16 h with furnace
3	1000 °C / 1 h	fast cooling on the air to ambient temperature

Experimental procedure

XRD patterns were obtained using $Cu_{K\alpha 1}$ radiation and Bragg-Brentano Philips PW 1050 X'Change diffractometer equipped with a graphite monochromator in the diffracted beam. The measurement were from 17.00 to 70.00 °(2Θ), step 0.02 °(2Θ), at 20 mA, 40 kV.

The obtained data were converted to ASCII format by PC-APD software and to DBWS 9006PC software format using PISARZ[††] software.

The inversion spinel with λ=0.45 was used as a starting model for the magnesioferrite structure (table 3.).

Table 3. The model of ordered magnesioferrite structure with λ=0.45.

name	magnesioferrite				
space group	Fd3m (227)				
lattice constant	a_0=8.375 Å				
fractional atomic coordinates	Mg^{2+}	Fe^{3+}	Mg^{2+}	Fe^{3+}	O^{2-}
Wyckoff number	8b	8b	16c	16c	32e
x:	1/2	1/2	1/8	1/8	0.2563
y:	1/2	1/2	1/8	1/8	0.2563
z:	1/2	1/2	1/8	1/8	0.2563
occupation number:	0.03750	0.04583	0.00417	0.03750	0.16667

The occupation numbers of cations were assigned the same codeword. Fe^{3+} in octahedral sites (1/2, 1/2,1/2) and Mg^{2+} in tetrahedral sites (1/8, 1/8, 1/8) had negative sign. Such way of assignment of codewords requires knowledge of accurate stoichiometry of the tested magnesioferrite because the sum of cations Mg^{2+} and Fe^{3+} will be constant during the refinement process. The result of chemical analysis which is necessary for estimation of starting occupation numbers values ensures correct results.

The results of refinement occupation numbers were recalculate into numbers of cations in tetrahedral and octahedral sites. The inversion parameter is equal to the percent of Fe^{3+} ions in tetrahedral sites. The degree of inversion was calculated accordingly to the definition given in the footnote [1].

[††] Autor software

Results

Table 4. The obtained results of inversion parameters of magnesioferrite after thermal treatment.

number of sample		Fe^{3+} in octahedral sites	Fe^{3+} in tetrahedral sites	Mg^{2+} in octahedral sites	Mg^{2+} in tetrahedral sites	λ	y [%]	S[1]
1	occ.number	0.04778	0.03556	0.0355	0.0061			
	number of atoms	9.17	6.83	6.83	1.17			
	x	1/2	1/8	1/2	1/8	42.69	85.38	1.35
	y	1/2	1/8	1/2	1/8			
	z	1/2	1/8	1/2	1/8			
2	occ.number	0.04609	0.03725	0.03725	0.00442			
	number of atoms	8.85	7.15	7.15	0.85			
	x	1/2	1/8	1/2	1/8	44.69	89.38	1.31
	y	1/2	1/8	1/2	1/8			
	z	1/2	1/8	1/2	1/8			
3	occ.number	0.04859	0.03475	0.03475	0.00692			
	number of atoms	9.33	6.67	6.67	1.33			
	x	1/2	1/8	1/2	1/8	41.69	83.38	1.33
	y	1/2	1/8	1/2	1/8			
	z	1/2	1/8	1/2	1/8			

Magnesioferrite annealed at 350 °C/8h and cooled very slowly has almost theoretic value of inversion parameter (sample 2) whereas samples as received (1) and fast cooled after short high temperature treatment (3) showed lower values of this parameter (table 4.). Small difference of inversion parameter between samples 1 and 3 may be explained by similar cooling rate. After short thermal treatment the sample 3 was removed from the furnace and free cooled in the air. In both cases the cooling rates were not sufficient to reach disordered structure. The measured and calculated x-ray diffraction pattern of the respective samples are presented in fig. 1,2 and 3.

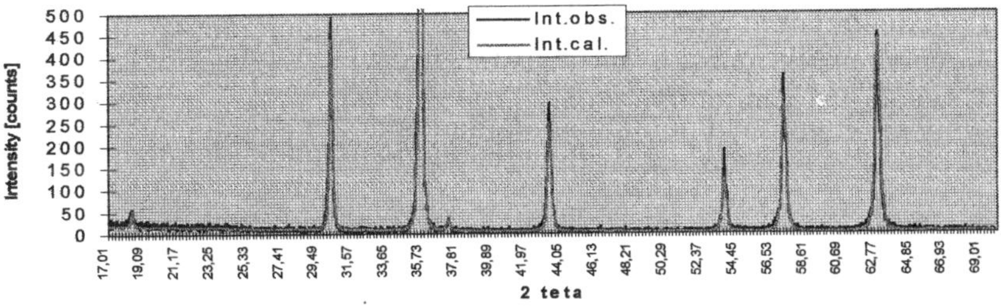

Fig.1. Experimental and calculated x-ray pattern of obtained magnesioferrite (sample number 1).

[1] S is the "goodness of fit", the ratio $R_{wp}/R_{expected}$

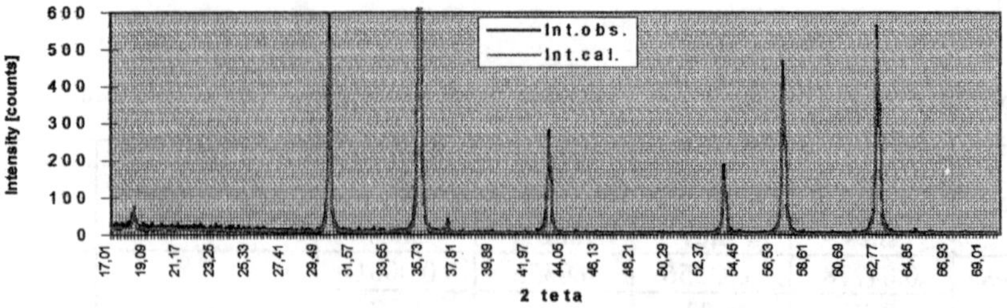

Fig.2. Experimental and calculated x-ray pattern of magnesioferrite after treatment at 350 °C / 8 h (sample number 2).

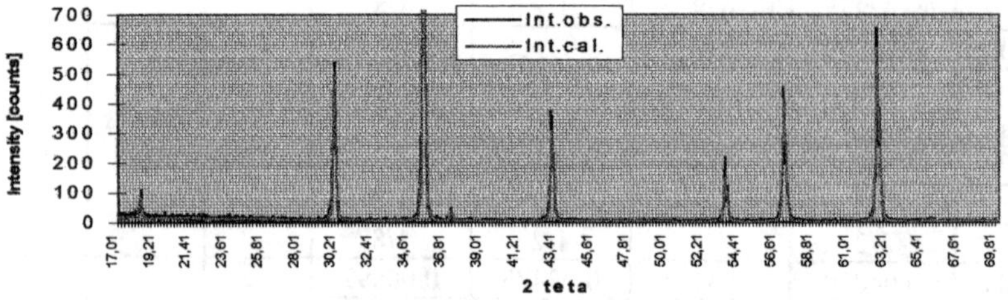

Fig.3. Experimental and calculated x-ray pattern of magnesioferrite after treatment at 1000 °C / 1 h (sample number 3).

Conclusion

- The proposed method requires knowledge of stoichiometry of tested magnesioferrite as a starting parameters (occupation numbers) for DBWS software.
- It is possible to use the Rietveld method to spinels which atomic scattering factors of cations differs significantly.
- As the continuation of this work the obtained results should be veryficated using another method.

Reference

1. Acta Crystallographica, 1953, **6**, 57.
2. PAGEOPH, vol. 141,No. 2/3/4 (1993).
3. American Mineralogist, vol. 68, 181-194, 1983.
4. Journal of Chemical Physics, 39, 435-439, 1963.

THE QUANTITATIVE PHASE ANALYSIS OF A COMMERCIAL TUBE OF THE DUPLEX STAINLESS STEEL

M. WRÓBEL
Academy of Mining and Metallurgy, Department of Metallurgy and Materials Engineering, 30 Av. Mickiewicza, PL-30059 Kraków, Poland

J. BONARSKI and K. PAWLIK
Aleksander Krupkowski Institute of Metallurgy and Materials Science, ul. Reymonta 25, PL-30059 Kraków, Poland

The applicability of the X-ray methods for determining the crystallographic texture and the amount of the phases in the commercial tube is estimated. The Schultz back reflection technique was used for the determination of incomplete pole figures of ferrite and austenite. The measurements were performed by means of the point counter (conventional) and by the pseudo position-sensitive (pPSD) detection technique. For each phase, the orientation distribution function (ODF) was calculated and then used for the reproduction of the complete pole figures. The complete pole figures were used in the Quantitative Phase Analysis (QPhA), according to the original method, free from the texture induced errors. The volume fraction of austenite and ferrite was determined also by means of the classical direct comparison method of Averbach and Cohen as well as the metallographic point method, for comparison.

1. Introduction

The duplex stainless steels offer an improved resistance to stress-corrosion cracking in chlorides and a higher resistance to pitting corrosion when compared with the standard austenitic stainless steels. Additionally, the higher tensile and yield strengths of the duplex stainless steels than those of the austenitic stainless steels and similar corrosion resistance make the duplex stainless steels a potential for many industrial applications.

The beneficial properties of the duplex stainless steels depend on the optimum balance of the constituent phases (approximately equal amounts of the bcc ferrite (α) and fcc austenite (γ)). The exact phase composition is a function of chemical composition (the balance between austenite-stabilising and ferrite-stabilising elements in the alloy) and the heat treatment conditions. The accurate and relatively quick experimental method of the Quantitative Phase Analysis (QPhA) is essential for the control of the material parameters.

Rolling and heat treatment lead to a preferred crystallographic orientation (texture). From the practical point of view it is reasonable to use the same method for the texture measurements and for the QPhA in the product examined. The X-ray diffraction techniques can be used for this purpose. Because of texture induced errors, the use of the classical direct comparison method of the QPhA by Averbach and Cohen[1] (A-C) can lead to some errors. The method of the QPhA proposed by Bonarski at al.[2,3] is free from the texture induced errors. The aim of the present paper was to evaluate the suitability of this method in practice to control both the phase composition and crystallographic textures of tubular products of the duplex stainless steel.

2. Experimental

The samples were cut off from a commercial tube supplied by the Metallurgical Plant „BUCZEK SA." after cold rolling from 88.9 x 5 mm to 25 x 2 mm (using the long-stroke cold pilger process with ring dies) followed by 1h annealing at 1050°C. The tube was made of the duplex stainless steel Bohler A903 with the chemical composition given in Tab. 1.

Tab. 1 Chemical composition of the investigated steel [wt.%].

C	Si	Mn	P	S	Cr	Mo	Ni	N
0.018	0.48	0.76	0.019	0.001	22.44	3.26	5.59	0.174

X-ray measurements were performed with the application of Philips X-Pert system equipped with a texture goniometer (ATC-3). The filtered X-ray radiation CoKα (λ=1.79026 Å) was used. The diffractometer patterns as well as the pole figures were measured. The sets of incomplete pole figures $P_{hkl}(\alpha, \beta)$ for each of the two phases ({111}γ, {200}γ, {220}γ and {110}α , {200}α, {211}α, respectively) were measured in the range of $\alpha = 0\pm75°$ and $\beta = 0\pm360°$ with $\Delta\alpha = 5°$ and $\Delta\beta = 5°$ by the Schulz back reflection technique[4]. To evaluate the influence of the measurement precision on the QPhA results, the measurements were performed using Θ - 2Θ scan technique[5] with *point counter* as well as with a pseudo-position detector (pPSD)[6] simulating a *position sensitive detector*. For the calculation of the orientation distribution function (ODF) and the complete pole figures ($\alpha = 0\div90°$) the arbitrary defined cells (ADC) method[7-9] was used. Basing on the complete pole figures, the phase volume fractions were calculated according to procedures described elsewhere[2,3]. Moreover, for comparison, the volume fractions of α and γ phases were also determined by means of the A-C method[1] as well as by the metallographic point method.

3. Results

The typical microstructures indicate that the investigated stainless steel consists of approximately equal amounts of ferrite and austenite. Using the metalographic point method the amount of ferrite was determined as being equal to 44%. On the other hand, the amount of ferrite, obtained by the A-C method, is far too high. It is assumed that the presence of crystallographical texture is a potential source of fault in the A-C method.

A moderately strong texture was found for the ferrite phase and a relatively weak one for the austenite phase. The ODFs maximum values in point counter measurements were higher than those by the pPSD techniques. For ferrite the ODF maximum values were respectively 23 and 13.5. For the austenite the ODF maximum values were 5 and 3, respectively. The decreased sharpness of the austenite texture in as-deformed and annealed duplex stainless steel was also found by Shek et al.[10]. Qualitative compatibility between the measured pole figures and the pole figures recalculated from ODF reveals the accuracy of the texture determination (Fig.1). The components: {011}<100>, {112}<111> and {213}<364> were identified in the ODF of austenite. The ODF of ferrite has the {112}<110> main component. Quantitative comparison of the

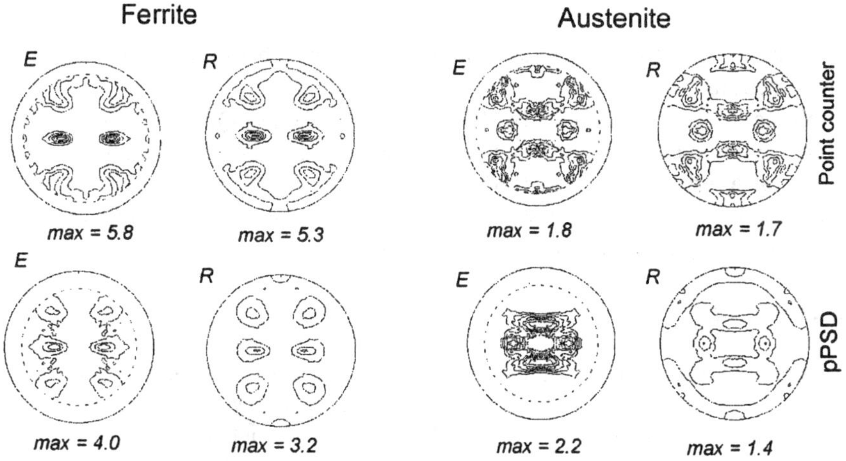

Fig. 1 {200}pole figures, experimental, normalised (E) and recalculated from ODF (R).

Tab.2 Results of austenite/ferrite volume fraction determination

Measurement technique		Bragg reflections		Phase volume fraction	
diffractometer pattern (*A-C method*)	pole figures (*proposed method*)	austenite	ferrite	austenite	ferrite
point counter	-	{200}, {220}	{200}, {211}	27.1 ± 1.4%	72.9 ± 1.4%
-	point counter	{200}, {220}	{200}, {211}	56.7 ± 0.7%	43.3 ± 0.7%
-	pPSD	{111}, {200}, {220}	{110}, {200}, {211}	55.8 ± 0.7%	44.2 ± 0.7%
Metallography		—	—	56%	44%

experimental and recalculated pole figures shows significant differences in the values of the peaks maxima. This indicates the existance of some significant errors of experimental pole figures. Because the result of the phase analysis depends on the precision of the ODFs determination, it has been improved by increasing the number of pole figures used for the calculation of the ODF. For the calculation of the ODFs, two and three pole figures were used for the classical point counter technique and the pPSD measurements, respectively.

From Tab. 2 one can conclude that the application of pPSD measurements does not greatly influence the results. It is clear, considering that the ratio of the ODF maximum for ferrite to that of austenite was not much influenced by the measurement technique.

4. Conclusions

- The proposed method enables to determine the texture of both phases as well as their volume fractions.
- The volume fraction determined by the proposed method of QPhA is much closer to that obtained by the metallographic point method, than that obtained as the results by the classical Averbach-Cohen method.
- Pseudo position-sensitive measurements improved the results of quantitative phase analysis of the duplex stainless steel only a little.

Acknowledgements

The authors are indebted to the Committee of Scientific Research (KBN) for financial support through grant No 7-T08A-064-08. The authors are also grateful to Eng. S. Starzykowski, Eng. J. Jaworski and Eng. T. Boksa (Metallurgical Factory BUCZEK SA.) for supplying samples of the tested tubes, and to Dr S. Dymek (AGH) for critical reading of the manuscript.

References

1. B.L. Averbach and M. Cohen, *Trans. Met. Soc. AIME* **176,** 401 (1948)
2. J. Bonarski, M. Wróbel and K.Pawlik, *Scripta Met. and Materialia* **25**, 1401 (1991)
3. J. Bonarski, M. Wróbel and K. Pawlik in *Textures of Materials, Proc. of the 11th Intern. Conf. on Texture of Materials (ICOTOM-11), Xi-an, China, 1996,* ed. by. Z. Liang et al. (Intern. Acad. Publish, Beijing, China, (1996), vol.2, p. 1507
4. L.G. Schulz, *J. Apply. Phys.*,20, 1030 (1949)
5. *X-Pert System User Guide, Philips Analitycal X-ray Customer Support,* Philips, Netherlandes, 2th edition, june 1992
6. J.Bonarski, W.Olek, K.Pawlik in *Textures of Materials, Proc. of the 11th Intern. Conf. on Texture of Materials (ICOTOM-11), Xi-an, China, 1996,* ed. by. Z. Liang et al. (Intern. Acad. Publish, Beijing, China, (1996), vol.2, p. 1215
7. K. Pawlik, *Phys. Stat. Sol.,* **B 134**, 477 (1986)
8. K. Pawlik and J. Pospiech, in *Theoretical Methods of Texture Analysis, Proc. Workshop*, ed. H.J. Bunge, Clausthall-Zellerfield 1986
9. K. Pawlik, J. Pospiech and K. Lücke, in *Texture of Materials, Proc. 8th Intern. Conf. on Texture of Materials (ICOTOM-8), Santa Fe, USA, 1987,* ed. J.S. Kallend and G. Gottstein (The Metallurgical Soc. Inc.) p. 105
10. C.H. Shek, G.J. Shen, J. K. L. Lai and B.J Duggan, *Materials Science Forum*, **157-162**, 853 (1994)

Paracrystallinity

PARACRYSTALLINITY

J. FIALA
ŠKODA Research Ltd, 31600 Plzeň, Czech Republic

A review is given on essential features of the structural disorder. A special attention is paid to the Hosemann's model of paracrystalline solids. Relationship between the structural defects and materials properties is demonstrated.

1 Introduction

According to the Webster [1],"paracrystal is a solid body with less than the three-dimensional order characteristic of a true crystal". So, at the first sight, paracrystallinity may seem to be something rare, just an exception from the generaly ruling crystalline order. In fact, deviations from this order are ubiquitous. Using parallel beams and dynamic difraction techniques, disturbances of order can be traced up to parts-per-billion levels in very, very "perfect crystals" [2]. Strictly speaking, there are no perfect crystals on Earth. The ideal structural order of a "true crystal" as mentioned by Webster is just an abstraction. Crystallographers, being aware of this fact, do not even mention anything on "order", "regularity" or "perfect periodicity" in the IUCr's statement of what crystalography is ("a branch of science concerned with the description and understanding of the structure and properties of condensed matter in terms of the spatial relationships of atoms and interatomic forces in an extended array").

2 The strange case of order and disorder

Paracrystallinity, defect of perfect crystalline order, is no exception; it is a general structural feature of condensed matter [3]. The structure of solid materials has a number of deviations from perfect three-dimensional periodicity. Different materials show different degree of order. Also, the number of ways in which the structure of real (para)crystalline materials differs from the perfect three-dimensional periodicity, is very large. There are areas in the structure of a given piece of matter which are quite regular. On the other hand, there are other regions which are pretty irregular. And we have many, many different kinds of defects, deviations on regularity. And these defects are arranged, combined in many, many different ways in the structure. E.g., high polymers consist of various regions, each with a characteristic degree of internal order, ranging continuously from something close to the perfect three-dimensional periodicity of an ideal crystalline state to the completely disordered structure of the amorphous state. Even in one and the same highly crystalline region of such a substance the arrangement of the atoms and molecules is not always perfectly regular, and even in amorphous regions there are often arrays in which the molecules are roughly parallel to one another, i.e. in which the arrangement is quasi crystalline. Regular distribution of atoms and molecules is a distinctive feature of the crystalline state of solid materials. On the other hand, the omnipresence of structural defects, their diversity and strong influence of materials properties recognized even Max von Laue, the discoverer of the diffraction of x-rays (diffraction by regular! structure of matter) that in the preface of the third edition of his book "Röntgenstrahlinterferenzen" [4] wrote on a "newly emerging science of structural defects and their consequences", in 1960.

The degree of structural order may be different in different directions, as is the case of liquid crystals. These are substances with perfect, crystalline order in one direction (two directions), while in the remaining directions the structure is quite disordered, resembling a liquid[5-7]. Arrangement (geometry) of the body is connected with the geometry of the molecules. So, elongated molecules form the so called calamitic liquid crystals which may be nematics, smectics and cholesterics[8-10]. Similar is the situation with discotic molecules (giving rise to discotic liquid crystals which reveal order in the direction normal to the molecular plane while two other directions are occupied randomly [11]). And we have, in addition to it, pyramidal liquid crystals [12] and phasmidic liquid crystals [13], where, again, assymetry of the molecules leads to asymmetry of the structure of the body in the sense that it has different degree of order in different directions.

Thirteen years ago, compounds have been discovered, the so called quasicrystals, which reveal orientational (long-range) order, while they have no translational periodicity[14-22]. The reason for such a strange behaviour was soon identified: Quasicrystals are composed of two types of building units, two types of unit cells which are assembled in the structure of the quasicrystal which, as a consequence of this, may reveal, e.g., five- or seven-fold axes of symmetry, something quite ruled out at crystals.

And we have a still another class of substances with reduced order, another class of paracrystals, the so called plastic crystals, which reveal translational (long-range) order, at the same time having no orientational (long -range) order [23,24]. In this sense, plastic crystals may be denoted as anti-quasicrystals. Their exotic behaviour is caused by rapid jump reorientations which undergo their structural units, that retain, at the same time, perfect positional order. We have cubic plastic crystals with sphere-like molecules and hexagonal plastic crystals with elongated molecules in which, at certain temperature, a "rotational transition" takes place and the whole molecules are allowed to rotate along the molecular long axis, which diminishes orientational order, but not completely as is the case of the cubic plastic crystals; here, at the hexagonal plasic crystals, the direction of the longer molecular axis remains conserved.

And there are also compounds that can be termed anti-glasses, as they retain long-range positional order, manifesting at the same time no short-range (positional) order (which is just the reverse of the case of glasses which reveal short-range order showing at the same time no long-range order). Anti-glasses are exemplified by $(SrTe_5)(O_{11}-)$, a ternary oxide defective in oxygen which belongs to the C1 (fluorite) structure type. While positions of calcium are occupied randomly by Sr and Te, positions of fluor are, again randomly, occupied by oxygen or vacancies [25].

3 Hosemann's paracrystal

In addition to the common meaning that we ascribe to the term of paracrystal (anything between an ideal crystal and a completely disordered amorphous substance), Rinne in 1933 began to use this term, paracrystal, to denote mesophases, liquid crystals [26]. But this practice was not universally accepted and nowadays it is almost abandoned. In 1950, Rolf Hosemann denoted as "paracrystal" his newly developed structural model of polymers and colloids [27]. Later, a large number of analyses demonstrated decisively that the model of Hosemann's paracrystal agrees perfectly with the diffraction patterns of not only polymers and colloids[28–35], but also many other materials including metals and alloys [36], graphite [37,38], glasses[39,40], biological materials[41,42], colloids[43] and catalysts[44–49]. The model became highly popular and the book of Hosemann and Bagchi "Direct Analysis of Diffraction by Matter" [50], which deals paracrystallinity according to the Hosemann's conception in detail, has been declared a "citation star"[51], having been cited 775-times since 1962 through 1990.

Structural distorsions in the Hosemann's paracrystal are such that only short-range order is retained. I.e., the distance between first neighbours does not vary very much, but there are no ideal fixed equilibrium positions throughout the paracrystal. In contrary to an ideal perfect crystal lattice embodied by a scaffold of equally shaped parallelepipeds, the lattice of paracrystals is distorted, being only topologically (but not metrically) equivalent to the perfect crystalline lattice. The "nearest neighbour" probabilities in the three directions corresponding to an ideal lattice ($\mathbf{a}_1$, $\mathbf{a}_2$, $\mathbf{a}_3$) may be written h_{100}, h_{010}, and h_{001}, respectively, where $h_{pqr}(\mathbf{x})$ denotes the probability that the vector $\mathbf{x}$ drawn from the origin where there is an atom, will locate another atom in the point which is the p-th in the $\mathbf{a}_1$ direction, the q-th in the $\mathbf{a}_2$ direction, and the r-th in the $\mathbf{a}_3$ direction. This probability can be expressed in terms of the p-fold convolution of h_{100}, the q-fold convolution of h_{010}, and the r-fold convolution of h_{001}. In case of an ideal perfect crystal, the probabilities h will be δ-distributions with zero widths, while in case of the Hosemann's paracrystal, the probabilities h may be approximated by a Gauss distribution the width of which is a measure of the paracrystalline distortion (standard deviation of the nearest neighbour interatomic distance from its ideal value in a perfect crystal). Although these fluctuations of the nearest neighbour distance are small, they cumulate when passing when passing from the nearest neighbour to the 2nd, 3rd and the 4th neighbouring atom, so that the position of the n-th atom with regard to the starting ("zeroth") atom will be eventually accidental[52–55]. Grouping the terms that correspond to the distribution of the first, second, third etc left and right neighbours, as well as the zero term, gives the pair distribution function $Z(x)$. Oscillations of this (selfcorrelation) function (of the distribution of atoms) die away with increasing interatomic distance and long-range order disappears.

Angular distribution of the intensity of the diffracted radiation which is, in general, the Fourier transform of the pair distribution function $Z(x)$ will be in case of Gaussian distribution of the nearest neighbour distances composed of a set of peaks the width of which increases quadratically with the order of reflection and with the variation coefficient of interatomic distances. According to the value of this distortion parameter, appearance of the diffraction pattern changes, and, in this manner, the Hosemann's model renders a broad spectrum of substances from ideal crystals to amorphous materials "... like Dante descending from the heaven of perfect crystalline order, through the purgatory of random displacements to the hell of utter chaos"[33]. There exists a number of different models describing the structure of real materials and their deviations from the perfect three-dimensional periodicity of an ideal crystalline state[56–90]. Each model leads to a more or less characteristic diffraction effetcs which allow to identify structural models by means of diffraction analysis. The characteristic feature of the Hosemann's model is that the width of diffraction peaks increases quadratically with the order of reflection. This feature has been found on diffraction patterns of many materials. So, many materials are paracrystalline in the Hosemann's sense.

E.g., liquid crystals may be described as Hosemann's paracrystals with anisotropic structural distortion in terms of a tensor distortion parameter[32]. The greater the distortion, the sooner will the structural coherency be lost in the respective direction. Thus, in the direction of large distortion, the substance behaves like a liquid, while in other directions, where the paracrystalline lattice distortion is small, the mesophase is like a crystal with structural order preserved up to long distances.

4 Disorder hinders the growth

Paracrystalline interatomic distance fluctuations increase with the square root of ther number of intermediary net planes. Therefore, paracrystalline distortion enthalpy increases with the fourth power of the dimension of the growing paracrystalline particle, while the bulk free enthalpy varies as the cube of the particle dimensions and the surface energy as their square. And this is exactly why paracrystalline distortions stop the growth of precipitate particles at a certain limit as was found empirically in many cases. This effect represents another, very important demonstration of the reality of the Hosemann's model[91–93].

When studying recrystallization of different materials, we find - from time to time - that secondary recrystallization (Ostwald ripening) stops at some limit. Then, suddenly, the grains go to pieces, and only after then their growth continues and even exceeds the limit reached during secondary recrystallization (we call this "tertiary recrystallization"). Besides other possible causes (like precipitation), paracrystalline distortions, that cummulate during the domain growth, may be the reason of such a behaviour. Disintegration of the large particles makes free the incorporated contaminants or frozen-in conformational singularities which do not fit in size and/or shape into the matrix and creates many channels of easy diffusion through which these causes of paracrystalline distortion run away. The root mean square of the deviations of interatomic distances from their mean values decreases in this way, so that, in the next run, the growing particles may reach greater size than in the former one.

5 α^* law

The greater the paracrystalline distortion, the smaller will be the size of particles (microparacrystals) that cannot be exceeded. Experiments with different materials demonstrated that the value of the product α^* of relative distance fluctuation of lattice bricks in the adjacent net planes and the square root of the average number of netplanes in a particle lies within a narrow range from 0.1 to 0.2[94–97]. There is no doubt that a fundamental law of solid state physics is hidden in this α^*-rule the meaning of which is not yet fully understood.

E.g., paracrystallinity and α^*-rule are connected with epitaxy. By statistical analysis of 3800 epitaxial pairs, the 90%-quantile of epitaxial misfit has been found to be 15%[98], which is exactly the value of α^*. Of course: α^* expresses the maximum local misfit at the extremity of a growing paracrystalline particle. The matter does not allow greater misfit and stops the growth of the particle at this instant. And, similarly, the maximum (homogeneous, epitaxial) misfit at which two crystals can grow together is 15%. These two values must be the same.

Another interesting respect is the relation between paracrystallinity and thermal vibrations. The value of α^*, some 15%, representing the maximum local structural misfit resulting from the paracrystalline distortions, is equal to the amplitudes of thermal vibrations (reduced to the interatomic distences) just below the temperature of the melting point, when interatomic bonds break.

Professor Friedel estimated that the misorientation of the separated boundary would have to be about 5° to cause fracture [99,100]. Which is not far from the maximum local misorientation induced by paracrystalline distortions at the surface of a particles where they have been accumulated during its growth up to the moment when they exhausted adaptability of matter and stopped its further emlargement (arctg $\alpha^* \approx 6$ - 11°). Further growth would break the bounds - cause fracture.

Paracrystallinity (here in the narrow, Hosemann's sense) is met everywhere: connected with fracture, thermal vibrations, epitaxy, recrystallization or precipitation.

6 Paracrystals in nature and technology

The phenomenon of order (and disorder) which is behind of all of this is still mysterious and recent calculations[101] showed that even at absolute zero point temperature a substance need not to converge to an ordered crystalline structure. Order-disorder processes accompany the birth and the death of galaxies in universe as well as of the living creatures on Earth. Crystallization of glasses, amorphization of crystals, acquisition of the native three-dimensional structure of proteins, settling suspensions of coloidal silica, forging and recrystallization heating of steel; from snowflake formation to the growth of bacterial colonies, it is an intricate interplay between ordering and disordering that takes the leading role. Handy use of structural imperfection may serve as a powerful tool in materials engineering. So, non- crystalline substances often function as promissing precursors for controlling the nano-scale structure of materials because a high degree of freedom for topological and chemical structure persists in amorphous materials. Ubiquity of liquid crystals in living systems and the central role of mesophases in structural biology [102] (as well as in electrooptics [103,104]!) is due to the right mixture of order and disorder in the structure of these substances for which Rinne coined the designation "paracrystalline" in 1933.

References

1. *Webster's Third New International Dictionary of the English Language Unabridged*, ed.P.B.Gove (G.& C. Merriam Company,, Springfield, 1968).
2. M.Deutsch, M.Hart and S.Cummings, *Appl.Phys.Lett.* **51,** 1410 (1987).
3. J.Fiala, *Ceramics-Silikáty* **36,** 49 (1992).
4. M. von Laue, *Röntgenstrahlinterferenzen,* 3rd ed. (Akademische Verlagsgesellschaft, Frankfurt am Main, 1960).
5. Stegemeyer, ed. *Liquid Crystals* (Steinkopf Verlag, Darmstadt, 1994).
6. H.Kelker, *Mol.Cryst.Liq.Crystals* **165,** 1 (1988).
7. T.Seideman, *Rep.Prog.Phys.* **53** , 659 (1990).
8. A.Adamczyk, *Mol.Cryst.Liq.Crystals* **249,** 75 (1994).
9. D.Frenkel, *Liquid Crystals* **5,** 929 (1989).
10. H.Sackmann, *Liquid Crystals* **5,** 43 (1989).
11. S.Chandrasekhar, *Liquid Crystals* **14,** 3 (1993).
12. H.Zimmermann, R.Poupko, Z.Luz and J.Billard, *Z.Naturforschung* **40a,** 149 (1985).
13. J.Malthete, A.M.Levelat and N.H.Tink, *J.Physique Lett.* **46,** 875 (1985).
14. D.Levine, *Journal de Physique* **46,** C8-397 (1985).
15. P.J.Steinhardt, *American Scientist* **74,** 586 (1986).
16. M.Ronchetti, *Philosophical Magazine B* **56,** 237 (1987).
17. D.R.Nelson and B.I.Halperin, *Science* **229,** 233 (1985).
18. T.Janssen, *Physics Reports* **168,** 55 (1988).
19. D.Schechtman, I.Blech, D.Gratias and J.W.Cahn, *Phys. Rev. Lett.* **53,** 1951 (1984).
20. V.Gramlich, *Fortschr.Miner.* **65,** 161 (1987).
21. A.Katz and M.Duneau, *Journal de Physique* **47,** 181 (1986).
22. D.Levine and P.J.Steinhardt, *Journal of Non-Crystalline Solids* **75,** 85 (1985).
23. J.Timmermans, *J.Chim.Phys.* **35,** 331 (1938).
24. M.-L.Saboungi and D.L.Price, *Journal of Non-Crystalline Solids,* **150,** 260 (1992).
25. M.Trömel, *Zeitschrift für Kristallographie* **183,** 15 (1988).
26. F.Rinne, *Transactions of the Faraday Society* **29,** 1016 (1933).
27. R.Hosemann, *Zeitschrift für Physik* **128,** 465 (1950).
28. W.Wilke, *Colloid and Polymer Science* **259,** 577 (1981).
29. W.Wilke, *Colloid and Polymer Science* **261,** 656 (1983).
30. R.Hosemann, *Polymer* **3,** 349 (1962).
31. R.Hosemann, *Berichte der Bunsen-Gesellschaft* **74,** 755 (1970).
32. M.Kakudo and N.Kasai, *X-ray Diffraction by Polymers* (Kodanska, Tokyo & Elsevier, Amsterdam, 1972).

33. B.K.Vainshtein, *Diffraction of X-rays by Chain Molecules* (Elsevier, Amsterdam, 1966).
34. L.E.Alexander, *X-ray Diffraction Methods in Polymer Science* (John Wiley & Sons, New York, 1969).
35. R.Hosemann and M.Hentschel, *Cellulose Chem.Technol.* **19**, 459 (1985).
36. M.S.Zei, *Physical Review A* **27**, 515 (1983).
37. W.Vogel and R.Hosemann, *Carbon* **17**, 41 (1979).
38. S.C.Benett and D.J.Johnson, *Carbon* **17**, 25 (1979).
39. R.Hosemann, M.P.Hentschel, U.Schmiesser and R.Brückner, *Journal of Non-Crystalline Solids* **83**, 223 (1986).
40. A.Burian, P.Lecante, A.Mosset and J.Galy, *Journal of Non-Crystalline Solids* **90**, 633 (1987).
41. E.D.Kuckuk and R.Hosemann, *Colloid and Polymer Science* **259**, 645 (1981).
42. E.J.Wheeler and D.Lewis, *Caldified Tissue Research* **24**, 243 (1977).
43. S.M.Kurilo, I.L.Glavati, T.P.Tautsyura, A.I.Marchenko and G.G.Kravchuk, *Colloid Journal* **48**, 137 (1986).
44. R.Hosemann, A.Preisinger and W.Vogel, *Berichte der Bunsen-Gesellschaft Phys.Chem.* **70**, 796 (1966).
45. F.Garbassi, G.Fagherazzi and M.Calcatera, *J.Catal.* **26**, 338 (1972).
46. G.Fagherazzi, F.Galante, F.Garbassi and J.Pernicone, *J.Catal.* **26**, 344 (1972).
47. C.J.Wright, C.G.Windsor and D.C.Puxley, *J.Catal.* **78**, 257 (1982).
48. W.S.Borghard and M.Bondart, *J.Catal.* **80**, 194 (1983).
49. J.M.Schultz, *J.Catal.* **27**, 64 (1972).
50. R.Hosemann and S.N.Bagchi, *Direct Analysis of Diffraction by Matter* (North Holland, Amsterdam, 1962).
51. E.Garfield, *Current Contents of the Physical, Chemical and Earth Sciences* **30**, No.12, 16 (1990).
52. A.M.Hindeleh and R.Hosemann, *J.Phys.C : Solid State Physics* **21**, 4155 (1988).
53. R.Hosemann, *Journal de Physique* **46**, C8-379 (1985).
54. R.Hosemann, W.Vogel and D.Weick, *Acta Cryst.A* **37**, 85 (1981).
55. W.Wilke, *Colloid and Polymer Science* **258**, 360 (1980).
56. M.Wilkens, *Phys.Stat.Sol.(a)* **2**, 359 (1970).
57. M.A.Krivoglaz, *Scattering Theory of X-rays and Thermal Neutrons by Real Crystals,* in Russian (Nauka, Moscow, 1967).
58. M.A.Krivoglaz, *Diffraction Theory of X-rays and Netrons by Non-Ideal Crystals,* in Russian (Naukova Dumka, Kiev, 1993).
59. H.Oettel, *Experimentelle Technik der Physik* **21**, 99 (1973).
60. C.N.J.Wagner, *Acta Metallurgica* **5**, 427 (1957).
61. R.L.Rothman and J.B.Cohen, *J.Appl.Phys.* **42**, 971 (1971).
62. B.I.Nikolin, *Multilayer Structures and Polytypism in Metallic Alloys,* in Russian (Naukova Dumka, Kiev, 1984).
63. P.Schattschneider and A.Wagendristel, *Zeitschrift für Naturforschung* **30A**, 1261 (1975).

64. E.J.Mittemeijer and R.Delhez, *J.Appl.Phys.* **47,** 1702 (1976).
65. R.Delhez and E.J.Mittemeijer, eds., *Diffraction Studies of Concentration Variations* (Delft University Press, Delft, 1978).
66. C.R.Houska, *J.Appl.Phys.* **41,** 69 (1970).
67. C.R.Houska, *J.Appl.Phys.* **49,** 2991 (1978).
68. B.E.Warren and B.L.Averbach, *J.Appl.Phys.* **21,** 595 (1950).
69. G.K.Williamson and W.H.Hall, *Acta Metallurgica* **1,** 22 (1953).
70. N.K.Misra and B.B.Mitra, *Acta Cryst.* **23,** 867 (1967).
71. J.I.Langford, D.Louer, E.J.Sonneveld and J.W.Visser, *Powder Diffraction* **1,** 211 (1986).
72. H.Toraya, *Powder Diffraction* **4,** 130 (1989).
73. D.Balzar, *Journal of Research of the NIST* **98,** 321 (1993).
74. S.Ergun, *Physical Review B* **1,** 3371 (1970).
75. E.Biedermann, *Acta Cryst.* **13,** 650 (1960).
76. S.Ariely, G.Kimmel, S.F.Dirnfeld, M.Bamberg and B.Priny, *Powder Diffraction* **6,** 66 (1991).
77. T.Ungár, M.Lambrigger and G.Kostorz, *Materials Science and Engineering A* **138,** 147 (1991).
78. M.F.Ebel, *Phys.Stat.Sol.* **5,** 91 (1971).
79. H.Mughrabi, T.Ungár, W.Kienle and M.Wilkens, *Philosophical Magazine A* **53,** 793 (1986).
80. T.Ungár, H.Mughrabi, K.Rönnpagel and M.Wilkens, *Acta Metallurgica* **32,** 333 (1984).
81. T.Ungár, *Materials Science Forum* **166-169,** 23 (1994).
82. P.Klimanek *in X-ray and Neutron Structure Analysis in Materials Science,* ed.J.Hašek (Plenum Press, New York, 1989), p.125.
83. J.G.M.Berkum, R. van Delhez, Th.H. de Keijser and E.J.Mittemeijer, *Phys.Stat.Sol.(a)* **134,** 335 (1992).
84. F.H.Cocks and S.F.Cogan, *Journal of Materials Science* **11,** 2061 (1976).
85. S.Rao and C.R.Houska, *Acta Cryst.A* **42,** 6 (1986).
86. S.Ciccariello, G.Fagherazzi and A.Benedetti, *Acta Cryst.A* **46,** 187 (1990).
87. G.Allegra and S.Brückner, *Powder Diffraction* **8,** 102 (1993).
88. P.H.Duvigneand and R.Deric, *Journal of Solid State Chemistry* **34,** 323 (1980).
89. A. de Bail in *Accuracy in Powder Diffraction II,* eds. E.Prince and J.K.Stalick (NIST, Washington 1992), p.142.
90. K.Miura, T.Ino and N.Minami, *Acta Cryst.A* **43,** 49 (1987).
91. J.Fiala and F.Filuš in the *Proceedings of the 12th Conference of Czech and Slovak Physicists,* eds. M.Lesňák, J.Luňáček and J.Pištora (JČMF, Ostrava, 1996), p.427.
92. R.Hosemann, *Zeitschrift für Kristallographie* **189,** 95 (1989).
93. R.Hosemann, *Coloid and Polymer Science* **260,** 864 (1982).
94. R.Hosemann, *Progress in Colloid and Polymer Science* **77,** 15 (1988).
95. R.Hosemann, W.Schmidt, A.Lange and M.Hentschel, *Colloid and Polymer Science* **259,** 1161 (1981).

96. R.Hosemann, M.Hentschel, F.J.Baltá-Calleja, E.Lopez Cabarcos and A.M.Hindeleh, *Experimentelle Technik der Physik* **33,** 135 (1985).
97. A.M.Hindeleh and R.Hosemann, *Polymer* **23,** 1101 (1982).
98. M.Malá, *Materials Chemistry and Physics,* to be published.
99. J.Friedel, *Les Dislocations*(Gauthier-Villars, Paris, 1956).
100. J.R.Low, *Progress in Materials Science* **12,** 31 (1965).
101. J.Mienkisz and Ch.Radin, *Physical Review B* **39,** 1950 (1989).
102. R.J.Hawkins and E.W.April, *Advances in Liquid Crystals* **6,** 243 (1983).
103. D.M.Walba, *Advances in Synthesis and Reactivity of Solids* **1,** 173 (1991).
104. D.M.Walba, *Science* **270,** 250 (1995).

PARACRYSTALLINE DISTORTIONS, INHERENT STRAINS AND CRYSTALLITE SIZE IN POLYMERS

W.WILKE
Department of Experimental Physics, University of Ulm,
D-89069 Ulm, Germany

The crystals of partially crystalline polymers are imperfect. The coherence length (crystallite size) in different crystallographic directions is in the order of some 10 nanometers and lattice distortions exist. Two types of distortions will be discussed: paracrystalline distortions and inherent strains. Paracrystalline distortions are a special type of lattice distortions, based on statistical fluctuations of the three cell edge vectors. With the assumption of statistical independence of the three coordination statistics, one arrives at the model of the ideal paracrystal. For this case an elegant mathematical formulation of the scattering theory of such a structure is possible (Hosemann). This type of distortions is found in a number of cases in polymer crystals. A structure with quasi long range order (QLRO) results. Paracrystalline distortions in polymers probably are the result of overlapping strain fields from dislocation like defects.
Inherent strains develop during cooling of partially crystalline polymers, caused by inherent stresses. The inherent strains, in contrast to paracrystalline distortion, increase with decreasing temperature. Crystallite size and lattice distortions are the reasons for line broadening in the wide angle X-ray scattering patterns. The angular dependence of the broadening for the two types of distortions is different. For inherent strains the broadening is proportional to the order p of reflexion (~ p) for paracrystalline distortions to the square of the order (~ p^2). Broadening by crystallite size is independent of p. The Warren-Averbach analysis cannot be applied, because no or only very few measurable reflexions of higher order exist. A method is presented, which, using all measurable line widths, allows the determination of the crystallite sizes (crystallite shape) and the magnitude of the distortions. In some cases, by combination with profil analysis, it is possible to measure also the size distribution

1. Introduction

The term "paracrystal" was used by Rinne[1] 1932 as a method of notation for liquid crystalline structures. Hosemann[2] introduced the "ideal paracrystal" 1950 as a special model for the description of statistical defined lattice structures. Later he discussed the "real paracrystal" and summarized the ideas and results about paracrystalline structures in a book [3]. Today paracrystals can be classified as systems with quasi long-range order (QLRO). For this systems the translational correlations decay according to a power law [4].
There are some problems, mainly in connexion with the ideal paracrystal, which is constructed from three independent coordination statistics by a folding process. To avoid these problems, models of real paracrystals were discussed, but also in this case problems with statistical correlations occur [5,6]. A reply to this objections is given in[7].

For the theoretical description of physical phenomena one has to use a model, having the limitations in mind. In the case of real polymer crystals with small lattice distortions and small crystallite sizes the above mentioned objections are not important and the model of the ideal paracrystal is a good approximation. Furthermore, it has the advantage that the scattering theory is formulated in a simple analytical way. The parameters are determined from the widths of the reflections in the X-ray powder diagrams.
In the following an overview of the theory and the method of parameter determination is given. The additionel line broadening by inherent strains is discussed also.
Some examples from the field of polymers illustrate the application.

2. Ideal paracrystal

2.1 Mathematical model: distance statistics, lattice factor

The distance (lattice) statistics z(**r**) gives the mean density and coordination of the adjacent points around any one and is sometimes also called coordination statistic. For the ideal paracrystal it is given by[3,8]

$$z(\boldsymbol{r}) = \delta(\boldsymbol{r} - \boldsymbol{0}) + \Sigma_m \Sigma_n \Sigma_o [H_{mno}(\boldsymbol{r}) + H_{-(mno)}(\boldsymbol{r})] \qquad (1)$$

$H_{100}(\mathbf{r}) = H_1(\mathbf{r})$ resp. $H_{010}(\mathbf{r}) = H_2(\mathbf{r})$ resp. $H_{001}(\mathbf{r}) = H_3(\mathbf{r})$ gives the probability of any given distance between first neighbours in
the direction $\mathbf{a}_1$ resp. $\mathbf{a}_2$ resp. $\mathbf{a}_3$ ($\mathbf{a}_i$ are the mean cell edge vectors).
The ideal paracrystal is constructed with the assumption that the edge statistic functions $H_k(\mathbf{r})$ are statistical independent and that adjacent cell edge vectors in direction k are uncorrelated. Then the expression for a peak mno of the lattice statistic is obtained by the folding of the edge statistics:

$$H_{mno}(\boldsymbol{r}) = H_1 \otimes..|m|..\otimes H_1 \otimes H_2 \otimes..|n|..\otimes H_2 \otimes H_3 \otimes..|o|..\otimes H_3 \qquad (2)$$

The resulting distorted lattice is statistically homogeneous. Because of the folding product in (2) the second moment (as measure for the width) of the peaks mno increases linear with the distance to the reference point.
To calculate the scattering intensity, one needs the Fourier transform of z(**r**). Let the transform of $H_k(\mathbf{r})$ be $F_k(\mathbf{b})$. The transform of $H_{mno}(\mathbf{r})$ is then given by

$$\mathscr{F}(H_{mno}) = F_1^m \, F_2^n \, F_3^o \qquad (3)$$

and that of z(**r**) by

$$Z(\boldsymbol{b}) = \Sigma_m \Sigma_n \Sigma_o \, F_1^m F_2^n F_3^o = Z_1 Z_2 Z_3 \qquad (4)$$

The vector **b** in reciprocal space is defined by

$$\boldsymbol{b} = \frac{\boldsymbol{s} - \boldsymbol{s_0}}{\lambda} \qquad (5)$$

where **s** resp. $\mathbf{s_0}$ are unit vectors in the direction of the scattered resp. incoming X-ray beam with wavelength λ. **b** and the diffraction angle 2ϑ are related by

$$|\boldsymbol{b}| = b = \frac{2 \sin\vartheta}{\lambda} \qquad (6)$$

Each summation in (4) gives the following result:

$$Z_k(\boldsymbol{b}) = Re\,\frac{1 + F_k(\boldsymbol{b})}{1 - F_k(\boldsymbol{b})} \qquad k = 1,2,3 \qquad (7)$$

For functions $H_k(\mathbf{r})$ having a center of symmetry, the statistical amplitude $F_k(\mathbf{b})$ can be expressed in the form

$$F_k(\boldsymbol{b}) = |F_k(\boldsymbol{b})|\exp(-2\pi i\boldsymbol{b}\cdot\boldsymbol{a}_k) \qquad (8)$$

and the lattice factor $Z(\mathbf{b})$ as

$$Z(\boldsymbol{b}) = \prod_{k=1}^{3} \frac{1 - |F_k(\boldsymbol{b})|^2}{1 + |F_k(\boldsymbol{b})|^2 - 2|F_k(\boldsymbol{b})|\cos(2\pi\boldsymbol{b}\cdot\boldsymbol{a}_k)} \qquad (9)$$

Each factor in (9) has maxima for $\mathbf{b}\cdot\mathbf{a}_k = h_k$ (integer).
In the following the formulae are given for orthogonal lattices (orthorhombic). The general case is discussed in[9].
For $H_k(\mathbf{r})$ a gaussian function is assumed:

$$H_k(\boldsymbol{r} + \boldsymbol{a}_k) = \frac{1}{(2\pi)^{3/2}\Delta_{k1}\Delta_{k2}\Delta_{k3}}\exp\left[-\frac{1}{2}\left(\frac{x_1^2}{\Delta_{k1}^2} + \frac{x_2^2}{\Delta_{k2}^2} + \frac{x_3^2}{\Delta_{k3}^2}\right)\right] \qquad (10)$$

The Fourier transform of H_k is

$$\mathcal{F}\,[H_k(\boldsymbol{r} + \boldsymbol{a}_k)] = F_k(\boldsymbol{b}) = \exp\left[-2\pi^2\left(\Delta_{k1}^2 b_1^2 + \Delta_{k2}^2 b_2^2 + \Delta_{k3}^2 b_3^2\right)\right] \qquad (11)$$

x_i are the cartesian components (axes parallel to $\mathbf{a}_i$) of $\mathbf{r}$ and b_i the related components of $\mathbf{b}$ in reciprocal space. The Δ_{ki} are the nine principal values of the three fluctuation tensors $\underline{\mathbf{T}}_k$. The first subscript to Δ denotes the cell edge vector, the second the direction of the fluctuation of this vector. This is illustrated in figure 1.
It is convenient to introduce the relative fluctuations g_{ki} by

$$g_{ki}^2 = \frac{\Delta_{ki}^2}{a_i^2} \qquad (12)$$

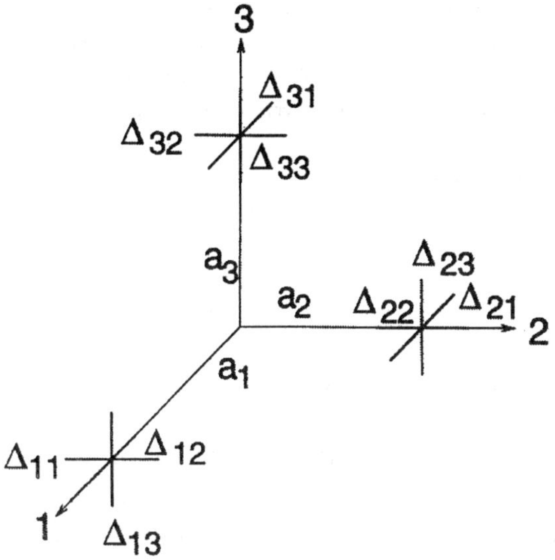

Fig.1: Principal values of the 3 fluctuation tensors

Insertion of (11) in (9) gives the explicit formula for the lattice factor of the ideal paracrystal.

2.2 Physical interpretation

Hosemanns conception of the paracrystal was, in brief, that it is a fundamental structure and that crystals and liquids are the limiting cases for condensed matter. Furthermore he wanted to explain the empirical relation between lattice distortions and crystallite size, the so called α^*-law. One attempt to solve this problems was the introduction of immobile defects (substitution and interstitial) during crystallization [10]. Later a distortion potential for the valence angles was defined to obtain an additional free enthalpy term [11,12]. It seems, that these attempts up to now are not real succesful. On the other side, in many cases it is possibel, especially for polymers, to describe the line broadening in X-ray diagrams in a quantitative way by paracrystalline distortions.

Therefore one can use the ideal paracrystal as a mathematical model for the describtion of a special type of distorted lattices.

Another point is the discussion about the physical origin of this distortion. There is a number of possibilities [13]. The most probable explanation (in the authors opinion), based on experimental results from polymers, is the following: paracrystalline distortions are the result of a superposition of strain fields, caused by dislocations or dislocation like defects [14,15,16]. For both cases the long range order is destroyed. If the dislocations are distributed randomly, the resulting strain field gives (in continuum approximation), within experimental errors, the same increase of line widths with the square of the reflection order as the paracrystal model. The average value $\langle g_{ki}^2 \rangle$ is a measure of the dislocation density. In polymer crystals dislocations are introduced e.g. during crystallisation, by incorporation of chain ends or branchings. Because of the weak van der Waals bonding between polymer chains, the strain field of kink defects is long ranged and acts, in connection with the small crystallite size, like a dislocation. Dislocations are also generated, as usual, during deformation of polymer crystallites. Examples will be given in chapter 4.

3. Scattering intensity of ideal paracrystals

3.1 General intensity formula

If the lattice distortions are independent of the crystallite size, the scattering power of the sample (N lattice cells with mean volume V) is given by [3]

$$I(\boldsymbol{b}) = N\left[<|F(\boldsymbol{b})|^2> - <|F(\boldsymbol{b})|>^2 \cdot |D(\boldsymbol{b})|^2 \right] + \\ + \frac{1}{V} <|F(\boldsymbol{b})|^2> \cdot |D(\boldsymbol{b})|^2 \cdot Z(\boldsymbol{b}) \otimes |S(\boldsymbol{b})|^2 \qquad (13)$$

$|D(\mathbf{b})|^2$ stands for the Debye-Waller factor, F(**b**) for the structure amplitude, Z(**b**) for the lattice factor (9) and S(**b**) for the shape amplitude. S(**b**) is the Fourier transform of the shape function s(**r**), which defines shape and size of the coherent scattering domains (crystallites).

The first term in (13) describes a diffuse background scattering, the second the broadened reflections.

The broadening is caused by paracrystalline distortions (Z(**b**)) and crystallite size ($|S(\mathbf{b})|^2$).

A good approximation for the shape factor is the shape ellipsoid. The crystallite shape and size (sample average) is characterized by the three principal axes D_i and the angle φ between D_1 and the crystallographic $\mathbf{a}_1$-axis (the chain direction of the polymer molecule is $\mathbf{a}_3$). In figure 2 a two dimensional section of the shape ellipsoid is shown.

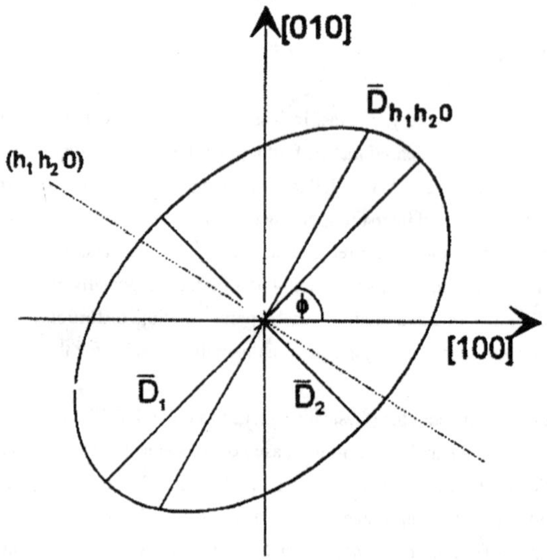

Fig.2: Shape ellipsoid (sectional drawing of the $\mathbf{a}_1$-$\mathbf{a}_2$ plane)

Within a reflection F(**b**) and D(**b**) are nearly constant and therefore the profile of the reflection is determined by the folding of the lattice- and the shape factor:

$$I(\boldsymbol{b}) \propto Z(\boldsymbol{b}) \otimes |S(\boldsymbol{b})|^2 \qquad (14)$$

3.2 Powder diagrams: integral widths of the reflections

For a powder diagram the intensity is proportional to the spherical average of (14) and is a function of $|\mathbf{b}| = b$. The integral width $\delta\beta_h$ of the profile of a reflection with Miller indices $h_1h_2h_3 \equiv h$ is defined as:

$$\delta\beta_h = \frac{\int I_h(b')db'}{I_h(b'=0)} \qquad b' = b - b_h \qquad (15)$$

The average size of the crystallites contributes to the integral width[17]:

$$\delta\beta_h(S) = \frac{2}{D_{h_1h_2h_3} + D_{\overline{h_1h_2h_3}}} = \frac{1}{D_{\{h_1h_2h_3\}}} \qquad (16)$$

D_{h1h2h3} is the averaged crystallite size perpendicular to the net plane $(h_1h_2h_3)$.
The lattice factor Z_h for a reflection h reads as follows[18]:

$$Z_h(b') \propto \frac{1}{1 + \frac{\pi^2}{\delta\beta_h^2} \cdot b'^2} \qquad (17)$$

The paracrystalline distortions give a contribution to the integral width which increases with the square of the order p of the reflection:

$$\delta\beta_h(Z) = \frac{\pi^2 g_s^2}{d_{h'}} \cdot p^2 \qquad (18)$$

where $d_{h'}$ is the netplane spacing (h' stands for the Miller indices without common measure, $h_i = h_i' \cdot p$) and g_s^2 is the resulting relative fluctuation perpendicular to the netplanes h:

$$g_s^2 = \frac{\sum_{i=1}^{3}\sum_{k=1}^{3} h'^2_i \cdot \frac{h'_k}{a_k^2} \cdot g_{ki}^2}{\sum_{i=1}^{3} \frac{h'^2_i}{a_i^2}} \qquad (19)$$

After correction of apparative broadening and broadening by sample thickness the integral width of the reflection h is obtained by adding the contributions from crystallite size and paracrystalline distortions:

$$\delta\beta_h = \delta\beta_h(S) \oplus \delta\beta_h(Z) \qquad (20)$$

If other types of distortions (e.g.strain distribution) also occur, further terms must be added.
The definite type of addition depends on the profiles of the contributions. The paracrystalline distortions contribute a Lorentz profile, as shown in (17). The profile of the shape factor can be determined approximately by measuring a reflection of first order (at low angles), because the influence of lattice distortions is small in this case. In many cases it is, within the experimental errors, a good approximation to work with the addition of two Lorentz profiles:

$$\delta\beta_h = \delta\beta_h(S) + \delta\beta_h(Z) = \frac{1}{D_{\{h_1h_2h_3\}}} + \frac{\pi^2 g_s^2}{d_{h'}} \cdot p^2 \qquad (21)$$

For some other profile combinations results are given in [19].

3.3 Determination of the parameters D_i, φ, g_{ki}

From the measured values of the integral width of n reflections the mean crystallite sizes D_i, the angle φ and the fluctuation parameters g_{ki} are calculated by minimizing the sum

$$F = \sum_{i=1}^{n} \left\{[\delta\beta_h^i(\exp) - \delta\beta_h^i(Z) - \delta\beta_h^i(S)] \cdot w^i\right\}^2 \qquad (22)$$

According to the accuracy of the measurement of the $\delta\beta_h^i(\exp)$ the values are weighted by the factor w^i. If the lattice distortion can be described by the model of the ideal paracrystal, the difference between measured and calculated values as a rule is less than 5% - 10% for all reflections.
One may be astonished at the success of the method, because of the relatively large number of parameters (4 + 9 = 13) and the limited number of reflections, especially in the case of polymers. But one should have in mind, that the

crystallite size and the relative fluctuations are smooth functions of the crystallographic direction. They cannot change abruptly between neighbouring directions. Therefore exist very strong constraints between the parameters. These constraints must be fulfilled by the simultaneous adjustment of the parameters .

3.4 Analysis of line profiles: Determination of crystallite size distribution

Without lattice distortions the shape factor $|S(b')|^2$ contains the information about the crystallite size distribution g(D). The line profile is proportional to the shape factor:

$$I_s(b') \propto |S(b')|^2 \qquad (23)$$

The autocorrelation function $p_s(r')$ of the shape function can evaluated from the line profile by Fourier transformation. The crystallite size distribution g(D) can be determined from the second derivative of $p_s(r')$ with respect to r' [20].
With lattice distortions the profile is influenced also by Z(b'). In this case in principle the Warren-Averbach method can be applied. For doing this, reflections of higher order are needed, which in the case of polymers in general are not useful (overlapping,background,low intensity). To solve this problem, a new method is proposed [21].
The method is a combination of the (above explained) line width analysis of a number of reflections, which results in the determination of the contribution of the lattice distortions (and the average values of the crystallite sizes) and an analysis of the profile of single reflections. In the following a short sketch of the method will be given.
The profile is corrected by the Stokes method.
The g_{ki} are determined by measuring the integral widths. Then, by (17)-(19) the function Z(b') (the subscript h is supressed in the following) is known. Fourier transformation of (14) gives (after normalization):

$$\mathcal{F}\,[I(b')] = \mathcal{F}\,[Z(b')] \cdot \mathcal{F}\,[|S(b')|^2] \qquad (24)$$

From this follows with (23):

$$p_s(r') = \mathcal{F}\,[I_s(b')] = \frac{\mathcal{F}\,[I(b')]}{\mathcal{F}\,[Z(b')]} \qquad (25)$$

and g(D) by

$$g(D) = D \cdot \left(\frac{d^2 p_s(r')}{dr'^2}\right)_{r'=D} \qquad (26)$$

The observed and instrumental line profiles are fitted with Pearson type VII functions [22]:

$$P(x) = \frac{f_0}{\left[1 + \frac{(x - x_0)^2}{ma^2}\right]^m} \qquad (27)$$

P(x) is Lorentzian (Cauchy) for m = 1, modified Lorentzian for m = 2 and Gaussian for m = ∞.
The Fourier transform of P(x) is known [23], but will not be given here.
All calculations are done by use of the computer algebra software 'MAPLE V'.

4. Examples of the application of the method to polymers

4.1 Deformation of polyethylene (PE)

Three PE samples (linear, branched and very high molecular weight) are investigated.For details see [16]. The D_i and g_{ki} are determined for different draw ratios $\lambda = l/l_0$ (l_0 initial length).
The volume V and surface S of the crystallites are given by

$$V = \frac{\pi}{6} \cdot D_1 \cdot D_2 \cdot D_3 \qquad (28)$$

$$S \approx \frac{\pi D_3}{2} \left[\frac{3}{2}(D_1 + D_2) - \sqrt{D_1 \cdot D_2}\right] \qquad (29)$$

During deformation the crystallites break up and new internal surfaces are formed.The volume of the new crystallites decreases to V_λ and their number increases to V_1/V_λ. The relative increase of the internal surface follows from

$$S_G = S_\lambda \cdot \frac{V_1}{V_\lambda} \qquad (30)$$

The internal surface is important for the analysis of creep processes [24]. The opposite process takes place during relaxation of the samples at higher temperatures in conjunction with recovery and recrystallization processes.
As an example the change of D_1 and V is illustrated in figures 3a and 3b.

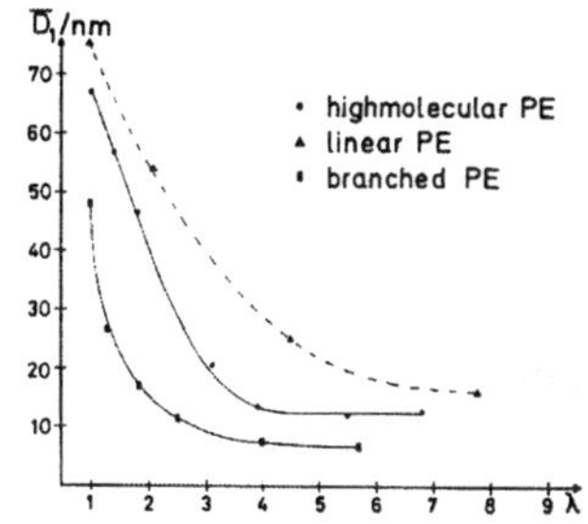

Fig.3a: D_1 as function of λ

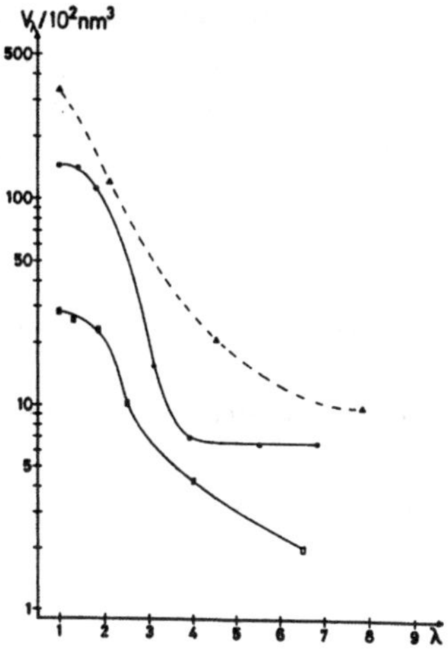

Fig.3b: V as function of λ

The crystallite size decreases drastically and approaches a limiting value at high draw ratios. In contrast the mean value $<g_{ki}>^2$increases, as shown in figure 4.

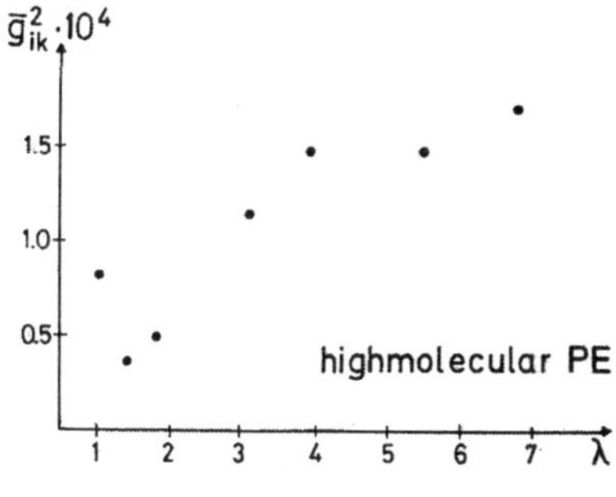

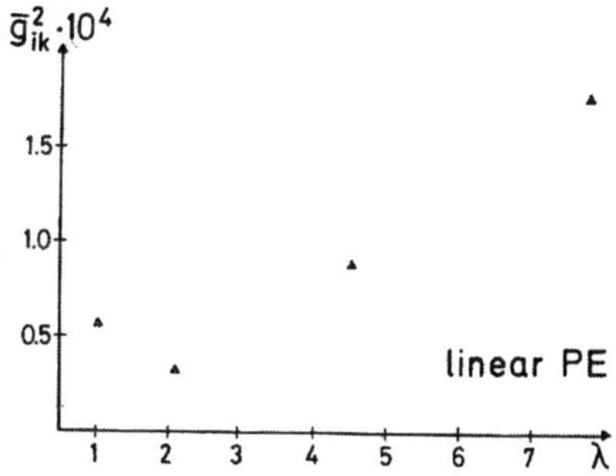

Fig.4: $<g_{ki}>^2$ as function of λ

The increase of $<g_{ki}>$ is attributed to an increase in dislocation density. The initial decrease is explained by the fact, that small, strongly distorted crystallites become unstable first and disappear. At the first stage of deformation the increase in dislocation density leads to the formation of new grain boundaries and consequently to the observed decrease of crystallite sizes. At high draw ratios the deformation process is mainly determined by grain boundary slip and no remarkable change of crystallite size occurs. But the crystallites are cross-linked by tie molecules and therefore stress induction and subsequently dislocation generation inside the crystallites occur.

4.2 Crystallization of fractions of (Ethylen-α-hexen)-copolymers

These substances are polyethylenes with definite side chain length. The fractionation was done with respect to side chain content ϵ (CH_3/1000 C)[25].

Fig.5 shows the volume V of the crystallites as a function of ϵ. V decreases with increasing ϵ to an asymptotic value, the volume of the smallest stable crystal.

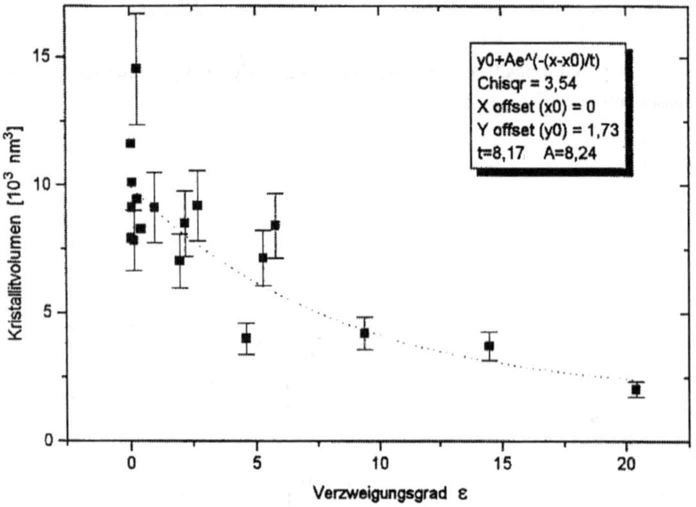

Fig.5: Mean crystallite volume in relation to side chain content

The averaged distortion parameters $<g_{ik}>$ do not increase with ϵ (see fig.6). This is an indication, that the side groups are included into the crystal lattice only to a limited amount.

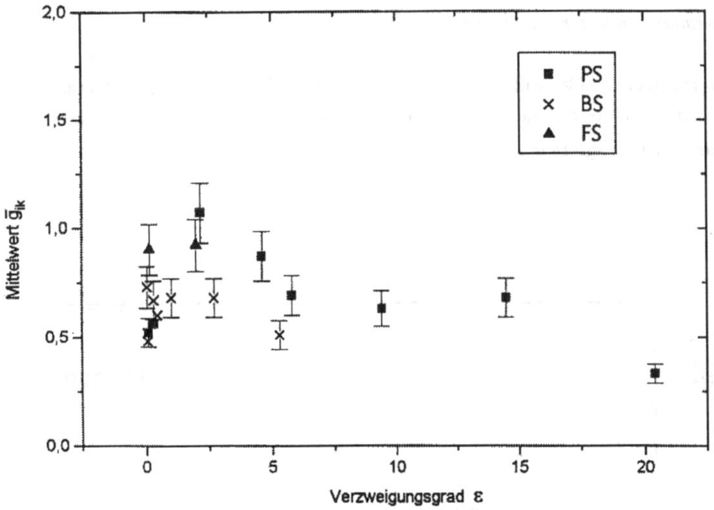

Fig.6: Averaged distortion parameters <g_{ik}> in relation to side chain content ϵ

The quotient of the lateral to the top surface areas of the crystals is nearly constant, as shown in figure 7. One possible explanation is the constancy of the related surface energies.

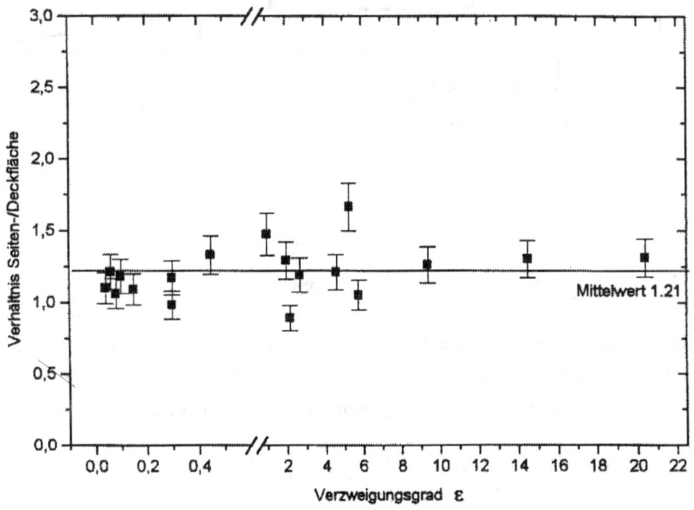

Fig.7: The qotient of lateral and top surface areas of the crystallites in relation to ϵ

4.3 Low temperature behaviour of polyethylene

Measurements at a fraction of PE with molecular weight 2000 (PE 2000)[26,27] and different samples of linear PE [28] at temperatures down to 10 K show a strong increase of the integral widths of the reflections with decreasing temperature (an example is given in figure 8)

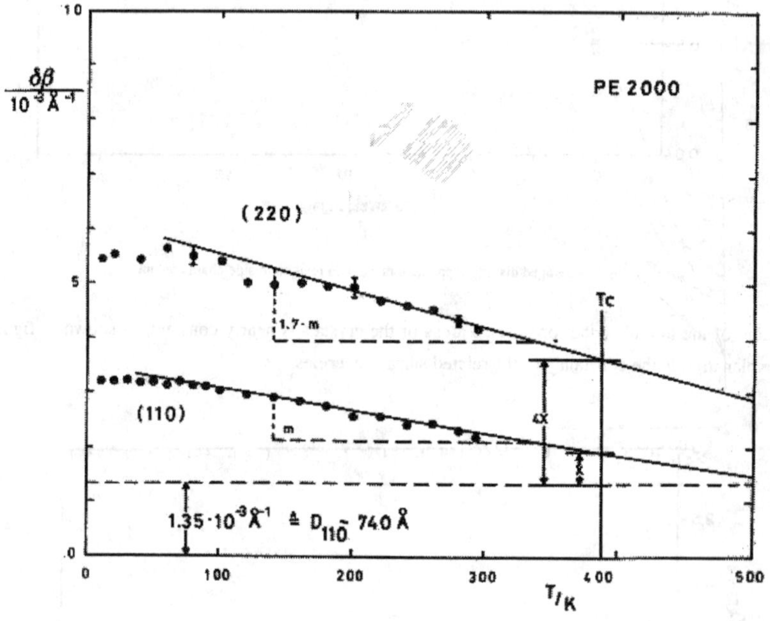

Fig.8: Integral widths of 110 and 220 as function of temperature

In these cases three contributions to δß must be taken into consideration: crystallite size and paracrystalline distortions, both independent of temperature, and temperature dependent inherent strains ϵ_h. The contribution of the strain distribution to $\delta\beta_h$ depends on $\langle\epsilon_h^2\rangle$ and is proportional to the order p of the reflections[17]:

$$\delta\beta_h(el)=\frac{\sqrt{2\pi}}{d_h}\sqrt{\langle \epsilon_h^2 \rangle}\cdot p \qquad (31)$$

If three orders of one reflection h' are available, it is possible to discriminate between paracrystalline distortions and broadening by strain distribution by measuring only the integral widths δß, because of the different dependence of δß on the reflection order p (proportional to p^2 resp.p). If only two orders are available, Fourier methods can be applied [29,30].

Another possibility presents itself, if $\delta\beta$ is measured at different temperatures. At $p = 0$ lattice distortions do not contribute to $\delta\beta$ and therefore in a plot of $\delta\beta$ - p^2 resp.p the extrapolated lines should have one point of intersection at $p = 0$. In figure 9 measurements are plotted for 200, 400 and 110, 220. Only the linear dependence results in one point of intersection, proving that the strain distribution dominates the temperature dependend part of the integral widths.

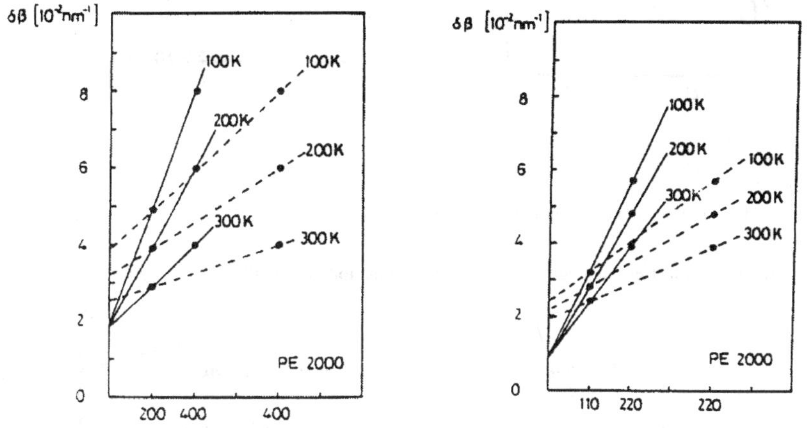

Fig.9: Integral widths vs. p resp. p^2 for different temperatures

There are no external stresses, so we can describe the increase of the integral widths with decreasing temperature due to the appearance of inherent stresses inside the sample. These stresses may be generated by small temperature inhomogeneities and by the different expansion coefficients of the lattice parameters of adjoining crystals with different orientation. They should disappear at the crystallization temperature T_c and therefore the integral widths of different orders of the same reflection should be the same. But fig.8 shows, that there remains a difference. This difference is interpreted as paracrystalline disorder, caused by dislocations generated during the crystallization[31], because we get significant crystal sizes only by quadratic separation (1:4) of this remaining integral widths. This is true for hk0-reflections. The analysis of the hk1-reflections gives the result, that stresses in chain direction are generated only below room temperature. The reasons may be the very large elastic constant in chain direction or the α-process which is effective only in chain direction and at temperatures greater than room temperature.

Another interesting result is the occurence of a minimum of the integral width for a number of reflections at very low temperatures (figure 10).

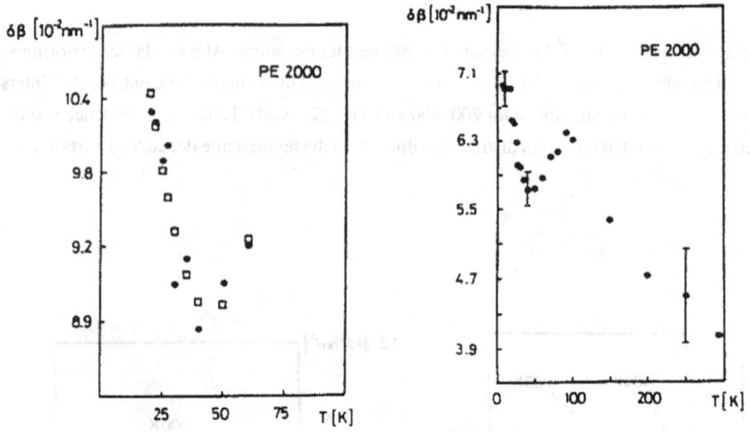

Fig.10: Integral width of reflection 211 resp.220 as a function of temperature

This effect is explained by a stress induced martensitic transformation of a small part of the crystals.

4.3 Change of crystallite size distribution of (Ethylen-1-hexen)-copolymers by heating

The above mentioned method of line profile analysis is applied to commercial polyethylene products of a special type (with uniform side groups).Measurements are done at four temperatures between 20°C and 121°C.The profiles of the reflections 200, 020 and 110 are investigated [21]. The following figures visualize the crystallite size distributions at different temperatures.

There are differences in the different crystallographic directions. With increasing temperature the distributions shift to larger crystallite sizes. This effect is more pronounced at higher temperatures (melting point $T_m = 133°C$) because of the higher chain mobility. Smaller crystals become unstable and melt and larger crystals appear as a result of recrystallization.

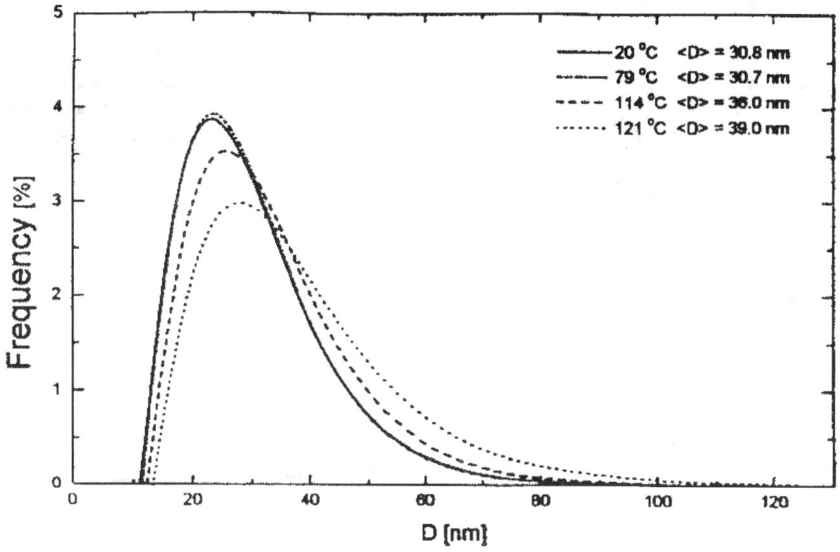

Fig.11: Crystallite size distribution normal to {110}-netplanes

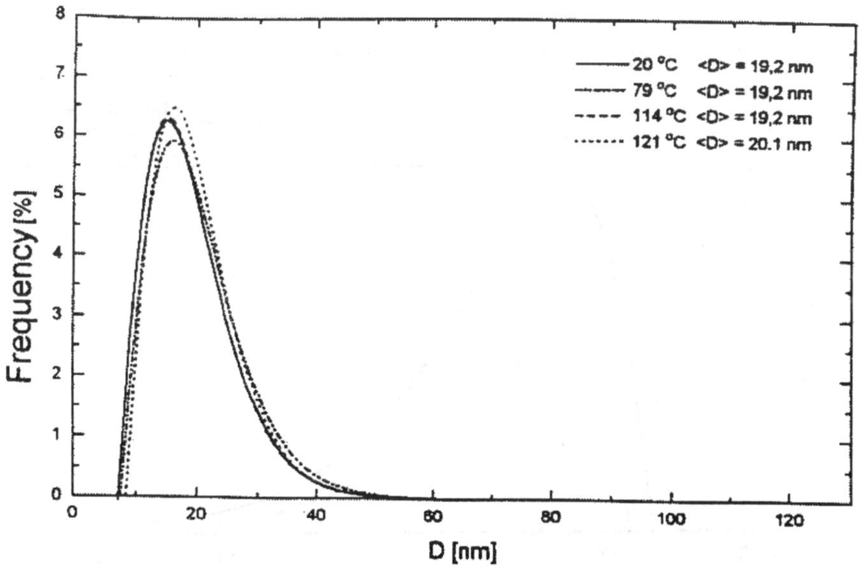

Fig.12: Crystallite size distribution normal to {100}-netplanes

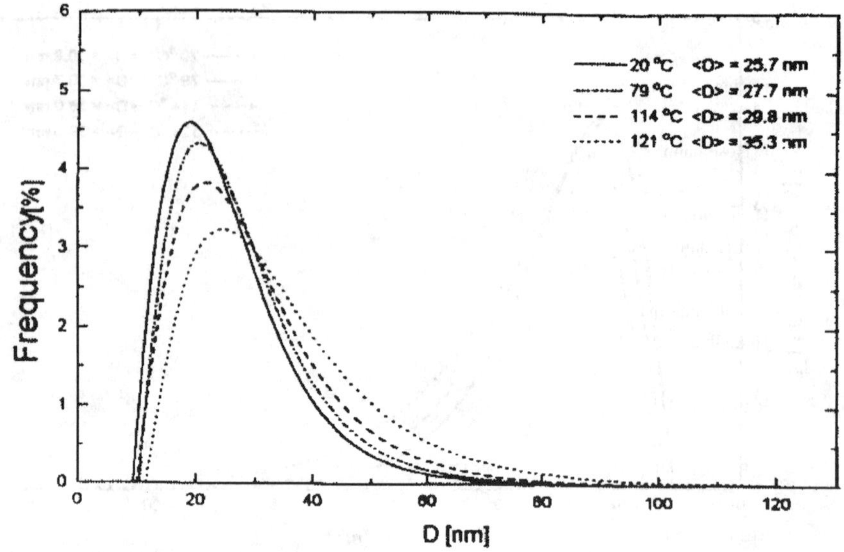

Fig.13: Crystallite size distribution normal to {010}-netplanes

In figure 14 the influence of the paracrystal correction is shown.

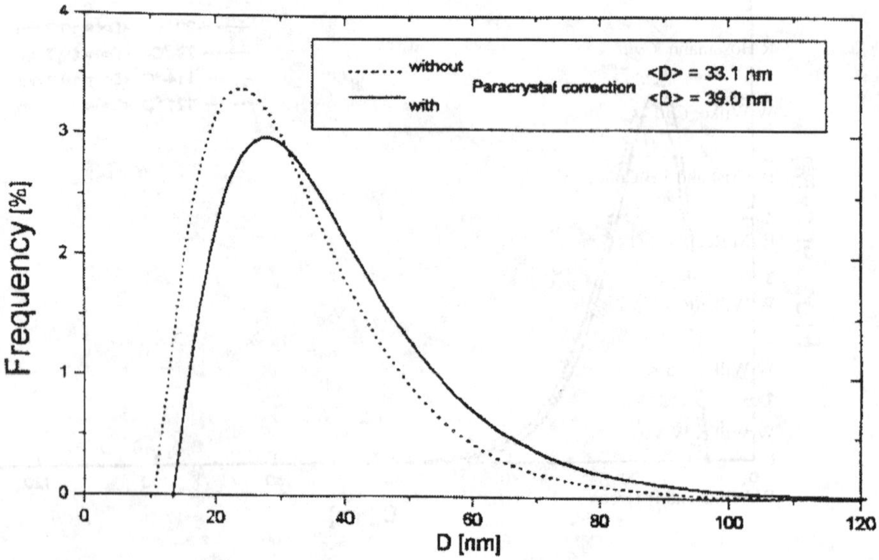

Fig.14: Crystallite size distribution (T = 121 °C,{110}) with and without paracrystal correction

References

1. F.Rinne, *Z.Krist.* **82**, 379 (1932)

2. R.Hosemann, *Z.Physik* **128**, 1 and 465 (1950)

3. R.Hosemann and S.N.Bagchi in *Direct analysis of diffraction by matter*, North-Holland Publ. Comp., Amsterdam (1962)

4. G.Venkataraman, D. Sahoo and V.Balakrishnan in *Beyond the crystalline State*, Springer-Verlag Berlin Heidelberg (1989)

5. R.Brämer, *Acta Cryst.* **A31**, 551 (1975)

6. R.Brämer and W.Ruland, *Makromol.Chem.* **177**, 3601 (1976)

7. M.Janke and R.Hosemann, *Progr.Coll. & Polym.Sci.* **64**, 226 (1978)

8. B.K.Vainshtein in *Diffraction of X-rays by chain molecules*, Elsevier Publ. Comp. Amsterdam, London, New York (1966)

9. W.Wilke, *Acta Cryst.* **A39**, 864 (1983)

10. P.H.Lindenmeyer, H.Beumer and R.Hosemann, *Progr.Coll. & Polym.Sci.* **64**, 232 (1978)

11. R.Hosemann, W.Schmidt, A.Lange and M.Hentschel, *Coll. & Polym. Scie.* **259**, 1161 (1981)

12. R.Hosemann, *Coll. & Polym. Sci.* **260**, 864 (1982)

13. W.Wilke, *Coll. & Polym. Sci.* **258,** 360 (1980)

14. B. Crist and J.B.Cohen, *J.Polym.Sci., Polym.Phys.Ed.* **17**, 1001 (1979)

15. E.-M.Reck, H.Schenk and W.Wilke, *Progr. Coll. & Polym. Sci.* **71**, 154 (1985)

16. W.Wilke, *Coll. & Polym. Sci.* **261**, 656 (1983)

17. W.Wilke and K.W.Martis, *Coll. & Polym. Sci.* **252**, 718 (1974)

18. W.Wilke, W.Vogel and R.Hosemann, *Coll. & Polym. Sci.* **237**, 317 (1970)

19. W.Wilke and R.Hosemann, *Faserforsch. u. Textiltechn.* **18**, 54 (1967)

20. A.Guinier in *X-Ray Diffraction in Crystals, Imperfect Crystals, and Amorphous Bodies*, W.H.Freeman and Comp., San Francisco and London (1963)

21. G.Bodor, S.Gall, K.D.Häberle, U.Lode and W.Wilke, *J.Polym.Sci. Part B: Polym.Phys.* **34**, 485 (1996)

22. K.Pearson, *Phil.Trans.Vol.A* **216**, 429 (1916)

23. M.M.Hall, V.G.Veeraraghavan, H.Rubin and P.G.Winchell, *J.Appl.Cryst.* **10**, 66 (1977)

24. J.Marosfalvi, G.Bodor and W.Wilke, *Kunststoffe German Plastics* **80** (10), 93 (1990)

25. G.Bodor, *Müanag és Gumi* **29**, 201-206 (1992)

26. W.Fronk and W.Wilke, *Coll. & Polym. Sci.* **259**, 797 (1981)

27. K.Schauer and W.Wilke, *Polym.Bull.* **34**, 477 (1995)

28. W.Fronk and W.Wilke, *Coll. & Polym. Sci.* **260**, 1107 (1982)

29. W.Vogel, J.Haase and R.Hosemann, *Z. Naturforsch.* **A 29**, 1152 (1974)

30. R.Hosemann, A.Lange and M.P.Hentschel, *Acta Cryst.* **A 41**, 434 (1985)

31. J.M.Peterson, *J.Appl.Phys.* **37**, 4047 (1966)

ON THE PARACRYSTALLINITY OF OPAL

Deane K. Smith
Department of Geosciences and Materials Research Laboratory
The Pennsylvania State University
University Park, PA 16802, USA

Gary P. Tomaino
Minerals Technologies Inc.
Easton, PA 18042, USA

Adrian C. Wright
J. J. Thomson Physical Laboratory, Reading University
White Knights, Reading, RG6 24F, United Kingdom.

Cristobalite and tridymite are distinct forms of crystalline silica, which along with quartz are encountered in industrial operations and industrial products. Because the International Agency for Research on Cancer has designated "crystalline silica" as a probable IARC type 2A carcinogen and quartz and cristobalite as a type 1 known carcinogen, it is important to properly identify and quantify the silica phase in all materials used in production and encountered in products. Opal is a form of hydrated silica, which is also encountered in industry. Although some forms of opal mimic cristobalite and tridymite, all forms of opal are not truly crystalline. Opal-A is disordered both in the short range and in the long range, but opal-CT and opal-C contain short-range ordered domains that mimic stacked sequences of cristobalite and tridymite sheets such that X-ray diffraction patterns show features similar to the crystalline cristobalite and tridymite. There is considerable debate as to whether the ordered regions have lost the water that many consider essential to the opals. Neutron diffraction has shown that opal-CT and opal-C yield similar patterns, which is distinctly different from the pattern obtained from cristobalite. DTA and TGA studies reveal that all opals show changes on heating characteristic of materials that lose water in the process. Opal-CT has proved stable to temperatures above 1500 °C even though the water has been removed by the heating.

1. INTRODUCTION

The term "silica" in the industrial sense is used to mean any material whose composition is SiO_2 whether it is crystalline or non-crystalline. Some people also consider silica to include hydrated SiO_2, which includes the opals and many new microsieve materials. The phases of SiO_2, both crystalline and non-crystalline, have long been very important materials in the manufacture of industrial products. Quartz and cristobalite are often encountered in the manufacture of a large number of products that involve natural rocks as source materials, and they may also form during processing when the product contains either "free" silica or high-alumina clays. Opal is also encountered when natural source materials are in use.

Quartz and cristobalite have been recognized for a long time as toxic to humans, in particular, as respirable particles, which are responsible for the disease silicosis. More recently, IARC, the International Agency for Research on Cancer, classified "Crystalline Silica" as a "type 2A: Probable Carcinogen" in 1986, and has classified quartz and cristobalite as a "type 1: Known Carcinogen" in 1997. Consequently, it is very important to identify and quantify properly the silica phases present in source materials, industrial environments and products and to establish which phases are crystalline and which are not. X-ray powder diffraction is the best method for confirming the phase identity, and it is used along with infrared and a few other methods for quantification.

2. Studies on Opal

Until 1964, the true nature of opal was poorly understood. Sanders (1964), Jones, Sanders and Segnit (1964) and Darraugh, *et al.* (1965) used replication techniques and the transmission electron microscope, TEM, to image the ordered arrangement of spheres of colloidal particles of hydrous silica in what is now called opal-A_G. Monroe, Sass and Cole (1969) confirmed this structure. Further details became available using the scanning electron microscope, SEM, (Segnit, Anderson and Jones, 1970). Individual colloid particles 30 to 40 nm in size cluster to form the spheres, which ranged in size from 150 nm to 400 nm in diameter. Some spheres showed a concentric shell structure, which was due to the mechanism of clustering. The spheres were most uniform in size in precious varieties of opal and non-uniform in other varieties. Where the size of the spheres were uniform, the packing of the spheres was most perfect, and the arrangement was essentially cubic close-packed. Void spaces existed between the spheres in most samples were filled with a silica cement. The density contrast due to the cement and the internal structure of the spheres along with the regularity of the packing created a diffraction grating with a periodicity with the magnitude that allowed the diffraction of incident light.

Jones, Sanders and Segnit (1964) and particularly Jones and Segnit (1971) expanded the X-ray diffraction studies first presented in Levin and Ott (1933), Dwyer and Mellor (1934), Floerke (1955) and Garavelli (1964) to classify opal into three types, opal-A, opal-CT and opal-C. Opal-A was described as "highly disordered, near amorphous"; opal-CT was as "disordered α-cristobalite, α-tridymite"; and opal-C as "well-ordered α-cristobalite. These descriptions were based on the interpretations of the characteristics and differences of the diffraction patterns. Opal-A showed only a broad hump around 22° 2θ for CuKα radiation; opal-CT showed more structure to the diffraction pattern; and opal-C showed many peaks which corresponded to the diffraction pattern of α-cristobalite", although they also stated that "it was not appropriate to refer to this material [opal-C] as α-cristobalite because all diffraction patterns show evidence of tridymitic stacking". No consideration was given to the role of the water in the structure models; in fact, they stated that the water played no essential structural role.

Jones and Segnit (1971) did equate the "lussatite" of Mallard (1890) to opal-CT because they re-examined equivalent material from the same locality and showed it agreed with their opal-CT. They state that many opal-CT samples showed the same optical fibrous structure Mallard attributed to cristobalite, many other opal-CT samples were optically isotropic. They do agree with Floerke's (1955) deduction that the structure of these opals "essentially result from α-cristobalite with varying degrees of stacking disorder leading to the appearance of maxima which can be attributed to tridymite".

Jones and Segnit (1971) consider the three forms of opal distinct from "their close relative" α-cristobalite. Infrared, differential thermal analysis, and dilatometric analysis also support the distinction of the three forms. They do discuss the water content as "[water] although structurally non-essential, is present in substantial amounts in all samples so far analyzed". This impression seems to be propagated because the water has no apparent effect on the diffraction trace, contributing only to the background. Heating studies showed that all forms showed water loss and changes in the diffraction patterns. The ultimate result of heating would always be α-cristobalite.

Floerke, Jones and Segnit (1975) discuss "opal-CT crystals". In this paper, they consider all opal other than opal-A as crystalline based on the structured X-ray diffraction patterns and show SEM images that illustrate faceting on the surface of lepidospheres of opal-CT. The surfaces are covered with thin blades oriented randomly giving the lepidospheres the appearance of growths of rosettes of crystals. The dimensions of the "crystallites" are about 100 x 1000 nm. These morphological structures are observed in many samples of opal-CT from many sources.

More recently, Langer and Floerke (1974) and Graetsch, Floerke and Miehe (1985) examined the water in the various forms of opal and showed both silane and molecular water existed at the internal and external surfaces. De Jong *et al.* (1987) have employed XRD and ^{29}Si MAS NMR to examine both opal-A and opal-CT. They showed that all natural opals show measurable chemical shifts characteristic of amorphous silica. The sharpness of the NMR spectra does increase with the presence of increased abundance of ordered domains. Graetsch, Gies, and Topalovic (1994) used XRD, NMR and IR to examine all types of opal. They showed that the shape of the signal for some opal-C samples suggested the presence of two components. The spectra for opal-C samples resemble but were not identical with the spectra for cristobalite. Guthrie, Bish and Reynolds (1995) have used a recursive calculational technique to model the diffracting component of opal-CT. It was shown that a model based on ordered and disordered intergrowths of planar cristobalite and tridymite units explained the shapes of the peaks observed in the X-ray pattern obtained from opal-CT. This model did not consider any role for the essential water nor the contribution from the disordered component of the sample. The fact that the model had to use a non-kinematic, non-three-dimensional approach is further evidence for the lack of true crystallinity in opal-CT. Elzea and Rice (1996) and Cady, Wenk and Downing (1996) have examined opal-CT and opal-C for the presence of lattice imaged domains which mimic the disordered stackings of cristobalite and tridymite using high resolution transmission electron microscopy, HRTEM, and have confirmed the presence of such domains in a matrix of disordered non-imaginable material.

3. NEW STUDIES ON OPAL-CT AND OPAL-C

This paper will report on some ongoing studies of opal-CT and opal-C by X-ray diffraction, neutron diffraction, and combined differential thermal analysis and thermal gravimetric analysis, DTA/TGA. In addition to the examination of several opal-A samples, two opal-CT specimens, one, known as "fire opal", from Queretara, Mexico and the other was opalized wood from an unknown locality in the Western United States were studied along with an opal-C specimen from Adrian, Oregon where it formed as an alteration product in layers of volcanic ash. Experimental data were compared with a sample of synthetic cristobalite prepared by heating a silica gel to 1450 °C. X-ray diffraction patterns obtained on a Siemens D500 instrument confirmed the identification of each specimen. Both the opal-CT specimens were similar having the characteristic double peak at 22° 2θ for CuK and the second peak at 36° 2θ. The opal-C specimen showed eight very broad rounded peaks, whereas the cristobalite reference showed similar peaks, they were slightly displaced and very narrow and sharp.

Neutron scattering experiments were carried out using beam line D1A at Institute Laue-Langevin, Grenoble, France. Although the ultimate goal of these studies is to obtain data to establish the radial distribution of electron density, the intercomparison of patterns is quite revealing at this time. Among the samples examined were the Mexican opal-CT, and the Oregon opal-C along with the cristobalite reference. The diffraction trace of the cristobalite is a typical cluster of sharp peaks characteristic of the crystallinity of the reference material. The diffraction traces of the opal-CT and opal-C were very similar to each other and very distinct from the cristobalite reference. They were composed of broadened peaks with only a partial similarity to the cristobalite pattern.

Two DTA/TGA instruments were employed, a Netzsch STA 409 unit and a Mettler TA-1 unit both capable of reaching 1600 °C. The Mettler unit was used to heat samples to 1325 °C with controled heating and cooling rates at 10 °C/min. The Netzsch unit was used for the higher temperature runs up to 1575 °C. Specimens were ground from –100 mesh starting material with about 0.5g used for each run.

Results of the DTA/TGA studies show that all opal samples lose 4 to 8 percent water on heating and that the water loss is in two or three stages indicative of different types of water in the structure. The evolved species was confirmed to be water using a Mitsubishi Instruments high-temperature water analyzer model VA-21. The lower temperature water loss is probably adsorbed water and zeolitic water in the gel structure. The fraction of free water is probably very small as these samples, and most opals, show little evidence for hygroscopic behavior. The higher temperature water is combined water possibly as Si-OH silane groups on the inner surfaces of the gel structure. This latter type of water is usually referred to as the essential water characteristic of the opaline state.

The DTA traces do show peaks corresponding to the several stages of water loss corresponding to the TGA changes in weight. All forms of opal, A, CT and C, show significant water loss. For opal-A, there is a typical exothermic peak for the crystallization to cristobalite usually in the 1250 °C to 1350 °C range. For both opal-C and opal-CT, there is no observation of this crystallization peak in the DTA traces. The opal-C sample shows a DTA peak on cooling in the 160 °C to 200 °C range that corresponds to the $\beta\leftrightarrow\alpha$ cristobalite inversion and indicates that cristobalite has indeed formed. X-ray diffraction patterns of the final product confirm the presence of crystalline cristobalite. The opal-CT samples also show no peak on heating to 1450 °C, but they also show no evidence of a $\beta\leftrightarrow\alpha$ inversion on cooling. The X-ray pattern of the cooled specimen shows that the opal-CT structure has been retained even with this drastic heating. In fact, it takes a heat treatment of 1575 °C to cause the crystallization of cristobalite from any of the opal-CT samples examined to date.

Figure 1 shows the DTA/TGA trace for the Mexican opal-CT. There are three distinct ranges for water loss, up to 240 °C, 240 to 600 °C, and above 600 °C. The specific type of reaction for these ranges still needs studies to confirm what is taking place. Figure 2 shows the X-ray diffraction pattern of the starting material and the specimen after heating to 1450 °C. The increase in intensity and decrease in FWHM of the peaks for the heated opal-CT are due an increase in the size of the ordered domains. The decrease in background is probably due to the water lost in the heating.

4. DISCUSSION

This study has shown that all opals do not respond to heat treatment in the same manner. It has long been believed that heating to 1400 °C was sufficient to convert all forms of opal to cristobalite. In fact, this procedure has been used for the quantification of non-crystalline silica in the analytical laboratory. Although all opal-A samples and Opal-C samples examined to date do convert to cristobalite with heating to1400 °C, none of the opal-CT samples so behave. It requires temperatures well over 1550 °C to initiate the crystallization of cristobalite from opal-CT.

The neutron diffraction data suggests that there is considerable difference between opal-C and cristobalite whereas the X-ray diffraction patterns would suggest a similarity. There is considerable debate on whether it is appropriate to apply the term crystalline to opal-C. The distinct difference in the neutron patterns and the DTA evidence for combined water in the structure would preclude such a designation. Actually, it is appropriate to apply the term *paracrystalline* to both opal-C and opal-CT. This term is often used on describing organic polymers and is certainly applicable to silica, which is also a polymeric structure with a three-dimensional network arrangement. In opal-C and opal-CT there are small domains where the order begins to develop in the short range similar to the ordering that occurs in organic polymers when the chains align.

The behavior of opal-CT is unique among all the opals examined. Gem type opals are generally considered very heat sensitive in the jewelry trade, but a few such opals appear to have unusual stability towards heat. Although not all such specimens have been studied by the methods used in this paper, X-ray diffraction has shown several of these anomalous opals to be of the opal-CT variety. The suggestion from all these studies is that the three categories of opal are indeed distinct forms and are also distinct from cristobalite.

5. References

Cady, S. L., Wenk, H. -R., and Downing, K. H. (1996) Amer. Mineral. V81, 1380-1395.
Darraugh, P. J., Gaskin, A. J., Terrell, B. C., and Sanders, J. V. (1965) Nature. V209, 13-16.
De Jong, B. H. W. S., van Hoek, J., Veeman, W. S., and Manson, D. V. (1987) Amer. Mineral. V72, 1195-1203.
Dwyer, F. P., and Mellor, D. P. (1934) J, Proc. R. Soc. N.S.W., V68, 47-50.
Elzea, J. M. And Rice. S. B. (1996) Clays and Clay Minerals, v44, 492-500.
Floerke, O. W. (1955) Neues Jahrb. Min. Mh. 217-233.
Garawelli, C. L. (1964) Atti. Soc. Tosc. Sci. Nat., V71, 3-56.
Graetsch, H., Gies, H. And Topalovic, I. (1994) Phys. Chem. Min. V21, 166-175.
Graetsch, H., Floerke, O. W., and Miehe, G. (1985) Phys. Chem. Min. V12, 300-306.
Guthrie, G. D., Bish, D. L., and Reynolds, R. C. Jr. (1995) Amer. Mineral. V80, 869-872.
Jones, J. B., Sanders, J. V., and Segnit, E. R. (1964) Nature. V204, 990-991.
Jones, J. B., and Segnit, E. R. (1971) J. Geol. Soc. Australia, V18, 57-68.
Langer, K., and Floerke, O. W. (1974) Fortsch. Minral. V52, 17-51.
Levin, I., and Ott, E. (1933) Z. Krist. V85. 305-318.
Mallard, E. (1890) Bull. Soc. Fran. Min. V13, 161-180.
Monroe, E. A., Sass, D. B., and Cole, S. H. (1969) Acta Cryst. A25, 578-580.
Sanders, J. V. (1964) Nature. V204, 1151-1153.
Segnit, E. R., Anderson, C. A., and Jones, J. B. (1970) Search. V1, 349-351.

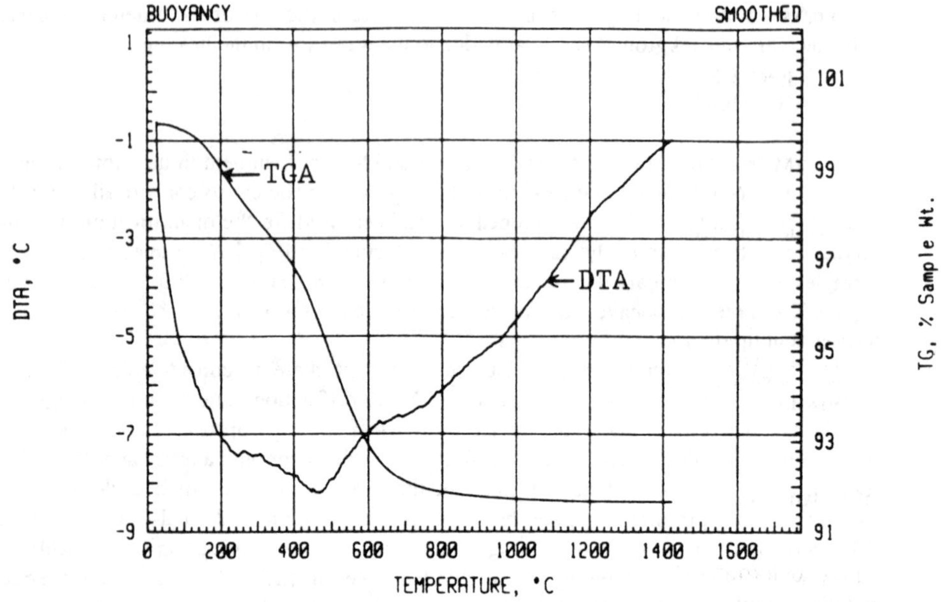

Figure 1. The differential thermal analysis trace of opal-CT from Queretara, Mexico showing water loss stages to over 600 °C but no crystallization to cristobalite up to 1450 °C.

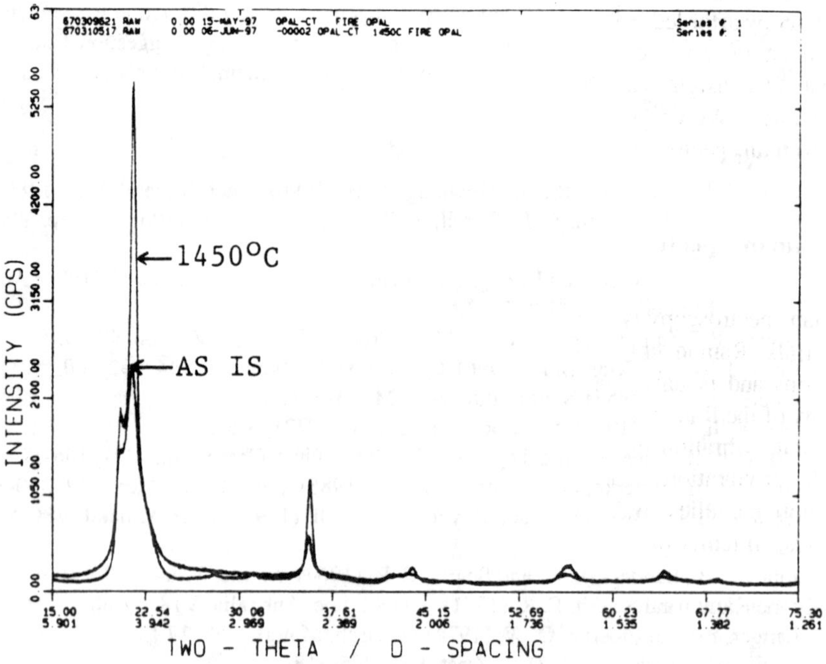

Figure 2. The X-ray diffraction pattern of opal-CT from Queretara, Mexico showing the sample prior to heating to 1450 °C and after cooling to room temperature.

CRYSTAL SIZE AND CRYSTALLINITY : X-RAY DIFFRACTION AND ELECTRON MICROSCOPY VERSUS RAMAN VIBRATIONAL SPECTROSCOPY

M. BOUDEULLE, B. CHAMPAGNON, C. CHEMARIN, E. DUVAL,
Y. GUYOT, J. MUGNIER, L. SAVIOT AND G. PANCZER

Laboratoire de Physico-Chimie des Matériaux Luminescents (LPCML), UMR CNRS 5620,
Université Claude Bernard LYON I,
69622 Villeurbanne, France

The interest of low-frequency Raman scattering for the determination of average particle size and the study of material organization at the nanometric scale is shown. The potentialities of high frequency scattering data for the assessment of short-range to long-range crystal order are emphasized.

1 Introduction

Particle size determination and evaluation of material crystallinity is still a big challenge for crystallography and specially now, with the development of nanotechnologies and the material use in extreme conditions (HT-HP). The possibilities and limitations of X-ray diffraction were largely debated in all the communities concerned with these problems and profited from the numerous methodological improvements, particularly those connected with the intensive use of computing. Conventional and high resolution electron microscopy afforded valuable observations, generally well correlated with diffraction data. However, in many cases, both the difficulties in the sample preparation and/or the statistically insignificant sampling impeded the extension of these methods to industrial laboratories. We propose to show, from our experience, how vibrational spectroscopies, like Raman spectroscopy, can bring some complements in such domains.

2 Raman spectroscopy : principle and method

Raman spectroscopy is used to measure the vibrational spectra of solid, liquid or gaseous materials. Raman emission results from the inelastic scattering of light by material phonons and is easily understood as a kind of energy-loss spectroscopy in which the energy of the Raman scattered radiation is equal to that of the incident beam minus that of a definite vibration in the sample. It results in a characteristic set of lines ("vibration bands" or vibration "frequencies", expressed in terms of wavenumber, cm^{-1}) for molecules and non metallic solids, dependant on the symmetry (point group) of the vibrating entities, in terms of local, medium and long-range order.

As the Raman effect is weak (approximately one photon out of a million will scatter from the sample at a wavelength shifted from the original wavelength), the monochromatic light of a polarised laser beam is used (Ar-laser, 514.5 nm). In microspectrometers, this parallel beam is focused by the objective (X50 - X100) of an optical confocal microscope which is used to select the studied zone (a few μm^3). The scattered light is dispersed by optical gratings and analysed with multi-channel charge-coupled device (CCD) detector. No sample preparation is required.

3 Low frequency Raman scattering

The low frequency Raman scattering corresponds to the frequencies below 100 cm^{-1}. It corresponds mainly to the interaction with the eigen vibration modes (breathing modes...) of nanostructures with sizes typically in the range 1 nm - 10 nm. Nanostructures can be amorphous (nanofluctuations in glasses) or crystalline (semiconductor nanocrystallites in glasses). We developed this method called LOFIS (Low Frequency Inelastic Scattering) for several types of nanostructures : glass ceramics, semiconductor quantum dots, nanostructures in glasses, polymers [1, 2, 3].

From a theoretical point of view, the method is based on the interaction with vibration modes of spheres in a continuous model. The resolution of the Navier equations and boundary conditions [1] lead to a general formula for the frequency ω of the normal mode as

$$\omega = S\frac{v}{dc} \tag{1}$$

v, c and d corresponding to sound velocity in the medium, velocity of light and diameter of the (spherical) nanostructures, respectively.

This formula has been explicitly calculated for Cd(S,Se) semiconductor quantum dots in a borosilicate glass matrix and a quantitative comparison with the other experimental results has been carried out. The relation between the frequency of the LOFIS modes and the inverse of the diameter of the nanostructures has been checked with TEM or SAXS [1, 2]. From formula (1) the average size of the crystallites ranged from 1.9 to 17 nm. The sensitivity of the method is high as it takes advantage of resonant effects which occur when the energy of the Raman excitation corresponds to transitions between energy levels of the nanocrystallites. The typical concentration is 1% of CdSe in the sample matrix.

For semi-cristalline polymers, the nature of the crystallites and eventually their size, perpendicular to the chains, can be determined from WAXS measurements, then using SAXS the long-period of the semicrystalline polymer is measured. Longitudinal acoustic modes LAM (zig-zag modes), localized in the chain segments of the ordered parts can be observed by low-frequency Raman scattering. In a simple model the wave-number of these modes is given by the following expression

$$\omega = \frac{(E/\rho)^{1/2}}{2\ell c} \tag{2}$$

where E is the Young modulus of the chain segment, ρ the density of the crystal and c, the light velocity. The wave-number ω ranged between 5 and 20 cm^{-1}, and the length ℓ of the polymer segment between 4 and 20 nm, depending of the sample and polymer. Furthermore the low-frequency Raman scattering allows to know the orientation of the crystallites by the determination of the scattered light polarization in comparison with the one of the laser. The effect of the molding conditions on the semicrystalline structure of polypropylene was shown when using in parallel WAXS and LOFIS which allowed to determine the crystal structure, the orientation and the size of the crystalline domains [3].

4 High frequency Raman scattering

As characteristic vibration bands, with precise wavenumbers and intrinsic band width can be assigned to definite vibrating entities, it can be easily understood that the spectrum of a crystal will be modified by small changes in crystal structure, like ion substitutions, vacancies, extended defects, as well as crystal size reduction.

Frequency shift and line broadening and dissymmetrisation are observed. As such changes can be induced by sample heating, applied pressure or residual stresses, one has to be cautious when explaining spectrum modifications in terms of crystal size.

In this case, the analysis of Raman lines is based mainly on the photon confinement model originally developed for microcrystalline silicon [4]. This approach was developed for various materials in which the finite crystallite size is taken into account by a weighting function on photon amplitudes. The q = 0 selection rule is thus relaxed due to the loss in long range order and the line-shape of the first-order photon peak becomes progressively asymmetrically broadened and shifted towards lower or higher frequencies, according to the concerned vibrations. The shift and the band width are correlated and related directly to the average size of ordered crystalline domains or nanoparticles.

The normal Raman spectra of *diamond* and *silicon* cubic crystal consist of a single line (1333 and 521 cm^{-1} respectively) assigned to the symmetric stretching of the covalent bond.

Raman scattering has become a standard tool in estimating the characteristic dimensions of silicon nanocrystals in porous silicon of which luminescence properties make it a potentially interesting material for opto-electronic devices. Crystallite sizes in the 20 - 100 nm range have been estimated using the model described above and correlated with TEM observations [5, 6].

The results indicate that in using the photon confinement model to estimate nanocrystal sizes in porous silicon greater attention should be paid to the polarization geometry, when the particular morphology of the porous layer might possibly favor certain scattering directions.

Line-shape analysis of the Raman spectrum was performed for diamond films grown using various methods like CVD [7] and nanocrystalline multi-twinned domains of "diamonds" formed during meteorite impact [8]. Broadening up to 19 cm^{-1} (compared to the intrinsic band width of ~2.2 cm^{-1}) were observed at the normal frequency, the original lorentzian profile of the band being preserved, when the confinement model predicts a large asymmetry.

A more plausible mechanism for Raman broadening than photon confinement has been suggested to be due to the photon scattering (phonon decay) at impurities, defects sites and grain boundaries, resulting in shortening of photon lifetime and Raman line broadening [7]. Such an interpretation is consistent with the other available structural data. X-Ray diffractometry was used to estimate the average crystallite sizes and lattice stresses from the broadening of the reflection profiles. A fairly good reciprocal dependance of the Raman peak width on the measured crystallite sizes was observed, suggesting that stress-induced broadening was of minor importance.

The different stages of nucleation and growth upon annealing of *titanium* and *zirconium oxide* particles from materials prepared with the sol-gel method was followed using in parallel low and high frequency Raman scattering.

The increase of crystal size with the temperature and annealing time was well imaged using Raman spectroscopy in the two frequency ranges. LOFIS provided the average diameters of the crystallites of the metastable ZrO_2 tetragonal phase which increase from 3.4 nm at 400°C to around 13 nm at 800°C, in good agreement with HRTEM observations, which show quasi-spherical particles embedded in amorphous material [9]. This size evolution confirms the FWHM variation of the Raman band at 148 cm^{-1}.

Similar observations were done for TiO_2 *anatase* samples. The broadening, shifting and slight asymmetry of the intense 144 cm^{-1} vibration band correlated well with the average particle size as deduced from LOFIS and measured using TEM in annealed sol-gel materials. Similar data were obtained for anatase in kaolin matrices. However, such a wide band is already present in low-temperature annealed sol-gel materials which give an amorphous signal using XRD. On the other hand, the particle size increase, as imaged by X-Ray reflection narrowing, can be compared to the last merging and strengthening of other vibration bands (195 cm^{-1}). In both cases, high frequency Raman bands would be recorded when the low frequency signal was weak and XRD was still blind.

It should be recalled that when distinct molecular or ionic units can be identified in a crystal, the vibration spectrum separates into two domains : at high wave-numbers, internal modes arising from the vibration of these groups themselves (stretching or bending modes) and lattice modes (internal modes) arising from motions of the molecules and ions against each other at lower frequencies. That means that some Raman emissions associated with distinct entities could be used, in appropriate cases, to check the local order in the early stages of nucleation and growth, like EXAFS ; while in growth experiments the emergence of the whole spectrum reveals the setting of the long-range order, which can be ascertained using conventional XRD.

As suggested by the previous data, an approach of the crystallinity or crystal order could be achieved when using Raman spectroscopy. Spectra were recorded for : i) brittle *quartz* samples amorphized during shock experiments with pressure up to 32 Gpa [10] ; ii) *quartz* grains showing fission tracks, altered at high temperature by U-bearing corrosive fluids [12]. The broadening, shifting, weakening of the vibration bands (464 cm^{-1}) and specially the one corresponding to the soft mode associated with the α-β transition (206 cm^{-1}), very similar for the two types of material, depicted the gradual loss of long range order.

4 Conclusion

The presented data put into evidence the interest of Raman spectroscopy in the estimation of the average size of nanoscaled particles and domains, even amorphous. They emphasized the potentiality of the method for checking the local, medium and long-range order in real crystals with a micrometric probe. A careful and quantitative interpretation of the spectra requires a complete assignation of the vibration bands and theoretical models, using photon dispersion curves have to be improved or developed.

However, in the present state, comparative studies can be done profitably. Works carried in parallel, using the different available methods, would contribute to set the possibilities and limits of such an approach. As a truly non-destructive method of characterisation, Raman vibrational spectroscopy, coupled with X-Ray diffraction and electron microscopy, could enlarge the fields of investigations of applied crystallography.

References

1. B. Champagnon, A. Andrianasolo, A. Ramos, M. Gandais, M. Allais and J.P. Benoit, *J. Appl. Phys.* **73**, 2775 (1993).
2. L. Saviot, B. Champagnon, E. Duval, I. A. Kudriavtsev and A. I. Ekimov, *J. Non-Cryst. Solids* **197**, 238 (1996).
3. A. Minardi, M. Boudeulle and E. Duval, *Polymer* **38**, 3957 (1997).
4. M. Fauchet in *Light scattering in Semiconductor structures and Superlattices*, ed. D. J. Lockwood and J. F. Young (Plenum, New York, 1991).
5. Y. Gregora, B. Champagnon and A. Halimaoui, *J. Luminesc.* **57**, 73 (1993).
6. Monin, L. Saviot, B. Champagnon, C. Esnouf and A. Halimaout, *Thin Solid films* **255**, 188 (1995).
7. H. Robins, E. N. Farabaugh and A. Feldman, *J. Mater. Res.* **5**, 11, 2456 (1990)
8. M. Boudeulle, C. Koeberl, G.I. Shafranovsky and V.L. Masaitis, submitted to *Meteoritics and Planetary Sciences* (1997).
9. C. Urlacher and J. Mugnier, *J. Raman Spectr.* **27**, 785 (1996).
10. B. Champagnon, G. Panczer, C. Chemarin and B. Humbert-Labeaumaz, *J. Non-Cryst. Solids* **196**, 221 (1996).
11. M. Boudeulle and C. Mosser, *Terra Nova* **9**, 677 (1997).

Amorphous Materials

EXAFS STUDIES OF $Fe_{66}Cr_8Cu_1Nb_3Si_{13}B_9$ AMORPHOUS AND NANOCRYSTALLINE ALLOYS

E. SOBCZAK, Y. SWILEM, R. NIETUBYĆ, A. ŚLAWSKA-WANIEWSKA
Institute of Physics, Polish Academy of Sciences, Al. Lotników 32/46, 02-668 Warsaw, Poland

M. TISCHER
HASYLAB at DESY, Notkestrasse 85, D-22603 Hamburg, Germany

The nanocrystalline alloys were obtained by annealing treatments at several temperatures ranging from 500°C to 530°C for time intervals 10, 20, 40 and 60 minutes. X-ray absorption spectra near the Fe K, Cu K and Cr K absorption edges for the studied alloys were measured in transmission mode at room temperature using synchrotron radiation in HASYLAB. The EXAFS oscillations and magnitude of Fourier transforms exhibited increasing volume fraction of the crystalline phase and ordering of the local bcc-structure around Fe and fcc-structure around Cu atoms. It was found that Cr atoms remained in amorphous matrix in the nanocrystalline alloys.

1 Introduction

Several years ago, it has been discovered that nanocrystalline Fe-based alloys exhibit excellent soft magnetic properties. Till now, most of the structural investigations were performed using X-ray diffraction and transmission electron microscopy. In the present work, we have chosen the EXAFS (Extended X-ray Absorption Fine Structure) method for study of the local ordering around Fe, Cu and Cr atoms in amorphous and nanocrystalline $Fe_{66}Cr_8Cu_1Nb_3Si_{13}B_9$ alloys, because it is sensitive for the element investigated and very well suited for study of changes in the local atomic environment occuring in alloys after devitrification. The EXAFS method examines the averaged distances from the absorbing atom to nearest atomic shells.

2 Experimental

The amorphous $Fe_{66}Cr_8Cu_1Nb_3Si_{13}B_9$ alloy was prepared by the single roller melt-spinning method. The nanocrystalline alloys were obtained by annealing treatments at several temperatures ranging from 500 ^{O}C to 530 ^{O}C for time intervals 10, 20, 40 and 60 minutes (see Table 1). The amorphous structure of as-quenched material and nanorystalline phase of heated alloys were confirmed by X-ray diffraction. The crystalline fraction was estimated from magnetization measurements.

The X-ray absorption measurements of the $Fe_{66}Cr_8Cu_1Nb_3Si_{13}B_9$ amorphous and nanocrystalline alloys were performed at HASYLAB using the synchrotron radiation from the DORIS storage ring. A Si (111) double-crystal monochromator was used. One of the crystal was slightly detuned from the Bragg conditions and its diffraction angle was piezoelectrically controled in order to stabilize the primary photon beam incoming to the first chamber.

Table 1: Heat treatment parameters of alloys studied, sample notation used in this work and crystalline fraction of α-Fe(Si) estimated from magnetization measurements.

No.	Sample notation	Parameters of heat treatment	Crystalline fraction
1.	QE	as quenched	0 %
2.	A	500 °C, 20 minutes	0 %
3.	B	520 °C, 10 minutes	5.5 %
4.	C	520 °C, 20 minutes	9.0 %
5.	D	530 °C, 10 minutes	11.9 %
6.	E	530 °C, 20 minutes	14.7 %
7.	F	530 °C, 40 minutes	19.4 %
8.	G	530 °C, 60 minutes	24.7 %

The absorption spectra at Fe K, Cu K and Cr K absorption edges were recorded at room temperature in a transmission mode, a foil of standard was placed and its absorption spectrum was measured simultanously with a sample.

The time of measurement was typically 2 s per point. In the EXAFS ranges the steps were 0.5, 1.0, 1.6, 2.4 and 3.2 eV depending on the distance from the absorption edge. The measured absorption spectra were of a good quality, even in the case of very small concentration of copper in alloys studied. Fe, Cu and Cr metallic foils were used as standards.

3 Results and discussion

Several scans of absorption measurements were performed for each sample and averaged spectra were calculated. Notation of samples is presented in Table 1. The X-ray absorption spectra were analyzed by using a PC computer and the program „EXAFS" made by Bonnin et al.[1] A linear function was fitted to the pre-absorption edge region, extrapolated along the whole absorption range and substracted from the spectrum.

The EXAFS oscillations were extracted in a standard way and then converted to photoelectron momentum k space. As usually, the EXAFS structure for an amorphous alloy had a sinusoidal shape and low amplitude, which rapidly decreased with momentum k due to the Debye-Waller factor. With increasing temperature of annealing the EXAFS amplitude increased and EXAFS shape becomes characteristic for a polycrystalline metal.

Due to a large Debye-Waller factor for amorphous alloys, the Fourier transforms (FT) were calculated in the limited range k from 3 to 10 $Å^{-1}$, applying a Hanning window function and weighting factor k^3. In Figure 1 the moduli of Fourier transforms of Fe EXAFS for all the alloys studied are compared with Fourier transform of Fe EXAFS for Fe metal. In Figure 2 and 3 the Fourier transforms of Cu EXAFS and Cr EXAFS for several alloys are compared with Fourier transforms for Cu and Cr metals, respectively.

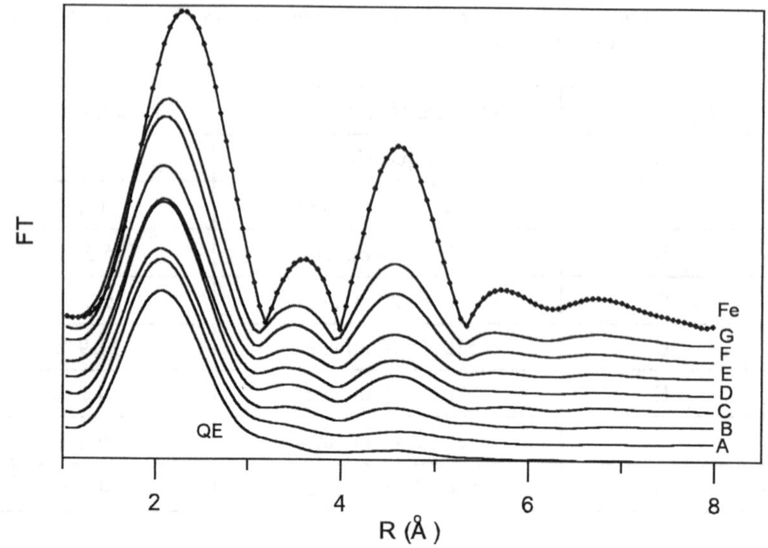

Figure 1: Fourier transform of Fe-K EXAFS for the $Fe_{66}Cr_8Cu_1Nb_3Si_{13}B_9$ amorphous alloy (sample QE), nanocrystalline alloys (samples A, B, C, D, E, F, G) and Fe metal. For clarity, successive curves are shifted on the vertical axis and the curve for Fe metal is reduced by factor 0.3.

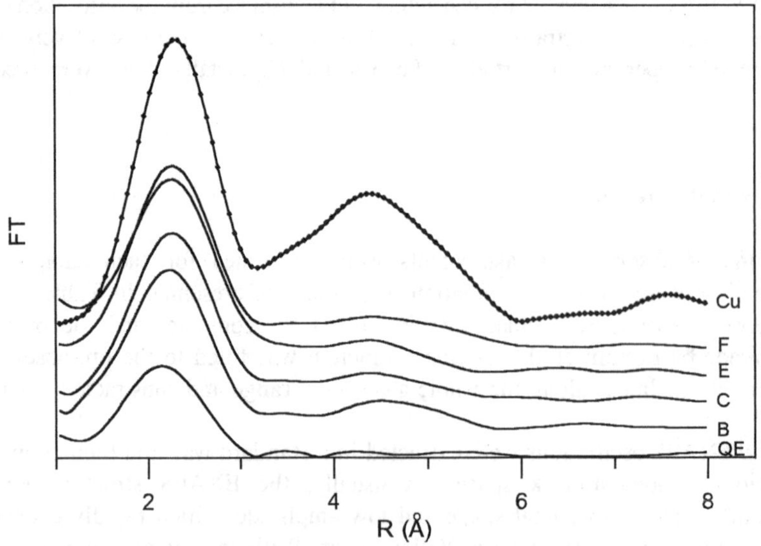

Figure 2: Fourier transform of Cu-K EXAFS for the $Fe_{66}Cr_8Cu_1Nb_3Si_{13}B_9$ amorphous alloy (sample QE), nanocrystalline alloys (samples B, C, E, F) and Cu metal. For clarity, successive curves are shifted on the vertical axis.

In Fig. 1 the FT of Fe EXAFS for nanocrystalline alloys are similar to that for Fe metal, however FT peaks for Fe metal are shifted to higher distances R than those for alloys, which can be easy explained by difference in lattice constants between α-Fe(Si) grains in alloys and α-Fe metal. A rapid crystallization of fcc Cu ultra-fine grains was observed already in the nanocrystalline alloys annealed at temperature 520 ^{O}C (curve B in Fig. 2).

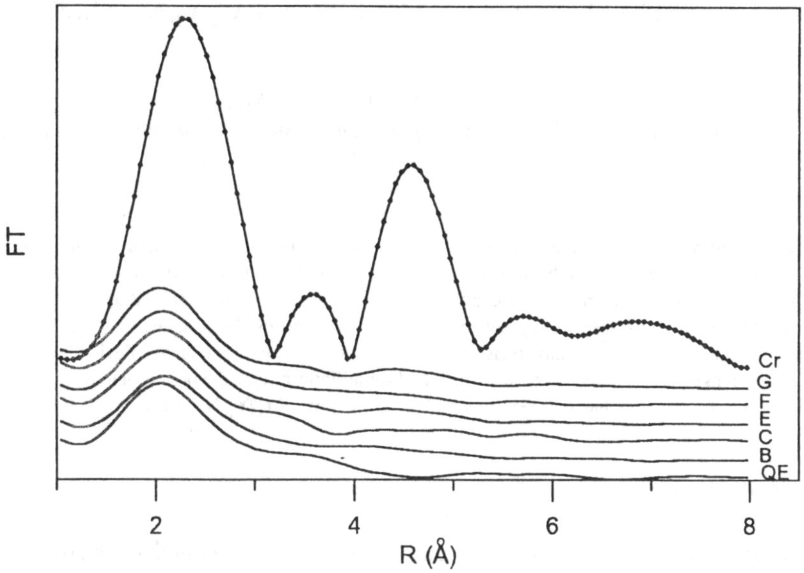

Figure 3: Fourier transform of Cr-K EXAFS for the $Fe_{66}Cr_8Cu_1Nb_3Si_{13}B_9$ amorphous alloy (sample QE), nanocrystalline alloys (samples B, C, E, F, G) and Cr metal. For clarity, successive curves are shifted on the vertical axis and the curve for Cr metal is reduced by factor 0.3.

The Fourier analysis of Cr K EXAFS of the nanocrystalline alloys has confirmed that Cr atoms are in an amorphous environment.

4 Conclusions

It was shown that EXAFS method is a useful method for studies of local atomic structure around Fe, Cu and Cr atoms in nanocrystalline alloys. EXAFS analysis of the alloys confirmed that in the as quenched $Fe_{66}Cr_8Cu_1Nb_3Si_{13}B_9$ alloy the Fe, Cu and Cr atoms are in an amorphous environment.

The most important result of the present EXAFS studies is that behaviour of Fe, Cu and Cr atoms is completely different during nanocrystallization of the $Fe_{66}Cr_8Cu_1Nb_3Si_{13}B_9$ alloy: nanocrystalline α-Fe(Si) grains and ultra-fine fcc Cu were observed, while Cr atoms remained in amorphous environment.

Acknowledgments

This study was partly supported by the KBN project No. 2PO3B 094 12.

Reference

1. D. Bonnin, P. Kaiser and J. Desbarres, *EXAFS, a program for PC IBM*, 1988.

CRYSTALLIZATION OF AMORPHOUS $Al_{85}Y_{10}Ni_5$ ALLOY

J. LATUCH, W. ZIELIŃSKI, H. MATYJA
Warsaw University of Technology, Deptartment of Materials Science & Engineering
Narbutta 85, 02-524 Warsaw, Poland

The crystallization behaviour and microstructural development during annealing of an amorphous melt-spun $Al_{85}Y_{10}Ni_5$ ribbon have been investigated by a combination of differential scanning calorimetry (DSC), transmission electron microscopy (TEM) and X-ray diffractometry (XRD). The processes of phase transformations were examined by continuous heating of the samples in the DSC at a rate of 40 K/min. As a result, many phases, mainly metastable, appeared in the devitrified glass. The compositions of most of them were identified. In the equilibrium state, the orthorhombic $Al_{16}YNi_3$ phase exists, which until now has not been reported to occur in the $Al_{85}Y_{10}Ni_5$ alloy.

1 Introduction

The recent discovery of Al-rich metallic glasses has generated considerable research interest because of their excellent mechanical and chemical properties. It has been found that these glasses can be obtained in the Al-Ln-TM (Ln is a lanthanide metal, TM is a transition metal) ternary systems[1,2]. These materials loose their interesting properties when undergo even partial crystallization. For this reason an understanding of the nature of the crystallization process is important from the point of view of possible applications of these alloys. The crystallization processes of several amorphous alloys of the Al-Y-Ni system have been described in previous reports[3,4]. Nevertheless, certain metastable phases occurring during annealing remain unidentified. The purpose of this study was to examine the crystallization process of the $Al_{85}Y_{10}Ni_5$ alloy using the DSC, TEM and XRD techniques.

2 Experimental methods

The ribbon to be investigated was produced by a melt-spinning device. The initial ribbon cross-section was 2 mm wide and 30 μm thick. The amorphicity of the as-quenched ribbon was assessed by XRD (CuK$_\alpha$ radiation) and TEM. The characteristic temperature peaks of the transformation processes (T_p) were determined from a thermal analysis using the DSC technique. The heating rate was 40 K/min.

In order to recognize the sequence and morphology of the metastable and stable crystalline phases present in the ribbons after isochronal heating, the microstructures of the alloy were examined by TEM. Precision measurements of the lattice parameters were performed by the zero-order central moment method; the error was estimated to be $\pm 2\times10^{-3}$ Å.

3 Results and discussion

During heating of the amorphous $Al_{85}Y_{10}Ni_5$ alloy, five exothermic peaks are observed in the DSC curve at the temperatures: 548 K, 604 K, 642 K, 677 K and 742 K.

The crystallization process was studied by heating the alloy continuously to the temperatures corresponding to the end temperatures of the subsequent transformations. Images of the microstructure and the corresponding electron diffraction pattern of the as-produced alloy are shown in Figure 1.

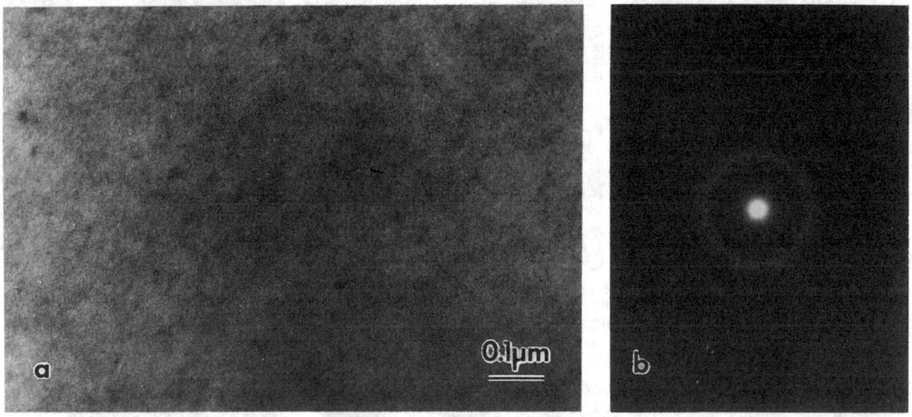

Figure 1: As-received state of $Al_{85}Y_{10}Ni_5$ alloy. a) image of microstructure, b) electron diffraction pattern

The featureless image of microstructure (Fig. 1a) and the diffuse ring in diffraction pattern (Fig. 1b) indicate that the alloy is in the amorphous state. The bright and dark field images of the microstructure obtained after heating through the first DSC peak (580 K) and the corresponding electron diffraction pattern are shown in Figure 2. At this temperature, numerous small crystals, uniformly distributed within the amorphous matrix, have appeared (Fig. 2a). The dark field image (Fig. 2b), obtained using the first spotty ring, clearly shows that the crystals have dendritic shapes. The corresponding electron diffraction pattern shown in Figure 2c, where a diffuse ring with several spotty rings can be seen, confirms this character of the microstructure. Knowing the TEM camera constant ($2\lambda L$) and the diameter of spotty rings, certain interplanar spaces were estimated. The calculated lattice parameter of this phase (a = 4.056 Å) is larger than that of pure aluminium. They fitted very well to the interplanar spaces of the solid solution of Y and Ni in Al (α-Al(Y,Ni)). What is interesting, is that the spotty rings are not continuous which implies a very strong crystallographic texture, - this may be related to the preferential crystallization.

Figure 3a shows a dark field image of the microstructure obtained after heating through the second DSC peak (625 K). Although no bright field image of the microstructure is available, the dark field image clearly shows that the crystals have lost their dendritic character and now are rather equaxial. The visible contrast of the crystals gives evidence of an internal disorientation due to the large number of structural defects. The electron diffraction pattern corresponding to this microstructure is shown in Figure 3b. Some new spotty rings appeared, which were identified to be related to the new β_2 and β_3 phases. The β_2 phase can be identified as orthorhombic (a = 5,99 Å, b = 7,66 Å, c = 5,17 Å) with a structure similar to that of the Al_3Ni phase. The β_3 phase can be described as cubic with the lattice parameter a = 4.274 Å. It has been reported[5] that the high temperature cubic modification of the Al_3Y phase exists. One can assume that the β_3 phase is disordered Al_3Y with some amount of nickel ($Al_3(Y_{1-x}Ni_x)$) which decreases the value of the lattice parameter. The character of the new rings is the same as that in the previous case

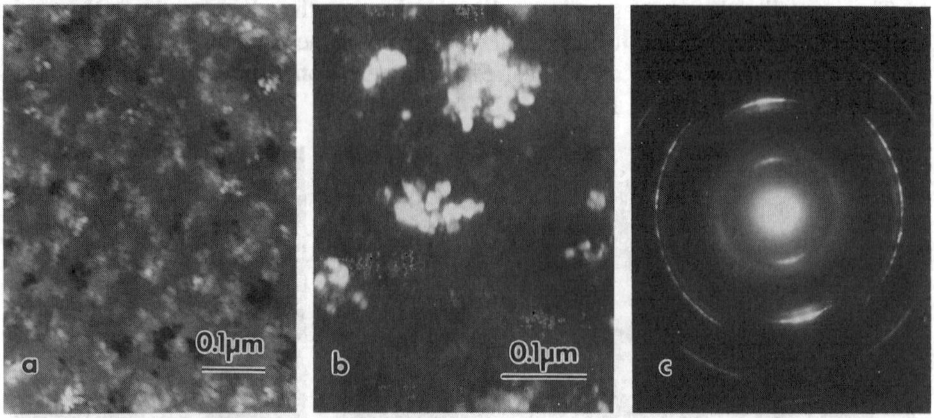

Figure 2: Microstructure of the $Al_{85}Y_{10}Ni_5$ alloy transformed at the first DSC peak. a) bright field image, b) dark field image, c) electron diffraction pattern

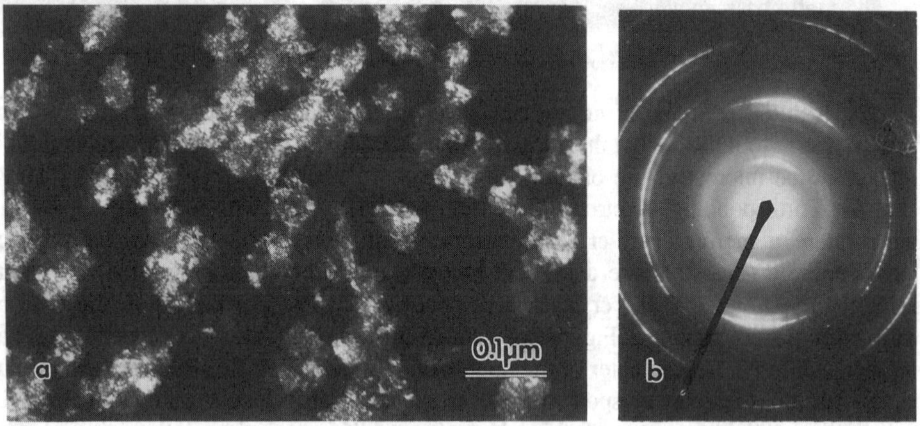

Figure 3: Microstructure of the $Al_{85}Y_{10}Ni_5$ alloy transformed at the second DSC peak. a) dark field image, b) electron diffraction pattern

indicating to the presence of a texture. However, in the case of the solid solution, the existing parts of the spotty rings are much longer than previously, which suggests that the texture has been weakened. In the electron diffraction pattern, we still have a diffuse ring which indicates, that some volume of the alloy has not been transformed yet. Heating above the third DSC peak (670 K) leads to the decomposition of the α-Al(Y,Ni) solid solution. The value of the lattice parameter decreases to 4.051 Å and becomes almost equal to that of pure aluminium. After heating the sample above the next DSC peak (705 K) we only observed the growth of β_3 phase crystals.

The major transformation of the microstructure has been registered after heating through the last DSC peak (760 K). Images of the microstructure in the bright and dark fields (using the most intense ring) as well as the corresponding diffraction pattern are shown in Figure 4. As can be seen in Figures 4a and 4b the microstructure is very complex. The presence of crystals of various kinds indicates to multiphase character of the microstructure.

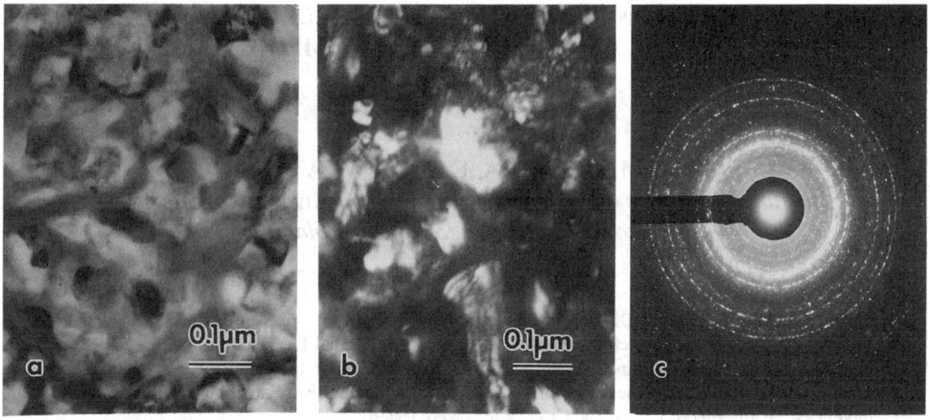

Figure 4: Microstructure of the $Al_{85}Y_{10}Ni_5$ alloy fully transformed. a) bright field image, b) dark field image, c) electron diffraction pattern

The multiphase character of the fully transformed alloy is confirmed by the electron diffraction pattern shown in Figure 4c. The following equilibrium phases have been identified: α-Al (cubic, a = 4.049 Å), Al_3Y (hexagonal, a = 6.195 Å, c = 21.129 Å) and $Al_{16}YNi_3$ (orthorhombic, a = 4.08 Å, b = 16.04 Å, c = 27.29 Å). The ternary $Al_{16}YNi_3$ phase is known from the literature[6]; the appearance of this phase during the crystallization of the $Al_{85}Y_{10}Ni_5$ alloy has not however been reported yet.

4 Conclusions

The crystallization process of amorphous $Al_{85}Y_{10}Ni_5$ alloy investigated by isochronal annealing includes five transformation steps at the DSC curves. This heat treatment reveals the presence of many phases, mainly metastable, in the devitrified glass. The sequence, composition and morphology of the phases that appear during the heating above the temperature of the successive DSC peaks were examined.

Acknowledgments

The financial support from the EC Copernicus Research Program under project no. ERBCIPA-CT94-0155 is gratefully acknowledged.

References

1. Y.He, S.J.Poon and G.J.Shiflet, *Science* **241** 1640 (1988).
2. A.Inoue, K.Ohtera, A.P.Tsai and T.Masumoto, *Jpn.J.Appl.Phys.* **27** L280 (1988).
3. V.Kwong, Y.C.Koo, S.J.Thorpe and K.T.Aust, *Acta Met.Mater.* **39** 1563 (1991).
4. R.Sabet-Sharghi, Z.Altounian and W.B.Muir, *J. Appl. Phys.* **75** 4438 (1994).
5. T.Dagerhamn, *Ark. Kemi* **27** 363 (1967).
6. H.Rosen, *J. Less-Common. Met.* **16** 285 (1968).

RADIAL DISTRIBUTION FUNCTION AND RIETVELD REFINEMENT ON MODELLING THE AMORPHOUS TWO-COMPONENT ALLOY STRUCTURE

M.NIEWIARA, E.ŁĄGIEWKA

Institute of Physics and Chemistry of Metals, University of Silesia, Bankowa 12, 40-007 Katowice, Poland

The idea of linking the Radial Distribution Function analysis and the Rietveld method in studies on amorphous materials is the subject of this paper. The results of Radial Distribution Functions: radii of coordination spheres and coordination numbers lead to the creation of a structure model and the preparation the Rietveld refinement input file. In order to examine the adaptability of the described methodology the amorphous Cu-Cd and Cu-Ti alloys were tested.

1 Introduction

The main idea of this work is using the connection between the RDF and the RR methods for describing the atomic structure of the two-component amorphous alloys which X-ray diffraction patterns have no typical diffraction peaks. In spite of the fact that the RDF theory has been simplified it is possible to obtain only a probable - approximate structure model. This fact leads to approximate coordination numbers and sphere radii, especially for two or more component materials. However the RDF theory gives good, basic information about structure model of nano- and micro-crystalline materials which have no diffraction peaks.

2 Material and methodology

The amorphous Cu-Cd and Cu-Ti alloys (Table 1) were tested in order to examine the adaptability of the described methodology. The electrodeposited Cu-Cd alloys were obtained in Institute of Physics and Chemistry of Metals in Katowice, at current density: D_K = 20, 15, 10, 7 A/dm^2 respectively, from etylenediamine electrolyte. The Cu-Ti alloy was obtained by the "melt spinning" method in Institute of Non - Ferrous Metals in Gliwice. The alloys are the samples of different kind compounds: Cu-Cd - with different atomic numbers: $Z_{Cu} = 29$, $Z_{Cd} = 48$ and Cu-Ti - with similar atomic numbers: $Z_{Cu} = 29$ and $Z_{Ti} = 22$. This fact is interesting from the interpretation of the RDF function point of view, calculation of the coordination numbers, the radii of coordination spheres and the structure determination.

The X-ray scattering intensities has been measured using the Philips PW 1130 type diffractometer with a diffracted beam graphite monochromator, the step scanning method and molybdenium radiation ($Mo_{K\alpha} = 0.7107$ Å). The samples were studied also by the JEM 3010 electron microscopy model.The calculation and analyses of the Radial Distribution Function were performmed using a computer program on an IBM/AT PC. The results of the RDF analysis have been leaded to the creation the Rietveld Refinement input files and the modelling of a structure model. Assuming that the "crystalline sizes" of searching samples are comparable to the size of the unit cell of the crystalline models, the line broadening and parameters defined the full at half maximum (FWHM) were modelled in the Rietveld Refinement [1,2,3,4]. From the structure deformation point of view, the unit cell parameters were also tested in the Rietveld Refinement. The R_{wp}, R_p, S agreement factors showed the evaluation of accuracy.

Table 1: Chemical composition, thickness and density of searching alloys.

Sample [at.%]	Thickness [μm]	ρ [kg/m^3] x 10^3
Cu87Cd13	31.44	8.35
Cu85Cd15	46.15	8.51
Cu74Cd26	31.20	9.18
Cu68.4Cd31.6	32.76	9.47
Cu64Ti36	69.33	4.66

3 Results and discussion

The electron microscopy analysis showed that the searching alloys have the nanocrystalline type of structure. Moreover, the chemical analysis performed by the EDS method using the electron microscopy JEM 3010 type showed that the main phases in materials are: the solid solution of cadmium in copper for Cu-Cd alloys and the solid solution of titanium in copper for Cu-Ti alloy.

The radii of coordination spheres obtained by RDF analysis showed that it was possible to obtain only the approximate structure model or a few models corresponding to real material structure. However, this way of analysis might be used in the case of materials, which nanocrystalline structure is confirmed by electron microscopy studies, for describing the input file to the Rietveld refinement. After comparing the preliminary RR patterns and the values of the agreement factors, it was concluded that the best fit was received for the *fcc* type of structure. The obtained values of the hypothetical "unit cell" parameters, atom positions and position occupancy parameters show a slight difference in the atom distribution for *fcc-type* of structure. Moreover, the obtained position occupancy parameters (which are: for *Cu* - atoms in the range of 0.779 - 0.210, for *Cd* - atoms: 0.201 - 0.879 and *Ti* - atoms: 0.724) show the presence of vacancies in the searching material and the creation of the *free volume* in the unit cell. The final *R -values* describing (in general) goodness of fit in the Rietveld Refinement are shown in the Table 2. The difference plots resulting from the RR are shown in the Figure 1. The obtained values of *R* - factors are typical to the results of the good refinement for other amorphous materials [3,4] also the values of radii of the coordination spheres obtained by the RDF and the RR methods (Table 3) are similar. All of that leads to the conclusion that the model of the hypothetical "unit cell" deduced from the Rietveld Refinement is good.

Table 2: The values of same *R - factors* [%] for Cu-Cd and Cu-Ti alloys.

Sample [at.%]	Values of *R - factors* [%] R_{wp}	R_p	R_{exp}	S
Cu87Cd13	20.07	15.67	2.80	7.17
Cu85Cd15	26.34	20.59	2.41	10.95
Cu74Cd26	23.33	17.81	2.65	8.79
Cu68.4Cd31.6	24.67	20.02	2.67	9.25
Cu64Ti36	42.76	34.60	2.05	20.89

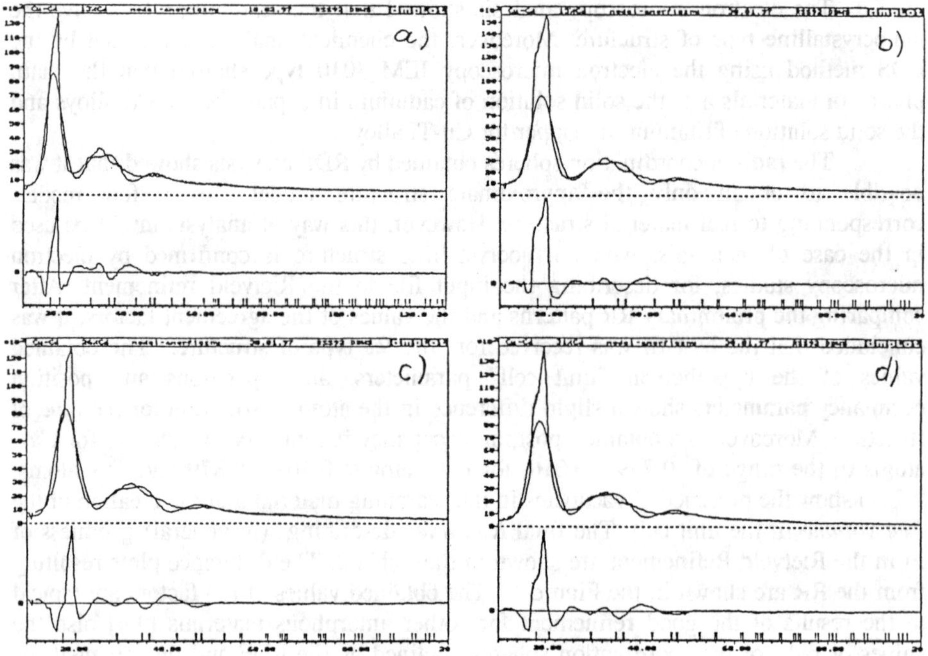

Figure 1: The experimental (dots) and calculated (line) X-ray diffraction patterns for Cu-Cd: - a - Cu87Cd13, - b - Cu85Cd15, - c - Cu74Cd26, - d - Cu68.4Cd31.6.
The difference plots are shown in the lower part of the figures.

Table 3: The values of radii of coordination spheres obtained by RDF method and calculated on the basis of RR results.

Sample [at.%]	Method	Radii of coordination spheres [Å] 1	2	3	4	5	6
Cu87Cd13	RDF	2.689	3.581	4.489	4.938	5.355	5.834
	Rietveld	2.63	3.72	4.56	4.92	5.26	5.88
Cu85Cd15	RDF	2.709	3.792	4.605	-	5.105	5.625
	Rietveld	2.74	3.80	4.67	-	5.22	5.72
Cu74Cd26	RDF	2.646	3.542	4.521	-	5.021	-
	Rietveld	2.54	3.59	4.40	-	5.08	5.68
Cu68.4Cd31.6	RDF	2.667	3.584	4.521	4.959	-	5.792
	Rietveld	2.56	3.62	4.44	-	5.13	5.73
Cu64Ti36	RDF	2.688	3.292	4.334	4.750	5.146	-
	Rietveld	2.66	3.76	-	4.61	5..32	5.95

4 Conclusion

- The described way of analysis of the materials might be used to structure model, especially for micro- and nano-crystalline materials.
- The Cu-Cd and Cu-Ti amorphous alloys, which X-ray diffraction patterns have no typical diffraction peaks, might be described by the hypothetical "unit cells" which are the modification of *fcc* model structure with different parameters and new atom positions.
- The R_{wp}, R_p, S agreement factors obtained for different kinds of material show different level of accurasy. E.g. for crystalline materials it is a range of 1 - 10 %, for nanocrystalline: 10 - 20% and for real amorphous materials: the range of 20 - 40%.
- The comparison of the theoretical (based on the Rietveld Refinement results) and experimental Radial Distribution Functions may show the level of the structure model accuracy much more better than the comparison of X-ray pattern intensities (like in the Rietveld method).

References

1. A.Le Bail, *Journal of Non - Crystalline Solids* **183** (1995) 39.
2. M.Niewiara, M.Socha, E.Łągiewka, *Archiwum Nauki o Materiałach* t. **18** nr **3** (1997) 205.
3. H.Krztoń, M.Niewiara, *Phisica Scripta* **5** (1995) 98.
4. H.Krztoń, M.Niewiara, *Proceedings of the XVI Conference on Applied Crystallography*, Cieszyn 1994, 403.

THE AMORPHOUS Fe-Si-B AND Fe-Cr-Si-B ALLOY STRUCTURES STUDIED BY RADIAL DISTRIBUTION FUNCTION AND RIETVELD REFINEMENT

M.NIEWIARA, D.GAWEŁ, E.ŁĄGIEWKA

Institute of Physics and Chemistry of Metals, University of Silesia, Bankowa 12, 40-007 Katowice, Poland

The idea of linking the Radial Distribution Function analysis and the Rietveld method in studies on amorphous materials is the subject of this paper. The results of Radial Distribution Functions - radii of coordination spheres lead to the creation of a preliminary structure model and the preparation the Rietveld refinement input file. In order to examine the adaptability of the described methodology the amorphous Fe-Si-B and Fe-Cr-Si-B alloys were tested.

1 Intoduction

The main idea of this work is using the connection between the RDF and the RR methods for describing the atomic structure of the amorphous Fe-Si-B and Fe-Cr-Si-B alloys. In earlier experiments [1,2], it was concluded that in the case of amorphous gels, this new method of analysis might be used. This paper shows that analysis of few first radii of coordination spheres obtained by the RDF method might be good enough step to determine the proper structure model and to create the RR input file. In fact, those first spheres are deformed the least, so in this way of refinemennt it is possible to receive the most probable structure model of amorphous material. In comparison to crystalline material, which is obtained after heating above the crystallization temperature our models show all "errors" and deformations in "the amorphous unit cell".

2 Material and methodology

The $Fe_{78}Si_{14}B_8$ and $Fe_{76}Cr_2Si_8B_{14}$ amorphous alloys were the material of studies. The searching materials were produced by the "melt spinning" method in the Institute of Non-Ferrous Metals in Gliwice and befor studing were annealed in vacuum trough 1 hour: $Fe_{78}Si_{14}B_8$ - in the range of 350 - 500°C and $Fe_{76}Cr_2Si_8B_{14}$ - in the range of 400 - 500°C. The alloys are the sample of compounds of different atomic number: $Z_{Fe} = 26$, $Z_{Cr} = 24$, $Z_B = 5$, $Z_{Si} = 14$, what is interesting from the interpretation of the RDF function point of view and the structure determination.

The X-ray scattering has been measured using a Philips diffractometer with a diffracted beam graphite monochromator, the step scanning method and molybdenium radiation ($Mo_{K\alpha} = 0.7107$ Å). The calculation and analyses of the Radial Distribution Function were performmed using a computer program on an IBM/AT PC. The results of RDF analysis have been leaded to the creation the Rietveld Refinement input files and the modelling of a structure model.

3 Results and discussion

The radii of coordination spheres were obtained as the results of the Radial Distribution Function (Figure: 1 - a, b) analysis. After the comparison of those values to the values characteristic of the typical crystalline structures of the compounds, which contain Fe, Si, Cr, B atoms, few initial structures were chosen as the imput files to the Rietveld Refinement. The results showed that only *Fe-α* - type structure might lead to the good resolution and giving the proper structure model of the amorphous alloys. The final *R -values* describing (in general) goodness of fit in the Rietveld Refinement obtained for the annealed alloys are shown in the Table 1, as well, the difference plot results from the RR are shown in the Figure: 1 - c, d. The obtained models are described by the hypothetical "unit cells". Their space groups, cell parameters, atom positions and position occupancy parameters are shown in Tables: 2 and 3. The values of *R - parameters* typical for the good refinement for other amorphous materials [3,4] and similarities in values of coordination spheres obtained by the RDF and the RR methods (Table 4) show that the RR model structure is good. Moreover, the results show that annealing temperatures don't have effect on the model structure.

Table: 1 The values of same *R - factors* [%] for the $Fe_{78}Si_{14}B_8$ and the $Fe_{76}Cr_2Si_8B_{14}$ annealed alloys.

Sample	Annealing temperature [°C]	Values of *R - factors* [%]		
		R_{wp}	R_p	S
$Fe_{78}Si_{14}B_8$	350	12.24	9.92	4.11
$Fe_{78}Si_{14}B_8$	500	13.90	10.78	3.87
$Fe_{76}Cr_2Si_8B_{14}$	400	10.51	7.98	3.21
$Fe_{76}Cr_2Si_8B_{14}$	500	12.58	9.32	4.06

Table 2: Space groups, cell parameters, atom positions and position occupancy parameters for $Fe_{78}Si_{14}B_8$ alloy.

Space group: I m3m , a = 3.0143 Å				
lp.	Atom	x	y	z
1	Fe	0.0	0.0	0.0
2	Fe	0.49	0.50	0.50
3	Fe	0.11	0.39	0.08
4	Fe	0.25	0.31	0.75
5	Si	0.26	0.33	0.42
6	Si	0.21	0.28	0.60
7	B	0.06	0.43	0.73
Position occupancy parameter (POP): Fe = 1, Si = 1, B = 1				

Table: 3 Space groups, cell parameters, atom positions and position occupancy parameters for $Fe_{76}Cr_2Si_8B_{14}$ alloy.

Space group: I m3m , a = 3.0143 Å				
lp.	Atom	x	y	z
1	Fe	0.0	0.0	0.0
2	Fe	0.50	0.55	0.50
3	Fe	0.25	0.30	0.73
4	Cr	0.27	0.30	0.39
5	Cr	0.26	0.39	0.09
6	Cr	0.45	0.58	0.56
7	Si	0.28	0.27	0.40
8	B	0.0	0.0	0.0
9	B	0.07	0.41	0.75
Position occupancy parameter (POP): Fe = 1, Si = 1, B = 1, Cr = 1				

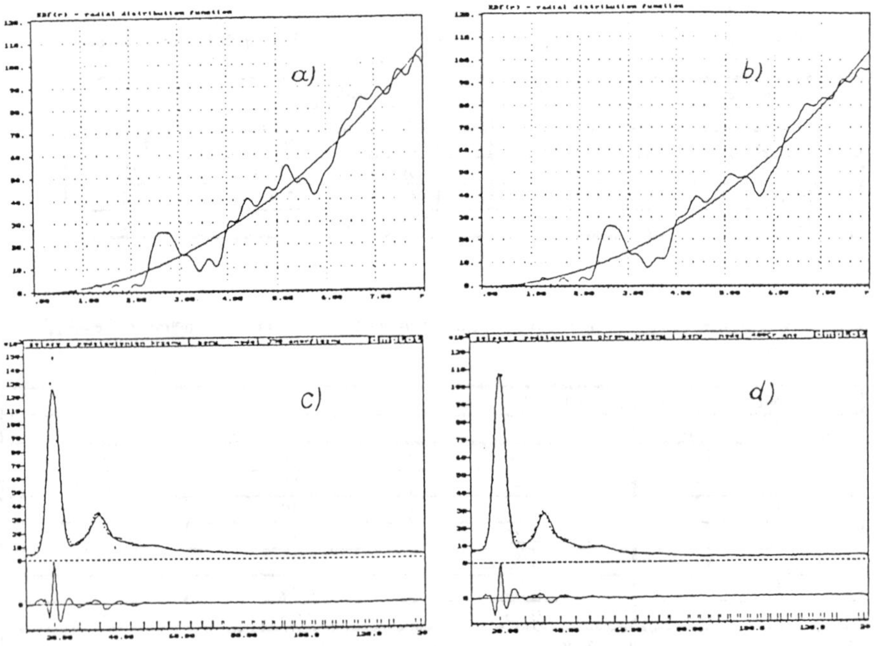

Figure:1 Radial Distribution Function for: - a - $Fe_{78}Si_{14}B_8$ and - b - $Fe_{76}Cr_2Si_8B_{14}$.
The experimental (dots) and calculated (line) X-ray diffraction patterns for:
- c - $Fe_{78}Si_{14}B_8$ and - d - $Fe_{76}Cr_2Si_8B_{14}$.
The difference plots are shown in the lower part of the figures.

Table: 4 The values of radii of coordination spheres obtained by RDF method and calculated on the basis of RR results.

Sample	Annealing temperature [°C]	Method	Radii of coordination spheres [Å] 1	2	3	4	5	6
$Fe_{78}Si_{14}B_8$	350	RDF	2.54	2.81	3.21	3.66	4.12	4.40
		Rietveld	2.67	2.89	3.30	3.70	4.09	4.39
$Fe_{78}Si_{14}B_8$	500	RDF	2.29	2.62	3.00	3.44	3.90	4.29
		Rietveld	2.31	2.64	3.09	3.51	3.99	4.30
$Fe_{76}Cr_2Si_8B_{14}$	400	RDF	2.56	3.19	3.60	4.00	4.38	5.05
		Rietveld	2.70	3.31	3.79	4.20	4.50	5.19
$Fe_{76}Cr_2Si_8B_{14}$	500	RDF	2.54	3.19	3.70	4.04	4.63	5.14
		Rietveld	2.61	3.17	3.79	4.23	4.76	5.20

4 Conclusion

- The amorphous alloys: $Fe_{78}Si_{14}B_8$ and $Fe_{76}Cr_2Si_8B_{14}$ might be described by the hypotetical "unit cell" which is a modification of the *Fe-α* unit cell with different parameters and new atom positions.
- The results of studies show that annealing temperatures don't have effect on the model structure.
- The addition of *Cr* - atoms to the Fe-Cr-Si-B alloys stabilize the hypotetical structure. This fact shows that this method might be used not only to structure modelling but also to studing of its stability.
- The described way of analysis of amorphous alloys might be used to model especially for micro- and nano-crystalline materials.

References

1. A.Le Bail, *Journal of Non - Crystalline Solids* **183** (1995) 39.
2. M.Niewiara, M.Socha, E.Łągiewka, *Archiwum Nauki o Materiałach* t. **18** nr **3** (1997) 205
3. H.Krztoń, M.Niewiara, *Phisica Scripta* **5** (1995) 98.
4. H.Krztoń, M.Niewiara, *Proceedings of the XVI Conference on Applied Crystallography*, Cieszyn 1994, 403.

CRYSTALLISATION OF AMORPHOUS ALLOYS OF Fe-X-Si-B (X=Cr,Mo) TYPE

J. LELĄTKO, L. PAJĄK, J. RASEK, P. KWAPULIŃSKI ,Z. STOKŁOSA
Institute of Physics and Chemistry of Metals, Silesian University
40-007 Katowice, Bankowa 12, Poland

Literature data and results of own investigations show that physical properties of Fe-X-Si-B amorphous alloys, obtained by melt spinning, depend on the type of alloying element X. A further changes of physical properties of the amorphous alloys may be achieved by heat treatment at elevated temperatures. The paper presents the results of investigations on the effect of heating on structural changes and electric properties occurring in $Fe_{78}Si_8B_{14}$, $Fe_{76}Mo_2Si_8B_{14}$ and $Fe_{76}Cr_2Si_8B_{14}$ alloys.

1 Introduction

Amorphous alloys of the FeSiB type obtained by melt spinning have interesting magnetic and electric properties. In the as-quenched state their magnetic and electric parameters depend mainly on chemical composition. However, in the materials obtained in this way there occur time and temperature instabilities leading to the thermodynamic equilibrium through structural relaxation and crystallisation [1,2]. Therefore it is possible to change the magnetic and electric properties of the alloy by a proper thermal treatment. It is also possible to modify the physical properties, to change the intensity of structural relaxation as well as to change the crystallisation temperature by introducing small amounts of additional elements X [3-5].
In the recent literature data there is no full data concerning the influence of the additional elements on the magnetic, electric properties and on intensity of structural relaxation and crystallisation process in Fe-Si-B alloys.

The aim of this work was to study the influence of the alloy additions such as Cr and Mo on the crystallisation process of the $Fe_{78}Si_8B_{14}$ type alloys.

2 Experimental procedure

The studies were carried out on the $Fe_{78}Si_8B_{14}$, $Fe_{76}Mo_2Si_8B_{14}$ and $Fe_{76}Cr_2Si_8B_{14}$ alloys obtained by melt spinning. The material was in form of 20 -30 μm thick and 10 mm wide ribbons. The X-ray phase analysis and electron microscopy were the methods used for the structural studies. The specimens studied were in the as-quenched state or after isothermal annealing in vacuum at the temperature range 300 - 875 K for every 25 K. X-ray studies were carried out on a Philips diffractometer using the CuK_α radiation. The electron microscopy studies were conducted on the JEM 3010 electron transmission microscope. The foils were polished by ion polishing. Also electric resistivity and magnetic permeability were measured.

3 Experimental results

The X-ray studies showed that the alloys in the as-quenched state have amorphous structure. On diffraction patterns from the basic $Fe_{78}Si_8B_{14}$ alloy annealed at 725 K sharp lines occurred on a background of diffusion reflexes from the amorphous phase.

These lines (the strongest lines from the αFe(Si) phase) prove that the crystallisation process has started in the alloy (Table 1). For the two other alloys the crystallisation temperatures are about 25 K higher. As it is seen from Table 1 the alloy additions also change the crystallisation start temperatures of other phases. The analysis of intensities of particular phase diffraction lines shows that the crystallisation process occurs with different intensity.

Table 1: Phase composition of the alloys after annealing at the temperature range 300 - 875 K determined by X -ray diffraction

Temperature [K]	$Fe_{78}Si_8B_{14}$	$Fe_{76}Cr_2Si_8B_{14}$	$Fe_{76}Mo_2Si_8B_{14}$
300	A	A	A
700	A	A	A
725	A, αFe (Si)	A	A
750	A, αFe (Si)	A, αFe (Si)	A, αFe (Si)
775	αFe (Si), Fe_2B	A, αFe (Si)	A, αFe (Si)
825	αFe (Si), Fe_2B	αFe (Si),	A, αFe (Si)
850	αFe (Si), Fe_2B	αFe (Si), Fe_3B,Fe_2B	αFe (Si), Fe_3B
875	αFe (Si), Fe_2B, Fe_3Si	αFe (Si), Fe_3B, Fe_2B	αFe (Si), Fe_3B, Fe_2B

A - amorphous phase

The high resolution images taken from the as-produced alloy showed that there exist only crystallisation nuclei quenched in the amorphous matrix [5]. The carried out studies proved that the crystallisation process in the amorphous matrix has a character of a primary crystallisation. In all of the studied alloys in the first stage of crystallisation crystals of the α-Fe(Si) phase precipitate. In the $Fe_{78}Si_8B_{14}$ alloy first crystals of this phase were observed after annealing at 750 K (Fig. 1). Annealing at 775 K causes intensive crystallisation of both αFe(Si) and Fe_2B phases. Complete vanishing of the amorphous areas was observed only after annealing at 825 K.

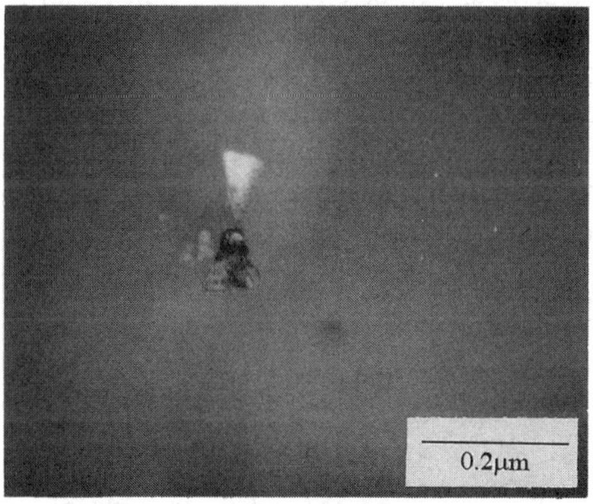

Fig.1 Electron microscope image obtained from alloy $Fe_{78}Si_8B_{14}$ annealed at 750 K

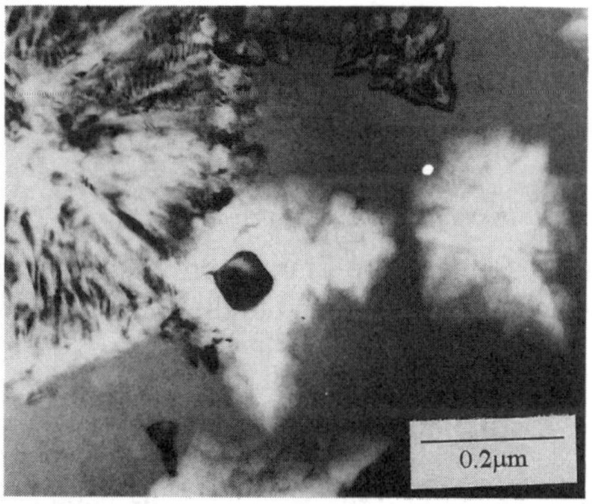

Fig.2 Electron microscope image obtained from alloy $Fe_{76}Cr_2Si_8B_{14}$ annealed at 800 K.

The crystallisation process in the alloy containing chromium had a similar character. In this case first crystals of the αFe(Si) were observed under microscope after annealing at 800 K (Fig. 2). In many cases the αFe(Si) phase in a dendrite form and some number of spherulites of the Fe_3B phase nucleate at fine (about 50 nm) nuclei of the αFe phase. Also for this alloy the amorphous phase vanished after annealing at temperatures higher than 825 K.

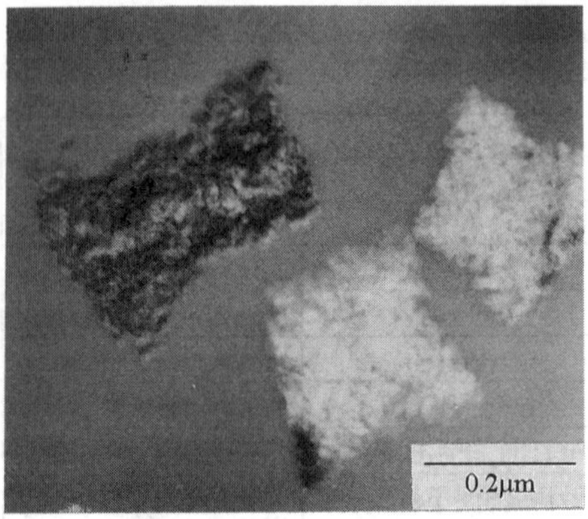

Fig.3 Structyre of the alloy $Fe_{76}Mo_2Si_8B_{14}$ annealed at 800 K.

Fig. 4 Structyre of the alloy $Fe_{76}Mo_2Si_8B_{13}$ annealed at 825 K

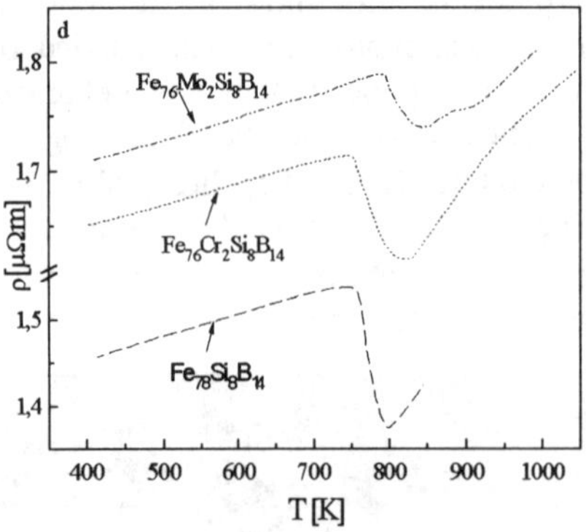

Fig. 5 Isochronous curves of electric resistivity in situ for the investigated alloy

The crystallisation process in the $Fe_{76}Mo_2Si_8B_{14}$ alloy had a different character. The microscopy method showed that small number of the first crystals of the αFe(Si) phase occurred after annealing at 800 K. The crystals had regular forms (Fig. 3). In figure 4 the structyre of the alloy $Fe_{76}Mo_2Si_8B_{14}$ annealed at 825K is presented. Teh electron diffraction patterns proved that expect for the Fe_3B and Fe_2B phases particles of the $Mo_5(B,Si)_3$ type phase precipitate. Increase in the annealing temperature to 875 K causes complete decomposition of the amorphous phase and a nanocrystalline structure occurs in the alloy. Some of the grains were the grains of the FeMo phase of a hexagonal structure which was not observed in the two previous alloys.

4 Discussion

According to [6] the crystallisation temperature T_x should increase when an addition of elements of higher atomic diameter than for iron is introduced to the basic FeSiB alloy. However, the carried out studies showed that this rule cannot be fully applied to the alloy studied. In this case the crystallisation temperatures for the $Fe_{76}Cr_2Si_8B_{14}$ and

$Fe_{76}Mo_2Si_8B_{14}$ were very close and differed from the crystallisation temperature of the basic $Fe_{78}Si_8B_{14}$ alloy of 25 - 50 K depending on an applied method. The investigations showed also different influence of these elements on the course of the crystallisation process. The differences consists in the crystallisation kinetics and in the amount of the particular phases occurring in an alloy. In the FeSiB alloy with the addition of chromium decrease of the amount of Fe_2B and Fe_3B borides comparing the amount of the αFe(Si) phase is observed.

On the other hand the addition of molybdenum causes significant stabilisation of the amorphous phase. Although the first crystals of the αFe(Si) phase are formed at the same annealing temperature as for the alloy containing chromium, they do not precipitate faster after increase in the annealing temperature, as it could be expected. Large amount of the amorphous phase exists even up to annealing at 850 K. This causes that at 850 K the crystallisation process is very intensive and in result a fine, nanocrystalline structure occurs. This structure proves that molybdenum serves as a modification (comparing to the $Fe_{78}Si_8B_{14}$ and $Fe_{76}Cr_2Si_8B_{14}$).

The described crystallisation courses of the studied alloys are in good agreement with measurements of physical properties (Fig. 5). Increase in the stability of the alloy with the molybdenum addition and the two-stage course of its crystallisation are well visible. The second maximum on the resistivity curve is connected with the intensive course of the crystallisation and creation of the fine, nanocrystalline structure. A small difference in the crystallisation courses occurring in the two first alloys are indicated only by small changes of the resistivity curves. Both chromium and molybdenum cause increase in the alloy resistivity.

5 Summary

The $Fe_{78}Si_8B_{14}$, $Fe_{76}Cr_2Si_8B_{14}$ and $Fe_{76}Mo_2Si_8B_{14}$ alloys obtained by melt spinning method have in the as-quenched state an amorphous structure.

The additions such as Cr and especially Mo cause increase in the crystallisation temperature T_x of about 25 - 50 K.

Presence of molybdenum in the $Fe_{76}Mo_2Si_8B_{14}$ alloy causes that the crystallisation process occurs in two stages. The two-stage course in the crystallisation process of the $Fe_{76}Mo_2Si_8B_{14}$ alloy is responsible for occurrence of two minimum of the resistivity versus temperature curve.

References

[1] Zieliński, A. Załuska, W. Dmowski, T. Kulik, H. Matyja, *Arch. Nauki o Mater.* **8**, 107 (1987).

[2] Zhou, K.Y.He, K.C.Bai, *J. Appl. Phys.* **80**, 5209 (1996).

[3] H.Matyja, T. Kulik, *Trends in Non Crystalline Solids* (World Scientific Publishing Co. Singapore 1992, p.107).

[4] Stokłosa, P.Kwapuliński, J. Rasek, J. Ilczuk, J. Lelątko, *Acta Phys. Polonica* **89A**, 437 (1996).

[5] Lelątko, M. Gigla, J. Rasek, *IX Conference on Electron Microscopy of Solids* 6-9 may 1996, Kraków - Zakopane, p. 363.

[6] J.L.Walter, *Mater.Sci.Eng.* **50**, 137 (1981).

AMORPHOUS COMPONENT IN QUANTITATIVE PHASE ANALYSIS

J. ŻELECHOWSKI
Institute of Non-Ferrous Metals, Light Metals Division,
ul.Pilsudskiego 19, 32-050 Skawina, Poland

Different results of quantitative phase analysis on Al-based alloy were discussed. Two methods were used: quantitative powder diffraction phase analysis (reference intensity ratio and calibration curve method) and quantitative metallography (point counting methods and image analyser system). The errors of quantitative powder diffraction phase analysis might be reduced by amount of amorphous component estimate.

Introduction

An aluminium alloy containing 12 weight percent of Si was subjected to phase analysis. The microstructure of this alloy consists of free silicon visible in figure 1 as big dark polygons and eutectic in aluminium matrix. From X-ray point of view there are only two phases in this alloy: Al and Si which are ascribed to the X-ray diffraction pattern illustrated in figure 2. The volume fraction of Al and Si in the alloy was determined by:

a) calculated chemical composition,
b) quantitative metallography:
- by a computer analysis of microstructure,
- by classic point method,

c) X-ray quantitative phase analysis:
- from the ratio of diffraction reflex intensity,
- from calibration curves.

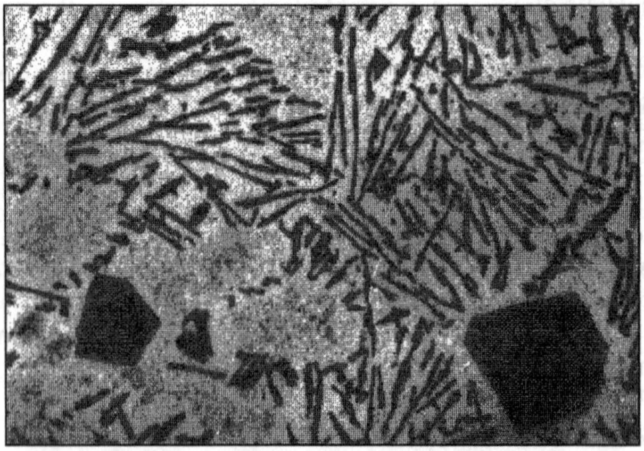

Figure 1: Microstructure of the investigated alloy at 400 × magnification presenting Si precipitates in the form of big dark polygons and eutectic in aluminium matrix.

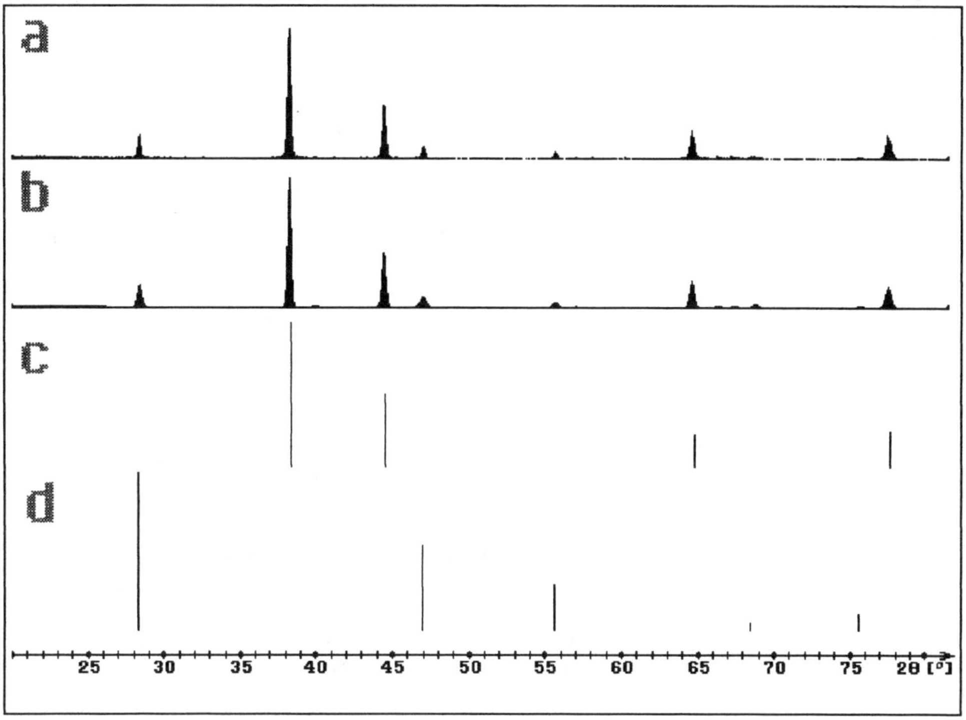

Figure 2: X-ray diffraction pattern (a) together with its optimal mathematical model (b) and a histograms of diffraction patterns from Al (c) and Si (d) phases.

Experimental

Cu Kα X-ray radiation and powder samples were utilised for X-ray quantitative analysis. The obtained diffraction patterns were transformed into optimised mathematical models [1] and in this form they were used for further calculations. The strongest reflection were measured: Al 111 from both: Al standard and from alloy and Si 111 reflections also from Si standard and the alloy. The intensity ratio for the sample and the standard, after taking into account the absorption correction, were used for independent calculation of each phase content [2÷5]. The alloy volume composition was also determined from a calibration curve. For this purpose standard samples were made as mixtures of powders of aluminium and silicon. Due to the texture and/or extinction [6] those standards provided diffraction patterns with heavily distorted intensities of Si phase reflex so that the required accuracy of the calibration curve was not reached. This effect was not observed in standards made of powdered AlSi50 master alloy diluted with powdered Al. Using the latter standards and applying the least square method a calibration curve presented in figure 3 was prepared. From microstructure photographs the volume fractions of Al and Si + eutectic phases were determined by computerised method of picture analysis [7]. Classical point method was also used [8, 9]. The phase composition obtained by different methods results are summarised in table 1.

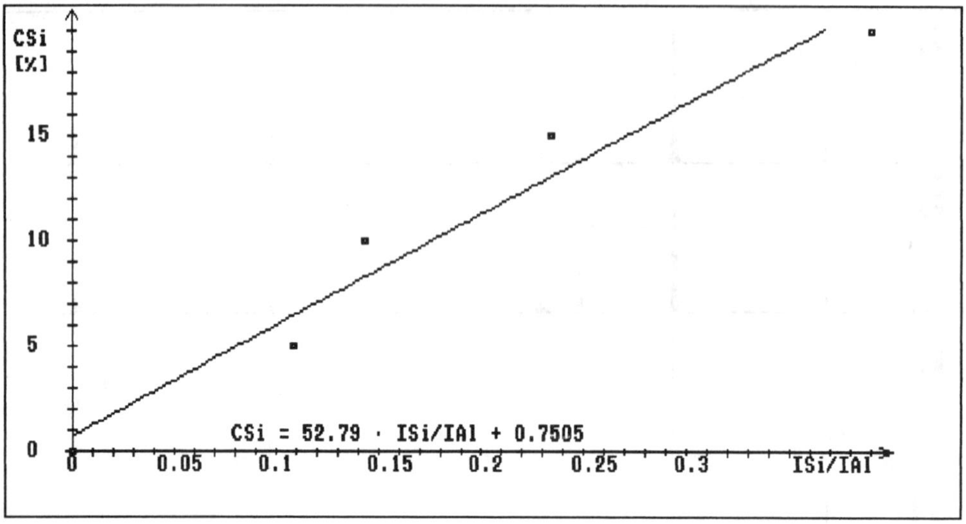

Figure 3: Calibration curve for determination of silicon content in Al-Si alloy.

Table 1: Summary of the calculated results of average volume fraction V obtained by different methods together with standard deviation given as a measure of the scatter.

phase	expected volume fraction	Experimentally determined volume fraction						
		metallographic method				X-ray method		
		computer microstructure analysis		classical point method		reflexes intensity ratio		calibration curve
	V [%]	V [%]	s [%]	V [%]	s [%]	V [%]	s [%]	V [%]
Al	86.36	79.22	3.39	90.07	1.94	80.15	7.44	-
Si	13.64	20.78	3.39	9.93 4.93*	1.94 1.51*	3.42	3.11	11.57

* This results reefers to the big particles only

Results and Discussion

The results of X-ray quantitative analysis and results of quantitative metallography differ by 2÷10% from expected values calculated from chemical analysis. The volume fraction of Si determined by X-ray analysis from calibration curve is lower then the expected value by only 2%. This is probably due to the advantages of the standards prepared from master AlSi50 alloy which is characterised by similar silicon distribution as the studied samples. Among the used metallographic methods the classical point method has yielded the more close results to the expected values. The fraction of Si determined by the point method was lower by about 4% (s≅2%) as compared to the expected value whereas the computer structure analysis gave values higher by about 7% (s≅3%). The higher values from the computer microstructure analysis as compared to the classical method may result from ineffectiveness of the applied program criteria for recognising the elements of the microstructure. The most unreliable results were obtained by the X-ray analysis based on the intensities ratio of diffraction reflection. This method gave the Si fraction lower by about 10% (s≅3%) and the Al fraction by about 6% (s≅7%) as compared to the expected values. The determination of Si fraction by the method of diffraction reflections intensities ratio (3.42%) yields a result similar to the point method (4.93%) when the latter is used for big Si particles exclusively. It means that the diffraction pattern is formed mainly by the silicon particles in the range of 10μm in size. The fraction of such particles is estimated to be about 3÷5% as compared to the expected 13.64% of silicon (by volume) in the alloy. Too small size of the eutectic crystals [10] has therefore become the reason for lower values of aluminium and silicon phase analysis (80.15% Al and 3.42% Si), totally by about 16% obtained from the diffraction reflections intensities ratio method.

Conclusions

The experiments demonstrate, that the results of phase composition analysis of a model aluminium alloy containing 12 weight percent of Si obtained by using several methods show significant discrepancies. The X-ray analysis as opposed to metallography allows for univocal determination of phase composition of a polycrystalline material provided the crystals of the investigated substance have sizes of 0.1÷10μm. The phenomena of texture and extinction are not desirable because they modify the diffraction reflections intensities and increase errors of X-ray quantitative phase analysis. Proper choice of methods and conditions for structural component analysis of aluminium alloy allows the accuracy of determination of the phase composition within few percent. The described example shows how inaccurate a phase composition determination can be when the size of crystals is not taken into account.

References

1. J. Żelechowski, Konferencja *Aluminium 96*, Mogilany 1996.
2. L. Alexander, H.P. Klug, *Powder Diffraction* 4 (1989), 66.
3. C.R. Hubbard, E.H. Evans, D.K . Smith, *J.Appl.Cryst.* 9 (1976), 169.
4. B.L. Davis, R. Kath, M. Splide, *Powder Diffraction* 5 (1990), 76.
5. C.R. Hubbard, R.L. Snyder, *Powder Diffraction* 3 (1988), 74.
6. Z. Bojarski, E. Łągiewka, Rentgenowska analiza strukturalna, Wyd. Uniwersytetu Śląskiego, Katowice 1995.
7. W. Szymański, *Rudy i metale nieżelazne*, 11 (1992), 330.
8. J. Ryś, Wstęp do metalografii ilościowej, Wydawnictwo „Śląsk", Katowice 1970.
9. J. Ryś, Metalografia ilościowa, Skrypty uczelniane AGH, Kraków 1983.
10. J. Fiala, *Archiwum nauki o materiałach* 14 (1993), 253.

Small Angle Scattering

SMALL-ANGLE SCATTERING STUDIES OF SOLIDS WITH STRUCTURE ON MORE THAN ONE LENGTH SCALE IN THE INTERVAL BETWEEN 5 Å AND 10^4 Å

Paul W. Schmidt, D. J. Voss, Randall D. Rice, Rainer Brückner, Laural Shulse
Physics Department, University of Missouri, Columbia, Missouri 65211 USA

Y. M. Kapoor
Department of Natural Sciences and Mathematics,
Lincoln University, Jefferson City, Missouri 65102 USA

T. P. Rieker
Sandia National Laboratories, Albuquerque, New Mexico 87185 USA

Francoise Ehrbuerger-Dolle
Institut de Chimie des Surfaces et Interfaces (CNRS)
B. P. nº 2478, F68057 Mulhouse, France

A new method has been developed for using small-angle X-ray or neutron scattering to investigate the structure of solids that have structure on more than one length scale in the interval from about 5 Å to 10^4 Å. These length scales are larger than the diameters of individual atoms or small molecules but still are too small to be resolved easily in an ordinary optical microscope. The new method is based on an exact equation developed by P. Debye [Annalen der Physik **46 (351),** Eq. (11), p. 816 (1915)] to describe the X-ray scattering from a system of randomly oriented interacting scatterers. The new technique has been employed in the analysis of the small-angle scattering data from a commercial carbon black, a commercial precipitated silica, a commercial catalyst carrier, a silica xerogel (dried gel), four soots obtained by the combustion of methane, and two carbon chars made from wheat.

I. Introduction

Almost everything in everyday life looks different when examined on different length scales (*i. e.,* when examined under different degrees of magnification or reduction). For example, when parts of the human body are viewed under an optical or electron microscope, their appearance is very different from what is seen when they are viewed with the naked eye from distances of a few centimeters or meters.

Similarly, many solids have different structures when examined on different length scales. This report reviews some the results that we obtained in our recent small-angle X-ray scattering studies of the structure of several solids on different length scales in the interval between about 5 and 10^4 Å.

In Section 2 we review some general properties of small-angle scattering that are useful in our studies. The small-angle scattering from solids with structure on more than one length scale is discussed in Section 3. Our new method for the analysis of the small-angle scattering data from these solids is presented in Section 4. In Section 5, we review the results that we obtained when we applied our new method in the small-angle X-ray scattering studies of a number of solids with structure on more than one length scale in the interval between 5 and 10^4 Å. Acknowledgements are given in Section 6, and references are collected in Section 7.

Although the techniques that we describe in this report apply to the small-angle scattering of both X-rays and neutrons, the studies that we describe used only small-angle X-ray scattering.

2. A Review of the Small-angle X-Ray and Neutron scattering Techniques

2.1 A General Description of the Scattering Process

Small-angle X-ray and neutron scattering [1-3] are important experimental techniques for investigating the structure of solids on length scales from about 5 Å to about 10^4 Å. This interval includes lengths that are larger than the diameters of individual atoms or small molecules but are too small to be able to be studied with an ordinary optical microscope.

Figure 1 is a schematic diagram of a typical small-angle scattering system. X-rays or neutrons from the source T are formed into a fine beam, often by slits, and strike the sample S that is being investigated. A small fraction of this beam that falls on the sample is re-emitted in directions other than that of the incident beam. In Fig. 1 the beam is deviated by an angle θ from the direction of the incoming beam. The deviated beam, which is called the scattered beam, is formed by slits or some other device and is recorded by the detector C.

X-rays or neutrons scattered from different parts of the sample travel different distances as they move from the source to the detector and thus arrive at the detector with different phases.[4,5] The intensity recorded by the detector is the diffraction or interference pattern that results from the waves scattered by different parts of the sample.

Calculations show [4,5] that the scattered intensity can conveniently be expressed in terms of the quantity

$$q = \frac{4\pi \sin(\theta/2)}{\lambda}; \qquad (1)$$

2.2 The Resolution Criterion

As we mentioned in the preceding section, qr is the maximum phase difference where λ is the wavelength, and θ, as we have mentioned, is the angle between the incident and deviated beams. The angle θ is called the scattering angle, and is equal to twice the Bragg angle.

A schematic diagram of the essential parts of a small-angle scattering system is shown in Fig. 1.

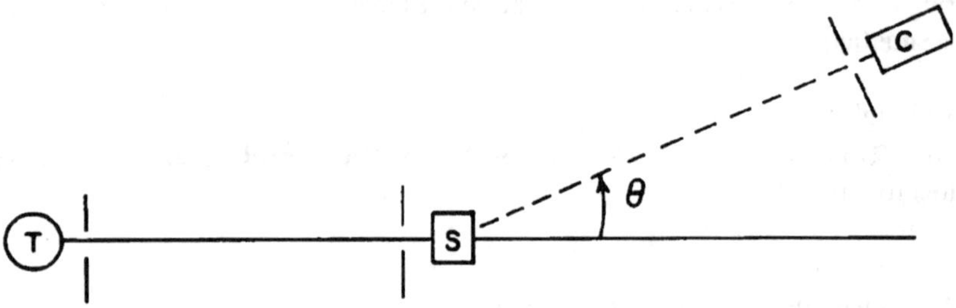

Fig. 1. A typical system for small-angle scattering. T is the source; S is the sample; C is the detector; and θ is the scattering angle.

When q is defined by Eq. (1), qr is the maximum phase difference between two points in the scatterer that are separated by a distance r.[4]

In discussions of small-angle scattering, we have found that it often is convenient to introduce a quantity that we call the diameter of a scatterer. We define the diameter A of a scatterer to be the largest distance between two points in the scatterer. For a sphere, this definition of the diameter is equivalent to the ordinary diameter.

The scattered intensity from almost all scatterers cannot be expressed in terms of simple functions. (One of the very few exceptions is a spherical scatterer with a uniform electron density.[6]) Nevertheless, for a scatterer with diameter A, simple approximations for the scattered intensity exist when

$$qA < 1$$

and

$$qA >> 1.$$

Expressions for the scattered intensity in these limiting cases are given in Sections 2.4 and 2.5, respectively.

The scattered intensity from a scatterer with a diameter A can often be conveniently expressed as a function of qA.

2.2 The Resolution Criterion

As we mentioned in the preceding section, qr is the maximum phase difference between two points separated by a distance r. As this result suggests, the quantity q*a* is useful for specifying the conditions that are necessary if scattering is to be employed to study structure on a length scale *a*. These conditions are expressed by a result that we call the resolution criterion, which states that if scattering methods are to be used to obtain information about a structure on a length scale *a*, the condition

$$0.1 \le qa \le 20. \tag{2}$$

must be satisfied.

Although the limits in Inequalities (2) cannot be sharply defined, we believe that these limits are convenient choices. As can be seen from Eq. (3) in Section (2.4) below, the intensity from a scatterer with diameter *a* will be measurably different from the intensity for q = 0 when qa is at least 0.1. Equation (6) in Section 2.5 below shows that when q*a* >> 1, no *new* information will be gained by measuring the intensity at still larger q, and so q need only be large enough to determine the power-law scattering reliably. These arguments suggest that the limits in Inequalities (2) are reasonable.

An important consequence of the resolution criterion is that large structures scatter primarily at small q, and measurements must be made at larger q in order to resolve smaller structures.

2.3 The Two-Phase Approximation

According to the resolution criterion, *small*-angle scattering cannot resolve structure on length scales as small as atomic diameters. Therefore, the atomic-scale structure normally can be neglected in small-angle scattering,[4] and the electron density of the sample often can be considered to have either one of the two constant values ρ_1 or ρ_0. This important and very useful result is often called the two-phase approximation.[7,8]

A consequence of the two-phase approximation is that the intensity of small-angle scattering is proportional to[7]

$$(\rho_1 - \rho_0)^2$$

for a scatterer of uniform electron density ρ_1 embedded in a medium of uniform electron density ρ_0.

Also, the same intensity is obtained for a scatterer of uniform electron density ρ_0 embedded in a medium of electron density ρ_1. Small-angle scattering experiments cannot distinguish between the two systems.

2.4 The Radius Of Gyration Approximation

When $qA < 1$, the intensity $I(q)$ from a system of N randomly oriented, independent scatterers of radius of gyration R can be at least reasonably well approximated by the equation[8]

$$I(q) = I_e N \rho^2 <V^2> e^{-(qR)^2/3}. \quad (3)$$

In this equation, I_e is the scattered intensity from one electron; ρ is the electron density; V is the volume of the scatterer, and the symbol < > represents an average over the N scatterers.

Usually

$$2R \leq A \leq 3.5R. \quad (4)$$

For a uniform-density sphere of diameter A,[9]

$$R = (20/3)^{-1/2} A \approx A/2.6. \quad (5)$$

2.5 Power Law Scattering

When $qA >> 1$, the intensity from a structure with a diameter A can usually be approximated by the equation[10]

$$I(q) = I_p q^{-\alpha}, \quad (6)$$

where I_p and α are constants.

Equation (6) shows that when I(q) is plotted as a function of q on logarithmic axes, the intensity will be a straight light with a negative slope. The value of the exponent α can be calculated from the slope.

Some values of α are listed in Table 1 at the end of Section 2.7.

Important information about the boundary surfaces of scatterers can often be obtained from the magnitude of the power-law-scattering exponent α in this equation [10]

Estimates of the average diameter of the scatterers can sometimes be obtained from I_p.[11]

2.6 *Fractals*

2.6.1 *Some General Properties of Fractals*

A fractal is a structure that looks the same at any magnification. This property is called self-similarity.[12,13]

Some properties of fractals are:

(1) There is no characteristic length scale.
(2) Fractal surfaces are rough.
(3) Real objects are fractal only within the interval of length scales l for which

$$L_{min} \leq l \leq L_{max}.$$

The lengths L_{min} and L_{max} are known as the inner and outer cutoffs, respectively.

2.6.2 *The Fractal Dimension D*

One of the most important properties, if not *the* most important property, of a fractal is its fractal dimension D.[12,13] The fractal dimension D can be evaluated by calculating the minimum number N of balls of diameter a that are necessary to cover a structure completely. If the structure is a fractal, N can be described by the equation

$$N = N_0 a^{-D}, \tag{7}$$

where N_0 is a constant. Pfeifer and Obert[13] discuss this method and other procedures for determining the fractal dimension.

Many disordered solids have structures that have fractal properties within a certain interval of length scales.[10(b)] Because the solids that we discuss in this report have fractal properties on certain length scales, we now review some of the relevant characteristics of fractals.

2.6.3 *Types of Fractal Systems*

Fractal solids can be described by one of three types of fractal systems: (1) Mass Fractals, (2) Surface Fractals, and (3) Pore Fractals.

A **Mass Fractal** has a mass that scales with (is proportional to) its surface area. For a mass fractal, $D_m < 3$.

A **Surface Fractal** consists of two non-fractal regions separated by a fractal surface. The values of the surface-fractal dimension D_s lie in the interval $2 \leq D_s < 3$.

A **Pore Fractal** can be considered to be the inverse of a mass fractal. That is, all the occupied regions in a pore fractal are empty in a mass fractal, and the occupied space in a mass fractal is empty in a pore fractal. For a pore fractal, $D < 3$. Pore fractals are much less common than mass or surface fractals. According to the two-phase approximation discussed in Section 2.3, scattering measurements cannot *by themselves* distinguish between mass and pore fractals.

2.6.4. *Some Properties of the Small-Angle Scattering from Fractal Systems*

The intensity of the small-angle scattering from solid fractal systems with a diameter A is appreciably different from the intensity scattered by other solids only when the intensity can be described by a power law, (*i. e.*, only when $qA >> 1$).[14] In the region of q in which the intensity follows a power law, however, there are characteristic values of the power-law-scattering exponent α. Table 1 lists some of these values.

Table 1
Some Values of the Magnitude α of the Power-Law-Scattering Exponent (d is the Euclidean dimension of the space in which the scatterer exists.)

Scattering System	α
Non-fractal scatterer ($d = 3$)	4
Thin plane lamina	2
Thin rod	1
Mass or pore fractal with fractal dimension D_m	D_m
Surface fractal with fractal dimension D_s in a space of Euclidean dimension d	$2d - D_s$
Porous solid with smooth pore boundary and a continuous power-law density transition with exponent β ($\beta > 0$)	$4+2\beta$

The last entry in Table 1 states that $\alpha > 4$ when the density at the boundary decays continuously to zero with a *positive* power-law exponent β.[15]

According to Table 1, for mass and pore fractals, $\alpha = D_m$. Also, $D_m < 3$ for mass and pore fractals, and so $\alpha < 3$. For surface fractals, $2 \leq D_s < 3$, and according to Table 1, $\alpha = 2d - D_s$, where d is the Euclidean dimension of the space in which the fractal exists. Therefore, $3 < \alpha \leq 4$ when $d = 3$. Thus, when a scatterer is known to be a fractal, evaluation of the power-law-scattering exponent α can be used to determine whether the scatterer is a surface or a mass or pore fractal.

3. Some Properties of the Small-Angle Scattering from Solids with Structure on More than One Length Scale

Figure 2 is a plot of the scattered intensity that might be expected from a solid with structure on *only one* length scale. The data points at smaller q in Fig. 2 can be fitted with Eq. (3), and the radius of gyration equation can be evaluated from the fit. According to Eq. (3), if this scatterer has structure on only one length scale, intensity data recorded at q values smaller than those in Fig. 2 will approach a constant value. Also, Eq (6) can be fitted to the data at larger q in Fig. 2.

The upper curve in Fig. 3 shows the complete intensity curve from which the points in Fig. 2 were obtained. The data at the smaller q in Fig. 3 clearly do not approach a constant intensity but instead follow a power law. Thus, in the upper curve in Fig. 3 there are *two* power laws. One power law is at the smallest q values in the plot, and the other is observed at the largest q.

In Fig. 3 the upper curve was obtained for carbon black N330, which was produced by Degussa. The middle curve shows the scattered intensity from Norton Chemical Process Products type SA6*73 catalyst carrier, and the lower curve is the intensity from a silica xerogel (dried gel) prepared by David Avnir and his colleagues at the Hebrew University of Jerusalem.

Since power laws in Fig. 3 are observed at the smallest q values for which data could be recorded, this large-length-scale structure is so large that the data do not allow its diameter A to the be evaluated. Nevertheless a lower limit to A can be calculated. Since the intensity follows a power law even at the smallest q value (3×10^{-4} Å^{-1}) at which data are reliable, the condition $qA >> 1$ is satisfied at this value of q. Our experience suggests that this condition can be satisfied at least reasonably well when $qA > 5$. Thus $A > 1.7 \times 10^4$ Å.

In Sections 4 and 5, we explain how we used the intensity data in the shoulder and at somewhat larger values of q to calculate the radius of gyration of the scatterers in solids with structure on more than one length scale. For carbon black N330, the silica-alumina catalyst carrier, and the silica xerogel, the radii of gyration were 2.7×10^2 Å, 34 Å, and 12.7 Å, respectively. (These values are listed in Table 2 in Section 5.)

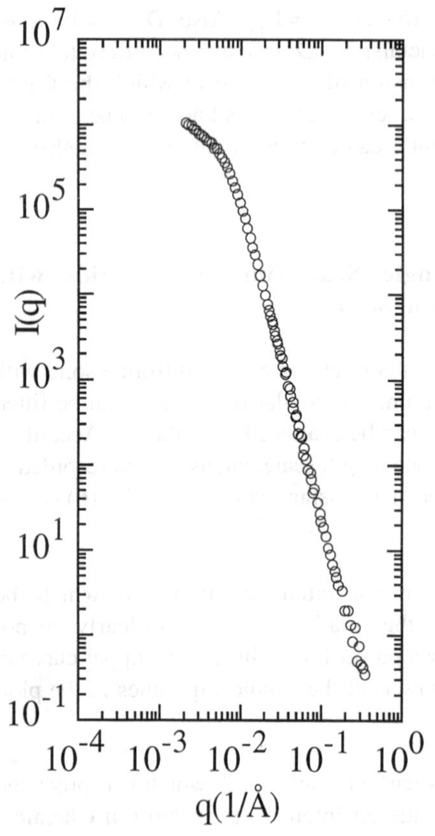

Fig. 2. (Above) The intensity of small-angle scattering that might be expected from a solid with structure on only one length scale.

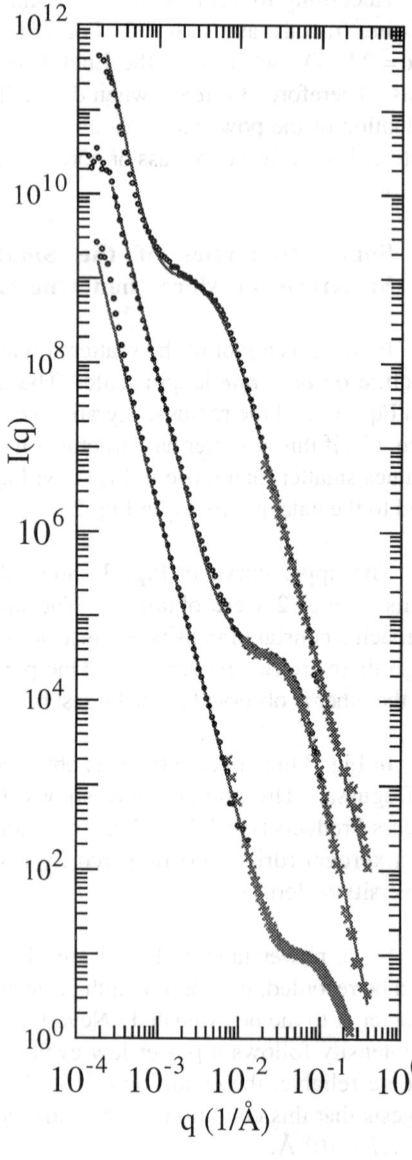

Fig. 3. (Right) The small-angle scattered intensity from carbon black N330 (upper curve) a silica-alumina catalyst carrier (middle curve) and a silica xerogel (lower curve). The lines are fits of Eq. (8). (See Section 5 and Table 2.)

The fact that the three curves came from very different kinds of solids suggests that structure on more than one length scale is very common in solids like those that we have investigated. This suggestion is confirmed by the intensity curves that we have recorded from many other solids. In fact, in our recent small-angle scattering studies of solids, we have observed *only* curves like those in Fig. 3 and have *never* seen a curve like Fig. 2, which is characteristic of a scatterer with structure on only one length scale.

The power laws and the shoulders in the curves like those in Fig. 3 have led us to believe that:

(1) Solids that give intensity curves like those in Fig. 3 are composed of systems of scatterers.

(2) The **systems** have average diameters of at least one or two microns.

(3) The average radii of gyration of the **scatterers** that comprise the systems can be calculated from the intensity in the region of the shoulders in the intensity curves.

(4) From the magnitudes of the power-law-scattering exponents in the intensity curves, information can be obtained about the boundary surfaces and often about other properties of the structure of the systems and the scatterers.

(5) The intervals of q at which power laws are observed often provide estimates of the diameters and other length scales that characterize the structure of the solid.

4. The Small-Angle Scattering from Systems of Scatterers

For N independent systems, each composed, on the average, of n randomly oriented, interacting scatterers separated by fixed distances, the average intensity I(q) can be expressed[17]

$$I(q) = nN\,P(q)\,S(q). \tag{8}$$

Here, the structure factor S(q) is

$$S(q)=1+\frac{1}{nNP(q)}\sum_{k=1}^{N}\sum_{i=1}^{n_k}\sum_{\substack{j=1\\ i\neq j}}^{n_k} f_{ki}(q)\, f_{kj}(q)\frac{\sin(qr_{kij})}{qr_{kij}}, \tag{9}$$

where the distance between primary particles i and j in system k, and n_k is the number of scatterers in system k. In Eq. (10), $f_{ki}(q)$ is the scattering amplitude of scatterer i in system k.

In Eq. (8), P(q), the form factor, which is the average scattered intensity from one of the nN scatterers, can be expressed

$$P(q) = \frac{I_e}{N}\sum_{k=1}^{N} P_k(q) = \frac{I_e}{nN}\sum_{k=1}^{N}\sum_{i=1}^{n_k} [f_{ki}(q)]^2, \tag{10}$$

where I_e is the scattered intensity from one electron.

The structure factor S(q) describes the effects of interactions between the scatterers. When S(q) = 1, interactions are negligible.

The equation for the intensity was obtained under the assumptions that the orientations of scatterers in the systems are not correlated.

The average diameters of the systems and the scatterers are defined to be A and b, respectively.

To analyze our small-angle scattering data, we found that useful expressions for S(q) and nNP(q) are

$$S(q) = \left[1 + \frac{C}{q^{\alpha}} + \frac{B}{q^{\beta}}\right]. \tag{11}$$

and

$$nNP(q) = nNP(0)\left[\frac{1}{1+f}\right]\left[T(qr, e, \gamma) + f\,T(qr_f, e_f, \gamma_f)\right], \tag{12}$$

where

$$T(x, e, \gamma) = \frac{1-e}{\left[1 + \frac{2/3}{\gamma + 2e}x^2\right]^{\gamma/2}} + \frac{e}{\left[1 + \frac{2/3}{\gamma + 2e}x^2\right]^{\gamma/2+1}}. \tag{13}$$

In Eq. (13), $-\gamma/2 < e < 1$, and $-\gamma_f/2 < e_f < 1$. The constant e is chosen to match the magnitude of the power-law scattering that is characterized by the exponent γ. The constant f is included in Eq. (12) to allow for the fact that sometimes there is structure *on length scales* much smaller than the average radius of gyration of the *scatterers* of which the system is composed. The *positive* constant f is a measure of the intensity of the scattering from this relatively small structure, which has an average radius of gyration r_f. The magnitude of the power-law-scattering exponent of the smaller structure is γ_f. (This structure is noticeable only at the larger values of q. Usually, *and in all curves discussed in this report,* f is equal to zero.)

The constants B, C, r, r_f, α, β, γ, e, nNP(0), e_f, and γ_f are determined by fits of Eqs (8) to the measured scattering data. In these fits, we substituted Eqs. (11), (12), and (13) for S(q), nNP(q), and T(x,e,γ), respectively.

Preliminary calculations suggest that the expression given by Eq. (12) for P(q) can can be obtained from a polydisperse system of independently scattering spheres of uniform density.[11]

5. Results and Discussion

5.1 Introduction

The scattering data were recorded at the scattering center operated jointly by Sandia National Laboratories and the University of New Mexico in Albuquerque, New Mexico (USA). We used a Bonse-Hart scattering system[17] for 2×10^{-4} Å^{-1} < q < 0.1 Å^{-1}, and a pinhole system[18] for q values between 10^{-2} Å^{-1} and 0.6 Å^{-1}. The data for the two regions of q were matched in the interval of q where the two sets of data overlapped.

We found that trial and error techniques were better than non-linear least squares methods for fitting Eq. (8) to the measured data. Perhaps the large number of parameters was responsible for many or all of the difficulties with least squares techniques.

The results of our fits of Eq. (8) to the data plotted in Figs. 3, 4, and 5 are collected in Table 2.

We used the condition qA > 5 when we estimated the lower limit A_{min} of the average diameter A of the systems discussed in Section 3. We now express A_{min} by the equation

$$A_{min} \approx \frac{5}{q_{min}}. \tag{14}$$

Table 2
Quantities Calculated from Fits of Eq. (8) to the Data in Figs. 3, 4, and 5

	Carbon Black N330	Catalyst Carrier	Silica Xerogel	Sipernat 22 LS
C	2.8×10^{-5}	6.0×10^{-5}	5.0×10^{-9}	9.3×10^{-3}
α	4.4	4.4	4.0	2.95
B	0	4.8	3.0×10^{-5}	2.1
β		1.8	3.6	2.1
nNP(0)	1.1×10^9	4.0×10^{-4}	11	2.3×10^4
r (Å)	2.7×10^2	34	12.7	87
γ	3.9	4.9	3.6	3.9
e	-1.4	-1.65	0.8	-1.5

In Eq. (14), q_{min} is the smallest value of q at which there is power-law scattering. We use Eq. (14) to discuss our results.

5.2 Carbon black N330

Figure 4 and the upper curve in Fig. 3 show our fit of Eq. (8) to the data from carbon black N330. The points in the lower part of Fig. 4 are the magnitudes of the relative errors in the fit. These errors are almost always smaller than 20%. Although these relative errors may seem large, we believe that they are quite satisfying in fits of Eq. (8) to intensity curves that vary by a factor of nearly 10^{10}.

According to Table 2, for carbon black N330, $\alpha = 4.4$ in the power law at smallest q in Fig. 3. Because $\alpha > 4$, the density at the boundaries of the systems decreases continuously to zero. From Eq. (14), the average diameter of the systems is at least one micron. Since $B = 0$ in Eq. (11), there is structure on *two* length scales: the diameter of the system and the average radius of gyration of the scatterers that make up the systems. The average radius of gyration r of the scatterers is 2.7×10^2 Å. Because γ is 3.9, the boundaries of the scatterers are fractal, with a fractal dimension 6 - 3.9 = 2.1. Thus the boundaries are nearly smooth.

5.3 The catalyst carrier

The middle curve in Fig. 4 and the fits of Eq. (8) to the data from Norton Chemical Products type SA6*73 catalyst carrier show that for 2.7×10^{-4} Å$^{-1} < q < 4.7 \times 10^{-4}$ Å^{-1}, $\alpha = 4.4$. Thus the systems scatter like solids with a density that approaches zero continuously at the boundary. The average diameter of the systems is at least one or two microns.

From our fits, we found that B in Eq. (11) was different from zero. There is a short region in which $\beta = 1.8$ for q between about 5×10^{-3} Å^{-1} and 10^{-2} Å^{-1}. This power-law region suggests that the systems are mass-fractal aggregates. From Eq. (14), the average diameter of the mass-fractal aggregates is at least 1×10^3 Å.

The fit to the intensity from the catalyst carrier showed that the scatterers have an average radius of gyration of 34 Å. Since $\gamma > 4$, the density of the scatterers approaches zero continuously at the boundary.

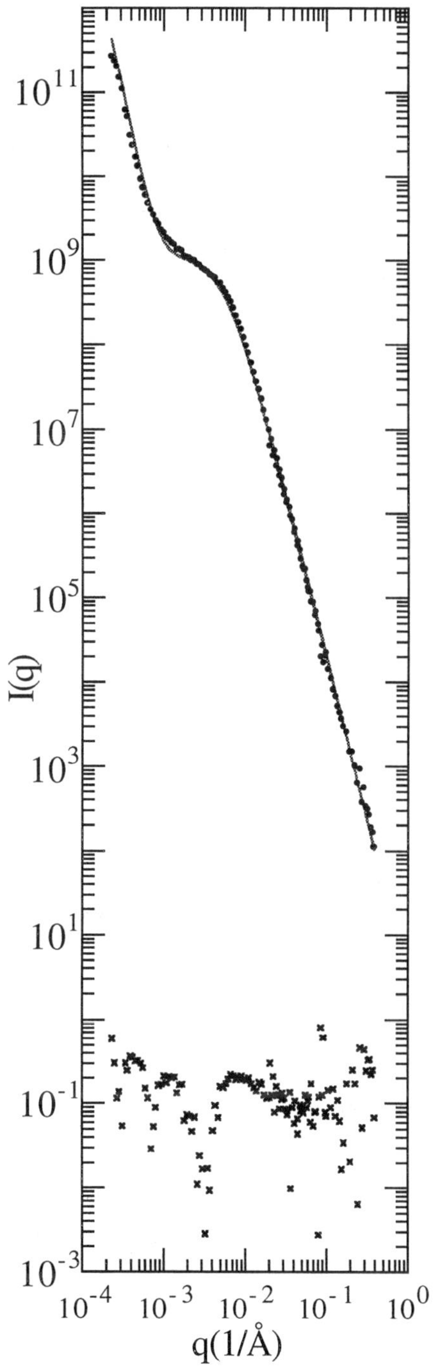

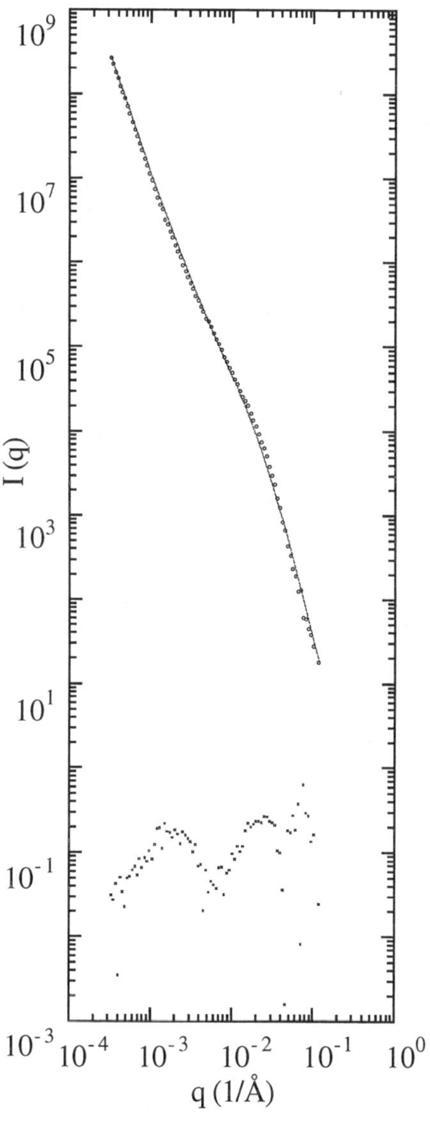

Fig. 5. (Above) The fit of Eq. (8) to the data from precipitated silica Sipernat 22 LS. The points in the lower part of the plot show the relative error in the fit.

Fig. 4. (Left) The fit of Eq. (8) to the data for carbon black N330. The points in the lower part of the plot show the relative error in the fit.

5.4. The silica xerogel.

For the silica xerogel, the fits of Eq. (8) gave the exponent $\alpha = 4.0$ for q values in the interval 4.7×10^{-4} Å^{-1} < q < 1×10^{-3} Å^{-1}. For 2×10^{-3} Å^{-1} < q < 2×10^{-2} Å^{-1}, we found that $\beta = 3.6$. We have tentatively interpreted these power laws to mean that the solid is composed of systems of *surface* fractals. The systems have smooth boundary surfaces. Since $q_{min} = 4 \times 10^{-4}$ Å^{-1} for the systems, Eq. (14) implies that the average diameter of the systems is at least 1 micron. Also, q_{min} is 2×10^{-3} Å^{-1} for the surface fractals, and so the average diameter the surface-fractals is at least equal to 2.5×10^3 Å. From Table 2, the surface fractals are composed of scatterers with an average radius of gyration of 12.7 Å. This radius of gyration is so small that the data do not extend to large enough q permit the exponent γ to be determined reliably. Thus, although we set γ equal to 4 in Eq. (8) for convenience in making the fits to the data for the silica xerogel, the value $\gamma = 4$ in Table 2 is *not* meaningful.

In the preceding paragraph, we said that our interpretation was tentative. Our reason for this uncertainty in our analysis is that when q < 4×10^{-4} Å^{-1}, the measured intensity lies above the power law from which we evaluated α. We have not discussed the data for this interval of q in more detail because there are only about ten of these points, and the data at these small q values are somewhat unreliable. Nevertheless, these data suggest that the systems that we have discussed may themselves be combined in other, larger systems.

5.5 Sipernat 22 precipitated silica

Our data for Sipernat 22 LS precipitated silica, which was produced by Degussa, are plotted in Fig. 5, where, as in Fig. 4, the absolute value of the relative errors are shown in the lower part of the plot. For 3×10^{-4} Å^{-1} < q < 1×10^{-3} Å^{-1}, our fits showed that α was 2.95. In the interval 2×10^{-3} Å^{-1} < q < 2×10^{-2} Å^{-1}, the intensity could be described by a power law with $\beta = 2.1$. We therefore have concluded that there are mass-fractal aggregates in Sipernat 22 LS, and from Eq. (14) the average diameter of these aggregates is at least 2.5×10^3 Å. The aggregates are made up of scatterers with average radius of gyration of 87 Å. Since $\gamma = 4.5$, the scatterers that make up the mass-fractal aggregates have a density that decreases continuously to zero at the boundary surface of the scatterer.

Thus, since Sipernat 22 LS consists of surface-fractal systems of mass-fractal aggregates of scatterers, in this solid there is structure on three length scales.

5.6. Soots produced by the combustion of methane.

Figures 6 and 7 show the scattered intensity measured from soots that C. M. Sorenson and C. Oh made by combustion of methane.[19] The work was performed in the Department of Physics at Kansas State University in Manhattan, Kansas (USA). There were four soot samples.

(a) The **Fluffy** soot was scraped from a plate that was placed in the flame.

(a) The **Ground** soot was obtained by grinding the fluffy soot in a mortar and pestle with light hand pressure for 30 seconds.

(c) The **Compressed Fluffy** soot was prepared by subjecting the fluffy sample to a pressure of 1.8×10^3 atmospheres in an isostatic press.

(d) The **Compressed Ground** soot was produced by subjecting the ground sample to a pressure of 1.8×10^3 atmospheres in an isostatic press.

Since electron micrographs show[19] that these soots consisted of very nearly spherical particles with diameters of approximately 300 Å, for the $[f_{ki}(q)]^2$ in Eq. (10), we used the expression for the intensity from a polydisperse system of spheres of uniform density. We also assumed the the distribution of diameters was the same in all systems. To evaluate nNP(q), in Eq. (10), we replaced the double sum, which extends over all nN scatterers in the N systems, with one integral. By this process, we obtained the equation

$$\mathrm{nNPq)} = \mathrm{nN}\int_0^\infty \rho_{\mathrm{m}}(a)\ \mathrm{I}(\mathrm{q}a, a)\ \mathrm{d}a. \qquad (15)$$

In Eq. (15), $\mathrm{I}(\mathrm{q}a, a)$ is the intensity from a scatterer that has a diameter a, and the diameter distribution $\rho_{\mathrm{m}}(a)$ is defined to be the probability that the diameter has a value between a and $a + \mathrm{d}a$. For $\rho_{\mathrm{m}}(a)$ we used the distribution

$$\rho_{\mathrm{m}}(a) = \frac{\mathrm{m}^{\mathrm{m}+1}\, a^{\mathrm{m}}}{\mathrm{A}^{\mathrm{m}+1}\, \Gamma(\mathrm{m}+1)}\, \mathrm{e}^{-(\mathrm{m}a/\mathrm{A})}. \qquad (16)$$

The distribution function $\rho_{\mathrm{m}}(a)$ is often called the factorial distribution or the Schultz distribution. In Eq. (16), A is the most probable diameter, and m determines the width of the distribution function. For $\mathrm{I}(\mathrm{q}a, a)$, we used the equation[6] for the *intensity* from a sphere of constant electron density ρ_e, and *diameter a*

$$\mathrm{I}(\mathrm{q},a) = \frac{\left(\frac{\pi}{2}\rho_{\mathrm{e}}\, a^3\right)^2}{\left(\frac{\mathrm{q}a}{2}\right)^6}\, \mathrm{I_e}\left[\sin\left(\frac{\mathrm{q}a}{2}\right) - \left(\frac{\mathrm{q}a}{2}\right)\cos\left(\frac{\mathrm{q}a}{2}\right)\right]^2. \qquad (17)$$

After Eqs. (16) and (17) were substituted into Eq. (15), we obtained an expression for nNP(q) in terms of elementary functions.[20] With this expression for nNP(q), we used a modification of a method developed by Kotlarchyk and Chen[21] to obtain an expression for the intensity that was analogous to Eq. (8). The results that we obtained in our fits of this equation to the scattering data from the soots are listed in Table 3.

For all four soots, the best fits were obtained for m = 35. When m is larger than 5, the full width of the w distribution function at half its maximum value can be obtained from the equation[11]

$$\frac{w}{A} = 2\left[\frac{2\ \ln(2)}{m}\right]^{1/2} \approx \frac{2.355}{\sqrt{m}} \,. \tag{18}$$

For m = 35, Eq. (18) shows that the full width of the distribution at half maximum is equal to 40% of the most probable diameter A.

When nNP(q) is calculated from Eq. (15), the power-law-scattering exponent γ of the spherical scatterers always equals 4, as is true for *any* scatterer with a smooth boundary surface and a uniform density that decreases discontinuously to zero at the boundary. Within the uncertainty of the fits, the data from the soots give $\gamma = 4.0$.

In Table 3, we have collected he results that we obtained in our analysis of the intensities plotted in Figs. 6 and 7 by use of the modified method of Katlarchyk and Chen.[21]

Table 3
Quantities Calculated from the fits to the Scattering Data from the Methane Soots

	Fluffy	**Ground**	**Compressed Fluffy**	**Compressed Ground**
m	35	35	35	35
A (Å)	3.4×10^2	3.4×10^2	3.4×10^2	3.2×10^2
C	0	1.0×10^{-10}	5.0×10^{-11}	1.7×10^{-8}
α		4.9	4.9	4.2
nNP(0)	2.5×10^4	4.2×10^4	3.7×10^4	4.5×10^4
B	3.4	2.5	1.8×10^2	0
β	1.85	1.8	1	

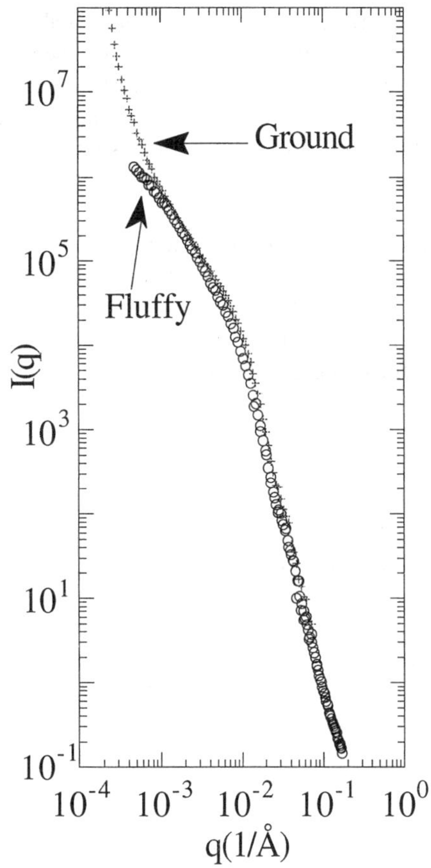

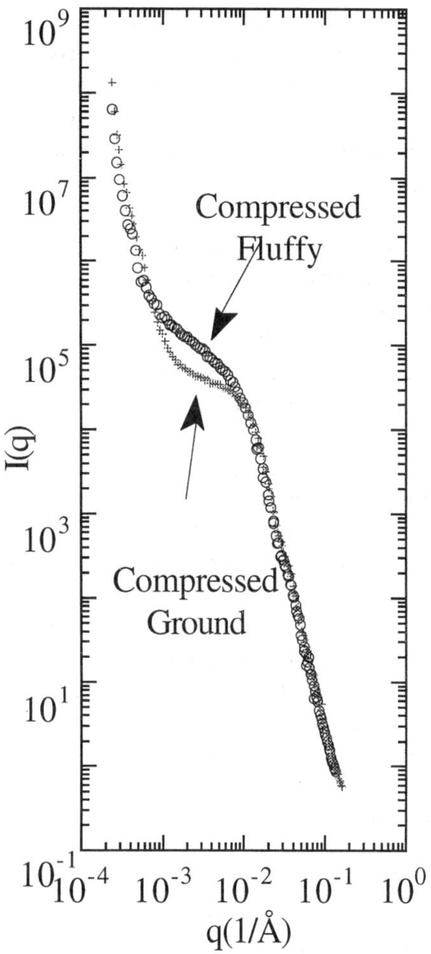

Fig. 6. (Left) The small-angle scattering intensity from the fluffy and ground soots. The intensity of the ground soot has bas multiplied by a constant that makes the two intensity curves in the plot very nearly equal for q greater than about 0.008 $Å^{-1}$.

Fig. 7. (Right) The small-angle scattering intensity from the compressed fluffy and compressed ground soots. The intensity of one of he soots has been multiplied by a constant that makes the two intensity curves in the plot very nearly equal for q greater than about 0.008 $Å^{-1}$.

The values of the most probable diameter A in Table 3 show that for all soots except the ground compressed soot, A = 3.4 x 10^2 Å. The best fit to the data from the ground compressed soot was obtained when A = 3.2 x 10^2 Å. The difference between these two A values is so small that we are not sure that it is significant.

The curves in Fig. 6 coincide nearly coincide for q greater than about 0.008 $Å^{-1}$. Also the two curves in Fig. 7 are nearly identical for these q values. Moreover, when the two curves in Fig. 7 are multiplied by a suitable constant, all *four* curves nearly coincide for $q > 0.008$ $Å^{-1}$. In the interval of q in which all four curves nearly coincide, the intensity is very nearly proportional to nNP(q), and so all the soots scatter like independent spheres.

From our analysis of the quantities computed from the fits to the data from the soots, we conclude that the fluffy soot consists of mass-fractal aggregates with fractal dimension 1.85. The four or five points for $q < 4 \times 10^{-4}$ $Å^{-1}$ lie below the power law, and so q_{min} is 4×10^{-4} $Å^{-1}$. The mass-fractal aggregates thus have an average diameter of at least one micron. The scatterers in these mass-fractal aggregates are spheres with most probable diameter 3.4×10^2 Å.

On the other hand, a comparison of the four curves in Figs. 6 and 7 show that the scattered intensity for $qA < 0.008$ $Å^{-1}$ is very much affected by grinding and compression. These differences show that grinding and compression changes the structure of the soots *only* on length scales *larger* than the diameters of the spherical scatterers.

A comparison of the scattered intensities from the ground soot, the fluffy compressed soot, and the ground compressed soots suggests that grinding and compression pack the carbon spheres into systems that have a density that is higher than the average density in the mass-fractal aggregates in that fluffy soot. The presence of these dense systems is suggested by the result that, as can be seen from Table 3, $\alpha > 4$ in all of these soots, while in the fit to the intensity from the fluffy soot, the constant C was zero.

The curve fitted to the intensity from the ground soot shows that S(q) contains terms with $\alpha = 4.9$ and $\beta = 1.8$. We therefore conclude that in the ground soot, there are both dense regions *and* mass-fractal aggregates.

The results in Table 3 for the compressed fluffy soot give $\beta = 1.0$ and $\alpha = 4.9$. Such low values of β have almost never been observed in mass-fractal aggregates. Because we believe that there may be mass-fractal aggregates with $\beta = 1.8$ in the compressed fluffy soot, we plan to make fits to the intensity from this soot in order to determine whether we can obtain fits with $\alpha > 4$ and $\beta = 1.8$.

Figure 7 shows that the intensity curve from the ground compressed soot qualitatively resembles the intensity curve recorded from carbon black N330. The data in Table 3 suggest that on length scales at least as large as one micron, this soot consists of systems of spherical particles with a uniform density and an average diameter of 3.2×10^2 Å. The systems have a density that approaches zero continuously at the boundary. There is no evidence of mass-fractal aggregates in the intensity from the ground compressed soot.

5.7 Wheat chars

In Fig. 8, we have plotted the intensity of the small-angle scattering from two chars made from hard red winter wheat by Venkatraman, Fan, and Walawender in the Chemical Engineering Department of Kansas State University in Manhattan, Kansas (USA).[22] Sample J was pyrolyzed in N_2 at 800° C for 5 min. Sample K was first pyrolyzed in N_2 at 300° C and then activated (oxidized) in CO_2 for 310 minutes at 875° C.

The pyrolyzed chars consist almost entirely of carbon. As amorphous carbon and graphite have no structure larger than a few Å, the *solid* carbon in the chars cannot be expected to produce appreciable small-angle scattering. We therefore have analyzed the intensity data from the wheat chars by assuming that the small-angle scattering from the solid carbon *itself* was negligible and that the measured scattered intensity thus was produced by pores in carbon.

The results of our fits of Eq. (8) to the measured intensities from Samples J and K are shown in Table 4.

For Sample J, in the interval 2×10^{-4} Å$^{-1} < q < 4 \times 10^{-3}$ Å^{-1}, α is approximately 4.0. This power law shows that the surfaces of the large pores are smooth. From Eq. (14), the average diameter of the large pores is at least two microns.

Also, β is about 2.95 for 8×10^{-3} Å$^{-1} < q < 4 \times 10^{-2}$ Å^{-1}. From this value of β, we conclude that Sample J is a pore fractal with fractal dimension 2.95. According to Eq. (14), the average diameter of the pore fractals is at least 6×10^2. Å. The fits showed that the radius of gyration of the small pores that make up the pore fractal is 2 Å. These pores are so small that the value assumed for γ used in the fits and listed in Table 4 has no appreciable effect on the other quantities in Table 4 and so is not meaningful.

We have no evidence that there is a relation between the large and the small pores in either char. We therefore will consider the large and small pores are independent.

For Sample K, α was approximately 3.9 for 4×10^{-4} Å$^{-1} < q < 3 \times 10^{-3}$ Å^{-1}, and β was approximately 2.3 in the interval 8×10^{-3} Å$^{-1} < q < 8 \times 10^{-2}$ Å^{-1}. Thus, the large pores in Sample K have fractals surfaces, with a surface-fractal dimension of 2.1. This fractal dimension is so near 2 that the surfaces of the large pore can be considered to be very nearly smooth. From Eq. (14), the large pores have an average diameter of at least one micron.

Since $\beta = 2.3$, Sample K is a pore fractal. Equation (14) states that the pore fractals have average diameters of at least 5×10^2 Å. The radius of gyration calculated from the fits is 6 Å. The pores are so small that the value of γ that we used in the fits does not affect the other quantities in Table 4 and thus has no significance.

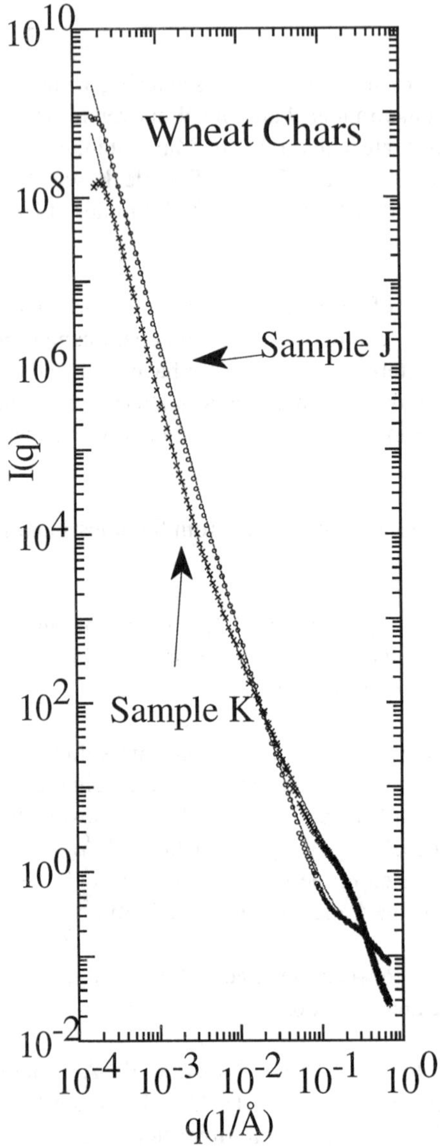

TABLE 4
Quantities Calculated from the Fits to the Scattering Data from the Wheat Chars

Sample	J	K
C	2.5×10^{-6}	1.0×10^{-6}
α	4.0	3.9
B	4.5×10^{-4}	6.7×10^{-3}
β	2.97	2.3
nNP(0)	2.0×10^{-1}	1.0
e	-1.1	-1.65
γ	3	5
r (Å)	2	6

Fig. 7. The scattered intensity curves for wheat chars J and K. The lines are fits of Eq. (8) to the measured intensity.

The result that the radius of gyration is larger for Sample K than for Sample J is not surprising, because the oxidation to which Sample K was subjected can be expected to remove carbon from the chars and thus will make the pores larger.

5.8 Concluding comments

The examples presented in this section illustrate how small-angle scattering can be used to investigate the structure of solids that have structure on more than one length scale. From our analysis of the scattering data, we have concluded that all of the solids except the wheat chars were made up of systems of scatterers. The diameters of the systems were at least one micron. Usually, the density of the systems decreased continuously to zero at the boundary of the systems. although in a few solids, the systems were surface fractals. The systems in Sipernat 22 LS were mass fractals. In some systems, the scatterers were mass-fractal aggregates of particles. Other systems were composed of particles that often had densities that decreased to zero continuously at the boundary of the particle. In other solids, the scatterers had smooth or fractal boundary surfaces.

We concluded that all of the scattered intensity from from the what chars was due to pores. On length scales at least as large as one micron, these surfaces of the pores in the chars were smooth or nearly smooth. In both chars, there also were pore fractals with with diameters of at least 5 x 10^2 Å. The radii of gyration of the small pores in the two chars were 2 Å and 6 Å.

6. Acknowledgements

We would like to express our thanks to Peter Pfeifer (Physics Department, University of Missouri, Columbia, Missouri 65211); Fred K. Ross (Chemistry Department, University of Missouri, Columbia, Missouri 65211) and J. S. Lin, (Oak Ridge National Laboratory, Oak Ridge, Tennessee 37831) for their assistance and advice in this work.

Acknowledgment is made to the Donors of the Petroleum Research Fund, administered by the American Chemical Society, for support of part of this work. Support for many of the scattering measurements was furnished by the Division of Materials Sciences, U. S. Department of Energy, and by the Sandia National Laboratories--University of New Mexico Scattering Center.

7. References

1. A Guinier, G. Fournet, C. B. Walker. Jr. and K. L. Yudowitch, *Small-Angle Scattering of X-rays,* Wiley, New York, 1955.
2. O. Glatter, and O. Kratky, *Small-Angle X-Ray Scattering.* Academic Press, New York, 1982.
3. P. W. Schmidt, in *Modern Aspects of Small-Angle Scattering,* pp. 1--56, edited by H. Brumberger [Proceedings of the NATO Advanced Studies Institutes C451, (Como, Italy, May, 1993). Kluwe, Drodrecht, the Netherlands, 1995.

4. Ref. 1, Chapter 1.
5. Ref. 3, Sections 1.2.1, 1.2.2, and 1.2.3, pp. 7--14.
6. Ref. 1, Eq. (31), p. 19.
7. Ref. 1, Sect. 2.4.3, pp. 78-81.
8. Ref. 1, Sect. 2.1.3, pp. 24-28.
9. Ref. 2, Sect. B, p. 28.
10. (a). Ref. 3, Sect. 1.2.7.2, pp. 24-27.
    (b) Ref. 3, Sect. 2.3, pp. 37-49.
11. P. W. Schmidt, unpublished research.
12. B. B. Mandelbrot, *The Fractal Geometry of Nature.* Freeman, New York, 1983.
13. P. Pfeifer and M. Obert, in *The Fractal Approach to Heterogeneous Chemistry,* edited by D. Avnir, Chapter pp. 11--13. Wiley, Chichester, England, 1989.
14. P. W. Schmidt and Xie D., Phys. Rev. A **33,** 560-566 (1986).
15. P. W. Schmidt, D. Avnir, D. Levy, A Höhr M. Steiner, and A. Röll, J. Chem. Phys. **94,** 1474, 1991
16. P. Debye, Annalen der Physik **46, (351),** Eq. (11), p. 816 (1915).
17. (a). U. Bonse and M. Hart, Appl. Phys. Lett. **6,** 155-156 (1965).
    (b). U. Bonse and M. Hart, in *Small-Angle X-Ray Scattering,* edited by H. Brumberger, pp. 121-130. Gordon and Breach, New York, London, 1967.
18. G. D. Wignall, J. S. Lin, and S. Spooner, J. Appl. Cryst. **23,** 241-245.
19. C. M. Sorenson and C. Oh. Unpublished research.
20 I. S. Gradshtein and I. M. Ryzhik, *Tables of Integrals, Series, and Products,* Academic Press, New York, 1965, p. 490, Integrals 5 and 6.
21 M. Kotlarchyk and S. H. Chen, J. Chem. Phys. **75,** 2461 (1983).
22. A Venkatraman, L. T. Fan, and W. P. Walawender, unpublished research.

INTERFACE CHARACTERIZATION BY SAS INTENSITY TAILS

Salvino CICCARIELLO
Dipartimento di Fisica 'G. Galilei' and I.N.F.M.,
via Marzolo 8, I-35131 Padova, Italy

The relations existing between the coefficients of the asymptotic expansion of the small angle scattered intensity and some geometrical features of samples' interfaces are reviewed. They have usefully been applied to characterize real samples as coated porous silicas, zeolite-supported metallic catalysts, demixing glasses. A further application to a polymer sample consisting of methylmethacrylate and polybutadiene is reported.

1 Introduction

The momentum transfers (h) explored in any small-angle scattering (SAS) experiment fall in the interval $[0.0001, 1]\text{Å}^{-1}$. Hence, the spatial resolution Δ attained by any SAS experiment cannot exceed 5Å. This means that, from the SAS point of view, a *real* sample is indistinguishable from any *hypothetical* sample characterized by an electron (or scattering-length) density-fluctuation, obtained by averaging that of the real sample over spherical regions with a diameter not exceeding 5Å, Hence, the quantities really measurable by SAS have to be left nearly invariant by the former average procedure. This condition is fairly met by samples which consist of homogeneous phases whose interfaces are separated by distances much larger than 5Å and whose curvature radii are also larger than 5Å. In these conditions, in fact, it becomes unimportant the fact that real interfaces are diffuse regions rather than geometrical surfaces and the area of the interfaces' area becomes a measurable quantity. Moreover, the particle shapes/sizes also could be observed within the reported uncertainty[1–3]. Another interesting class of samples refers to the case of power-law behaved scattered intensities[4,5]. In these cases, the observable quantity is the power law exponent $(-\gamma)$. Concerning the sample's texture it results that the former conditions on the sample interfaces are no longer true. However, they are violated in such a way that, depending on whether it is observed $\gamma < 3$ or $3 < \gamma \leq 4$, the amount of matter or the area of the interface surface, falling inside any sphere with diameter L, scale as L^{γ} or as $L^{6-\gamma}$, respectively.

In the following we shall confine our attention to the first class of samples. In particular, our first task will be that of reminding the conditions which have to be met for the asymptotic analysis of the scattered intensity to make sense[6]. Following Debye and Bueche[1], $\eta(\mathbf{r})$, the electron density fluctuation of a sample, will be written as

$$\eta(\mathbf{r}) = \eta_{D}(\mathbf{r}) + \nu_{D}(\mathbf{r}) \tag{1.1}$$

where $\eta_D(\mathbf{r})$ is a discrete valued function whose values will be denoted by $\eta_1, \eta_2, .., \eta_N$. The regions where $\eta_D(\mathbf{r}) = \eta_j$ define the jth phase of the sample. In this way, the number of the different phases will be N. Decomposition (1.1) implies that $\eta_D(\mathbf{r})$ and $\nu_D(\mathbf{r})$ have finite discontinuities, with opposite sign, in the same space points. The discontinuities are located on geometrical surfaces which will depict the sample's interfaces. If $\eta_D(\mathbf{r})$ approximates $\eta(\mathbf{r})$ rather closely, then $\nu_D(\mathbf{r})$ will fluctuate around zero and it will be everywhere rather small except in the sample's regions located around the discontinuities. These regions, however, are rather narrow and their typical width will be denoted by δ_{sp}. Reminding that the scattered intensity is simply given by the square modulus of the Fourier transform (FT) of the sample electron density-fluctuation and confining ourselves to isotropic samples, from eq. (1.1) it follows that[6]

$$I(h) = \langle|\tilde{\eta}(\mathbf{h})|^2\rangle = I_D(h) + I_{bck}(h) \tag{1.2}$$

with

$$I_D(h) \equiv \langle|\tilde{\eta}_D(\mathbf{h})|^2\rangle, \tag{1.3}$$

$$I_{bck}(h) \equiv \langle|\tilde{\nu}_D(\mathbf{h})|^2\rangle \; + \; 2\langle\Re e\big[\tilde{\nu}_D(\mathbf{h})\overline{\tilde{\eta}_D}(\mathbf{h})\big]\rangle, \tag{1.4}$$

where the tilde and the bar over a function symbol respectively denote the FT and the complex conjugate, while the angular brackets denote the angular average over all the possible direction of $\mathbf{h}$. Owing to the properties of $\nu_D(\mathbf{r})$ noticed above and to those of the FT, it results[6,7] that $\tilde{\nu}_D(\mathbf{h}) \approx 0$ for the h-values such that $h\delta_{sp} << 2\pi$, while the discontinuity of $\nu_D(\mathbf{r})$ implies that $\langle|\tilde{\nu}_D(\mathbf{h})|^2\rangle \propto h^{-4}$ in the h-range: $h\delta_{sp} >> 2\pi$. On the other hand, if the (typical) minimal distance between the sample's interfaces is denoted by D_{min}, the discontinuous behaviour of $\eta_D(\mathbf{r})$ implies that $\langle|\tilde{\eta}_D(\mathbf{h})|^2\rangle \propto h^{-4}$ for the h's such that $hD_{min} >> 2\pi$. By these results and provided that $D_{min} >> \delta_{sp}$, the h-range can be split into three parts: the inner one where $hD_{min} < 1$, the intermediate part where $hD_{min} > 1$ and $h\delta_{sp} < 2\pi$ and the outer part where $h\delta_{sp} > 2\pi$. Since $\delta_{sp} \leq 5$Å, the last region falls outside the h-range explored in SAS experiments and can be ignored. In the inner part, $I_{bck}(h)$ is quite negligible, while in the intermediate part it should satisfactorily be described by the two parameter expression

$$I_{bck}(h) = \beta_0 + \beta_1 h^2, \tag{1.5}$$

where β_0 and β_1 must be determined by a best-fit procedure. Moreover, when $D_{min} > 50$Å, the intermediate region is sufficiently wide to make $hD_{min} > 2\pi$ fulfilled in its outer part. It is now reminded that, from the analysis of the intensities relevant to some particular particle shapes as concentric spheres, coaxial cylinders and tori, it has been found that the asymptotic expansion is already quite satisfactory in the region $hD_{min} > 2\pi$ (instead of $hD_{min} >> 2\pi$). Then, in the outer part of the intermediate region it is possible to use the asymptotic expression of $I_D(h)$ and, in this way, to determine the parameters present in this expression and related to the sample's interface. These relations will be reported in the next section, while their application to get structural information on real samples will be illustrated in sect. 3. The conclusive remarks are reported in sect. 4.

2 Intensity asymptotic expansion

First it is recalled that the scattered intensity is the FT of the so called sample correlation function[1] $\gamma_D(r)$, i.e.

$$I_D(h) = 4\pi < \eta_D^2 > V/h \int_0^\infty r\gamma_D(r)\sin(hr)dr, \tag{2.1}$$

where

$$\gamma_{_D}(r) \equiv \left(4\pi < \eta_{_D}^2 > V\right)^{-1} \int d\hat{\omega} \int_{R^3} \eta_D(\mathbf{r}_1)\eta_D(r\hat{\omega} + \mathbf{r}_1)dv_1. \tag{2.2}$$

and

$$\eta_{_D}(\mathbf{r}) \equiv \sum_{i=1}^{N}(n_i - < n >)\rho_i(\mathbf{r}). \tag{2.3}$$

In these equations, $\rho_i(\mathbf{r})$ is defined as being equal to 1 inside the region occupied by the ith phase and to zero elsewhere, n_i and $< n >$ respectively denote the electron density of phase i and the average electron density of the sample, whose volume is V, and

$$< \eta_{_D}^2 > = \sum_{1 \le i < j \le N} (n_i - n_j)^2 \phi_i \phi_j, \tag{2.4}$$

where ϕ_i is the ith phase volume fraction. It is easy to show that[3,8–10]

$$\gamma_D(r) = \sum_{1 \le i < j \le N} \eta_{ij}\gamma_{ij}(r), \tag{2.5}$$

where

$$\gamma_{ij}(r) \equiv 1 - P_{ij}(r)/(\phi_i\phi_j), \tag{2.6}$$

$$\eta_{ij} \equiv (n_i - n_j)^2 \phi_i \phi_j / < \eta_{_D}^2 >, \tag{2.7}$$

and

$$P_{ij}(r) \equiv (4\pi V)^{-1} \int d\hat{\omega} \int_V \rho_i(\mathbf{r}_1)\rho_j(\mathbf{r}_1 + r\hat{\omega})dv_1. \tag{2.8}$$

Eq.s (2.6) and (2.8) make it evident that the correlation function (CF) depends only on the geometry of the phases. According to a general theorem on FT[11–13], the asymptotic behaviour of $I_{_D}(h)$ is simply determined by the discontinuities of the derivatives of $\gamma_{_D}(r)$. In particular, a non-null value of each odd derivative of the CF, at the origin, can be considered as a discontinuity. Thus, the derivative of order $(2m-1)$ contributes to the $I_{_D}(h)$ asymptotic expansion with the monotonically decreasing term

$$8m\pi V < \eta_{_D}^2 > (-1)^m \frac{\gamma_{_D}^{(2m-1)}(0^+)}{h^{2(m+1)}}, \quad m = 1, 2... \tag{2.9}$$

The famous Porod law results from that fact that

$$\gamma_{_D}^{(1)}(0^+) = -\sum_{1 \le i < j \le N} \frac{(n_i - n_j)^2 S_{ij}}{4V < \eta_{_D}^2 >}, \tag{2.10}$$

where S_{ij} is the area of the interface between phases i and j. For the 3rd derivative, Kirste and Porod[14] showed that

$$\gamma_{_D}^{(3)}(0^+) = \sum_{1 \le i < j \le N} \frac{(n_i - n_j)^2}{4V < \eta_{_D}^2 >} \int_{S_{ij}} dS \frac{2(R_m + R_M)^2 + (R_m - R_M)^2}{(4R_m R_M)^2}, \tag{2.11}$$

where R_m and R_M are the principal curvature radii. From this expression, valid only in the case of smooth surfaces, it is evident that $\gamma_D^{(3)}(0^+) \geq 0$. However, in presence of vertices or curvilinear edges, a correction to (2.11) has to be accounted for and this correction can make $\gamma_D^{(3)}(0^+)$ negative[15,16]. The integral expressions of $\gamma_D^{(5)}(0^+)$[17] and of $\gamma_D^{(7)}(0^+)$[18] are also known in the case of smooth surfaces. Moreover, it has been shown that the interfaces' smoothness ensures that all the even derivative values of the CF are null at the origin, i.e. $\gamma_D^{(2m)}(0^+) = 0$ for m=1,2,.. These last results, however, can hardly be used in the practice, since they require an accuracy in the intensity measurement far beyond the one experimentally achievable. In fact, a practical application is possible only for the sum-rule related to the $\gamma_D^{(2)}(0^+)$ value[19]. Thus, in the following, it will be assumed that the monotonically decreasing terms, present in the asymptotic expansion of $I_D(h)$, are simply given by

$$\frac{\mathcal{P}}{h^4} + \frac{\mathcal{K}_{KP}}{h^6} \tag{2.12a}$$

where

$$\mathcal{P} \equiv 2\pi \sum_{1 \leq i < j \leq N} (n_i - n_j)^2 S_{ij} \tag{2.12b}$$

$$\mathcal{K}_{KP} \equiv 2\pi \sum_{1 \leq i < j \leq N} (n_i - n_j)^2 S_{ij} \left[3\langle H^2 \rangle_{S_{ij}} - \langle \kappa \rangle_{S_{ij}}\right]. \tag{2.12c}$$

In the last relation, H and κ denote the mean and the gaussian curvatures of the interfaces, while the angular brackets remind that the quantities are the average values evaluated throughout the S_{ij} interface. Moreover, the quantity inside square brackets refers to the smooth contribution only.

By developing the procedure, first considered by Schmidt and Wu[13,17], it has been shown[20] that: when the interfaces are parallel at a relative distance δ, the second derivative of the CF shows a finite or a logarithmic singularity at $r = \delta$. More definitely, it is reminded that two portions Σ_1 and Σ_2 of a generic surface S are said to be parallel when, considering Σ_1 as a wave-front portion, Σ_2 turns out to be the wave-front after that the first has travelled a distance δ. By the Huyghens principle, Σ_2 is the surface resulting from the envelope of the spheres of radius δ centered on Σ_1. Two cases are now possible: the spheres lie on one side of Σ_2 or the spheres cut Σ_2. In the first case, Σ_1 and Σ_2 are said to be elliptically parallel, and it results that $\gamma_D^{(2)}(r)$ has a finite discontinuity at $r = \delta$. In the second case, Σ_1 and Σ_2 are said to be hyperbolically parallel and $\gamma_D^{(2)}(r)$ has a logarithmic discontinuity at $r = \delta$. The two discontinuities respectively contribute to the asymptotic expansion of $I_D(h)$ with

$$\frac{\mathcal{A}\cos(h\delta)}{h^4} \quad \text{or} \quad \frac{\mathcal{B}\sin(h\delta)}{h^4}, \tag{2.13}$$

and the general integral expressions of coefficients $\mathcal{A}$ and $\mathcal{B}$ are given by eq.s (3.13) and (3.14) of ref. [20]. The former geometric conditions imply that oscillatory deviations from a horizontal behaviour have generally to be expected in the Porod plot of the scattered intensity. Of course, the general asymptotic expansion of $I_D(h)$

can contain further contributions behaving as

$$\frac{\mathcal{C}\cos(h\delta+\varphi)}{h^{4+\alpha}},$$

with $\alpha > 0$. For instance, a contribution of this kind with $\alpha = 0.5$ is known to exist in the case of right circular cylinders[13]. However, it is presently unknown how $\mathcal{C}$, α and φ are related to the geometry of the interface. For this reason and for the fact that these contributions, as h increases, fade in the Porod plot owing to their faster decrease, it will be assumed that the expression

$$I_{asy,D}(h) \approx \frac{\mathcal{P}}{h^4} + \frac{\mathcal{K}_{KP}}{h^6} + \sum_l \frac{\mathcal{A}_l\cos(h\delta_l)}{h^4} + \sum_m \frac{\mathcal{B}_m\sin(h\delta_m)}{h^4} \tag{2.14}$$

is asymptotically quite accurate. For later reference, eq. (2.14) will now be specified in the case of coated porous materials as well as in the cases of monodisperse samples consisting of particles having the form of two concentric spheres (hollow spheres) or two coaxial cylinders (hollow cylinders), respectively. For the first kind of samples, under the assumption that the pore size (D) is much larger than δ, the coating film thickness, eq. (2.14) can be written (neglecting the $\mathrm{O}(h^{-6})$ contribution) as[21]

$$I_{asy,D}(h) \approx \frac{2\pi}{h^4}\left[\hat{\mathcal{S}} - \mathcal{A}\cos(h\delta)\right] \tag{2.15a}$$

with

$$\hat{\mathcal{S}} \equiv \left[(n_1-n_3)^2 + (n_2-n_3)^2\right]S, \tag{2.15b}$$

$$\mathcal{A} \equiv \left[(n_1-n_3)^2 + (n_2-n_3)^2 - (n_1-n_2)^2\right]S, \tag{2.15c}$$

where S denotes the area of the porous material, while n_1, $n_2(=0)$ and n_3 are respectively the electron density of the support, of the voids and of the film. For hollow spheres, after denoting by n_1, n_3 and n_2 the electron densities of the inner sphere, of the spherical shell and of the external medium and by R_1 and R_2 the inner and outer radius, the oscillatory contributions turn out to be given by[22]

$$\frac{8\pi^2 N}{h^4}\Bigg[R_1^2(n_1-n_3)^2\cos(2R_1h) + R_2^2(n_2-n_3)^2\cos(2R_2h) +$$

$$R_1R_2\Phi(n_1,n_2;n_3)\Big\{\cos\big((R_1-R_2)h\big) + \cos\big((R_1+R_2)h\big)\Big\}\Bigg] \tag{2.16a}$$

where

$$\Phi(n_1,n_2;n_3) \equiv (n_1-n_2)^2 - (n_1-n_3)^2 - (n_2-n_3)^2 \tag{2.16b}$$

while N denotes the number of the particles present in the sample. For hollow cylinders, characterized by the radii $R_1 < R_2$ and by the heights $2H_1 < 2H_2$, the explicit expression of the oscillatory contributions is[22]

$$\frac{4\pi^2 N}{h^4}\Big[-(n_1-n_3)^2\left[R_1^2\cos(2H_1h) + 2R_1H_1\sin(2R_1h)\right] -$$

$$(n_2-n_3)^2\left[R_2^2\cos(2H_2h)+2R_2H_2\sin(2R_2h)\right]+$$
$$\Phi(n_1,n_2;n_3)R_1^2\left[\cos\left((H_2-H_1)h\right)-\cos\left((H_2+H_1)h\right)\right]+$$
$$2\Phi(n_1,n_2;n_3)H_1\sqrt{R_1R_2}\left[\cos\left((R_2-R_1)h\right)-\sin\left((R_2+R_1)h\right)\right]\Bigg] \quad (2.17)$$

where the meaning of the remaining symbols is as in eq. (2.16). It is also noted that by setting $R_1=0$ in eq.s (2.16a) and (2.17), the resulting expressions yield the oscillatory asymptotic terms for monodisperse spheres and cylinders, respectively.

3 Numerical results

By best-fitting the expression of $I_{asy,D}(h)+I_{bck}(h)$, resulting from (2.14) and (1.5) to the observed pin-hole intensity in the outer part of the explored h-range, in principle it would be possible to determine all the parameters present in the former equations. For the procedure to work satisfactorily and unambiguously, it is necessary that the intensities are accurately measured in the outer h-range. To this aim, intensity measurements at fixed (and rather high) number of counts are highly desirable.

So far, the pin-hole intensities[23] scattered by some coated porous silicas, have been analyzed by using eq. (2.15a) and eq. (1.5), with β_1 set equal to zero in order to reduce the parameter number. The analysis results are reported in ref. [21]. They look rather satisfactory, despite the fact that the considered h-range is weakly asymptotic, in the sense that with the resulting film thickness values the condition $h\delta>2\pi$ turned out to be fulfilled only at the border of the explored h-range in the case of the thickest film, only.

However, the asymptotic analysis can be performed also with slit intensities. Assuming that the slit can be considered as ideal, i.e. very long and quite narrow, the resulting intensity is

$$J_D(h)=2\int_0^\infty I_D(\sqrt{h^2+t^2})dt \quad (3.1)$$

By substituting here expression (2.14), the asymptotic expansion of the 'infinite-slit' smeared intensity is immediately recovered. This reads

$$J_{asy,D}(h)\approx\frac{\pi\mathcal{P}}{2h^3}+\frac{3\pi\mathcal{K}_{KP}}{8h^5}+\sum_l\frac{\pi\mathcal{A}_l\phi_{cos}(h\delta_l)}{2h^3}+\sum_m\frac{\pi\mathcal{B}_m\phi_{sin}(h\delta_m)}{2h^3} \quad (3.2)$$

where it has been put

$$\phi_{cos}(h\delta)\equiv\frac{4}{\pi}\int_1^\infty\frac{\cos(h\delta t)dt}{t^3\sqrt{t^2-1}}dt \quad (3.3a)$$

and

$$\phi_{sin}(h\delta)\equiv\frac{4}{\pi}\int_1^\infty\frac{\sin(h\delta t)}{t^3\sqrt{t^2-1}}dt \quad (3.3b)$$

The integrals defining the functions $\phi_{cos}(h\delta)$ and $\phi_{sin}(h\delta)$ can be evaluated and then parameterized as explained in ref. [24]. At this point, the asymptotic analysis of infinite-slit intensities becomes straightforward. Analyses of this kind have already been carried through for demixing glass samples[24,25] and for coated porous silicas[21,26] and the results turned out to be satisfactory. The procedure can also be applied to the SAS intensity scattered by supported metallic catalysts. In this case, the contribution to the intensity resulting from the only metallic particles is obtained by the procedure described in ref. [27] and then this contribution is analyzed by the asymptotic procedure. The resulting particle radii turn out to be close to the ones reported in [27].

3.1 An application to a polymer sample

Styrene and butadiene block copolymers have undergone a great deal of experimental and theoretical work in order to understand how the morphology of these materials change as their volume fractions are varied. In particular it has been found that the samples' texture evolves from spheres to cylinders and then to lamellae as the styrene's volume fraction increases[28,29]. It is believed that the same phenomenon occurs for mixtures of methylmethacrylate (MA) and polybutadiene (BD). Provided the particle sizes turn out to be rather large and the polidispersity degree rather small, the texture evolution ought to be observed by SAS. For this reason, Dr. R. Sobry and coworkers of Liège University have collected the small-angle X-ray intensities scattered by a rather large number of these samples. Since their preliminary analyses of these intensities appear to confirm the reported evolution, a more refined analysis has now been undertaken on the basis of the procedure expounded above. The first results obtained for sample MA06BD85, the most dilute in MA among the measured ones, are now reported. The *log-log* plot of the SAXS intensity (○) is shown in fig. 1a, while the corresponding Porod plot is shown in fig. 1b. (This plot refers to an infinite-slit collimation, since the intensities have been collected with a Kratky camera. The finite-slit corrections as well as all the details on the intensity collection and on the numerical analysis will reported in a forthcoming paper[31].) Some peaks and shoulders appear evident in the figures. The h-range where the intensity can be considered asymptotic was chosen on the following basis. Being the first peak observed in fig. 1b likely related to interference effects between different particles[30], it was assumed that the asymptotic region lies beyond the value $h_{min} = 0.034\text{Å}^{-1}$. The monotonically decreasing terms present in (3.2) and the background contribution given by (1.5) were fitted to the observed intensity by choosing different upper bounds (h_{max}) for the fit interval. Fig. 1c shows the corresponding Porod plots for

$$J_{osc}(h) \equiv J^{obs}(h) - \left[\pi\mathcal{P}/2h^3 + 3\pi\mathcal{K}_{KP}/8h^5 + \beta_0 + \beta_1 h^2\right]$$

(where subscript *osc* is a reminder of the fact that, by (1.5) and (3.2), $J_{osc}(h)$ must be fitted with the oscillatory terms present on the right side of (3.2)). The dotted, the continuous and the broken lines refer to $h_{max} = 0.25$, 0.3 and 0.35Å^{-1}, respectively. The figure makes it evident that: i) whatever the h_{max} choice, $J_{osc}(h)$ does not appreciably change in the h-range $0.034 - 0.2\text{Å}^{-1}$; ii) since the oscillations fade quickly away as h increases, a polidispersity in the particle sizes has to be expected; iii) the presence of the noise (unfortunately, the intensity were measured

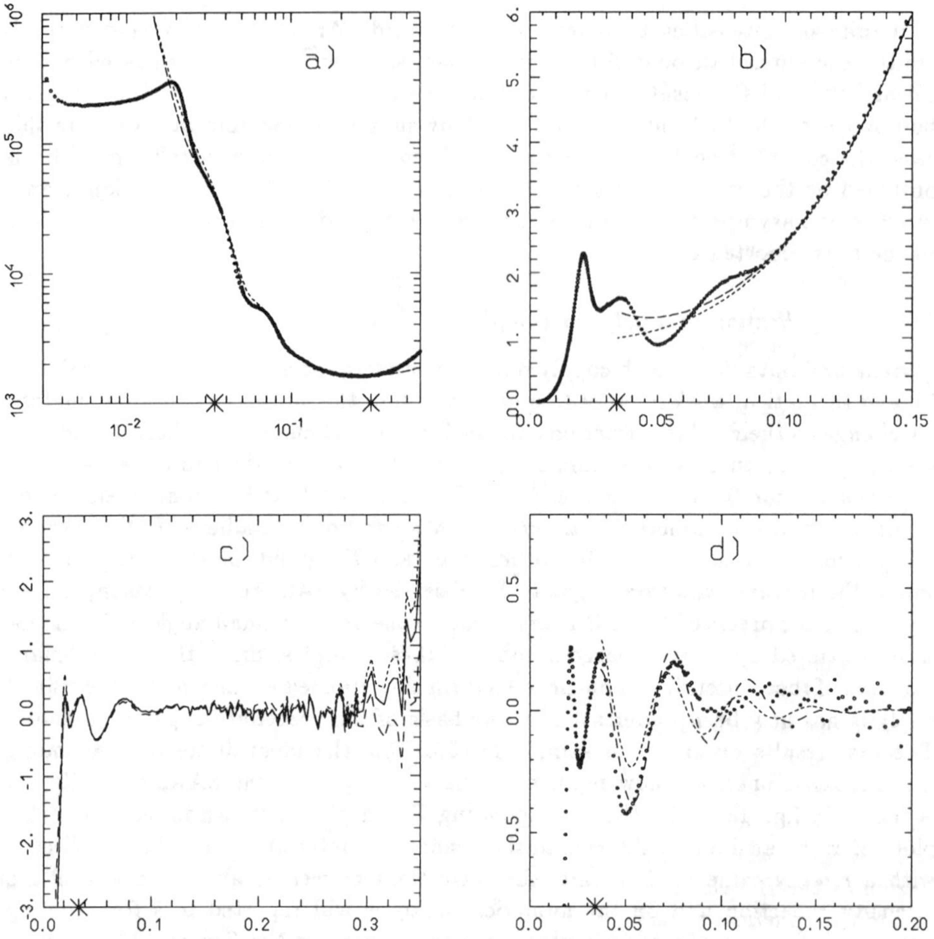

Figure 1: The horizontal axes give the h-values in Å^{-1}. The asterisks denote the lower and, when present, the upper bound of the fit-interval. **a)** Symbols: plot of the observed intensity [in arbitrary units]; dotted line (broken line): best-fitted intensity with a set of polidisperse spheres (polidisperse hollow spheres). **b)** Porod plots of the observed intensity (symbols), of the background plus $\pi\mathcal{P}/2h^3$ (dotted line), of the background plus $\pi\mathcal{P}/2h^3$ plus $3\pi\mathcal{K}_{KP}/8h^5$ (broken line). (Note that the Kirste-Porod contribution becomes appreciable only in the inner part of the best-fitted h-interval.) **c)** Porod plot of $J_{osc}(h)$ [the difference between the observed intensity and the best-fitted monotonic contributions]. Dotted, continuous and broken lines: h_{max} = 0.25Å^{-1}, 0.3Å^{-1} and 0.35Å^{-1}. The three choices are practically equivalent since the difference in the outer h-range is pure noise. **d)** Symbols: $h^3 J_{osc}(h)$ for the case h_{max} = 0.3Å^{-1}. Dotted and broken lines: best-fitted contributions resulting from a set of spheres and from a set of hollow spheres, respectively.

at fixed time) does not allow one to choose among the considered h_{max} values; iv) as a compromise, it was chosen h_{max}=0.3Å^{-1}and the resulting $(J_{osc}(h))$ has been fitted by eq. (2.16a), after converting the latter to the infinite-slit collimation, i.e. by substituting each $cos(h\delta)/h^4$ with $\pi\phi_{cos}(h\delta)/2h^3$ (see eq.s (3.2) and (3.3a)). The final plots are shown in fig. 1d, where symbols o represent $J_{osc}(h)$, while the continuous and the dotted lines are the results obtained with a polidisperse distribution[31] of spheres and of hollow spheres, respectively. Concerning the fit-goodness value, this turned out to be equal to 0.16 after best-fitting the only monotonic and background contributions, and then it reduced to 0.12 and to 0.08 by considering a set of spheres or a set of concentric spheres, respectively. In the case of the spheres the average diameter turns out to be equal to 165Å, while in the case of concentric spheres the outer and inner average resulting diameters are 158Å and 38Å, respectively. In both cases, the size-dispersion is 10%, while the values of $\mathcal{P}$, $\mathcal{K}_{KP}$, β_0 and β_1 are, respectively, 0.595Å^{-3}u, $0.407 \cdot 10^{-3}$Å^{-5}u, 1474u and 1069Å^2u, u denoting an unspecified unit related to the fact that the observed intensity is in relative units. For this reason, it is not possible to obtain from the $\mathcal{P}$ values the specific surface areas. However, some consistency check can be performed. In particular, if one assumes that all the interface comes out from spherical particles, the $\mathcal{K}_{KP}/\mathcal{P}$ ratio, equal to $0.68 \cdot 10^{-3}$Å^{-2} by the best-fit values just reported, can also be determined by the resulting radii's values. In this way, for spheres and for hollow spheres $\mathcal{K}_{KP}/\mathcal{P}$ is respectively equal to $0.18 \cdot 10^{-3}$ and $0.24 \cdot 10^{-3}$Å^{-2}. The comparison with the value reported above shows that the order of magnitude is satisfactory in both cases, though the concentric sphere morphology appears to be more satisfactory. The smaller fraction of the MA in comparison to that of the BD suggests that the particles consist of MA in a BD surrounding. Having found, in the case of concentric spheres, that $\sqrt{8\pi^2 N}(n_2 - n_3) = -0.0050u^{1/2}$Å$^{-5/2}$ and $\sqrt{8\pi^2 N}(n_1 - n_3) = 0.0038u^{1/2}$Å$^{-5/2}$, allows one to conclude that n_2 (the outer-region or the BD electron-density) is smaller than n_3 (the electron density present inside the spherical shell) which in turn is smaller than the inner sphere's electron density (n_1). This result qualitatively can easily be explained. Since the electron density of the MA is known to be higher than that of the BD, then either the inner sphere is made up of pure MA and the intermediate region of MA and BD, or the intermediate region is made up of amorphous MA and the inner sphere of crystalline MA. Finally, it is reminded that electron microscopy confirms the spherical shape as well as the (outer) radius value, since it yielded a diameter of 156Å.

4 Conclusion

The results reported show that the asymptotic analysis of SAS intensity tails yield valuable information on the samples' texture. One might rightly wonder of the advantages of performing the best-fit analyses on the basis of eq.s (2.16) and (2.17) instead of directly using the form-factors of hollow spheres or cylinders. Clearly, when the involved typical distances (D_i) are well resolved, in the sense that they obey the condition $hD_i > 2\pi$ throughout the best-fitted h-range, the results will coincide, because eq.s (2.16) and (2.17) quite accurately describe the form-factors. However, the asymptotic analysis presents the advantage of using simpler analytical expression, and this advantage is particularly evident in the case of cylinders whose exact form-factor expressions require the use of elliptic functions[32]. This conclusion also applies when the smaller distances are weakly resolved. On the contrary, when most of the typical lengths are weakly or even not resolved, the asymptotic analysis fails. However, the best-fit results obtained by using the form-factors cannot be confidently accepted, In fact, the relevant analyses unavoidably considered small h values, where interparticle interference effects could be important. (These considerations are confirmed by the detailed numerical analysis of Pedersen and Schurtenberger[33]. For instance, by looking at the results reported in their figure 7, it appears evident that, with $q_{max} = h_{max} = 0.125\text{Å}^{-1}$, the smaller radius ($R_{min}$) of 10Å they considered cannot be resolved because $h_{max}\delta_{min} = 2h_{max}R_{min} = 4.50 < 2\pi$.) It should also be noted that eq.s (2.16) and (2.17) allow one to appreciate immediately whether an oscillatory contribution can be observed or not. For instance, in the case of rather long cylinders the prefactor $2H_2R_2$ dominates the remaining ones. This conclusion also applies to the case of long bent cylinders, provided the length of the curved parts of cylinders be negligible in comparison to the sum of the linear parts. Finally, it is remarked that, in carrying through the asymptotic analysis, parameters $\mathcal{P}$ and $\mathcal{K}_{KP}$ are treated independently from the radii' or heights' values. In this way, at the end of the numerical analysis, it is possible to check whether the results are consistent or not. The last case takes place when the interface area resulting from the heights and/or radii is larger than that resulting from the $\mathcal{P}$ value. On the contrary, when it results smaller, it can be concluded that only a fraction of the interface arises from the assumed particle shape.

In conclusion, the most delicate point of any asymptotic analysis consists in finding out the asymptotic region through the subtraction of an appropriate background contribution (i.e $I_{bck}(h)$). Finally, it is stressed that disposing of intensities which have been accurately measured in absolute units, a goal easy to achieve by a further small experimental effort, is highly rewarding in order to assess the accuracy of the final results.

Acknowledgments

I am grateful to Dr. Roger Sobry for useful discussions and for the permission of presenting the results on sample MA06BD85. Financial support from MURST through 40% and 60% Funds is gratefully acknowledged.

References

1. P. Debye and A.M. Bueche, *J. Appl. Phys.* **20**, 518 (1949).
2. G. Porod, *Kolloid-Zeit.* **124**, 83 (1951); *ibid.* **125**, 51 and 109, (1952).
3. P. Debye, H.R. Anderson and H. Brumberger, *J. Appl. Phys.* **28**, 679 (1957).
4. P.W. Schmidt, *J. Appl. Cryst.* **24**, 414 (1991).
5. J.E. Martin and A.J. Hurd, *J. Appl. Cryst.* **20**, 61 (1987).
6. S. Ciccariello, J. Goodisman and H. Brumberger, *J. Appl. Cryst.* **20**, 117 (1988).

7. S. Ciccariello, in *X-Ray investigations of Polymer structures*, ed. A. Włochowicz, Proc. SPIE 3095, 142, (1997).
8. A. Peterlin, *Makromol. Chem.* **87**, 152 (1965).
9. J. Goodisman and H. Brumberger, *J. Appl. Cryst.* **4**, 347 (1971).
10. S. Ciccariello and A. Benedetti, *J. Appl. Cryst.* **18**, 219 (1985).
11. A. Erdélyi, *Asymptotic expansions*, (Dover, New York, 1958).
12. D.S. Jones and M. Kline, *J. Math. Phys. (Cambridge, Mass.)* **37**, 1 (1958).
13. P.W. Schmidt, *J. Math. Phys.* **6**, 424 (1965).
14. R. Kirste and G. Porod, *Kolloid-Zeit* **184**, 1 (1962).
15. R. Sobry, J. Ledent and F. Fontaine, *J. Appl. Cryst.* 24, 516 (1991).
16. S. Ciccariello and R. Sobry, *Acta Cryst.* *A***51**, 60 (1995).
17. H. Wu and P.W. Schmidt, *J. Appl. Cryst.* **4**, 224, (1971); *ibid.* **6**, *ibid.* 66 (1971); *ibid.* **7**, 131 (1974).
18. S. Ciccariello, *J. Math. Phys.* **36**, 219 (1995).
19. D. Tchoubar and J. Méring, *J. Appl. Cryst.* **2**, 128 (1969); S. Ciccariello, *J. Appl. Phys.* **56**, 126 (1984).
20. S. Ciccariello, *Phys. Rev* **A44**, 2975 (1991).
21. A. Benedetti and S. Ciccariello, *J. Appl. Cryst.* **27**, 249 (1994).
22. S. Ciccariello, *J. Appl. Cryst.* **24**, 509 (1991).
23. P.W. Schmidt, D. Avnir, D. Levy, A. Höhr, M. Steiner, and A. Röll, *J. Chem. Phys.* **94**, 1474 (1991).
24. S. Ciccariello and A. Benedetti, *J. Appl. Cryst.* **19**, 195 (1986).
25. A. Benedetti, S. Ciccariello and G. Fagherazzi, *Phys. Chem. Glasses* **29**, 173 (1988).
26. A. Benedetti, S. Ciccariello, F. Pinna and G. Strukul, *J. Phys. IV (France)* **3C8**, 499 (1993).
27. H. Brumberger, J. Goodisman, R. Ramaya and S. Ciccariello, *J. Appl. Cryst.* **29**, 526 (1996).
28. B.R.M. Gallot, *Adv. Polym. Sci.* **29**, 85 (1978).
29. E.L. Thomas, D.B. Alward, D.J. Kinning, D.C. Martin, D.L. Handlin and L.J. Fetters, *Macromolecules* **19**, 2197 (1986).
30. J.B. Hayter and J. Penfold, *Mol. Phys.* **42**, 109 (1981).
31. S. Ciccariello and R. Sobry, *in preparation.*
32. S. Ciccariello, *Acta Cryst.* *A***45**, 86 (1989). See also the errata reported in the note at page 511 of ref. [22].
33. J.S. Pedersen and P. Schurtenberger, *J. Appl. Cryst.* **29**, 646 (1996).

SAXS STUDIES ON POROUS MATERIALS

L. PAJĄK

Institute of Physics and Chemistry of Metals, Silesian University, Bankowa 12, 40-007 Katowice, Poland.

Porosity is a common feature of many materials. The pores are very desirable in catalysts, filters, membranes, adsorbents, etc. The knowledge of the specific surface area, pore size, roughness of the pore-solid interface is of the prime importance.

On the whole the structure of the most porous materials is quite complex. In last two decade the characterization of porous materials made a milestone progress owing to the use of fractal geometry.

The methods based on the scattering effect (small-angle X-ray scattering (SAXS), small-angle neutron scattering (SANS) and light scattering (LS)) have already been found very useful for studies of complex, disordered, porous materials.

In the present paper the possibilities of the study of complex, disordered, porous materials by SAXS method are shown. The three types of the structure of such materials are presented.

1. Introduction

The pores are very desirable in materials applied as catalysts, filters, membranes, adsorbents. For these materials the knowledge of the specific surface area, pore size, roughness degree of pore-solid interface is of prime importance. These parameters can be estimated by methods based on the scattering behaviuor like SAXS, SANS and light scattering.

Electron microscopy can yield some general information on the structure of disordered, porous materials but generally the overlapping of the structural feature does not permit precise conclusion to be drawn. Moreover, the preparation of the sample of porous material for electron microscopy investigation without deformation of its primary structure is extremely difficult. Nevertheless, electron microscopy can help in the choice of more realistic model of the structure of porous materials which is applied in the interpretation of SAXS results.

The porosity of materials it means the total volume fraction of material occupied by pores can vary in a very broad range, from much less than 1% up to more than 99% as in the case of aerogels.

2. SAXS from complex, disordered, porous materials

When discussing the problem of SAXS from complex, disordered, porous materials at least three, two-phase, well defined types of pore structure can be distinguished (Fig. 1). These types of pore structure can be effectively described by fractal geometry provided the property of self-similarity is observed. The self-similarity or dilation symmetry means that the objects look the same under transformation of scale of observation [1].

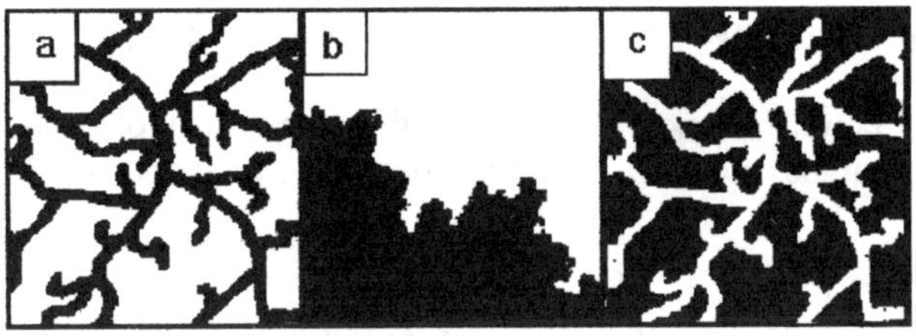

Fig.1 Types of pore structure of complex, diordered materials.

The first type (Fig. 1a) represents materials of high porosity (usually much greater than 50 vol.%). In this case very often the solid phase is an aggregate of primary particles (subunits, blocks) formed during aggregation or gelation processes. The aggregate of mass objects can be described by mass fractal.

The second type of pore structure (Fig.1b) corresponds to rough pore-solid interface. The knowledge of the roughness degree is very important because many chemical and physical processes depend on surface morphology. The region bounded by rough pore-solid interface can be described by surface fractal.

The third type of pore structure (Fig. 1c)corresponds to the network of connected, branched pores which can be described by pore fractal [2]. Mass and pore fractals are sometimes called "volume fractals", because their fractal properties extend through the fractal object. For surface fractal the fractal properties are met only in the interface region.

At present it is generally accepted that the fractal models of pore structure describe the real structure better than the classical models which assume the existence of macro-, meso- and micropores of simple geometry.

The mass M of mass fractal object scales with size R and mass fractal dimension D_m by the relation

$$M = M_0 R^{D_m} \quad (1)$$

where M_o and D_m are constants and the values of mass fractal dimension lie in the interval

$$1 < D_m < 3. \quad (2)$$

The surface area A of surface fractal object which is bounded by a rough, fractal surface scales with size R and a surface fractal dimension D_s by relation

$$A = A_0 R^{D_s} \quad (3)$$

where A_o and D_s are constants and the values of D_s lie in the interval

$$2 < D_s < 3. \quad (4)$$

Self-similarity of real objects can be satisfied only in statistical sense and real objects can reveal fractal properties over a limited range of length

$$a < l < \xi. \quad (5)$$

Ratio ξ/a can be regarded as the range of fractality. Is is often postulated that physical fractals ought to show a fractal properties over at least one order of magnitude for the ratio ξ/a.

The SAXS intensity from fractal objects has a simple power-law form [3,4]

$$I(q) = I_0 q^{-\alpha} \quad (6)$$

where I_o and α are constants, $q=(4\pi\sin\Theta)/\lambda$, 2θ is scattering angle and λ is the X-ray wavelength.

The values of power-law exponent α (and thus the values of mass and/or surface fractal dimension) can be calculated from the slope of the linear parts of log $I(q)$ vs. log q plots (Fig. 2).

For mass fractal

$$\alpha=D_m \quad \text{so} \quad 1<\alpha<3 \quad \text{since} \quad 1<D_m<3, \tag{7}$$

whereas for surface fractals

$$\alpha=6-D_s \quad \text{so} \quad 3<\alpha<4 \quad \text{since} \quad 2<D_s<3\,. \tag{8}$$

The values of exponent α for mass and surface fractals don't overlap. Thus from the value of α it is possible to distinguish whether the structure of scatterer can be described by a mass or a surface fractal. Above consideration are valid when scatterers are of narrow, "bell" shape polydispersity. The polydispersity of power-law type considerably affects the value of exponent α [4].

3. Experimental

SAXS studies were performed on silica aerogels and electrodeposited copper foil. Fig.2 shows examples of SAXS curves from porous materials with different fractal morphology. Mass and surface fractality (Fig.2 a,b) was observed for silica aerogels obtained in a single-step base catalyzed process through hydrolysis of TEOS (TetraEtoxyOrtoSilane)

$$Si(OC_2H_5)_4 + 4H_2O \rightarrow Si(OH)_4 + 4\ C_2H_5OH \tag{9}$$

and polycondensation of silicic acid

$$n\ Si(OH)_4 \rightarrow (SiO_2)_n + 2n\ H_2O \tag{10}$$

Wet alcogels were dried in supercritical conditions. Both presented aerogels (Fig. 2 a, b) were obtained at the same water/TEOS molar ratio which was equal to 2 but at different concentration of TEOS in ethanol solution which was equal to 5 and 60 vol.% for the sample with mass and surface fractality respectively [5,6].

4. Discussion

At low concentration of TEOS in ethanol solution (5 vol.%) the hydrolysis is complete and polycondensation leads to formation of branched, polymeric type structure which results in aerogel with mass fractality and mass fractal dimension D_m equal to 2.30 (Fig.2a).

At very high concentration of TEOS in ethanol solution (60 vol.%) the complete hydrolysis of TEOS is less probable what means that reacting monomers and clusters have unhydrolyzed sites. In this case polycondensation is not complete leading to the growth of the rough colloidal particles with surface fractal dimension D_s=2.8 (Fig. 2b).

To my best knowledge up to now there are no generally accepted example of porous material with pore fractal morphology. Fig. 2c shows SAXS curve from electrodeposited Cu foil obtained from ethylenediamine electrolyte in galvanostatic conditions at current density 0.07 A/dm^2. The pore structure of this sample can be assigned to pore fractal morphology, but unfortunately up to now these results are not verified by the other independent method.

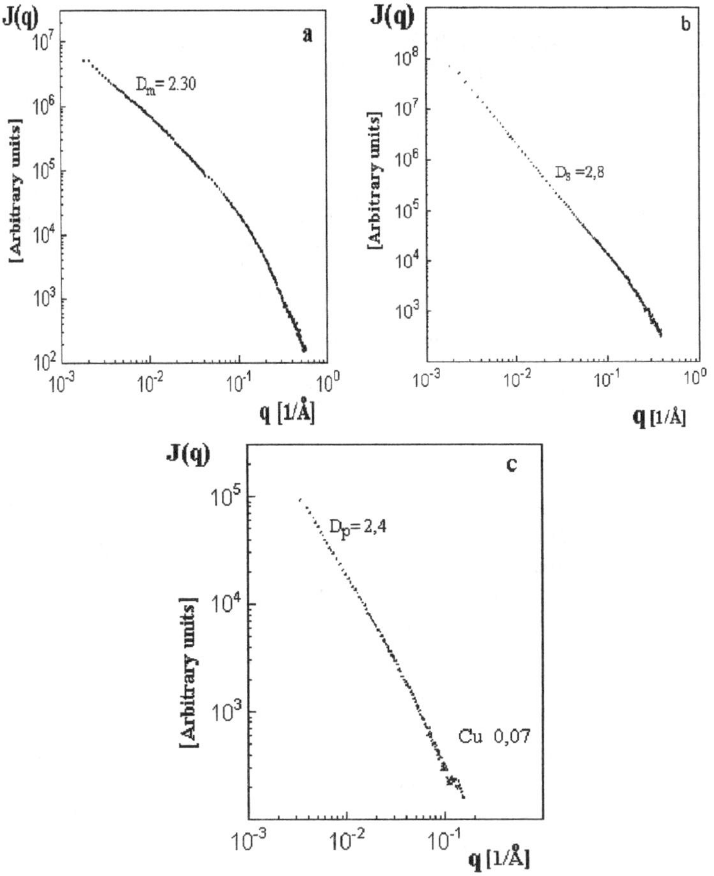

Fig. 2 Examples of SAXS curves from porous materials with different fractal morphology : a) mass fractal, b) surface fractal, c) pore fractal

From presented results it can be concluded that SAXS method can be very useful in the studies on comlex, disordered and porous materials.

[1] B.B. Mandelbrot, *The Fractal Geometry of Nature,* New York : Freeman (1983)
[2] P. Pfeifer, M. Obert, In *The Fractal Approach to Heterogeneous Chemistry,* edited by D. Avnir., pp. 11-43. Chichester, England: Wiley (1989)
[3] J.E. Martin, A.J. Hurd, *J.Appl.Cryst.* **20**, 61 (1987)
[4] P.W. Schmidt, *J.Appl.Cryst.* **24**, 414 (1991)
[5] A.B. Jarzębski, L. Pająk, *J. Non-Cryst. Sol.* **204**, 172 (1996)
[6] A.B. Jarzębski, J. Lorenc, L. Pająk, *Langmuir.* **13**, 1280 (1997).

SAXS INVESTIGATION OF THE SILICEOUS MATERIALS WITH SURFACE POLYMER LAYERS

S. PIKUS, A.L. DAWIDOWICZ, E. KOBYLAS, D. WIANOWSKA and S. RADKIEWICZ
Faculty of Chemistry, Maria Curie-Sklodowska University, M. Curie-Sklodowska Square 3, 20-031 Lublin, Poland

The paper presents the results of the SAXS investigation of the materials in which a polymer layer obtained by cross-linking a dextran-polimine mixture is deposited on the surface of controlled porosity glasses (CPGs). The power-law scattering conditions are fulfilled for all investigated samples in a broad range of the vector scattering q values. The slope of logI vs logq plots for the examined materials changes from 3.87 to 4.49. The value exceeding 4 suggests the existence of the diffuse profile of the electron density in the boundary layer (transition layer). The correlation between the amount of dextran or polyimine or cross-linking agent and the thickness of the transition layer is observed. The similar correlation is also observed between the amounts of the polymer layer and values of the specific surface area and the changes of the volume pore size distribution function.

1 Introduction

Among a wide range of sorbents applied as column packings for chromatography, siliceous materials (silica, gels, controlled porosity glasses - CPGs) are the most frequently used. Forming a polymer layer on the surface of siliceous materials is one of the methods for protecting the silica skeleton from dissolution in the alkaline mobile phases as well as eliminating the negative influence of silanol groups on separated (by means of chromatography) molecules e.g. proteins. Siliceous supports with the surface polymer layer composed of cross-linked polysaccharide deserve a special attention [1, 2].

During the past 10 years, considerable progress has been made in the study of disordered systems and small angle X-ray scattering (SAXS) and neutron scattering (SANS) have been found to be especially well suited for studying these systems [3]. The intensity of SAXS from the complex, disordered and porous systems can often be described by the power law scattering equation:

$$\mathbf{I(q) = I_o q^{-\alpha}}, \qquad \text{where } I_o,\ \alpha \text{ are constants and } q = \frac{4\pi \sin\theta}{\lambda}$$

(2θ is the scattering angle, λ is the X-ray wavelength).

From the magnitude of the α exponent very interesting information can be pointed out e.g. a fractal character of porous systems, an existence of the diffuse profile of the electron density in the boundary layer etc.. Besides, the SAXS method allows to estimate also other values characteristic for porous materials e.g. specific surface areas, volume pore size distribution function etc.. Thus, the question appears whether the SAXS method could be useful for examination of so complicated sorbents which are prepared by cross-linking a dextran-polyimine mixture deposited on the surface of controlled porosity glass.

2 Experimental

2.1 Materials

Controlled porosity glass (CPG) was obtained from Vycor glass according to the procedure described elsewhere [4, 5]. The procedure of CPG coverage by means of dextran or dextran polyimine polymer layer was described in detail in the paper [6].

2.2 Small angle X-ray scattering measurements

Measurements were performed on a slit-collimated Kratky camera using a filtered Cu radiation. A scintillation counter and a pulse-height analyzer were used to measure the scattered intensity. The correction of the scattering curves for equal absorption was calculated for all samples based on the absorption coefficients.

3 Results and Discussion

The first columns of Table 1 summarize the amounts of the reagents (calculated per 1g of CPG) applied for the formation of polymer layer on the CPG surface.

Table 1. The amounts of dextran, triethylenetetraamine (TECZA) and diethyleneglicol diglycidyl ether (EDG-2) used for preparing the polymer layer and data obtained from SAXS measurements.

Sample	Dekstran (g)	TECZA (g)	EDG-2 (g)	α	Ruland [9] (Å)	Vonk [8] (Å)	Specific surface area (m^2/g)
III_1	0.1	0.2	0.03	3.8705	-	-	48
III_2	0.1	0.2	0.06	4.0137	-	-	44
III_3	0.1	0.2	0.12	4.0624	3.9	5.943	40
III_4	0.1	0.2	0.24	4.2738	8.9	10.6	36
IV_4	0.3	-	0.24	4.1280	6.4	7.7	41
II_4	0.2	0.1	0.24	4.4080	13.7	13.6	31
I_4	0.25	0.05	0.24	4.4912	14.7	14.6	33
IV_2	0.3	-	0.06	4.0142	6.9	6.4	46
0	CPG not covered by polymer layer			3.8868	-	-	64

Fig. 1 presents only the exemplary scattering curves in the form of log I vs log q plot but

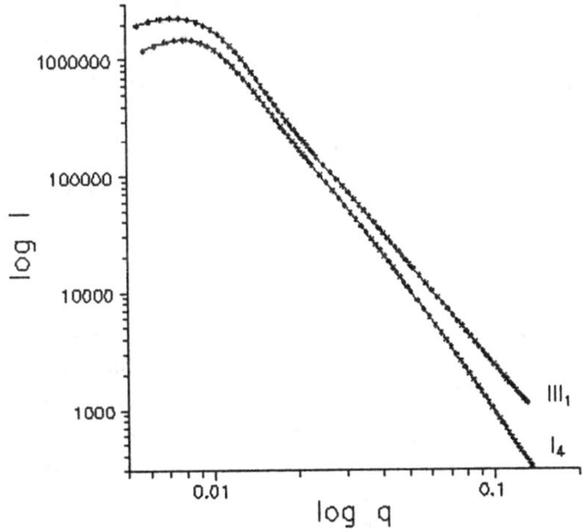

Fig. 1. The log-log plots of the SAXS intensity vs q for the samples I_4 and III_1.

for the all investigated samples the power-law scattering conditions are satisfied in a broad range of q values. The α values for the desmeared SAXS data are collected in the next column of Table 1. As a results, the α values for some materials exceed 4. According to Schmidt interpretation [3, 7] the α value in the range $4 < \alpha < 6$ can result from the diffuse profile of the electron density in the boundary layer. Taking this into consideration, a transition layer in which the electron density continuously (not sharply) decreases to zero has to exist in the polymer layer deposited on the CPGs surface. There are two methods for the calculation of the transition layer thickness - Vonk's [8] and Ruland's [9]. Fig.2 presents the exemplary run of the Ruland plot ($I(q)*q$ vs $1/q^2$) which is typical for the investigated samples.

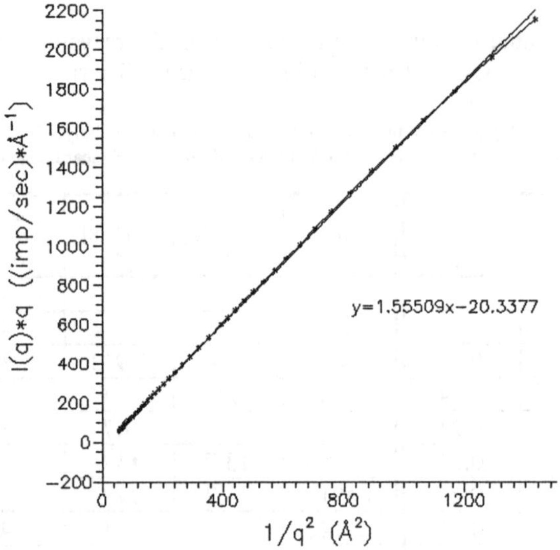

Fig. 2. The Ruland plot $I(q)*q$ versus $1/q^2$ for sample III_4.

The excellent linearity of the presented dependence is noticeable. Thus, the precise value of the transition layer thickness can be estimated from the SAXS data (see next columns of Table1).The thicknesses of the transition layers calculated by two independent methods are very similar in the case of the same material. There can be also seen a correlation between the amount of chemicals (especially dextran or EDG-2) used for the surface layer formation and the thickness of the transition layer.

Fig.3 presents the Porod plot for the chosen samples. In the q range which satisfies the power-law scattering conditions the dependencies can be described using the following equation:

$$I(q) = k_1/q + k_2/q^3 \qquad I(q) \text{ - slit collimated intensity}$$

Taking the k_2 coefficient value as the classical "Porod value", the specific surface areas were calculated for all samples (the last column of Table 1) [10]. As appears from the presented data the specific surface area of the samples with polymer layer is lower compared to that of the initial CPG. Besides, the specific surface area decreases with the increase of EDG-2 which was used for the polymer layer cross-linking. The pore size distribution function $F_V(D)$ (not presented in this paper) for the initial support (CPG) shows a main peak with the maximum close to 400Å and small peak with the maximum at 200Å. After polymer layer formation the position of main $F_V(D)$ maximum shifts towards the lower values of D whereas the second small one disappears.

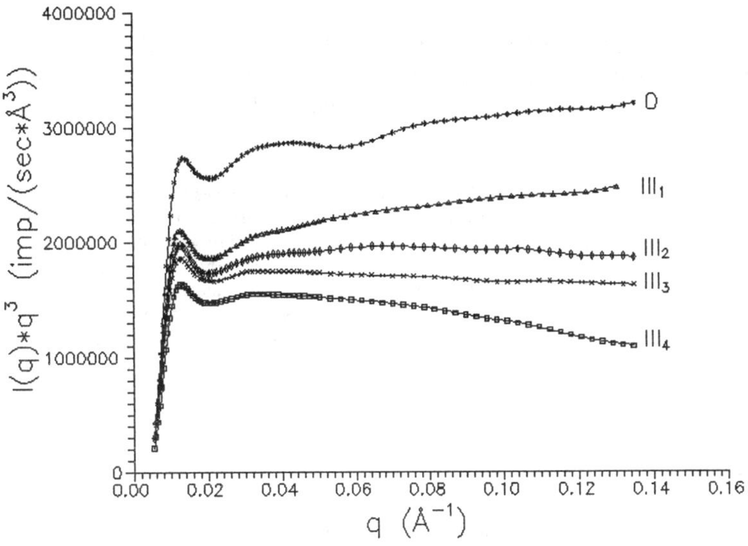

Fig.3. The "Porod value" curve for series samples III_1-III_4 and pure glass "0".

4 Conclusions

The obtained results clearly show that the SAXS method is well suited for studying siliceous supports with the surface polymer layer, because a few important parameters which characterize the investigated materials can be estimated by this method. Especially interesting information results from the magnitude of the α exponent. For many investigated samples the value of α exponent is greater than 4 and hence the thickness of the translation layer on the surface of the polymer can be calculated. The thickness of such a layer can be a very important value for characterization of the chromatographic and adsorption features of the support materials with such a layer. The SAXS method allows additionally to show the changes of the porous structure (specific surface areas, pore size distribution function) resulting from the polymer layer formation on the surface of siliceous supports.

References

1. M. Petro, D. Berek, I. Novak, Reactive Polymers, **23**, 173 (1994).
2. F.E. Regnier, US Pat., 3 983 299 (1976).
3. P.W. Schmidt, NATO ASI Ser., **C 451**, 1 (1995).
4. W. Haller, J. Phys. Chem. **42**, 686 (1965).
5. A.L. Dawidowicz, Polish J. Chem. **66**, 313 (1992).
6. A.L. Dawidowicz, D. Wasilewska, S. Radkiewicz, Chromatographia **42**, 49 (1996).
7. P.W. Schmidt, D. Avnir, D. Levy, A. Höhr, M. Steiner and A. Röll, J. Chem. Phys. **94**, 1474 (1991).
8. C.G. Vonk, J. Appl. Cryst. **6**, 81 (1973).
9. W. Ruland, J. Appl. Cryst. **4**, 70 (1971).
10. L. Pajak, S. Pikus, D. Wasilewska, Archiwum nauki o materialach **17**, 259 (1996).

SAXS INVESTIGATION OF THE SILICA GEL WITH CARBON DEPOSIT

S. PIKUS, R. LEBODA, E. KOBYLAS and B. CHARMAS
Faculty of Chemistry, Maria Curie-Sklodowska University, M. Curie-Sklodowska Square 3, 20-031 Lublin, Poland

A series of carbon-silica gels (carbosils) differing in carbon deposit contents were investigated by SAXS and gas adsorption (nitrogen adsorption). For the investigated series of carbosils the satisfaction of power-law scattering conditions in a broad range of values of q was observed. The estimated values of exponent α were changed from 4.6 to 4.3 according to the contents of carbon in the sample. The volume pore size distribution function $F_V(D)$ and specific surface areas were calculated from the SAXS data. A fair agreement between the results obtained from the SAX scattering data and from the adsorption method (BET) was confirmed. The SAXS data suggest that carbon fills some of the pores completely but it does not deposit on the surface of all pores evenly.

1 Introduction

The complex carbon-mineral adsorbents (CMA) have become recently very attractive materials for many analytical and technological processes as well as in many research works [1, 2]. This new type of adsorbents proves to be very useful in adsorption processes on a technical scale e.g. in water and sewage purification, in petroleum derivative products removal from polluted waters etc.. Because of great possibilities of CMA application, determination of carbon deposit morphology and topography is one of the most important tasks.

The SAXS method allows to estimate many different values (specific surface areas, the pore size distribution function etc.) which characterize the investigated samples. The intensity of SAXS from complex, disordered and porous systems can often be described by the power law scattering equation:

$$\mathbf{I(q) = I_o q^{-\alpha}} \quad \text{where } I_o, \alpha \text{ are constants and } q = \frac{4\pi \sin\theta}{\lambda}$$

(2θ is the scattering angle, λ is the X-ray wavelength).

From the magnitude of α exponent very interesting information can be drawn e.g. fractal character of porous systems, existence of the diffuse profile of the electron density in the boundary layer etc. Thus, the SAXS method should be well suited for studying some properties of the CMA materials.

2 Experimental

2.1 Materials

Silica gel Si-60 (Schuchardt München, Germany) was the parent substance used in the studies. Carbon-coated silica gels (carbosils) were obtained according to the procedure described elsewhere [1].

2.2 Small angle X-ray scattering measurements

Measurements were performed on a slit-collimated Kratky camera using a filtered Cu radiation; a scintillation counter and pulse-height analyser were used to measure the scattered intensity. For all samples the absorption coefficient was estimated and the correction of the scattering curves for equal absorption was also made.

3 Results and Discussion

The first three columns of Table 1 give a few characteristic values of investigated carbosils.

Table 1. The values of time coating (h), the amount of the carbon deposit (%) and the data obtained from the SAXS or BET measurements.

Sample	Coating time (h)	C (% wt)	S_{BET} (m/g^2)	S_{SAXS} (m/g^2)	α	Ruland [5] (Å)	Vonk [6] (Å)
Sil	-	-	336	327	4.62	6.02	8.07
Bas1	0.5	0.77	329	355	4.53	5.88	8.10
Bas2	1	4.37	324	366	4.44	5.63	6.75
Bas3	2	14.89	275	365	4.39	4.87	7.23
Bas4	3	20.32	254	301	4.56	5.96	8.78
Bas5	4	26.70	213	236	4.29	3.56	5.39
Bas6	6	35.00	149	163	4.32	6.04	5.51

Fig.1 presents only the exemplary scattering curves in the form of log I vs log q plot, but for all investigated samples the power-law scattering conditions are satisfied in a broad range of q values. The α values for the desmeared SAXS data are collected in sixth column of Table 1. Because, the α values for all the investigated materials exceed 4 the thickness of the transition layer [3, 4] was calculated using two methods - Ruland's [5] and Vonk's [6], (see the last two columns in Table 1).

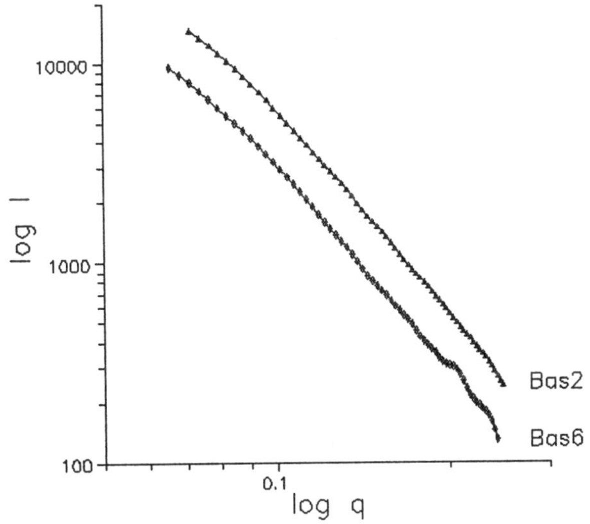

Fig.1. The log-log plots of the SAXS intensity vs q for samples Bas2 and Bas6.

Fig.2 presents the Porod's plot for the investigated samples. In the q range which satisfies the power-law scattering conditions the dependencies can be described using the following equation

$$I(q) = k_1/q + k_2/q^3 \qquad I(q) \text{ - slit collimated intensity}$$

Taking the k_2 coefficient value as the classical "Porod values", the specific surface areas were calculated for all samples using as a standard porous glass with known specific surface area [7] - see fifth column of Table 1.

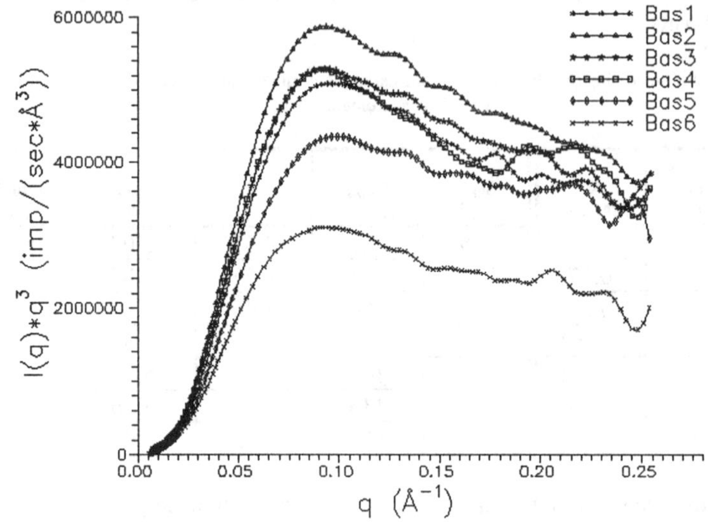

Fig.2. The "Porod value" curves for series samples Bas1-Bas6.

Fig.3 presents the volume pore size distribution function $F_V(D)$ calculated from the SAXS data (a) and $F_V(D)$ from gas adsorption data (b). The $F_V(D)$ calculated from the SAXS data is not normalized for amounts of the pores, in order to improve showing the difference between different $F_V(D)$. As shown in Fig.3 the $F_V(D)$ (SAXS) have almost the same run for all samples; they show one sharp maximum situated at the same value of the diameter (50**Å**). The $F_V(D)$ (BET) show also one maximum for all carbosils (D= 42**Å**). No change of pore sizes with the increase of carbon amounts in the sample, stated by means of SAXS method, indicate the way of carbon deposition on the silica gel. It seems that carbon fills some of the pores completely but it does not deposit on the surface of all pores evenly.

The values of specific surface areas determined using the SAXS and BET methods are similar though some differences are observed (Table 1). The values of specific surface areas determined by BET decrease with the increase of carbon content in the sample while those determined by SAXS increase at first and next decrease for larger content of carbon. The above inconsistency can be due to the fact that the carbon which depositing at the beginning has probably a structure (has some inhomogeneties in electron density distribution) causing additional SAX scattering.

As mentioned above, for all samples (also for pure silica gel) the values of α coefficient exceed 4 (Table 1); it means that the boundary layer between the solid and the air has diffuse profile of electron density (transition layer). The thickness of this transition layer for carbosils and pure silica gel has a value of several **Å**. There is no correlation between the thickness of this layer and the carbon content in the sample (Table 1). As follows from the above the presence of the transition layer results rather from the silica gel surface properties and complete filling of some pores by the deposited carbon does not have any effect on it.

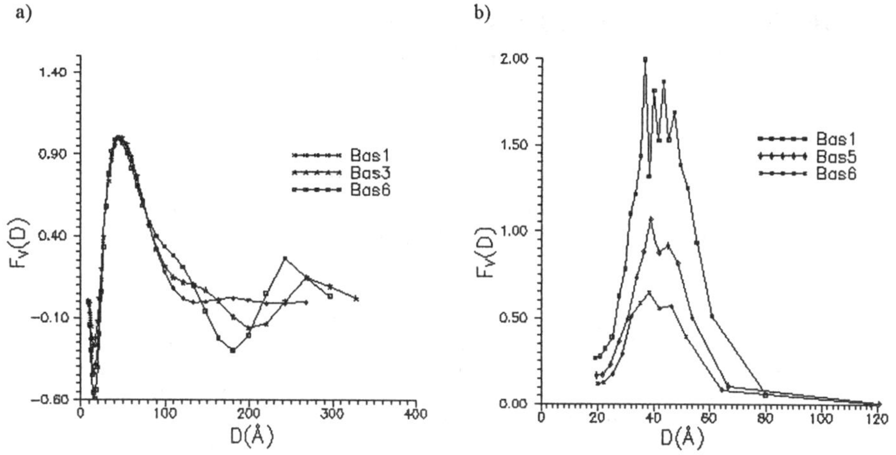

Fig.3. The volume pore size distribution function F_V (D) for some samples
a) SAXS method
b) BET method.

4 Conclusions

The obtained results show clearly that the SAXS method is well suited for studying carbosils, because a few important values which characterize the investigated materials can be estimated by this method. The values of specific surface areas determined using the SAXS methods, the range of q values for which the power-law scattering is obeyed and also the values of the α coefficient give information about structure of the investigated materials. However, especially interesting information can be obtained from the run of the volume pore size distribution function $F_v(D)$. No differences between $F_v(D)$ for pure silica gel and the samples with different amounts of carbon deposit suggest a mosaic topography of carbosils.

Acknowledgments

Financial Support from the State Committee for Scientific Research (KBN, Warsaw). Project No 3 TO9A 03611, is gratefully acknowledged.

References

1. R. Leboda, Mater. Chem. Phys., **34**, 123, (1993).
2. R. Leboda and A. Dabrowski, "Complex Carbon-Mineral Adsorbents: Preparation, Surface Properties and their Modification" in: Adsorption on New and Modified Inorganic Sorbents, Eds A. Dabrowski and V.A. Tertykh, Elsevier, Amsterdam 1996, pp. 115- 147.
3. P.W. Schmidt, NATO ASI Ser., **C 451**, 1 (1995).
4. P.W. Schmidt, D. Avnir, D. Levy, A. Höhr, M. Steiner and A. Röll, J. Chem. Phys., **94**, 1474 (1991).
5. W. Ruland, J. Appl. Cryst., **4**, 70, (1971).
6. C.G. Vonk, J. Appl. Cryst., **6**, 81, (1973).
7. L. Pająk, S. Pikus, D. Wasilewska, Archiwum nauki o materiałach, **17**, 259, (1996).

THE INFLUENCE OF THE CATALYST COMPOSITION ON THE STRUCTURE OF SILICA AEROGELS

B.BIERSKA, L.PAJĄK, E. ŁĄGIEWKA.
Institute of Physics and Chemistry of Metals, University of Silesia, Bankowa 12, 40 - 007 Katowice, Poland

The structure of aerogels can be controlled by the conditions of their preparation and especially by the choice of catalyst. The structure of seven silica aerogels synthesized at different concentration of catalyst (NH_4OH and NH_4F) was investigated by small-angle X-ray scattering (SAXS) method. From SAXS results it can be concluded that fluoride anions promote the formation of more ramified, self-similar, polymeric structure with mass fractal dimension $D_m \approx 2.5$, whereas hydroxyl anions give somewhat less ramified structure ($D_m \approx 2.15$) with narrower range of fractality.

1. Introduction

SAXS was applied for the study of the influence of the chemical composition of the catalysts (NH_4OH, NH_4F) on the structure of silica aerogels. The methods based on the scattering behaviour are particularly useful for the study of porous materials. The SAXS method gives information on the pore structure in the length scale 10 - 1000 **Å**.

The fractal geometry introduced by Mandelbrot [1] was applied for the description of the structure of complex, disordered, porous materials [2]. The fractal geometry is based on self-similarity property which means that the objects looks the same under transformation of scale of observation.

The intensity of SAXS from system of scatterers which exhibit the property of self-similarity can be described by the equation [3]:

$$I(q) = I_0\, q^{-\alpha} \qquad (1)$$

where I_0 and α are constants, $q=(4\pi\sin\theta)/\lambda$ is the scattering vector modulus, 2θ is the scattering angle and λ is the x-ray wavelength.
For mass fractal,

$$\alpha = D_m \quad \text{so} \quad 1 < \alpha < 3 \quad \text{since } 1 < D_m < 3$$

whereas for surface fractals

$$\alpha = 6 - D_s \quad \text{so} \quad 3 < \alpha < 4 \quad \text{since } 2 < D_s < 3$$

where D_m and D_s are mass and surface fractal dimensions respectively.

The correlation between the structure of disordered system which is the aggregate built up from smaller primary particles and the shape of the scattering curve is shown in fig.1. In real space the mass fractal behaviour extends from the size of primary particles a up to the size of aggregate in dilute solution ξ. In the case of greater, connected aggregate, such as these encountered in gel structures ξ can by regarded as a minimum length above which the system appears to be homogeneous. In the reciprocal space the mass fractal behaviour extends in the interval of q values from $1/\xi$ up to $1/a$.

The properties of surface are considered on relatively short lenght scale in the region of pore-solid interface. Thus the information on surface fractals is contained in the outer part of the scattering curve.

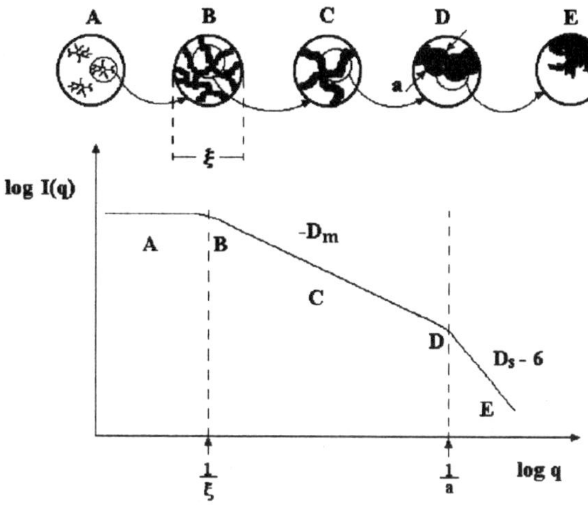

Fig.1. The correlation between the structure of disordered system and the shape of scattering curve.

2.Experimental Conditions

The silica aerogels were obtained in a single-step, base catalyzed process through the hydrolysis of TEOS (TetraEtoxyOrtoSilane)

$$Si(OC_2H_5)_4 + 4H_2O \rightarrow Si(OH)_4 + 4\ C_2H_5OH \qquad (2)$$

and polycondensation of $Si(OH)_4$

$$n\ Si(OH)_4 \rightarrow (SiO_2)_n + 2n\ H_2O \qquad (3)$$

applying different chemical composition of catalyst (tab.1). Water/TEOS molar ratio in hydrolysis reaction was equal to 4, whereas the volumetric concentration of TEOS in ethanol solution was equal to 5 %. Wet alcogels were dried under supercritical conditions.

The silica aerogels preparation and specific surface area S_{BET} measurements (Tab.1) were performed at the Institute of Chemical Engineering in Gliwice, Poland [4].

The SAXS measuments were performed with a Kratky type camera made by Jeol. To cover a wider range of the scattering vector modulus q two set of data were collected. The experimental set-ups corresponded to a geometry of "infinite" slit length. Therefore, a desmeared power-law exponent α corresponding to point collimation of the primary beam was estimated by adding 1 to the actual exponent α^* resulting from the smeared, experimental scattering curve, so

$$\alpha = \alpha^* + 1 \qquad (4)$$

3.Results

The structure of aerogels can be controlled by the conditions of their preparation and especially by the choice of catalyst. Depending on the catalyst composition the gelation of alcogel lasts from 0.23 to about 100 hours. The addition of NH_4OH considerably decreases the gelation time (Tabl.1). The specific surface area S_{BET} for all aerogels is generally similar what can indicate similar structure of aerogels. The SAXS curves for aerogels obtained at constant concentration of NH_4F and NH_4OH and different concentration of the second component are shown in Fig.2 and Fig.3 respectively.

From the slope of the linear parts of *log J(q)* versus *log q* plots the values of mass fractal dimention were estimated. From the slope of the outermost parts of scattering curve ($\alpha=4$) it can be concluded, that the pore-solid interface is smooth [3].

Table 1. Characteristic parameters of the silica aerogels

aerogel	c_{NH_4OH} [mol/dm^3]	c_{NH_4F} [mol/dm^3]	*gelation time [h]*	S_{BET} [m^2/g]
1K05	0	0.0016	1.30	770
2K05	0.0075	0.0016	0.55	796
3K05	0.015	0.0016	0.51	842
4K05	0.030	0.0016	0.45	741
5K05	0.015	0	100.00	757
6K05	0.015	0.0008	0.43	743
7K05	0.015	0.0032	0.23	725

c - concentration of catalyst components
S_{BET} - specific surface area determined by BET method

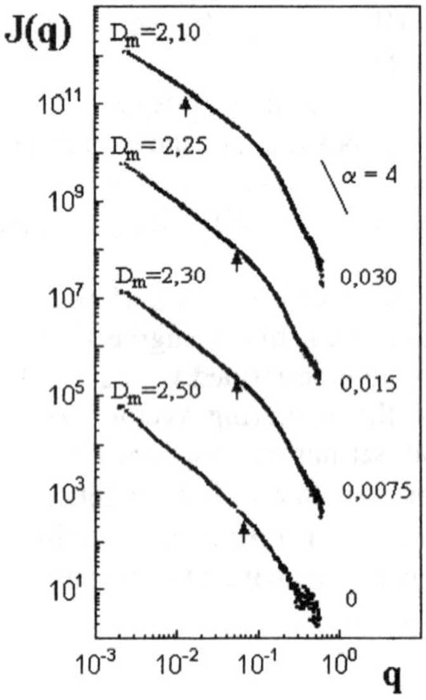

Fig.2. The SAXS curves from the silica aerogels obtained at the constant concentration of NH_4F (0.0016 mol/dm^3) and the different concentrations of NH_4OH (the numbers in the labels are in mol/dm^3).

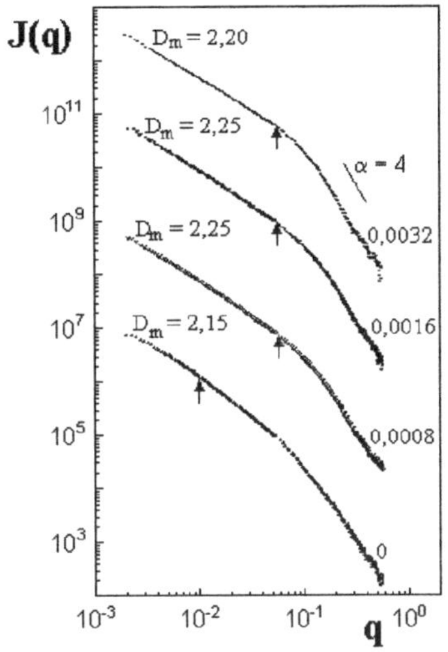

Fig.3. The SAXS curves from the silica aerogels obtained at the constant concentration of NH_4OH (0.015 mol/dm^3) and the different concentrations of NH_4F (the numbers in the labels are in mol/dm^3).

4. Conclusion

- Morphology of silica aerogels depends on the composition of catalysts applied during preparation. Fluoride and hydroxyl anions differently affect the process of structure formation.
- Fluoride anions promote formation of a branched self-similar polymeric structure with mass fractal dimension D_m=2.5. At the constant concentration of fluoride anions the increase in the concentration of hydroxyl anions leads to decrease in D_m what means the formation of a less ramified structure self-similar over a narrower range.
- Application of hydroxyl anions as catalyst results in sparser structure of silica aerogels. Additions of fluoride anions gives more ramified structure but change of concentration of fluoride anions has little effect on the values of D_m and the range of mass fractality.
- From the present and previous [5] works it can be concluded that the water/TEOS molar ratio in hydrolysis reaction and TEOS concentration in ethanol solution have greater influence on the structure of silica aerogels than the composition of catalysts.

[1] B.B.Mandelbrot. *The Fractal Geomerty of Nature*, New York, Freeman (1983)
[2] D.W.Schaefer, *Science* **243** , 1023 (1989)
[3] P.W.Schmidt, *J. Appl. Cryst.* **24**, 414 - 435 (1991).
[4] A.B. Jarzębski, J.Lorenc, Y.I. Aristov, N. Lisitza, *J. Non-Cryst. Sol.* **190**, *198* (1995*)*.
[5] A. B. Jarzębski, J. Lorentz, L. Pająk, *Langmmir* **13**, *1280* (1997).

Texture Analysis

NEW CONCEPTS OF TEXTURE ANANLYSIS AND THEIR CONTIRBUTION TO OUR UNDERSTANDING OF MATERIALS PROPERTIES

H.J. BUNGE
Department of Physical Metallurgy
Technical University of Clausthal

The texture of a polycrystalline material, in the *classical sense*, is defined by the orientation distribution function $f(g)$ of the crystallites irrespective of their local arrangement. The *new concept* of texture considers crystal orientation $g(x)$ as a function of location x in the material. Additionally, also the substructure (lattice defects) may be considered as a function $s(x)$ of location. Several integrals over the aggregate function called generalized textural quantities have also been used. The classical texture function $f(g)$ can be obtained from pole figures measured with a texture goniometer and followed by mathematical pole figure inversion. More advanced experimental equipment uses one-dimensional position sensitive detectors, area detectors, parallel beam technique or location resolved texture measurement. The greatest step forward was, however, achieved by automated *orientation-location scanning* on the basis of Kikuchi diagrams. This technique yields a two-dimensional section of the aggregate function $g(x)$. Model calculations of physical properties of polycrystalline materials show that the classical texture $f(g)$ defines only the Voigt and Reuss bounds which deviate up to 25% from each other. In order to reach a higher accuracy the aggregate function $g(x)$ must be used. This can be done in terms of the "Cluster Model" containing $\sim$ 1000 crystals. Using this model it was shown that grain shape, grain packing, and orientation correlation have a definite influence on the properties of polycrystals. By particular choices of $g(x)$ the whole range between the Voigt-Reuss bounds can be reached.

1 The classical concept of texture

The texture of a polycrystalline material is defined (in the classical sense [1]) by the **O**rientation **D**istribution **F**unction (ODF) of the volume fraction of crystallites having the orientation g of their crystal axes with respect to the chosen sample coordinate system

$$\frac{dV_g/V}{dg} = f(g) \quad ,g = \{\varphi_1, \phi, \varphi_2\} \tag{1}$$

where g may be specified by the Euler angles $\{\varphi_1, \phi, \varphi_2\}$ or in many other ways. The size, shape, and mutual arrangement of crystallites is not taken into consideration. The classical method of texture measurement is pole figure measurement by which the direction dependence of a chosen crystal direction $h \perp (hkl)$ is obtained [1]

$$\frac{dV\,/\,V}{d\Omega} = P_{(hkl)}(\alpha\,\beta) \quad , \quad \begin{array}{ccl} \{\alpha\,,\beta\} & = & \text{sample direction} \\ (hkl) & \perp & \text{crystal direction} \end{array} \tag{2}$$

Pole "figures" (or pole density distribution functions) are two-dimensional projections of the ODF

$$P_{(hkl)}(\alpha\,\beta) = \frac{1}{2\pi} \int\limits_{\{\alpha\beta\}\perp(hkl)} f(g) \cdot d\psi \tag{3}$$

If several pole figures $P_{(hkl)}(\alpha\,\beta)$ for different (hkl) are available then eq(3) can be solved for $f(g)$ (pole figure inversion) [2]. The classical concept of texture is adequate to the classical measuring technique i.e. the texture goniometer based on X-ray or neutron diffraction.

2 New concepts of texture - The aggregate function

The structure of a polycrystalline material can be considered on four different levels:

1. Crystal structure - position of atoms in the unit cell
2. Phase structure - size, shape, arrangement of the phases
3. Grain structure - size, shape, *orientation*, arrangement of crystallites
4. Substructure - all kinds of lattice defects.

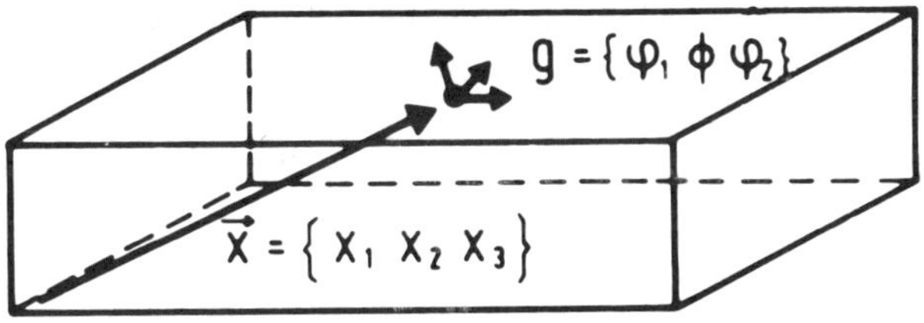

Fig.1
The aggregate function $g^i(x)$ specifies crystal orientation g of the phase i in a small volume element located at x.

If we assume the crystal structures of the phases to be given then the polycrystalline aggregate can be described completely by the aggregate function [3]

$$G(x) = \begin{matrix} g^i(x) & \text{ideal part} \\ s(x) & \text{real part} \end{matrix} \qquad \text{Aggregate function} \tag{4}$$

with:
$x = \{x_1 x_2 x_3\}$ Location coordinates
$g = \{\varphi_1, \phi, \varphi_2\}$ Orientation coordinates
$i = 1 \ldots N$ Number of the phase
$s = \{s_1, s_2, \ldots s_m\}$ Substructure parameters
The ideal part of the aggregate function is illustrated in Fig.1. The real part is extremely complex and will not be considered here in detail. The aggregate function $g^i(x)$ "operates" in the six-dimensional space $\{x_1 x_2 x_3 \varphi_1 \phi \varphi_2\}$. In order to reduce the "dimensionality" several integrals of $g^i(x)$ over this space, called *generalized textural quantities* may then be considered:

1.) **Texture** $f(g)$ (the classical ODF) only *orientation* coordinates

↑

2.) The Mis-Orientation Distribution Function MODF of grain and phase boundaries, Fig.2

$$\frac{dA^{ij}_{\Delta g}/A^{ij}}{d\Delta g} = F^{ij}(\Delta g) \quad , \qquad \Delta g = g_2 \cdot g_1^{-1} \qquad \text{MODF} \tag{5}$$

3.) The local texture - Texture fields

$$\frac{dV_{g,x}/V_x}{dg} = f(g, x) \tag{6}$$

:

Many others

:

↓

4.) **Stereology** ("Gefüge", in the classical sense) only *location* coordinates

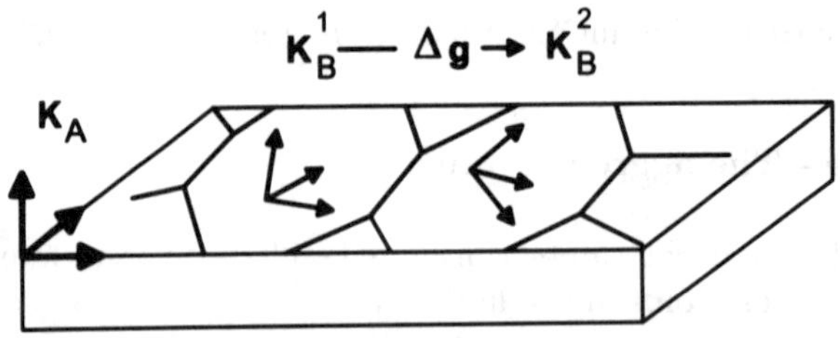

Fig.2
The Mis-Orientation Distribution Function specifies the area fraction of grain (or phase) boundaries having the misorientation $\Delta g = g_2 \cdot g_1^{-1}$.

3 New experimental techniques

Two different types of texture measuring techniques may be distinguished:

1. Single crystal techniques $\longrightarrow g$
2. Polycrystal techniques $\longrightarrow h \perp (hkl)$

as is illustrated schematically in Fig.3. They allow to measure the full orientation g respectively one crystal direction h.

3.1 New polycrystal techniques

Position Sensitive Detector PSD

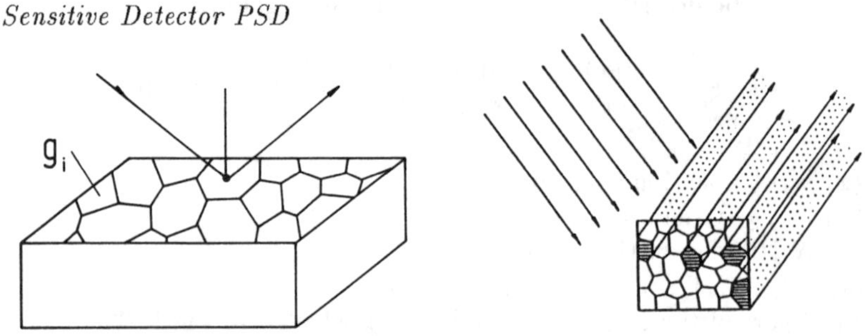

Fig.3 Texture measurement can be done by single crystal or polycrystal techniques.

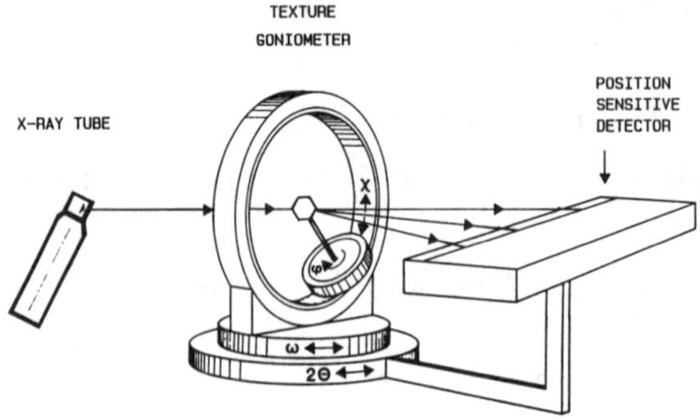

Fig.4
Texture goniometer (schematic) equipped with a one-dimensional position sensitive detector.

The classical texture goniometer is equipped with a single detector e.g. a scintillation detector using a wide entrance slit (e.g. $1^\circ \ldots 3^\circ$) which receives the integral intensity of a chosen Bragg-reflection. This detector can be replaced by a (one-dimensional) position sensitive detector, Fig.4. This way the complete diffraction spectrum $I(2\vartheta, \alpha, \beta)$ can be obtained for each sample orientation α , β with a resolution $\Delta 2\vartheta \sim 0.01°$.

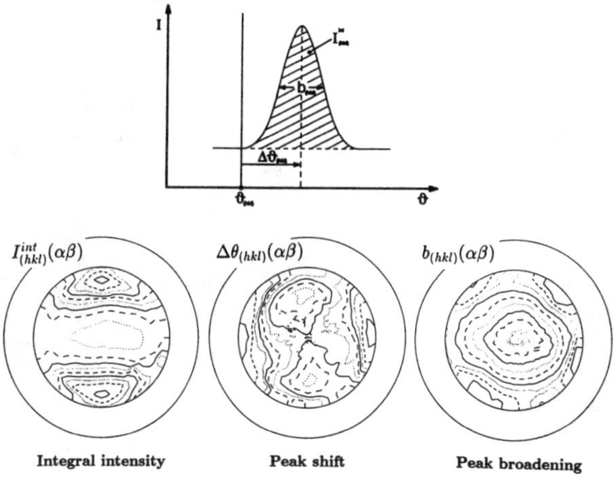

Fig.5
Peak profile analysis applied to all sample orientations $\{\alpha\,\beta\}$ yields peak-position and peak-width pole figures additional to the conventional integral intensity pole figures.

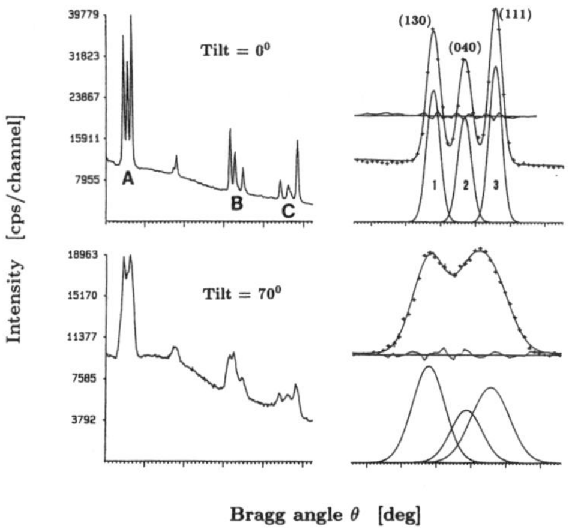

Fig.6
Peak profile analysis allows to seperate overlapping diffraction peaks and hence allows texture analysis in materials with peak-rich diffraction diagrams.

This method has two main advantages compared with the classical texture goniometer [4]:

1. Peak-shift and peak-broadening pole figures can be obtained along with the classical integral intensity pole figure, Fig.5, containing information about $s(x)$ eq(4).
2. Overlapped peaks can be deconvoluted, Fig.6. Hence, texture analysis can be done in materials with complex, peak-rich diffraction diagrams.

Area detector

The detector can be 2-dimensional, Fig.7. It registers, in each sample orientation, a two-dimensional diffraction diagram as is shown for instance in Fig.8. The essential advantage of this method is its high angular resolution in the pole figure angles $\{\alpha\beta\}$ e.g. $\sim 0.1° \times 0.1°$. This high resolution has mainly two applications [5]:

1. The correct measurement of sharp textures e.g. Fig.9.
2. Individual grain resolution with respect to:

- grain orientation g
- grain volume V

as well as changes of these quantities in the course of physical processes e.g. plastic deformation, recrystallization, phase transformation.

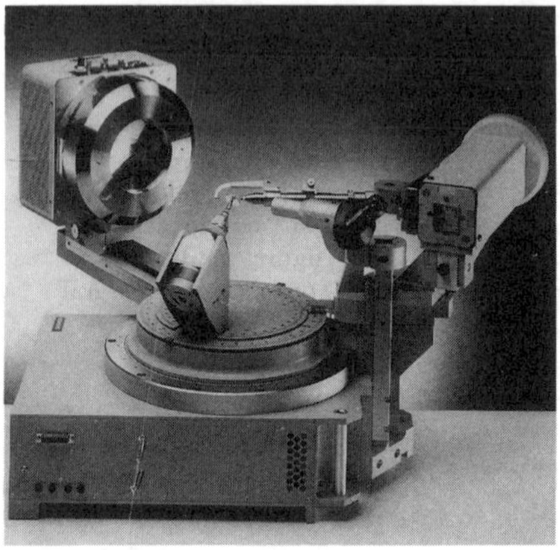

Fig.7
Texture diffractometer with a two-dimensional (area) detector.

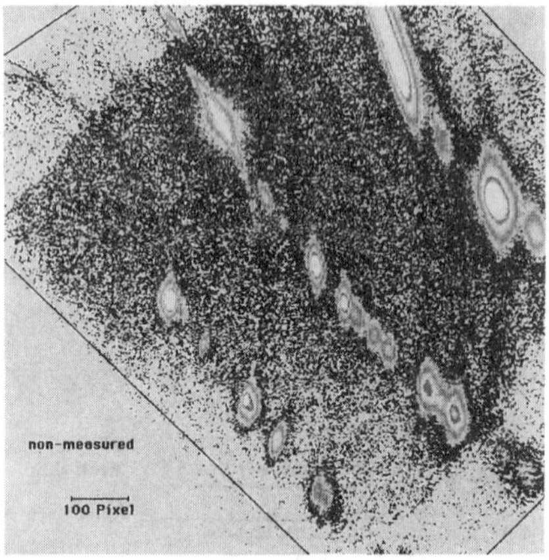

Fig.8
Diffraction diagram obtained with an area detector. The sample was a lithium disilicate glass-ceramic with very sharp texture.

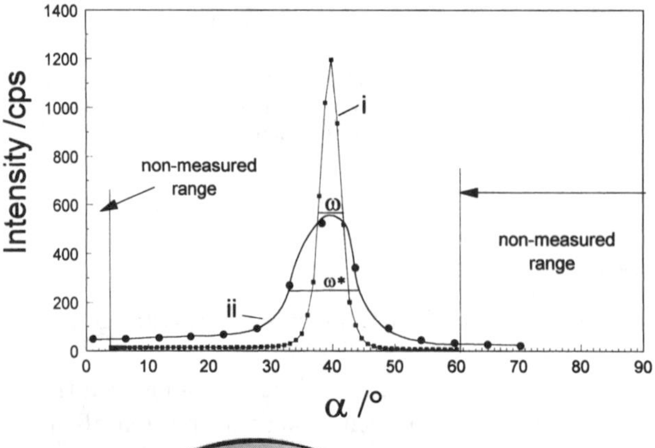

Fig.9
Section through the pole figure of cold rolled nickel measured with a conventional texture goniometer and an area detector. The conventional measurement shows considerable instrumental broadening of the sharp texture peak.

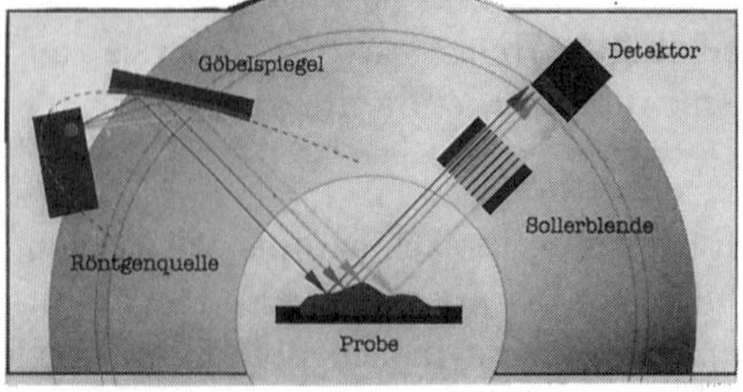

Fig.10
The parallel beam technique, here implemented by a Göbel mirror, is insensitive to the sample geometry, particularly to sample tilt.

Parallel beam technique

In all methods mentioned above the necessary sample tilt through the angle α leads to a decrease in the resolution of 2ϑ (instrumental peak broadening). This is not the case in the parallel beam technique which can be realized in two ways:

1. Long distance between source and sample, synchrotron radiation
2. Parallelization by Göbel mirrors [6]

This technique has many different implementations one of which is shown in Fig.10. Angular resolution in the pole figure of $0.01° \times 0.01°$ or better can be reached this way.

Texture topography $\longrightarrow$ Texture fields

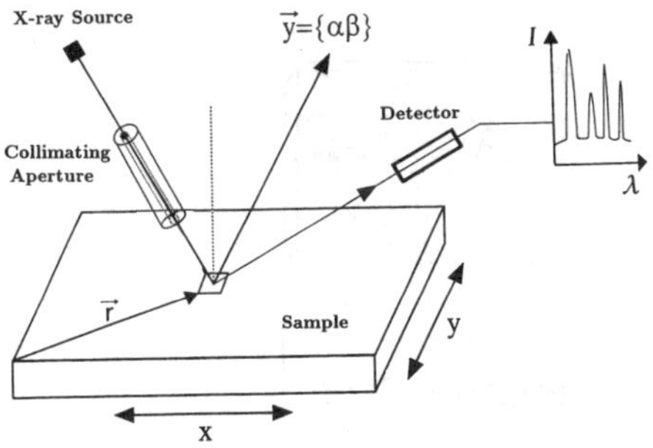

Fig.11
Principles of texture topography, here using an energy dispersive detector.

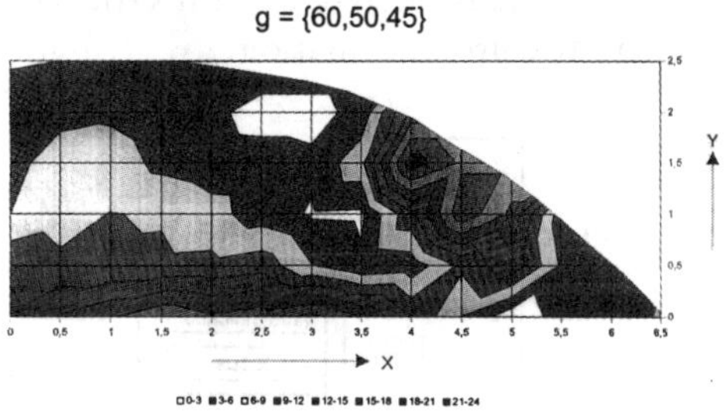

Fig.12
The texture field (texture topograph) of a "pofile rolled" sample. Here distribution of one particular crystal orientation.

By using a narrow entrance apperture e.g. a glas-capillary, the local texture can be measured with a resolution in the order $50\mu m \times 50\mu m$ (or even better) [7]. If the sample is then scanned in two directions Fig.11 the two-dimensional texture field eq(6) can thus be obtained. This is particularly interesting in the case of inhomogeneous deformation e.g. profile rolling Fig.12 or relief-stamping e.g. in coins Fig.13.

3.2 New single crystal techniques

Single crystal techniques as shown in Fig.3a require a high location resolution. This can be obtained particularly by electron diffraction with a resolution in the order of
$1\mu m \times 1\mu m$ in back-reflection
$1nm \times 1nm$ in transmission

Fig.13
Texture topograph of a coin of which the stamped relief was removed by grinding and polishing. The stamped relief can be made visible again by the local texture beneath the surface introduced by the inhomogeneous deformation of the "coining" process.

The used diffraction diagrams are wide angle diagrams (Kikuchi diagrams). These are evaluated by image analysis, automatic indexing followed by orientation analysis, and automatic location scanning [8] as is illustrated schematically in Fig.14 [9]. As a result a two-dimensional section of the aggregate function $g^i(x)$ eq(4) is thus obtained with an orientation resolution of$\sim$ 0.5°. Since crystal orientation g is a three-dimensional quantity which must be plotted in each point x, two two-dimensional colour plots are often used, each of which "codes" two orientation angles according to the "colour triangle". An example (though only in grey shades) is shown in Fig.15. This technique, known as **O**rientation **I**maging **M**icroscopy (OIM) [8] or **A**utomated **C**rystal **O**rientation **M**apping (ACOM) [9] gives the most complete texture information.

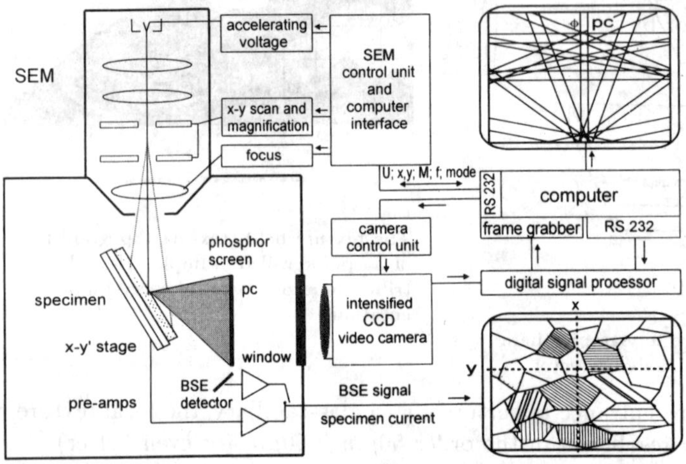

Fig.14 Principles of the Automated Crystal Orientation Measurement (ACOM technique) also known as EBSP, BKD, or OIM technique.

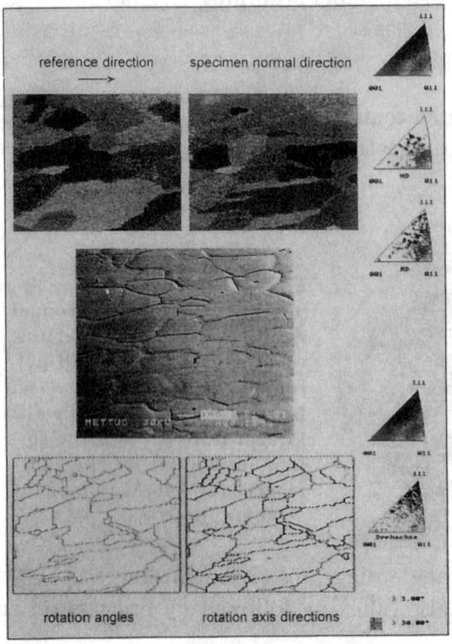

Fig.15
The result of an ACOM-measurement, i.e. the two-dimensional section of the aggregate function $g(x)$ represented in different ways:

a) The local orientation is represented by the crystallographic orientation of sample normal and in-plane reference direction, both colour-coded by the "colour-triangle" superposed to the inverse pole figure.

b) The misorientation Δg across the grain boundaries, colour-coded in a similar way.

4 Orientation versus location resolution

Crystal orientation is defined by the orientation of lattice planes or lattice directions. Since we must admit an uncertainty in the order of an atom diameter Fig.16 orientation resolution $\Delta\alpha$ and location resolution Δx are not independent of each other, rather they are reciprocal to each other [10]

$$\Delta x \cdot \Delta\alpha \sim 1\,\mathring{A} \qquad \text{Uncertainty Relationship} \qquad (7)$$

In this equation Δx may be determined by the used location resolution e.g. electron diffraction or by the size of the crystallites e.g. nano-crystalline materials as is shown in Fig.17.

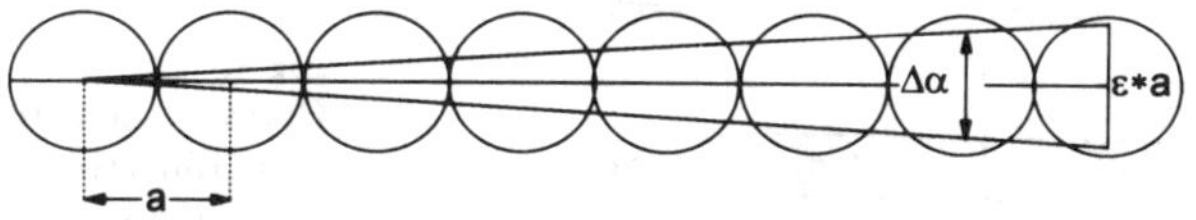

Fig.16 Crystal orientation is defined by crystallographic directions the uncertainty $\Delta\alpha$ of which is the greater the smaller the crystal.

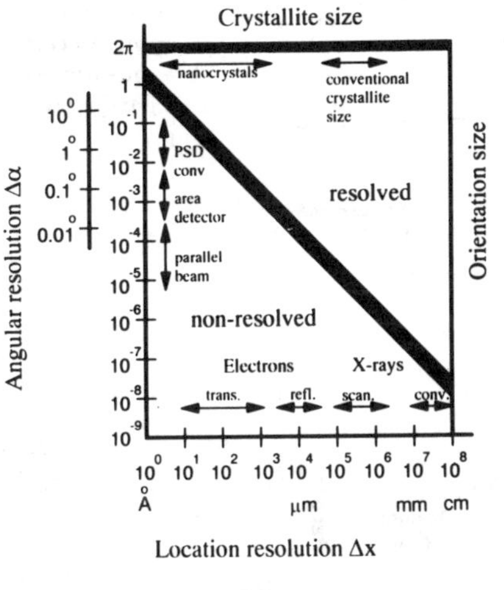

Fig.17.
The uncertainty relationship defines the boundary of simultaneous high resolution with respect to location and orientation.

5 Physical properties of polycrystalline materials

According to the classification of structure levels the physical properties of polycrystalline materials depend on:

1.) Crystal structure $\Longleftarrow$ Chemical composition
2.) Aggregate structure $\Longleftarrow$ Physical treatment

Crystal structure is described by the position of a (limited) number of atoms. Hence, the number of structure parameters is often < 100. Furthermore, the structures are often assumed to be known. So, these parameters are fixed values.

Aggregate structure $G(x)$ or even $g^i(x)$ eq(4) involves a number of parameters in the order of $[L/\Delta x]^3$ where L is sample size and Δx is location resolution respectively crystallite size. This may easily be in the order of $10^6 \ldots 10^{12}$ or even more and these parameters vary from sample to sample, hence they must be treated as variables. The described generalized textural quantities are attempts to reduce this number of parameters. This makes a thorough physical treatment of properties of polycrystalline materials so much more difficult than that of single crystals.
A physical property of a material is defined by the "response" of the material to any external "action" exerted upon it.

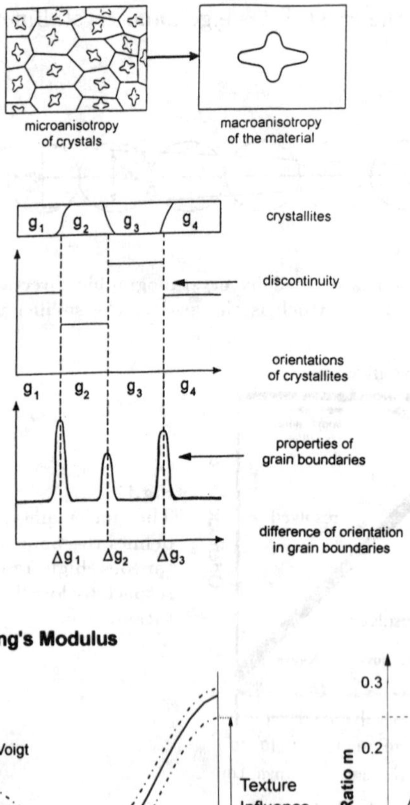

Fig.18
Crystal anisotropy together with different crystal orientations gives rise to three types of effects
1) volume average of the anisotropy,
2) discontinuity across the grain boundaries and hence grain interaction,
3) misorientation dependent particular grain boundary properties.

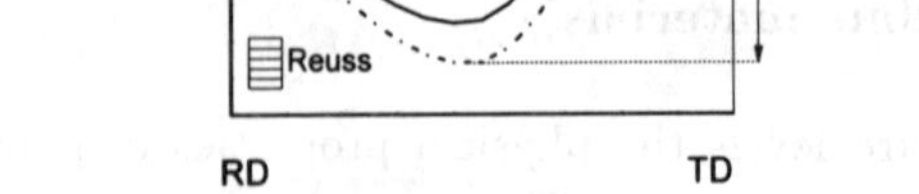

Fig.19
Anisotropy of Young's modulus in the sheet plane (schematic) showing the "texture effect" and the "geometric effect".

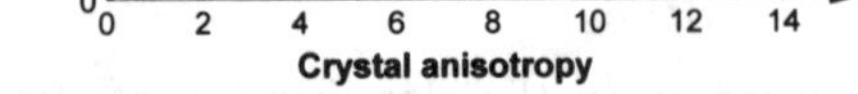

Fig.20
The ratio of maximum geometric anisotropy to maximum texture effect as a function of crystal anisotropy (single-phase material).

We may distinguish two kinds of properties:

1. Static properties — no structure change
   e.g. elasticity
2. Dynamic properties — with structure change
   e.g. plasticity

If the latter ones are considered up to strong structure changes then they may also be looked at as the *structure forming processes.*
Properties of polycrystalline aggregates are influenced by *crystal orientation* via *anisotropy* of the property in a single crystal. Thereby at least three levels of approximation may be distinguished

1. Volume average — 1. Order
2. Grain interaction — 2. Order
3. Boundary properties — 3. Order

This is illustrated in Fig.18. Here, it is admitted that the properties of the crystallites may be modified by substructure described by the function $s(x)$ in eq(4).

5.1 Example: Elastic properties

The elastic properties of a single crystal of orientation g are described by Hooke's Law:

$$\begin{aligned} \sigma_{ij} &= C_{ijkl}(g) \cdot \varepsilon_{kl} \\ \varepsilon_{ij} &= S_{ijkl}(g) \cdot \sigma_{kl} \end{aligned} \qquad [S_{ijkl}] = [C_{ijkl}]^{-1} \qquad \text{Hooke's Law} \tag{8}$$

For a *single-phase* polycrystal with texture $f(g)$ we obtain the volume averages of the two kinds of property parameters

$$\begin{aligned} \overline{C}^{V}_{ijkl} &= \oint C_{ijkl}(g) \cdot f(g) \cdot dg \\ \overline{S}^{R}_{ijkl} &= \oint S_{ijkl}(g) \cdot f(g) \cdot dg \end{aligned} \qquad \left[\overline{S}^{R}_{ijkl}\right] \neq \left[\overline{C}^{V}_{ijkl}\right]^{-1} \qquad \begin{matrix}\text{Voigt}\\ \text{Reuss}\end{matrix} \tag{9}$$

These two averages are different from each other as is illustrated schematically in Fig.19 (Each of them is valid for a very particular grain structure as is also shown in this figure). The influence of texture and grain structure respectively may be estimated by the maximum anisotropy due to both structural parameters [11].

1. Volume average $\longrightarrow$ texture (Maximum = crystal anisotropy)
2. Grain interaction $\longrightarrow$ grain structure (Distance between the Voigt-Reuss bounds)

Both increase with increasing crystal anisotropy but also the ratio of the two increases from zero to a maximum of 0.25 at highest crystal anisotropy as is shown in Fig.20.

5.2 Model calculations of the elastic properties with the aggregate function $g^i(x)$

As a discretization of the aggregate function $g^i(x)$ a "Cluster" of grains embedded in a continuous medium with average properties Fig.21 can be used. Each grain at the location x can belong to any phase i and have any orientation g. All boundary conditions inside and at the surface of the cluster are taken into consideration [12]. It was shown that the error due to discretization was $< 1\%$ at ~ 1000 grains (and it can be further decreased if the number of grains is further increased). This may be compared with the uncertainty of up to 25% if only the texture $f(g)$ is taken into consideration.

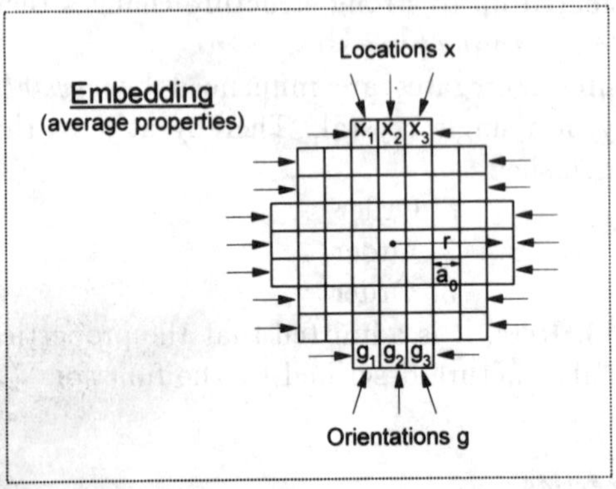

Fig.21
A cluster of grains of identical shape is embedded in a homogeneous matrix which has the averaged cluster properties. Each grain can belong to a different phase and can have its own orientation g.

Fig.22 Interaction strains across the grain boundaries obtained by the cluster model (a). The displacement is continuous in the grain boundaries (b).

The following structural parameters were considered in the model [13].

1. Grain shape by the axis ratios $a : b : c$
2. Grain packing. The centres of the grains formed a cubic primitive, *bcc*, or *fcc* lattice.
3. Orientation correlation MODF
4. Texture
5. Two-phase structure

As an example Fig.22 shows the interaction strains in the grains (the displacements are continuous in the grain boundaries). Fig.23 shows the influence of one axis ratio of the grains in a fibre-textured material. Finally, Fig.24 shows the influence of grain packing and orientation correlation (i.e. three different MODFs) in a material with random texture.

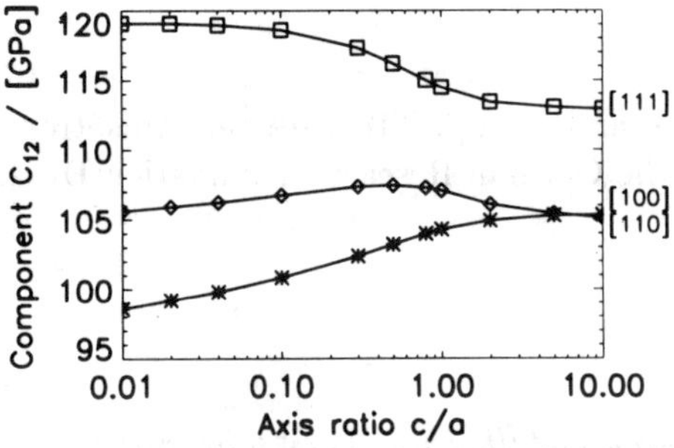

Fig.23 Influence of axis ratio of the grains on the elastic properties (here stiffness component c_{12}) for copper with sharp fibre textures, obtained by the cluster model.

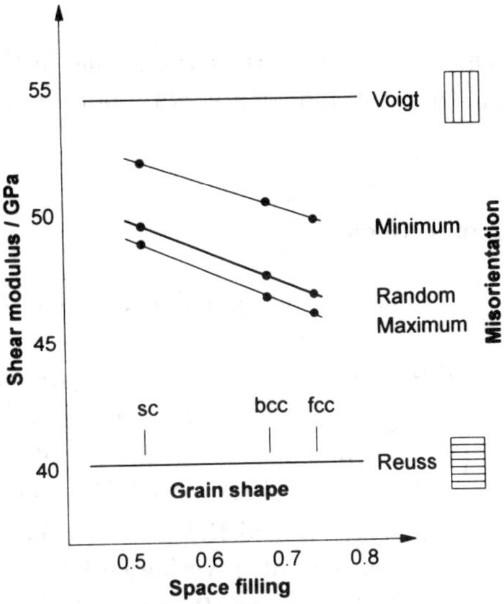

Fig.24 Influence of grain packing according to cubic primitive *sc*, *fcc*, or *bcc* lattice and misorientation on the shear modulus of copper, obtained with the cluster model. Grain packing is particularly plotted versus the "space-filling" factor of the three types of lattices.

6 Conclusions

- The geometrical part of the aggregate function $g(x)$ (single-phase material) may influence materials properties by up to 25% of the maximum influence of texture.
- This part can be satisfactorily modelled by a cluster model containing $\sim$ 1000 crystals. The polycrystal properties must lie between the Voigt-Reuss bounds given by the texture. With particular grain structures they can vary in this whole range. Particularly, the properties may violate the narrower Hashin-Shtrikman bounds [14].
- The knowledge of the aggregate function (in access to the *classical* texture) is necessary for a satisfactory understanding of the properties of polycrystalline materials.
- The aggregate function (at least in two dimensions) is available by the OIM or ACOM technique.
- Advanced polycrystal techniques allow the measurement of the classical texture with much higher orientation - as well as location resolution.
- They also allow to measure some orientation dependent location averages of the substructure function $s(x)$.

Acknowledgment

This paper summarizes work done in the research group: "Texture and Anisotropy of Polycrystalline Materials" sponsored by the German Research Foundation DFG.

References

[1] G. Wassermann and J. Grewen, *Texturen metallischer Werkstoffe*, Springer Verlag Berlin 1969.

[2] H.J. Bunge, *Texture Analysis in Materials Science*, Butterworths Publ. London 1982, 2nd Edition Cuvillier Verlag Göttingen 1993.

[3] H.J. Bunge, Statistical Crystallography of the Polycrystal, in: *Textures of Materials ICOTOM-10*, Materials Science Forum **157-162** 13-30 (1994).

[4] L. Wcislak and H.J. Bunge, *Texture Analysis with a Position Sensitive Detector*, Cuvillier Verlag Göttingen 1996.

[5] H.J. Bunge and H. Klein, Determination of Quantitative, High-resolution Pole Figures with the Area Detector, *Z. Metallkde.* **87** 465-475 (1996).

[6] H.F. Göbel, Planar Graded Multilayer Reflectors as Selectable Wavelength Monochromators in Bragg-Brentano Diffractometers, *Proc. EPDIC-5*, Materials Science Forum (in print).

[7] R.A. Schwarzer and M. Wehrhahn, X-ray Scanning Apperatus for Mapping Texture and Element Distribution, *Textures and Microstructures* **29** 65-76 (1997).

[8] B.L. Adams, D.J. Dingley, K. Kunze, and S.I. Wright, Orientation Imaging Microscopy: New Possibilities for Microstructural Investigations using Automated BKD Analysis, in: *Textures of Materials ICOTOM-10*, Materials Science Forum **157-162** 31-42 (1994).

[9] R.A. Schwarzer, The Study of Crystal Texture by Electron Diffraction on a Grain Specific Scale, *Microscopy and Analysis* **45** 35-37 (1997).

[10] H.J. Bunge, Physical Versus Mathematical Aspects in Texture Analysis, *Textures and Microstructures* **25** 71-108 (1996).

[11] H.J. Bunge, Textur- und Gefügeanisotropie vielkristalliner Stoffe, in: *Materialwissenschaftliche Grundlagen*, DGM Informationsgesellschaft - Verlag Frankfurt 289-293 (1997).

[12] H. Kiewel and L. Fritsche, Calculation of Effective Elastic Moduli of Polycrystalline Materials including Non-textures Samples and Fibre textures, *Phys. Rev.* **B 50** 5 (1994).

[13] H. Kiewel, H.J. Bunge and L. Fritsche, Calculation of Linear and Non-Linear Elastic Properties of Polycrystalline Materials by Use of a Cluster Model, *Textures and Microstructures* **28** 261-271 (1997).

[14] Z. Hashin and S, Shtrikman, A Variational Approach to the Theory of Elastic Behaviour of Polycrystals, *J. Mech. Phys. Solids* **10** 343 (1962).

QUANTITATIVE APPROACH TO ORIENTATION DIFFERENCE DISTRIBUTION ANALYSIS BASED ON MODEL FUNCTIONS

J. JURA, J. POSPIECH
Instytut Metalurgii i Inżynierii Materiałowej, Polska Akademia Nauk, PL -30059 Kraków, Poland

G. GOTTSTEIN
Institut für Metallkunde und Metallphysik der RWTH D-52056 Aachen, Germany

In numerous cases orientation distribution function (ODF) can be interpreted as a superposition of the Gauss-type distributions in the surrounding of the distinguished single orientations. These model distributions describe the main texture components with their parameters: the volume fractions and the scattering widths. Orientaion difference distribution function (ODDF) depends only on the ODF. Because of the simple, unambiguous relation defining ODDF, which links it with ODF, it is possible to carry out the analysis of quantitative relations of both distributions. Such an analysis may constitute a starting point of investigations into the orientation difference distributions. The considerations presented in the paper, which are based on model representation of the orientation and orientation difference distribution functions, allow for more closer study of the quantitative relations linking them as well as the possibility of ODDF analysis in practice.

1 Introduction

In the quantitative description of the microstructure of polycrystalline material the orientation characteristics being functions of orientations or functions of orientation differences are of great importance (Bunge[1], Pospiech[2]). Among these the most important is the orientation distribution function (ODF) describing the material texture, and in the microstructure investigations the misorientation distribution function describing the relations between the nearest neighbourhood orientations (MDF).

From among the orientation difference distributions the global orientation difference distribution function ODDF (also called "uncorrelated" misorientation function (Bunge *et al.*[3]) appears interesting. ODDF provides information about the distribution of all possible orientation differences, it may also be used as a normalizing factor for other orientation characteristics; e.g. reducing effect of texture in MDF (Bunge *et al.*[3,4], Plege[5]).

On account of the simple, unambiguous relation defining ODDF which links it with ODF, it is possible to carry out the analysis of quantitative relations of both distributions, e.g. on the basis of ODF model as superposition of Gaussian type distributions, (Pospiech *et al.*[6], Lücke *et al.*[7], Jura *et al.*[8]).

2 ODF and ODDF Components

Let assume that the orientation distribution around the position g_i is described by the function (Bunge[1])

$$f_i(g) = H_i \exp(-\phi^2/\phi_i^2) \tag{1}$$

$$H_i = V_i\, 2\sqrt{\pi} / \{\phi_i [1 - \exp(-\phi_i^{\,2}/4)]\},$$

where ϕ_i - is the scattering width around the position g_i ,
V_i - defines the volume fraction of the component g_i in the texture (ODF).

The relation defining ODDF has the form (Bunge[1])

$$F_i(\Delta g) = \oint f_i(\Delta g g) f_i(g) dg \,. \qquad (2)$$

F(g) function is thus completely defined by f(g).
From Eq. 2 there follows directly the relation

$$F(\Delta g = e) = \oint f^2(g)\, dg = J \,, \qquad (3)$$

where J is the so-called texture index - a measure of the texture sharpness.

Hence there arises the question in what way the transformation defined by Eq. 2 affects the form (the parameters ϕ_i and V_i) of the "representative" of the component g_i in the global orientation difference distribution function $F_i(\Delta g)$

To examine this problem we will use the expansion of both functions into a series of generalized spherical functions $T_l^{mn}(g)$. The coefficients of a series expansion of the function ODF - C_l^{mn} and those of the function ODDF - $C_l^{m_1 m_2}$ are linked by the known relation (Bunge *et al.*[3] , Zhao *et al.*[9])

$$C_l^{m_1 m_2} = \frac{1}{2l+1} \sum_{n=-l}^{l} C_l^{m_1 n} C_l^{*m_2 n} \,. \qquad (4)$$

Let consider a model ODF composed of I components of the parameters $\{g_i, V_i, \phi_i\,;\ i = 1,I\}$ described by Eq. 1. The coefficients of series expansion of such a function are defined by the formula (Pospiech *et al.*[6])

$$C_l^{mn} = \sum_{i=1}^{I} V_i \frac{e^{-l^2\phi_i^2/4} - e^{-(l+1)^2\phi_i^2/4}}{1 - e^{-\phi_i^2/4}} T_l^{*mn}(g_i) \,. \qquad (5)$$

C_l^{mn} can be written down in approximate form applying the approximation (Jura *et al.*[8])

$$C_l^{mn} \cong (2l+1) \sum_{i=1}^{I} \exp[-l(l+1)\frac{\phi_i^2}{4}] V_i T_l^{*mn}(g_i) . \qquad (6)$$

Putting above relation into Eq. 4 we get

$$C_l^{m_1 m_2} = (2l+1) \sum_{i,i'}^{I} c_{ii'} V_i V_{i'} \exp[-l(l+1)\frac{\phi_i^2 + \phi_{i'}^2}{4}]\, T_l^{*m_1 m_2}(\Delta g_{ii'}) \,, \qquad (7)$$

where

$$\Delta g_{ii'} = g_i g_{i'}^{-1} \,, \quad c_{ii'} = 1 \ \text{for}\ i = i' \ \text{and} \ c_{ii'} = 2 \ \text{for}\ i \neq i' .$$

Thus if ODF is described by the components of the parametrers: $\{g_i, V_i, \phi_i; i = I\}$, the components of a corresponding ODDF (considering $\Delta g_{ii'} = \Delta g_{i'i}$) are described by the parameters: $\{\Delta g_{ii'}, c_{ii'} V_i V_{i'}, \sqrt{\phi_i^2 + \phi_{i'}^2}\ ; i, i' = 1, I\}$.

3 Model of ODDF Function

If the texture (ODF) contains only one component (we neglect, for the present, the possibility of the occurance of statistical symmetry of the sample), then Eq. 7 refers to the Gaussian type distribution, analogous to Eq.1 of a height equal to J_{ODF} and the scattering width $\phi_{ODDF} = \sqrt{2}\, \phi_{ODF}$ (ϕ_{ODF} is the scattering width of the peak in the orientation space, while ϕ_{ODDF} is the scattering width after the transformation into the misorientation space) . Thus, independently of the position of g_i in the orientation space the peak "coupled" with it in the misorientation space will be in the position $\Delta g = e = \{0,0,0\}$ and will have the width $\sqrt{2}\,\phi_{ODF}$.

Let now consider the case of ODF composed of K constituents (components and their complementary subcomponents) in the positions $\{g_k ; k = 1,K\}$ with the volume fractions V_k and the scattering widths ϕ_k . Each complementary subcomponent (generated by the statistical symmetry of the sample) of the component g_i must be regarded as an independent constituent of the ODF model. Analysis of Eq. 7 leads to the following conclusion with reference to the ODDF model: K constituents of ODF are transformed into $1+\frac{1}{2}K(K-1)$ components of ODDF. In the position $\Delta g = e$ there occurs the superposition of the "peaks" from each of the K constituents of ODF with the volume fractions V_k^2 and the scattering width $\sqrt{2}\,\phi_k$. On the other hand, in the positions $g_{kk'} = g_k\, g_{k'}^{-1}$ ($k, k' = 1, K ; k \neq k'$) there occur the components (local maxima) of ODDF - when considering the equivalence $\Delta g_{kk'} = \Delta g_{k'k}$ - there are $\frac{1}{2}K(K-1)$ of them, with the volume fractions $2V_kV_{k'}$ and scattering widths $\sqrt{\phi_k^2+\phi_{k'}^2}$.

If $\sum_{k=1}^{K} V_k = 1$ then there also holds: $\sum_{k,k'=1}^{K} c_{kk'}V_kV_{k'} = 1$, where: $c_{kk} = 1, c_{kk'} = 2, k \neq k'$.

The presented transformation of an ODF model resulting from the dependence defining ODDF allows to conclude that the structure of ODDF is more complex than that of ODF which generates it. ODDF is also more "scattered" around its local maxima ($\phi_k, \phi_{k'} < \sqrt{\phi_k^2+\phi_{k'}^2}$), and its local maxima are additionally suppressed on account of the dependence $V_k, V_{k'} > V_k * V_{k'}$. A special case is the point $\Delta g = e$, at which the peaks from all K constituents of ODF (components together with subcomponents generated by the statistical symmetry of the sample) with the scattering widths $\sqrt{2}\,\phi_k$ and the volume fractions V_k^2 overlap.

Low symmetrical multiplicity of the misorientation $\Delta g = e$ (S_A - on account of crystallographic symmetry) in comparison with the symmetrical multiplicity of general misorientation ($S_A \times S_A$ to the maximum) additionally exposes the local maximum in the position e, which is of no interest during ODDF analysis.

Let us note that in the case of a coincidence, i.e. equal $\Delta g_{kk'}$ for some pairs of ODF constituents g_k , $g_{k'}$; different ODFs may generated identical ODDF.

Figure 1a shows a model of ODF containing two constituents; a corresponding model of ODDF is shown in Figure 1b.

4 Conclusions

- The structure of the ODDF is more complex than the ODF generating it. Interesting properties of the ODDF can be, as shown, defined on the basis of a model presentation of the ODF.
- On account of "close packing" of peaks forming the ODDF in misorientation space there occurs their overlapping. If the symmetry of the crystallographic lattice is higher than that of the sample symmetry, then the basic region in misorientation space is

smaller than the basic region in the orientation space. This makes the differentiation of the peaks, i.e. the analysis of the ODDF difficult.

- Considerable differences in the symmetrical multiplicity of Δg (S_A for $\Delta g = e$ and S_A^2 at arbitrary Δg) additionally expose the peak in the position $\Delta g = e$. Simultaneously (on account of the normalization of ODDF), the heights of the remaining peaks - essential for the orientation difference distribution analysis - become relatively lowered.

- A primary assessment of a possibility of an ODDF analysis (identification of its local maxima) may be obtained by determininig for an ODF model: $\{g_k, V_k, \phi_k; k = 1, K\}$ all possible constituents: $\{\Delta g_{kk'}, c_{kk'} V_k V_{k'}, \sqrt{\phi_k^2 + \phi_{k'}^2}\ ; k, k' = 1, K\}$.

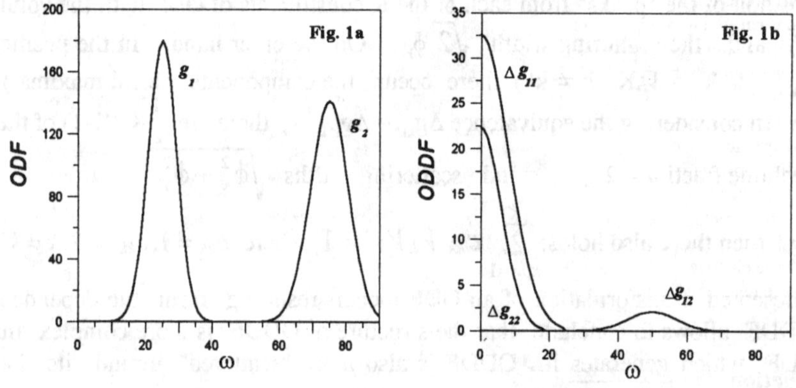

Figure 1a) Model ODF for the constituents $g_1 = \{\ \omega = 25°, V = 35\%, \phi = 6°\ \}$ and $g_2 = \{\ \omega = 75°, V = 65\%, \phi = 8°\ \}$. b) The corresponding ODDF of $\Delta g_{11} = \{\ \omega = 0°, V = 12.25\%, \phi = 6°\ \}$, $\Delta g_{22} = \{\ \omega = 0°, V = 42.25\%, \phi = 8°\ \}$ and $\Delta g_{12} = \{\ \omega = 50°, V = 45.5\%, \phi = 10°\ \}$ constituents (Jura *et al.*[8]).

References

1. H.J.Bunge, Texture Analysis in Materials Science, 1982, London: Butterworths.
2. J.Pospiech, Acta metall. mater. **41,1,** 305 (1993).
3. H.J.Bunge and H.Weiland, Textures and Microstructures, **7, 231** (1988).
4. H.J.Bunge and E.Dahlen, Texture changes and growth anamalies during continous grain growth. In : Annealing Processes, Recovery, Recrystallization and Grain Growth, eds. N.Hansen, D. Juul-Jensen, T.Lefers, B.Ralph, Riso 1986, p. 255.
5. B.Plege, On Orientation Correlations and their Relationship to Textures. In: Theoretical Methods of Texture Analysis, ed. H.J.Bunge, DGM Informationsgesellschaft Oberursel 1987, p. 315.
6. J.Pospiech, W.Truszkowski, J.Jura and J.Król, Proc. 4th European Conference on Textures in Metals, Cambridge UK 1975, p.23.
7. K.Lücke, J.Pospiech, J.Jura and J.Hirsch, Z.Metallkd. 77, **5**, 312 (1986).
8. J.Jura, J.Pospiech and G.Gottstein, Proc.11th International Conference on Textures of Materials, Xi'an, China 1996 ed. Z.Liang, L.Zuo, Y.Chu, International Academic Publishers, p.1351.
9. J. Zhao, B.L.Adams and P.R.Morris, Textures and Microstructures, **8&9**, 493(1988).

DEVELOPMENT OF THE DEFORMATION AND RECRYSTALLIZATION TEXTURES IN Cu-6.5Al SINGLE CRYSTALS

W. RATUSZEK, K. CHRUŚCIEL
University of Mining and Metallurgy,
Al. Mickiewicza 30, 30-059 Kraków-Poland

Microstructure and texture of cold rolled and annealed Cu-6.5Al single crystals with $(017)[57\bar{1}]$, $(207)[\bar{7}\bar{7}2]$, $(616)[\bar{5}66]$ and $(110)[\bar{3}32]$ orientations were investigated. With increasing deformation shear bands and the areas with high density of twins in the form of lenses were observed. Simultaneously crystal orientations were changed to final „alloy type", described by $(110)[1\bar{1}2]$ and $(110)[001]$ orientations, however, the course of these orientation changes was different for individual single crystals. Crystallographic planes of $\{111\}$ type took the positions parallel to the rolling plane in the range of intermediate deformations.

During annealing new orientations were developed. Orientations of the recrystallization nuclei were mostly from the spread of deformation texture. Only small changes of the orientation were observed in the areas with homogeneous orientation in the early stage of recrystallization and this process was caused by recovery or „in situ" recrystallization. Rapid recrystallization and crystal orientation changes occurred in the areas of inhomogeneous orientation. Prolongated annealing times led to some orientation changes too.

Mechanism of the recrystallization texture development in fcc alloys with low stacking fault energy is complex and could not be explained neither by the oriented growth nor oriented nucleation theories. The presence of some common <111> axes and rotation directions suggest that the „selective" choice of nuclei (orientation) occurs at the very beginning and then their growth leads to the formation of final texture.

1. Introduction

Single and multiple slip or twinning proceed at the onset of yielding (plastic deformation) depending on the initial orientation and the value of SFE (stacking fault energy). Initial orientations play a special role in the range of intermediate deformations (50-80%) when the component with $\{111\}$ planes parallel to the rolling plane is formed. Shear bending develops with increasing deformation and at high strains the structure and texture are nearly the same as in the polycrystalline material.

In the course of annealing the processes of recovery and recrystallization occur. Migration of vacancies and dislocations leads to their annihilation or rearrangement without orientation and texture changes during recovery. Continuous recrystallization „in situ" includes dislocation rearrangement in the deformed matrix. Subboundaries are formed and the growth of subgrains proceeds by slow migration of small-angle boundaries. Continuous recrystallization is not connected with the migration of recrystallization fronts therefore does not lead to the local orientation changes. Hence, the recrystallization texture should not be distinct from the texture after deformation. Discontinuous recrystallization occurs in two stages of nucleation and growth. Nucleation is connected with diffusional rearrangement of dislocations. The growth of nuclei proceeds by migration of recrystallization fronts which incorporate point defects and dislocations.

Development of recrystallization textures is strongly dependent on nucleation and the subsequent growth of nuclei. There are two theories describing formation of recrystallization textures, namely, the theory of the oriented nucleation and theory of the oriented growth

Theory of oriented nucleation [1] assumes that the nuclei formed in deformed material, have the structure and orientation dependent on the previous deformation mechanism. Hence, the decisive factor for recrystallization texture is the structure of deformed material.

Theory of oriented growth [2,3] assumes, that the recrystallization textures develop due to the favourable growth of those nuclei only, for which the difference of orientation with respect to the deformed matrix is 40° around <111> axes.

Recrystallization textures in fcc metals and alloys have different orientations (components) which depend on structure and texture after deformation as well as alloying additions i.e. stacking fault energy (SFE) [4].

2. Experimental and results

Single crystals of Cu-6.5wt%Al alloy were produced by zone melting method. Initial crystal orientations were determined by Greninger-Laue method [5]. Investigations were performed on single crystals after deformation (cold-rolling) and annealing. Pole figures of {111} planes were measured by Schulz method [5] (figs. 1÷4).

For the (207)[$\bar{7}\bar{7}$2] single crystal the fibre τ (<110> || TD) occurs after 75% of strain which includes (112)[11$\bar{1}$] and (552)[$\bar{1}\bar{1}5]_{Tw}$, (332)[$\bar{1}\bar{1}$3], (556)[$\bar{3}\bar{3}$5], (556)[33$\bar{5}$], (221)[$\bar{1}\bar{1}$4] and (110)[001] orientations, as well as two components (111)[$\bar{1}\bar{1}$2] and (111)[11$\bar{2}$], (fig. 1a). After initial annealing the weakening of (111)[11$\bar{2}$] and (111)[$\bar{1}\bar{1}$2] orientations was observed. Simultaneously sharpening of (110)[001] occurred and the appearance of some-new orientations, (fig. 1b). Further annealing led to the disappearance of deformation texture (figs. 1c, 1d). The recrystallization texture is described by the following main orientations: {123}<112> (four components), {265}<432> (four components), (274)[32$\bar{5}$] and (724)[23$\bar{5}$] (fig. 1e).

In the (017)[57$\bar{1}$] single crystal two strong components were observed in deformation texture after 90% strain, namely; (110)[001] and orientations with {111} planes parallel to the rolling plane: (111)[$\bar{4}\bar{3}$7], (1$\bar{1}$1)[$\bar{5}\bar{3}$3] and (111)[$\bar{1}\bar{1}$2], (fig. 2a). The recrystallization texture includes {527}<234>, {427}<235> and {123}<112> orientations with four components in each case (fig. 2c, 2d).

Relations between orientations within deformation texture and orientations after annealing may be described by rotation of ~{111}<112> and ~(110)[001] components. The rotations angles are respectively: 20-28° around the common <111> poles (fig. 2e) and ~40° about <111> axis laying close to the TD (fig. 2f).

Deformation texture of $(6\bar{1}6)[\bar{5}66]$ single crystal after 75% of strain contains strong component $(110)[11\bar{2}]$ and small contribution of orientations having $\{111\}$ planes parallel to the rolling plane (fig. 3a). Initial annealing caused disappearance of the $\{111\}$ components, sharpening of the $(110)[1\bar{1}2]$ and the appearance of weak component $(110)[\bar{1}12]$, (fig. 3b). The recrystallization texture may be described by the following orientations:
$(\bar{2}\bar{1}3)[\bar{2}1\bar{1}]$, $(\bar{2}\bar{3}1)[2\bar{1}1]$, $(265)[1\bar{2}2]$, $(265)[\bar{1}22]$, $(\bar{2}74)[3\bar{2}5]$, $(2\bar{7}5)[\bar{3}25]$, (fig. 3c, d). Orientations after annealing and the orientations laying within the spread of the $(110)[1\bar{1}2]$ component may be related by the rotation of 36-42° around the poles situated on or near the circumference of the projection circle.

The $(110)[\bar{3}34]$ single crystal preserved nearly the same orientation during deformation. Only after 70% of strain the traces of the $\{111\}$ component appeared. The initial stadium of annealing did not change the orientation substantially (fig. 4a, b). It may be the result of recovery or recrystallization „in situ" only. However, after longer time of annealing the growth of new grains changed the orientation (fig. 4c). The main components of recrystallization texture, namely: $(2\bar{1}3)[\bar{5}\bar{7}1]$ and $(213)[\bar{5}42]$ as well as $(3\bar{2}7)[\overline{23}0]$ and $(3\bar{2}7)[230]$ are described by rotation $\sim 38 \div 40^{\circ}$ (fig. 4e) and $25 \div 35^{\circ}$ (fig. 4f) about the <111> poles common with deformation texture respectively.

3. Conclusions

The results of Cu-Al single crystal investigations after deformation and recrystallization lead to the following remarks:

- orientations of recrystallization nuclei were not random but laying within the spread of deformation textures;
- within the homogeneously oriented areas the recovery or recrystallization „in situ" proceeded at the beginning, accompanied by slight change of orientation, however, the longer annealing time caused the grain growth and development of new orientations;
- within the inhomogeneously oriented areas the recrystallization proceeded rapidly with complete change of orientation and the prolongation of annealing time resulted in further partial orientation changes due to grain growth,
- relation between the deformation and recrystallization textures could not be described by the „typical" rotation of 40° around <111> directions; the choice of <111> poles as well as the angle and direction of rotation occurred by selective way in each case.

4. References

1. W.G.Burgers: Z.Physik, **599,** 651, (1920).
2. P.A.Beck, H.Hu: Trans AIME, **194,** 83, (1952).
3. K.Lücke: Canadian Metallurgical Quaterly, **13,** 261, (1974).
4. W. Ratuszek: Dissertation and Monographies No.27, Wydawnictwa AGH, Kraków 1995.
5. A.B.Greninger : Trans. AIME, **117,** 61, (1935).
6. L.G.Schulz: J. Appl. Physics, **20,** 1030, (1949).

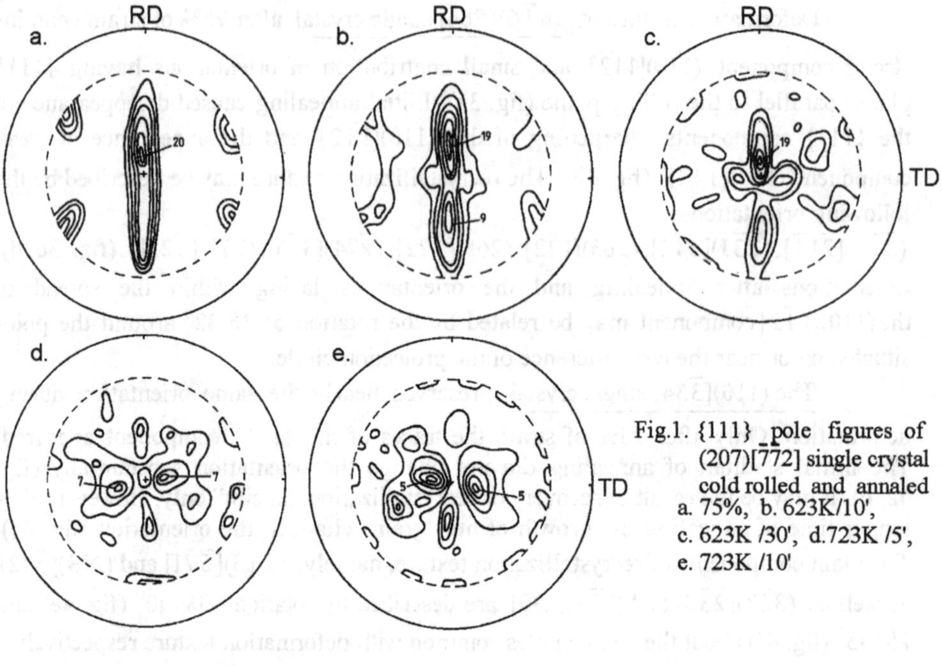

Fig.1. {111} pole figures of (207)[$\bar{7}$72] single crystal cold rolled and annaled
a. 75%, b. 623K/10',
c. 623K /30', d.723K /5',
e. 723K /10'

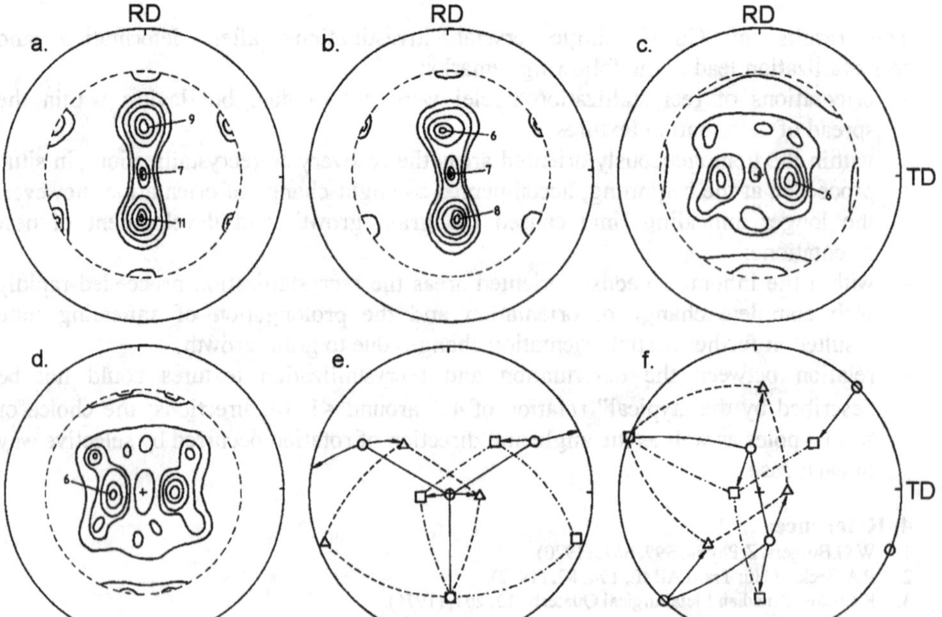

Fig. 2. {111} pole figures of (017)[57$\bar{1}$] single crystal, a. 90%, b. 623K/5', c. 723K/5', d. 723K/30',
e. (274)[32$\bar{5}$] □ (724)[23$\bar{5}$] △ ,
The orientation ~(111)[11$\bar{2}$] rotated ±26° around [111] axis which is near RD,
f. (13$\bar{2}$)[$\bar{5}\bar{3}$7] □ (31$\bar{2}$)[357] △ ,
The orientation (110)[$\bar{1}$17] rotated ±40° around axis [111] which is placed near TD.

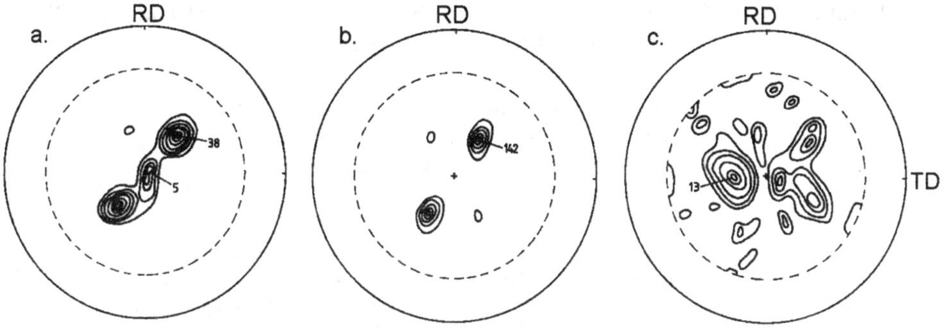

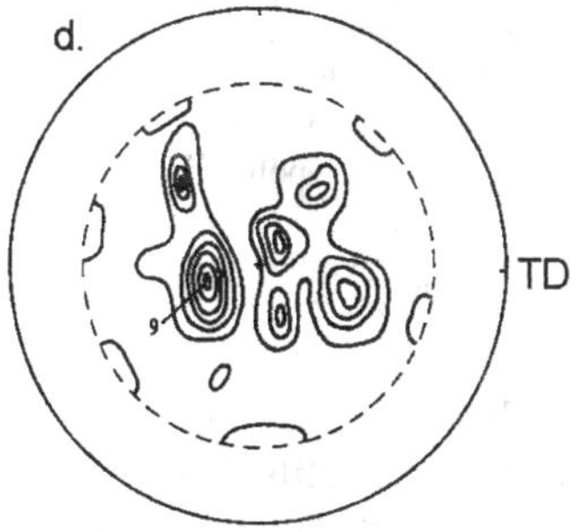

Fig. 3. {111} pole figures of single crystal (6$\bar{1}$6)[$\bar{5}$66]
after rolled a. 75 %,
after annealed b. 623K /30', c. 723K /5', d. 723K /10'

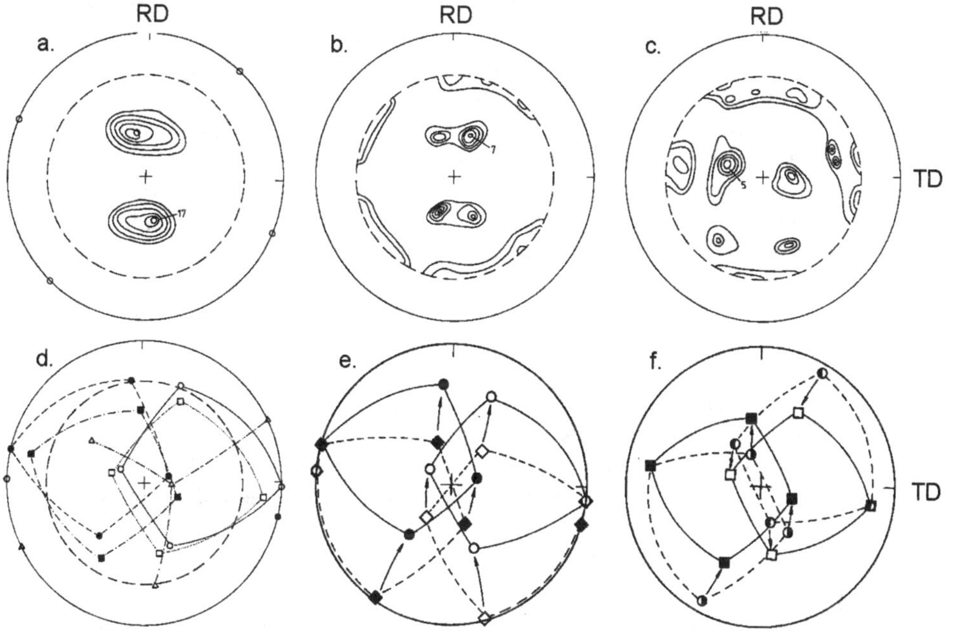

Fig. 4. {111} pole figures of near ~(110)[$\bar{3}$32] single crystal
after rolled: a. 98%, annealed : b. 723K/ 2', c. 723K /150'
d. Ideal orientations describing recrystalization texture coresponding to fig c
e. (2$\bar{1}$3)[$\bar{5}\bar{7}$1] ○ (213)[$\bar{5}$42] ● ,
Rotation of two components of 40° around axes <111> situated near TD
f. (3$\bar{2}$7)[$\bar{2}\bar{3}$0] □ (3$\bar{2}$7)[230] ■ ,
Rotation of angles 25° - 35° around <111> axes near TD

RELATION BETWEEN DEFORMATION AND RECRYSTALLIZATION TEXTURES IN Cu-Al ALLOYS

W. RATUSZEK, A. BUNSCH, K. CHRUŚCIEL
Faculty of Metallurgy and Materials Science
University of Mining and Metallurgy,
Al. Mickiewicza 30, 30-059 Kraków - Poland

Texture development during cold rolling and subsequent annealing were studied in Cu-Al alloys with different aluminium contents. Experimental results were compared with computer aided texture transformation. It is suggested, that development of the recrystallization texture components could not be explained only on the base of the oriented growth theory. Most of the ideal orientations corresponding to recrystallization texture components are related not to the ideal orientations corresponding to main deformation texture components but to the orientations within the spread of these components. This relation is described by rotation around <111> axes, however, very often rotation angle differs from 40° and additionally selective choice of defined <111> direction takes place.

1. Introduction.

For many years recrystallization texture formation was explained on the base of two theories: oriented nucleation [1] and oriented growth [2], however, still there are no evidence which of the theory better describes texture changes in recrystallization of real materials with different stacking fault energy (SFE) [3-4]. The aim of this work is to find out if the oriented growth theory can describe texture changes during recrystallization of alloys with different stacking fault energy (SFE). For this purpose copper alloys with different aluminium contents were investigated.

The oriented growth theory assumes that nucleus with defined orientations to the matrix (usually 40° around common <111> axes) grow up faster and the rate of nucleus growth determines recrystallization texture [2-3]. So using a computer aided procedure texture transformation was done and experimental recrystallization textures were compared with the picture of textures obtained according to the oriented growth theory by a theoretical transformation of deformation texture.

2. Material and experimental procedure

Copper alloy ingots containing 2,41; 5,24 and 6,5wt.% of aluminium were annealed at 450°C for 1 hour, to obtain samples not exhibiting texture (random oriented material). This material was cold rolled to 97% deformation and annealed at 450°C. Texture was measured by the Schulz method for samples on each stage of the material processing [4].

3. Deformation and recrystallization texture

Texture for low deformed samples are very the same. With increasing deformation „alloy type" texture was developed in all materials, however, some differences in texture development were observed in particular for deformation in a range from 70% to 90%. In all samples deformed to 97%, textures were similar, with α fibre limited τ fibre and strong S orientation which is close to {146}<211> and {123}<412>. Only different were the dominant orientations in α fibre and there was slight difference in orientations describing S component. Texture after 97% deformation of all investigated alloys are given in form of ODF cross section in Figure 1.

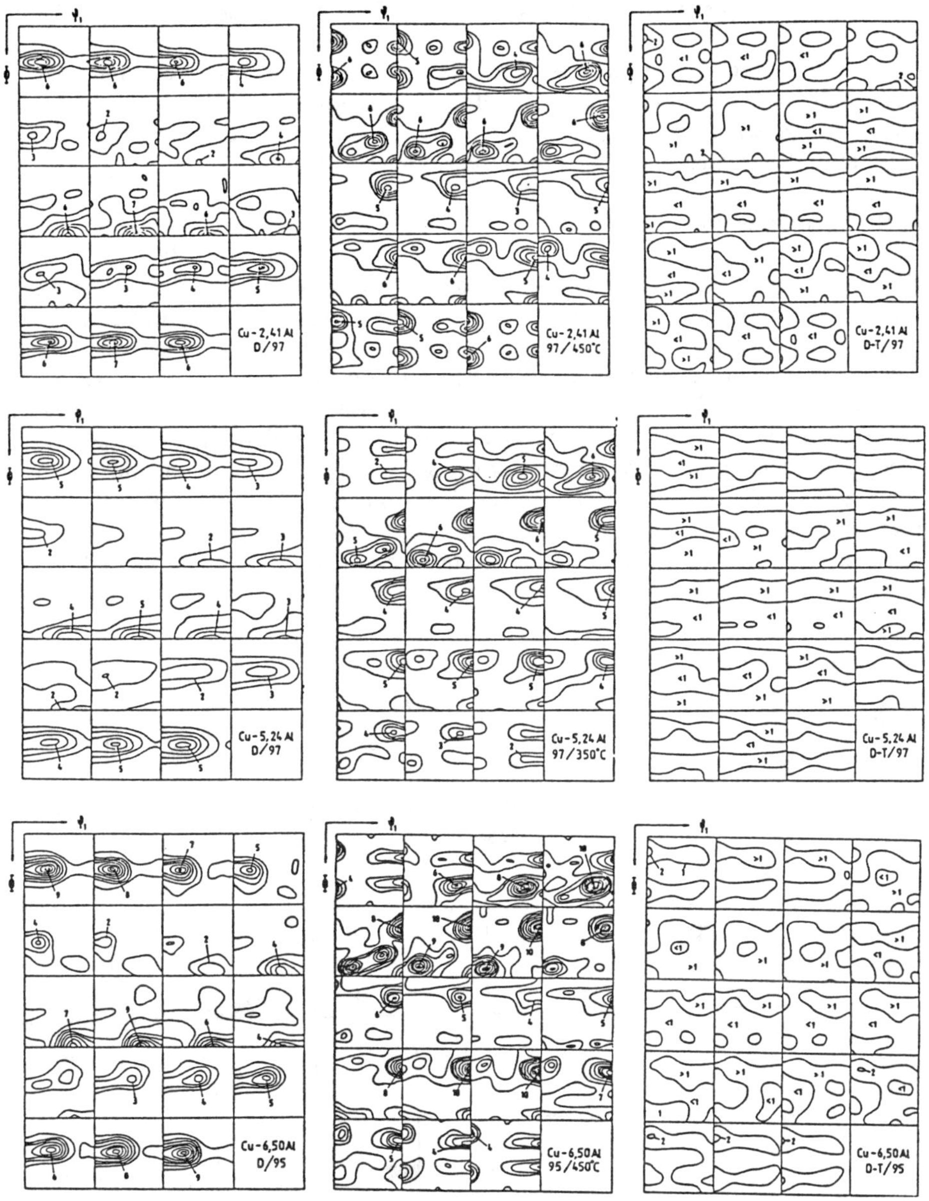

Fig. 1. ODF cross section showing texture of the Cu-Al alloy after 97% deformation (D/97), cold rolled 97% and annealed at 450°C for 10 minutes (97/450°C) and picture of the texture obtained by transformation according to the oriented growth theory (D-T/97).

For all 97% deformed and annealed at 450°C samples the major recrystallization texture components were {263}<358> and {142}<253> - Fig. 1b. Volume fraction of other components varied with aluminium contents in alloys. In Cu-2,41% Al alloy {530}<001> component was very strong in Cu-5,24% Al orientations {113}||RP and in Cu-6,5% Al <113>||ND and <001>||TD.

In Figure 1. also pictures of textures obtained by a theoretical transformation from deformation texture according to the oriented growth theory are given for each alloy. This presentation, one close to the other, makes it easy to compare experimental recrystallization textures with picture of transformed texture.

4. Result analyse and discussion.

When compare experimental recrystallization textures with picture of transformed texture one can see that they do not fit well one to another - Fig. 1. Transformed textures were obtained by theoretical transformation, which was done with assumption, that all deformation texture components are rotated 40° around all <111> type axes in both directions. This leads to very weak and spread texture and it is evident, that such a transformation does not describe real changes occurring in materials during recrystallization. However, some transformations effects are worth to mention. There are some differences in picture of the texture obtained by transformation for each alloy and these differences correspond to experimental recrystallization textures. As a conclusion it could be stated, that although deformation textures differ only slightly, these differences influence on the picture of transformed texture.

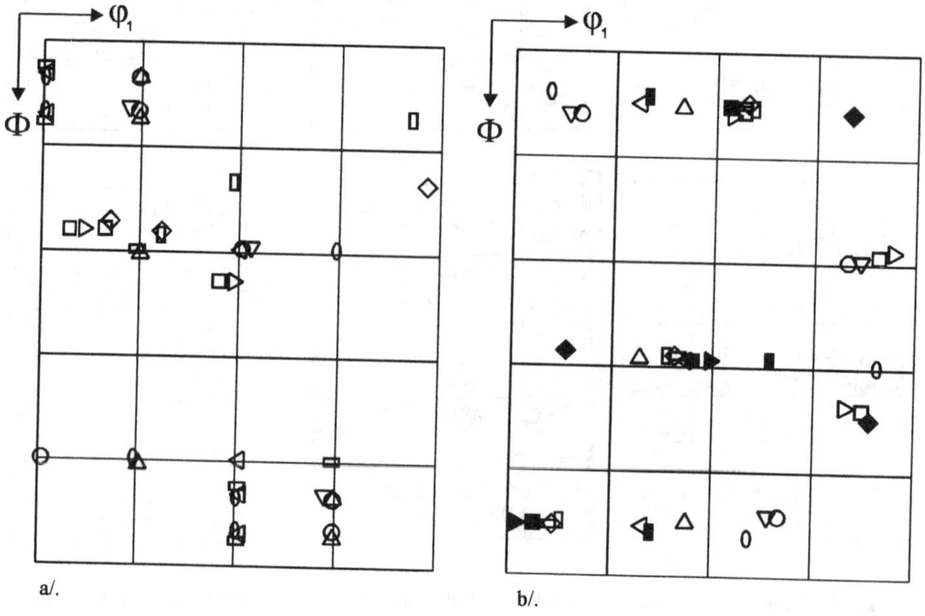

Fig. 2. Ideal orientations describing experimental recrystallization components (a) and orientations corresponding to deformation texture components (b) which were obtained by theoretical transformation (shape of symbols exhibit mutual relation - see Table 1.).

Table 1. Relations in-between ideal orientations given in Euler space in Fig. 2.

Components in recryst. texture	Relation		Components obtained by transformation
	rotation angle	rotation axis	
◇ $(326)[83\bar{5}]$	38°- 40°	$[11\bar{1}]$ $[\bar{1}\bar{1}\bar{1}]$	◇ ~$(011)[21\bar{1}]$ ◆ ~$(101)[12\bar{1}]$
▯ $(124)[25\bar{3}]$	38°- 40°	$[111]$ $[1\bar{1}\bar{1}]$	▯ ~$(011)[21\bar{1}]$ ▮ ~$(101)[12\bar{1}]$
□ $(113)[21\bar{1}]$	38° 60°	$[\bar{1}\bar{1}\bar{1}]$ $[\bar{1}\bar{1}\bar{1}]$	□ $(416)[12\bar{1}]$ ■ ~$(616)[16\bar{2}]$
▷ $(113)[33\bar{2}]$	38° 60°	$[\bar{1}\bar{1}\bar{1}]$ $[\bar{1}\bar{1}\bar{1}]$	▷ $(416)[2\ 10\ \bar{3}]$ ▶ ~$(101)[010]$
◁ $(210)[001]$	32°	$[\bar{1}\bar{1}\bar{1}]$	◁ $(99\bar{1})[4\bar{3}9]$
△ $(210)[\bar{1}20]$	32°	$[\bar{1}\bar{1}\bar{1}]$	△ ~$(110)[\bar{3}32]$
0 $(530)[001]$	38°- 40°	$[11\bar{1}]$	0 ~$(110)[\bar{1}12]$
▭ $(520)[001]$	38°	$[\bar{1}\bar{1}\bar{1}]$	▭ $(66\bar{1})[3\bar{2}6]$
○ $(053)[0\bar{3}5]$	38°- 40°	$[\bar{1}\bar{1}\bar{1}]$	○ $(034)[3\bar{4}3]$
▽ $(053)[1\bar{3}5]$	38°- 40°	$[\bar{1}\bar{1}\bar{1}]$	▽ ~$(011)[3\bar{2}2]$

Because described previously transformation of the experimental deformation texture does not describe real changes occurring in materials during recrystallization, other computer aided procedure was done. Ideal orientations corresponding to the main recrystallization components were rotated according to relation assumed by the oriented growth theory and obtained orientations were compared with experimentally measured deformation textures. Results of these procedure are given in Euler space in Figure 2 and Table 1. When comparing these results with previously measured deformation texture it is evident that locations of the obtained orientations are always within the spread of the main deformation texture components.

5. Conclusions

As a final conclusion it could be stated, that development of recrystallization components could not be explained only on the base of the oriented growth theory. Orientations describing recrystallization texture components are related to deformation texture components by rotations around <111> axes, but these relations are usually not to the orientations describing main rolling texture components, but to the orientations laying within the spread of these components. Also very often rotation angle differs from 40° and additionally selective choice of defined <111> axes takes place [4].

References

[1]. G. Burgers - Z. Physik **599**, 651 (1920)
[2]. P. A. Beck, H. Hu - Trans. AIME **194**, 83 (1952)
[3]. C. A. Verbraak - ICOTOM 6, 98 (1981)
[4]. W. Ratuszek - Wydawnictwa AGH, Kraków - **27**, (1995)

INFLUENCE OF GRAIN COMPOSITION AND PRESSING PROCESS ON THE ANISOTROPY OF THE GRAPHITE PRODUCTS.

W. RATUSZEK, A. DŁUGOSZ, K. CHRUŚCIEL, S. BUDZYŃ
University of Mining and Metallurgy,
Al. Mickiewicza 30, 30-059 Kraków - Poland

Electrode cokes of different grain composition (0-40μm, 100-320μm and 630-1000μm) were used as a charge for electrodes and test blocks. Using mentioned solid compounds graphite electrodes were produced by pressing in passage press and graphite blocks were produced in matrix (die). Anisotropy (texture) index $I^{||}_{00.2} / I^{\perp}_{00.2}$ was calculated and grain size was measured and additionally $I_{00.2}/I_{10.1}$ index were calculated to indicate the texture degree on particular cross sections of the graphite products. Electrode anisotropy depends on pressing method and grain composition. Electrodes obtained from solid compounds with the smallest grains were the most textured. In these electrodes with increase of grain size, anisotropy index decreases. In case of graphite blocks, grain composition does not influence on their texture

1. Introduction.

Anisotropy of the properties of graphite products results on several factors, among which the most important are: grain composition of solid components, forming method and heat treatment parameters.

Polycrystalline graphite products are built up from many graphite crystallites, which have characteristic anisotropy of properties. Final products exhibit anisotropy when they are textured what means that constituent graphite crystallites have preferred orientations. Graphite has layer structure. Distances between carbon atoms in layer are 0.1415 nm and they are much smaller than the distance in-between layers, which is 0.34nm. Bonds within the layer are atomic ones, whereas in-between layers the bonds are similar to the van der Waals bonds. Weakness of interlayers bonds explains perfect cleavage of the graphite along (00.1) crystallographic planes. Crystals with layer structure exhibit anisotropy of physical properties. Direction of the strongest bonds usually corresponds to the direction of the high electric and heat conductivity. In graphite the ratio of the heat conductivity in direction of the highest and the smallest value of this conductivity is equal 4:1 [1, 2]. This indicates the existence of high anisotropy in graphite.

Texture and microstructure of graphite products results on chemical and physical properties of the solids and liquid components of the charge, as well as on methods of plastic mass preparation, greens product formation and finally heat treatment conditions.

2. Material and experimental procedure

Electrodes and test blocks were produced from solid components with different grain composition (Table 1.). Electrodes 115 mm in diameter were formed in a passing press. Test blocks were prepared by pressing in matrix (pressing die) which dimension was 200x200x75 mm. This way formed electrodes and test blocks were fired and graphitized in the same conditions. First they were fired in tunnel furnace at 1200°C during 10 days and than they were graphitized at 1850 and 2500°C in the Tamman furnace in argon. Electrodes and blocks were also graphitized at technological conditions in the Acheson furnace at 2500°C.

Table 1.
Grain size composition of solid filler for the production test electrods and blocks

Grain size [mm]	Percentage fraction of grains in a set [%]					
	Series I			Series II		
	Set 1	Set 2	Set 3	Set 1	Set 2	Set 3
> 4	-	-	2	-	-	2
4 - 2	-	-	12	-	3	33.7
2 - 1	-	29	22	-	38.6	22.2
1 - 0.5	-	24	18	-	24.8	13.8
0.5 - 0.2	1	23	14	2	17.8	11.2
0.2 - 0.06	11	11	8	70.5	11.2	15.8
< 0.06	88	13	24	27.5	4.6	1.3

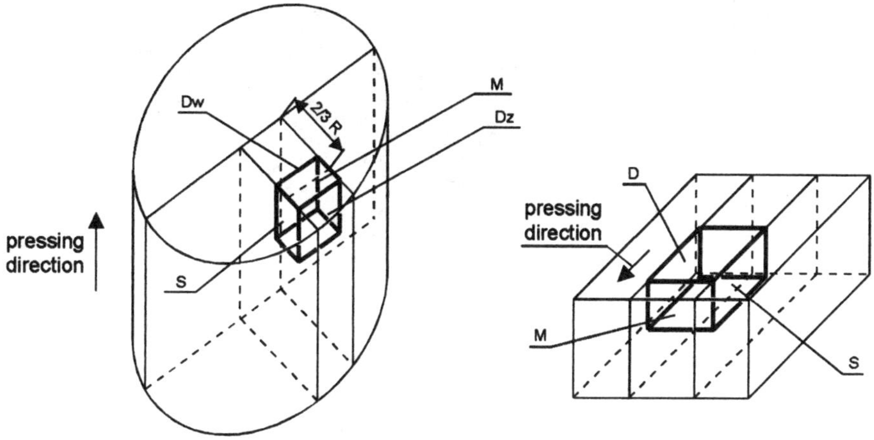

Fig. 1. Scheme of samples cutting for structure investigations of the electrodes and blocks by X-ray measurements (D_z , D_u , D, S, M - sections under examination)

Structure measurements were done on diffractometer with HZG4 goniometer using the X-ray cobalt tube radiation. The texture index for electrodes and test block was estimated on the basis of simplified anisotropy index $I^{\parallel}_{00.2} / I^{\perp}_{00.2}$. This index for electrodes was taken as a ratio of the total intensity of 00.2 diffraction line measured for the sample intersection parallel to the electrode axis and the total intensity of 00.2 diffraction line measured on intersection perpendicular to this axis [3]. For test blocks, which were pressed in the matrix, as the anisotropy index, ratio of the same intensities were taken for the surface parallel and perpendicular to the pressing axis. Additionally, measurements of the other ratios were done for particular intersections. Namely, the ratio of the total diffraction line intensity for two pairs of lines 00.2 and 10.0 and also 00.2 and 10.1 . The dependence of the anisotropy index on heat treatment conditions for electrodes and blocks is given in Figure 2 and 3.

3. Experimental results.

On the basis of the obtained experimental results it can be stated, that anisotropy of the investigated electrodes and test blocks depends on grain composition of solid components and method which was applied for pressing. Results presented in figure 2 and 3 suggest that the main factor which influences the products anisotropy (estimated on the basis of mentioned above anisotropy index) is the forming method of intermediate green products, which is prepared from plastic mass containing solid components and binding agent. The anisotropy index decreases with the increasing of the grain size in solid components for both the electrodes and test blocks, no matter what was the heat treatment stage. The higher value of anisotropy posse electrodes formed by the passing pressing.

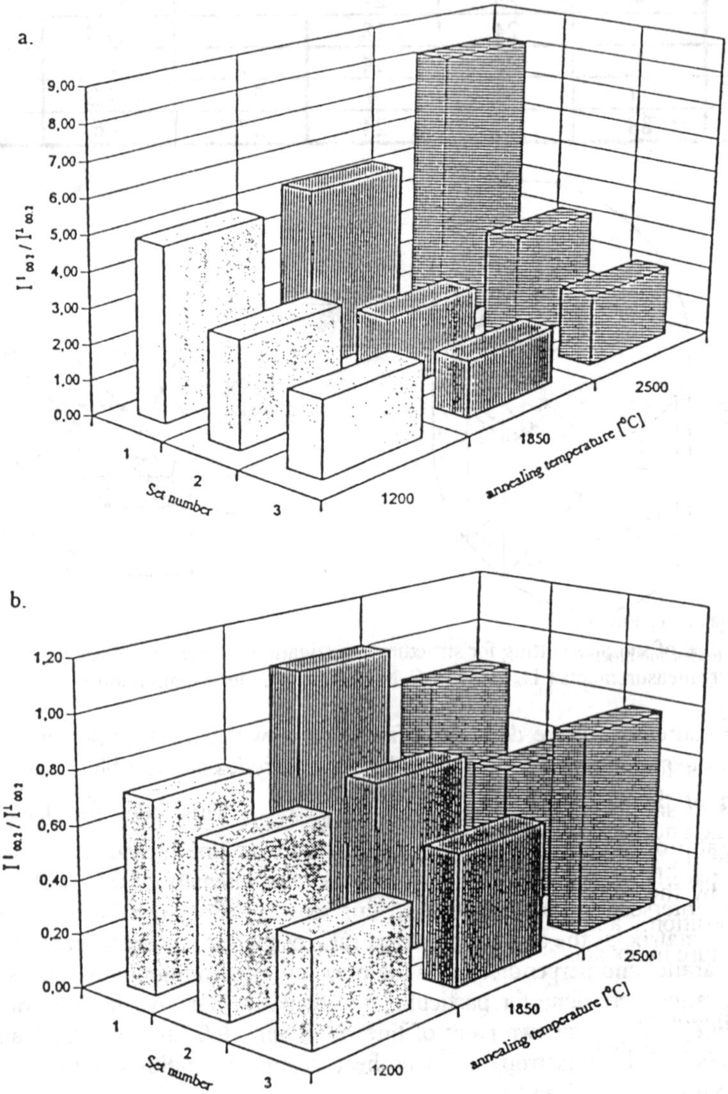

Fig. 2. Graphical representation of the reflection intensity coefficient - $I^{\parallel}_{00.2} / I^{\perp}_{00.2}$ (the planes parallel and perpendicular to the pressing direction) for electrodes (a) and blocks (b) produced from the three different sets annealed of 1200°, 1850° and 2500°C - series I (Table 1)

For electrodes produced in the passing press this value is the highest for the finest grain size of the charge and was increasing with the heat treatment temperature [Fig. 2 and 3]. Products formed as test blocks in matrix (die) from the same charge have the value of anisotropy index equal close to 1 or smaller. These indicate, that they were almost isotropic. Forming in matrix does not lead to creation of preferred arrangements of grains in the blocks structure and also the grain size does not influence on the products anisotropy (Fig. 2 and 3). It should be stated that no matter what is the grain composition in charge and what is the pressing method, the average graphite (grain) crystallite size in electrodes and in blocks are almost the same.

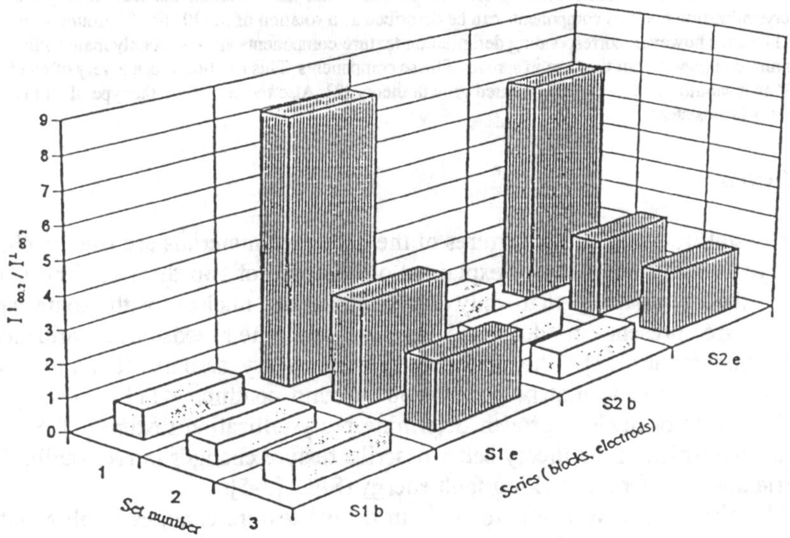

Fig. 3. Graphical representation of the reflection intensity coefficient - $I^{\parallel}_{00.2} / I^{\perp}_{00.2}$ for the industrial electrodes and blocks annealed at 2500° in Acheson furnace

Conclusions

It was found in this work that by the right choice (selection) of the grain composition of the charge (solid components), the graphite products with expected texture index and anisotropy can be obtained. Anisotropy and texture are specially influenced by the smallest grains in the charge and also the applied pressing method.

So by the selective grinding of the solid components, by selection of proper grain composition and choice of pressing method, electrodes and blocks with the expected texture index and anisotropy can be produced.

Acknowledgements

This work has been founded by the KBN, Grant # N7 S201 069 07

References

1. A. Taylor, X-RayMetallography, London, 1961
2. S. Szarras, Budowa ciała stałego, Warszawa 1974
3. A.Długosz, W.Ratuszek, S.Budzyń, C.Czarnek, Karbo-Energochemia-Ekologia **41**, 73, (1996)

COLD ROLLING AND ANNEALING TEXTURES IN Cu-3.0Si ALLOY.

W. RATUSZEK, A. BUNSCH, K. CHRUŚCIEL
Faculty of Metallurgy and Materials Science
University of Mining and Metallurgy,
Al. Mickiewicza 30, 30-059 Kraków - Poland

Rolling and recrystallization textures and their mutual relation were studied in Cu-Si3.0% alloy. Relation between ideal orientation describing rolling and recrystallization texture components were analysed by computer aided procedure. It was found that relations between rolling and recrystallization textures components can be described as a rotation of 30-40° or 60° around some <111> axes however, corresponding deformation texture components are not exactly main rolling texture components but they are in spread of these components. This rotation was not very often of 40° as it should be according to oriented growth theory [2]. Also the choice of the type of <111> axes is inconsistent.

1. Introduction.

During recrystallization process, textures of the deformed materials are usually changed. For many years this changes were explained on the base of two theories. The **oriented nucleation** theory assume, that only recrystallization nucleus with some specific orientations are present in deformed matrix, and their existence influence the recrystallization texture [1]. The **oriented growth** theory assume, that nucleus with defined orientation to the matrix (usually 40° around common <111> axes) grow up faster and the rate of nucleus growth determine recrystallization texture [2]. Still there are no evidence which of the theory better describe texture changes in recrystallization of real materials with different stacking fault energy (SFE) [3-5].

The aim of this work was to study in details texture changes in alloy with low stacking fault energy. For this purpose the copper alloy containing 3,0% silicon was chosen. This work is a part of much wider investigations, which were done for materials with different chemical composition and because of it different SFE and electron concentration [5].

2. Material and experimental procedure

Copper alloy ingots containing 3.0 wt.% of silicon were annealed at 450°C for 1 hour to obtain samples not exhibiting any texture (random oriented material). This material was cold rolled. Set of samples after 75% and 97% deformation were annealed at 350°C and 450°C. Texture was measured by the Schulz method for samples on each stage of the material processing.

3. Deformation and recrystallization texture

Textures of the deformed Cu-Si cold rolled samples (Figure 1.a.) can be described as „alloy type", with two orientation fibres α and τ and also strong S component. In details it could be stated that:

- in fibre α major components are {110}<112> , {110}<001> and fibre α contains all components, which are placed in-between {110}<001> and {110}<111> orientations
- in fibre τ , dominating components are {332}<113>, {552}<115>, {110}<001> and fibre τ is limited by {111}<112> and {110}<001> orientations.

- the S type component is described by the strongest {146}<211>, {135}<112> orientations and some others, which can be defined as {123}<uvw> type.
- weak γ fibre with <111> direction parallel to ND is present in deformation texture after intermediate deformation to 75%.

With increasing deformation, all rolling texture components in α and τ fibres are becoming stronger, with exception of γ fibre components which disappears.

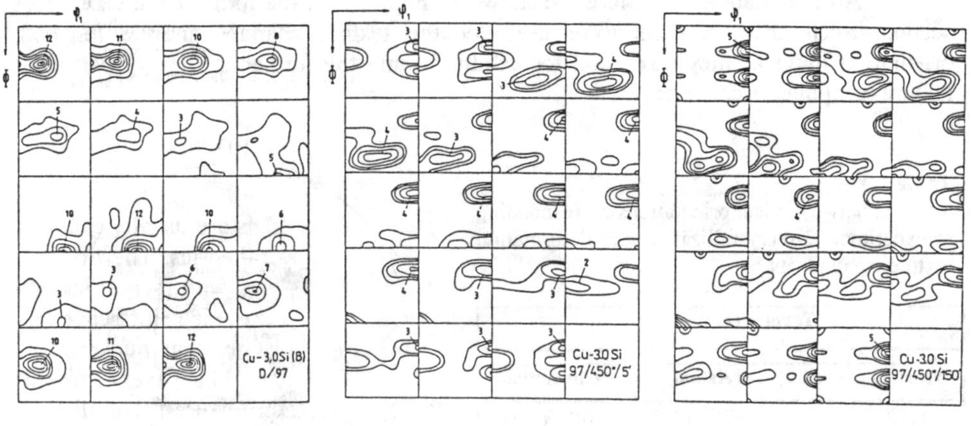

Fig. 1. ODF cross section showing texture of the Cu-Si samples a/. after 97% deformation and subsequent annealing at 450°C for b/. 5 and c/. 150 minutes.

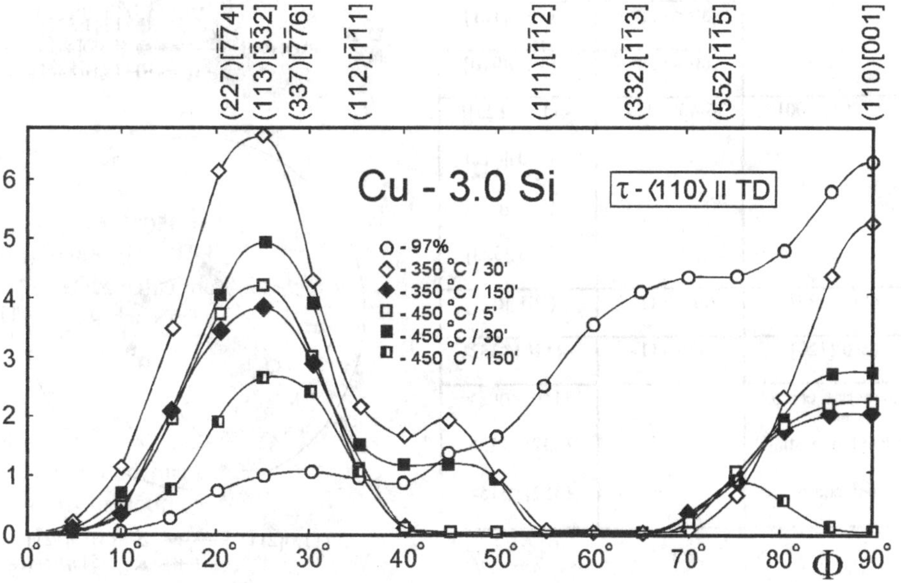

Fig. 2. ODF cross section of the orientation fibre τ, for 97% deformed samples and annealed at different conditions.

When annealing at 350°C and 450°C , texture changes observed during recrystallization are very evident (Figure 1.) and can be described as follow:

- fibre α - is gradually shortened and finally fibre α disappears -
- fibre τ - disappears in part near orientations {332}<113> and {110}<001>
- strong components {113}<332> elongated to {227}<774> and {337}<776> are created (these components lie also in fibre τ).
- other components like {113}<112>, {326}<835>, {012}<021>, {100}<001> and {124}<235> are created, too.

All the changes in τ fibre are shown in Figure 2 in the form of the ODF cross section. Such cross sections were used for the texture changes analyse for both orientation fibres as they are good for exhibiting and the further analyse of particular texture components changes.

Table 1.
Relation between ideal orientations corresponding to experimental recrystallization and deformation texture components.

Texture components		
recrystallization	**relation**	**deformation**
(326)[835]	36-40° <111>	~(011)[211]
(124)[253]	35-40°<111>	~(011)[211]
(113)[211]	38°-40° <111>	~(416)[121]
	30°<111>	~(351)[121]
	60°<111>	~(101)[010]
~(001)[100]	36-40° <111>	~(124)[823]]
		~(123)[412]
		~(123)[634]
		~(123)[523]
(113)[332]	60° <111>	~(101)[010]
(210)[120]	32°<111>	~{110}<332>
no recryst. comp.		{110}<001>
related to existing		{332}<113>
def. comp.		{552}<115>

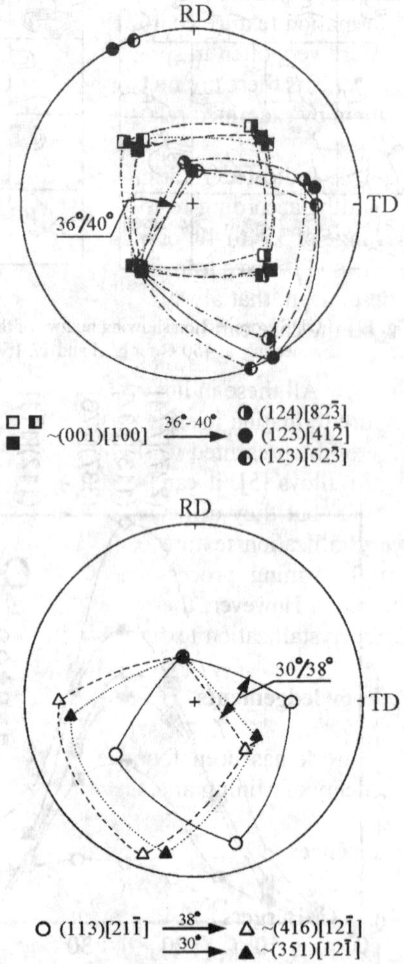

Fig. 3. Stereographic projection presenting mutual relations between some orientations listed in Table 1.

4. Analyse and results discussion

Mutual relations between experimentally measured recrystallization and deformation texture components were analysed. The aim of this analyse was to find, if ideal orientations describing these experimental texture components, can be obtained one from the other by rotation of the type similar to assumed by oriented growth theory [2]. This was done by the computer aided procedure. The relations, which were found for some ideal orientations describing main texture components, experimentally measured in investigated Cu-Si alloy are given in Table 1. Some examples of the ideal orientation rotations are shown on stereographic projection in Figure 3.

From Table 1. it is evident, that relations between ideal orientations corresponding to recrystallization and rolling textures components can be described as a rotation of 30-40° or 60° around respective <111> axes. However, corresponding deformation texture components are not always the main rolling texture components but they are very often in spread of some of these components. For some deformation texture components there are no components in recrystallization texture which are exactly related to them by examined relation (one could not be obtained from another by this rotation).

Indicated rotation around <111> axes was not very often precisely 40° as it should be according to the oriented growth theory [2]. In many cases these angles were in range of 35° to 40° or was equal 60° [5]. Additionally, only rotation around some types of the <111> axes transforms recrystallization components into deformation components. These mean, that always choice of the type of <111> axes are different and no fixed rule for all the case can be found.

All these indicate that oriented growth theory could not explain recrystallization texture formation itself and additionally some other factors should be considered. On the base of the presented work and some previous investigations of Cu-Zn, Cu-Al, Cu-P and Cu-Si alloys [5], it can be stated that deformation textures in all these alloys are very similar, but they differ significantly in recrystallization textures. These differences in recrystallization texture could be caused by: difference in stacking fault energy of the alloys, twining process occurring in recrystallization or segregation of the alloying elements. However, there is not evident explanation how these parameters influence on the recrystallization textures formation.

Acknowledgements

This work has been founded by the Metallurgical and Material Science Department, Academy of Mining and Metallurgy, Krakow, Poland - Grant # 11 110 230

References

[1]. G. Burgers - Z. Physik, **599**, 651 (1920)
[2]. P. A. Beck, H. Hu - Trans. AIME **194**, 83 (1952)
[3]. I. Slakhorst - Acta Metall., **3**, 301 (1975)
[4]. K. Lücke - ICOTOM 7, 195 (1984)
[5]. W. Ratuszek - Wydawnictwa AGH, Kraków - **27**, (1995)

DEFORMATION AND RECRYSTALLIZATION TEXTURES AFFECTED BY THE CHANGE OF THE DEFORMATION PATH IN CuGe8 ALLOY.

J. POSPIECH*, I. KOPACZ*, J. GRYZIECKI**

**Polish Academy of Sciences, Institute of Metallurgy and Materials Science, 30-059 Krakow, ul. Reymonta 25, Poland*

***Academy of Mining and Metallurgy, Department of Metal Forming, 30-059 Krakow al. Mickiewicza 30, Poland*

The development of the deformation and the primary recrystallization textures of samples CuGe8 alloys subjected to unidirectional and additional slight transverse rolling were analysed and compared with corresponding textures of unidirectionally and cross rolled materials. The additional rolling strain imposed by a small reduction of the transverse rolling causes relatively great changes in the textures especially after recrystallization. Regularities were found and explained by using the orientation distribution function (ODF).

INTRODUCTION

If a metal with a stabilized texture caused by unidirectional rolling is additionally deformed with the rolling direction changed (e.g. 90° around ND), then there appear tendencies constraining the new stabilization conditions thus destroying the effects of the former. This additional rolling, introducing a change in the deformation path, leads to the destabilization of the earlier formed structure. Then there become activated such systems which have not taken part in the deformation or have participated in it to a minimal extent. This again leads to changes in the distribution of the dislocation densities and, as a consequence, to unloading of destabilities in the form of deformation localization such as micro and macro shear bands. The phenomena occurring in the microstructure caused by changes of the deformation path (e.g. by additional transverse rolling) have been described and discussed by Korbel [1]. It has been shown there that structural alterations are accompanied by texture effects in the deformed as well as in the recrystallized materials. However, there exists, as yet, no systematic analysis of the texture effects caused by change of the deformation path. There is not much information especially with regard to metals of low stacking fault energy (SFE). It can be expected that the change of the deformation path will have a destructive effect on the texture but, at the same time, there should appear new orientations corresponding to the new rolling geometry. More complicated is the case of the recrystallization texture. Great gradients of dislocation densities, numerous twin intersections and local lattice rotations in areas of shear bands determine the locations of nucleation during the recrystallization annealing. The change of the deformation path will disturb the earlier formed distributions of potential recrystallization nuclei and it may introduce significant changes in the recrystallization texture. Effects destroying the earlier formed recrystallization texture are then expected. Texture alterations induced by transverse rolling will be the subject of a more detailed analysis in the present paper. Deformation textures as well as textures after primary recrystallization of the CuGe8 alloy which is characterized by a very small stacking fault energy will be considered.

EXPERIMENTAL

The investigated material was a copper alloy with 8 wt % of germanium. Such alloy is characterized by small SFE amounting to about $4 mJ/m^2$ [2]. The alloy prepared from components of 99.99% purity was melted in an inductive furnace in argon atmosphere. The obtained ingot was homogenized in 800°C during 20 hours in argon atmosphere. After hot rolling up to thickness of 3 mm the alloy was annealed at 400°C for 15 minutes. The alloy was next cold rolled in a triple way: unidirectional rolling up to 70% reduction, cross rolling up to 70% (alterating the rolling direction by 90° after each pass) and with a 5% transverse rolling (after an unidirectional rolling up to 65% reduction). Part of the CuGe8 alloy (from each deformation mode) has been annealed at 400°C for 20 min. For each sample after deformation and after recrystallization three pole figures {111}, {100}, {110} were measured. The ODFs as the basis for quantitative texture analysis have been calculated from pole figures using the ADC method [3]. The ODFs are presented in the space of Euler angles (φ_1 ϕ φ_2) in 5 degree section of φ_2-angle.

RESULTS AND DISCUSSION

1. Rolling texture.

1.1 The texture of a sample unidirectionally rolled up to 70%. The observed texture shown in fig.1 is typical for fcc metal alloys of low SFE from the range of medium reduction (see e.g. [4]). The brass component Bs:{110}<112> connected along the α-fibre with the weak Goss:{110}<001> component dominates in the texture. Also the β fibre is visible connecting the brass component with the weakly represented components S:{213}<364> and C:{112}<111>. On the section φ_2=45° one can recognize the consequences of twinning of C-oriented grains through the occurrence of the {552}<115> twin orientation and its scattering region in the direction of the position {111}<112>, giving evidence of rotation of the twin planes to parallelism with the rolling plane. The shear bands intersecting the twin panels {111}<112> introduce their own intrinsic orientation near the Goss position [5].

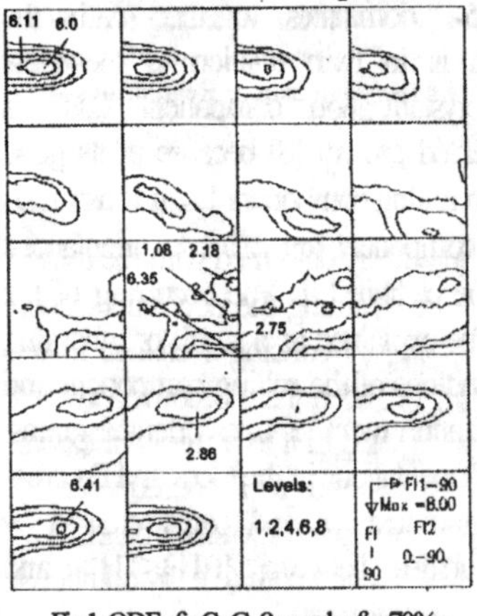

Fig.1 ODF of a CuGe8 sample after 70% unidirectional rolling.

1.2 The texture of 5 % transversely rolled sample after unidirectional rolling up to 65%. The small amount of additional transverse rolling introduces disturbing alterations into the earlier formed texture. One can see in fig.2 section φ_2=45° that orientations near C together with the twin positions from {111}<112> in the direction to the Goss position are most sensitive to the destabilizing effects of transverse rolling. A general lowering of the orientation density is observed. The orientation densities along the β fibre decrease from the Bs-position through S-position till to a complete disappearance near the C-position. These alterations are accompanied by the occurrence of new orientations as a consequence of the activation of new twinning processes and shear band formation [1]. New orientations become visible as a weak incomplete fibre running through the positions {012}<021>] and through twin positions {124}<021> till the orientation {156}<432>. These position can result, according to the investigation on single crystals [6], from orientations induced by the newly formed shear bands rotated about 90° around the normal direction.

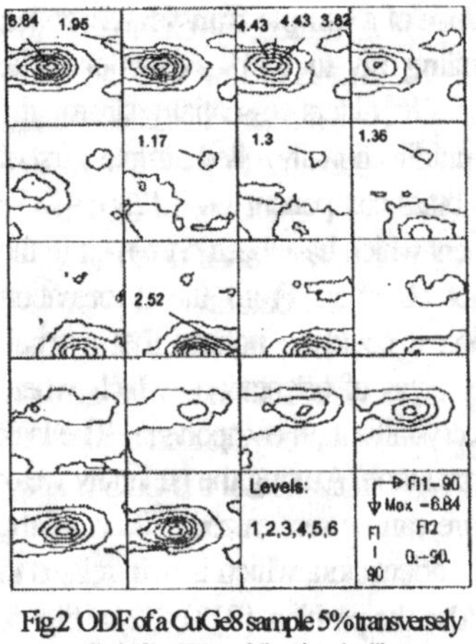

Fig.2 ODF of a CuGe8 sample 5% transversely rolled after 65% unidirectional rolling.

1.3 The texture of a cross rolled sample up to 70%. The texture presented in fig.3 formed by cross rolling shows similarity to the earlier described texture after a small additional transverse rolling. It is less complex due to the four-fold symmetry induced by the cross rolling [7] and due to the reduction of a number of components to one widely smoothed around the {011}<322> orientation (the shifted Bs orientation). Strongly represented twin positions of the main component appear. These are {114}<3 12 0> (fig. 3 section

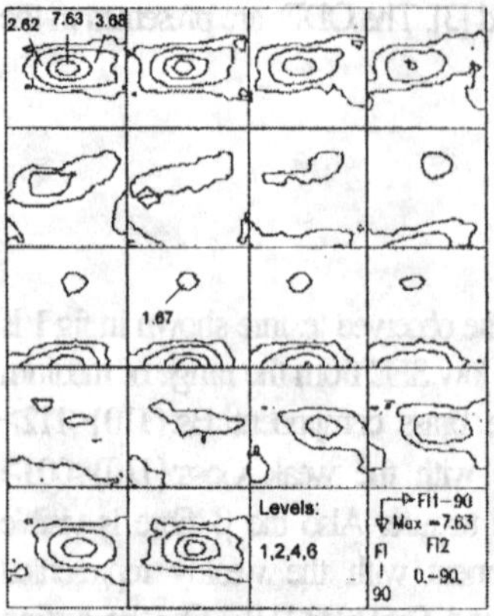

Fig3 ODF of a CuGe8 sample after 70% cross rolling.

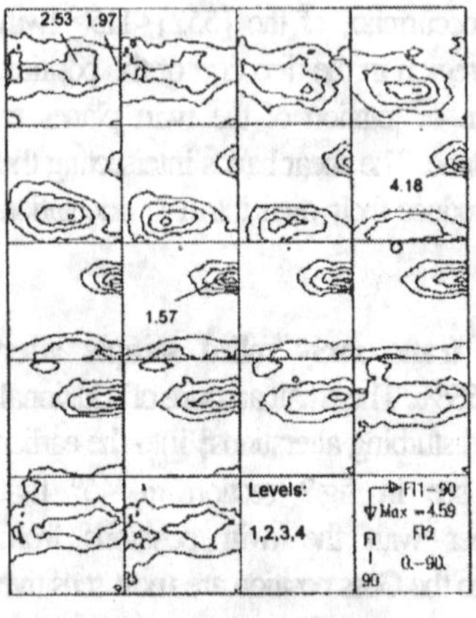

Fig4 ODF of a CuGe8 sample recrystallized after 70% unidirectional rolling.

φ_2=0°) and the positions around {011}<311> and {011}<233>, which extend the scattering region of the {011}<322> component.

2. *Recrystallization texture.*

2.1 Recrystallization texture of a sample rolled up to 70% reduction. In the recrystallization texture the brass recrystallization component Bs/R:{236}<385> dominates together with the {114}<221> position which is in twin relation to the Goss component. The brass recrystallization component Bs/R is interpreted as a result of oriented growth [8] because of its good relation 40°<111> to the main rolling component Bs. It is assumed that the nuclei of the Bs/R component are twins to orientations from the area Goss/Bs of the α fibre [9]. The occurring of the {114}<221> orientation can be explained as a result of inheriting the orientations of the surroundings of the rolling component and also as positions of recrystallization twins of Goss oriented grains. There are also small shares of orientations in the Goss and Bs areas inherited from the deformation textures and partly created by annealing twins in the subarea between {011}<111> and {011}<112> orientations (fig. 4 section φ_2=0°).

2.2 The recrystallization texture of a sample transversely rolled 5% after unidirectional rolling up to 65% reduction. The texture is shown in fig. 5. It differs essentially from the recrystallization texture of the unidirectionally rolled sample. First of all, the main recrystallization brass component Bs/R:{236}<385> and the {114}<221> component which have been dominant in the unidirectional sample are not observed. From the deformation textures before recrystallization it could be noticed that a small additional transverse rolling reduction leads to the disappearance of areas of orientations which where interpreted as orientations of nucleus of the above mentioned mean recrystallization components. The lack of these components in fig.5 may confirm the correctness of this interpretation. Among the generally weak components visible in fig.5 the effects of inheriting the rolling texture can be recognized. Effects of the formation of annealing twins are also observed where the {011}<122> orientation, which is twin related to the Goss orientation, is the most pronounced (fig. 5, section φ_2=0°). Also the position {012}<321> (fig. 5, section φ_2=20°) which has a twin relation to {012}<021>, is visible. In the recrystallization texture the occurrence of the {136}<310> position (fig. 5, section φ_2=20°), which creates a good oriented growth relation with the {011}<655> position should be mentioned. This position will become dominant in the recrystallization texture after cross rolling. The nuclei of this orientation can result from recrystallization twins of the {034}<243> orientation (fig. 2, section φ_2=0°).

2.3 Recrystallization texture of a cross rolled sample up to 70% reduction. Similarly as in the previous case, in the recrystallization texture the phenomenon of inheriting the main rolling component which occurs together with its twins from its wide surroundings can be noticed. In the recrystallization texture the most

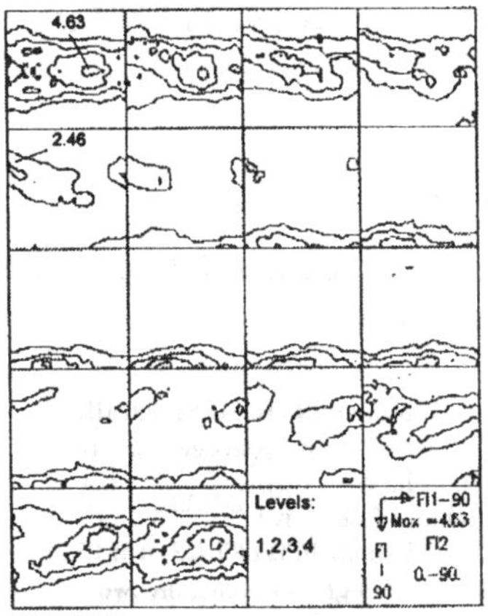

Fig.5 ODF of a CuGe8 sample recrystallized after 65% unidirectional and 5% transversely rolled.

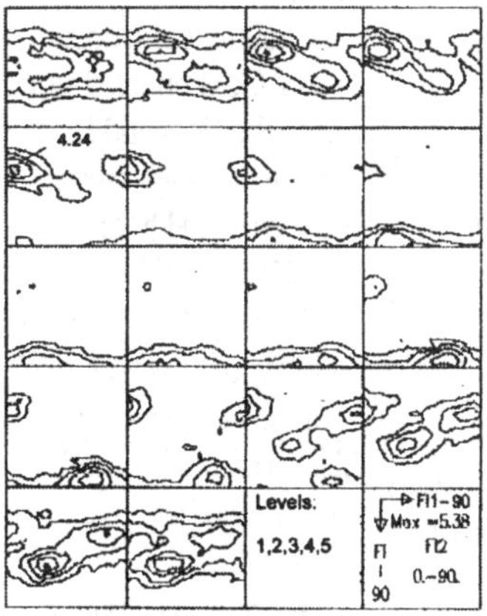

Fig.6 ODF of a CuGe8 sample recrystallized after 70% cross rolling.

pronounced is the {136}<310> component which was observed as a weak one in the previous described case (fig. 5). The great share of this component is justified by the great share of the {034}<243> component (fig. 3, section φ_2=0°) which is the source of nuclei due to the mechanism of twinning.

CONCLUSIONS

Transverse rolling leads to a destabilization of the unidirectional rolling texture, but at the same time it introduces orientations from newly activated twin systems and new shear bands. In the action of destabilization disappear all texture components except the brass component which shifts in the direction of {110}<233>, symmetrical in respect to the four-fold axis parallel to the rolling plane vector. This position is achieved during cross rolling. The main recrystallization brass component Bs/R and the {114}<221> component formed in the recrystallization texture of an unidirectionally rolled sample are not observed in the recrystallization texture of a additionally transverse rolled sample. The recrystallization texture after a small additional transverse rolling reduction does not differ in general from the recrystallization texture after cross rolling. The newly formed component can be explained as being due to the oriented growth and to the oriented nucleation mechanism.

This research was supported by the Committee of Scientific Research (KBN) Poland Grant T08B05910.

REFERENCES

1. A. Korbel: Proc. Of the Z. S. Basiński Intern. Symp. On Crystal Plasticity (1992) p. 42.
2. P. C. J. Gallagher: Met. Trans. **1**, (1970) 2429.
3. K. Pawlik, J. Pospiech, K. Lücke: Textures & Microstructures **14-18** (1990), 25.
4. J. Gryziecki, W. Truszkowski, J. Pospiech, J. Jura: Textures & Microstr. **14-18** (1991), 1061.
5. H. Paul, J. Jasieński, A. Piątkowski, L. Lityńska: Proc. IX Conf. Electron. Microscopy of Solids, Kraków-Zakopane 1996, p. 239.
6. J. Jasieński, A. Piątkowski: Arch. of Metallurgy **38** (1993), 279.
7. A. Böcker, H. J. Bunge: Textures & Microstructures **14-18** (1991), 1037.
8. J. Pospiech, K. Lücke: Z. Metallkde **70** (1979), 567.
9. H. Eichelkraut, J. Hirsch and K. Lücke: Z. Metallkde **75** (1984), 113.

DETERMINATION OF GRAIN ORIENTATIONS IN RECRYSTALLIZED ODS ALLOY INCOLOY MA956

A. WASILKOWSKA, W. RATUSZEK
A. CZYRSKA-FILEMONOWICZ
University of Mining and Metallurgy, Faculty of Metallurgy and Materials Science
Al. Mickiewicza 30, PL-30059 Kraków, Poland

The crystallographic grain orientations in ferritic ODS alloy INCOLOY MA956 were systematically established by Laue-Greninger method. The recrystallized bar exhibited a pronounced texture inhomogeneities in the cross section: a sharp texture around <111> into the core, and randomly grain orientations near the outer surface. The calculations of a maximum Schmid factor give an information about distributions of preferred slip systems: about 60% of {112}<111> and about 30% of {123}<111> were determined. Only in the outer zone of the bar cross section, where random grain orientations were found, slip systems {110}<111> are preferable. The variations of the maximum Schmid factor in the cross section of the bar might be responsible for plastic instabilities of the material during creep deformation.

1 Introduction

Oxide Dispersion Strengthened alloys (ODS) have been developed for applications at temperatures above 1000°C, mostly for advanced power stations and gas turbine components [1]. The coarse elongated grain structure, obtained by extrusion and recrystallization of the mechanically alloyed powder, provides very good uniaxial creep properties. ODS alloys possess directionally dependant elastic and creep/fracture properties, because of the existence of a strong material texture [2].

For the quantitative description of high temperature creep behaviour of ODS alloys, the evaluation of texture is very important. For obtaining statistically representative data of an orientation distribution, the experimental method used for texture measurements is significant.

The aim of the present work was to evaluate the nature of crystallographic orientations in commercial ferritic ODS alloy INCOLOY MA956. An intention of such investigations was improving our understanding of the deformation mechanisms in ferritic ODS alloys.

2 Experimental details

INCOLOY MA956 was supplied by Inco Alloys International as hot extruded and recrystallized (1350°C/1h) 20mm diameter bar. Chemical composition of the bar was as follows: Fe-19.9Cr-4.6Al-0.4Ti-0.5Y_2O_3 (in wt %).

Three specimens (numbered 1, 2 and 3) cut from different places of the bar cross section were polished and etched. One longitudinal specimen with well defined orientation, was taken very close to the specimen nr 1.

The specimens were examined using light microscope and X-ray diffractometer. Crystallographic orientation measurements were performed by Laue-Greninger method using a tungsten lamp and 0.5mm collimator. The Laue measurements were taken systematically along defined lines (every 1mm), or over concentric circles (every 15-30°) on the etched surface.

The crystallographic orientations were determined using a computer programme based on the Greninger method. The calculations of the maximum Schmid factor distribution lead to the determination of the preferable slip systems.

3 Results and discussion

The macrostructure of the as-received bar MA956 exhibits large columnar grains with up to few millimetres size (Figure1). The grains were elongated in extrusion direction (grain aspect ratio, GAR, exceeds 30:1). Two perpendicular sections of specimen nr 1 (Figure1) were oriented relative to each other in a way, which enables the change in crystallographic orientation along the bar length to be detected. The grain orientations of three different specimens were measured for the statistical analyses of the bar.

The results of X-ray Laue diffraction measurements are presented in Figure 2. It was found, that there are pronounced differences in the grain orientations on the specimen cross section. The columnar grains near the bar surface bar exhibited a random orientation (high angle grain boundaries). The specimen core consisted of a few grains oriented about 4-10° with respect to the <111> direction. Orientations spread from <667> to <334> and from <757> to <323> (Figure 3a).

Similar results were obtained for the other MA956 specimens investigated (Figure 3 b,c). The sharp texture was present in the core of the bar (about 12mm in diameter); it means in about 65% of the cross section. The activation of different slip planes in the bar cross section might be a reason for the plastic instability previously observed in MA956 alloy [3,4].

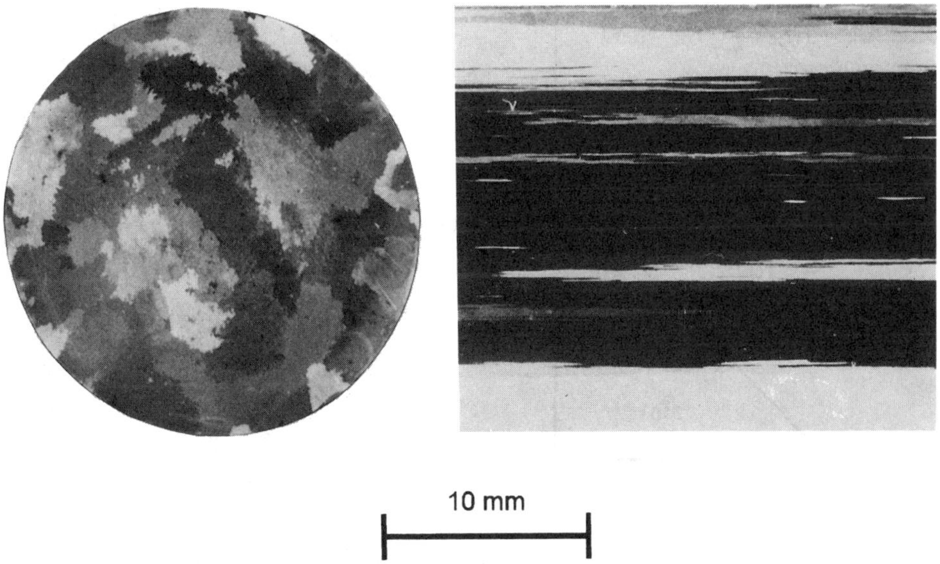

Figure 1: INCOLOY MA956 macrostructure; cross - (left) and longitudinal (right) sections.

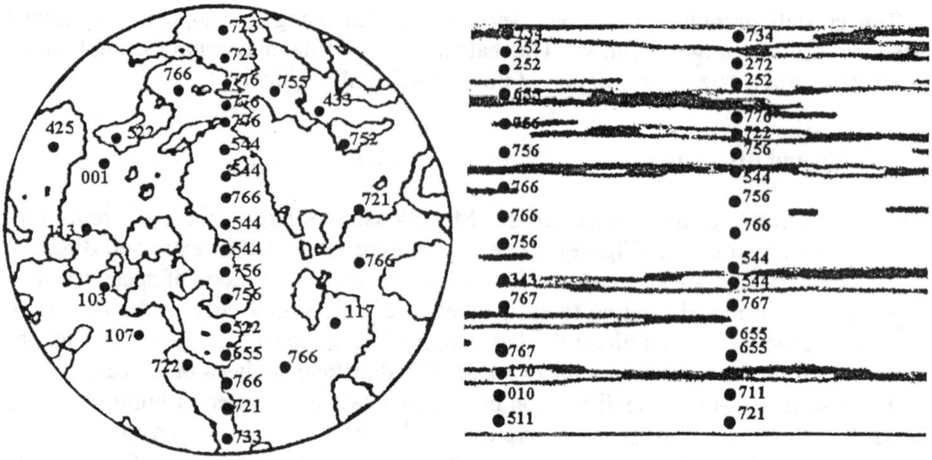

Figure 2: X-ray Laue measurements. Grain orientations in longitudinal direction of the bar: plane indexes (left) and direction indexes (right).

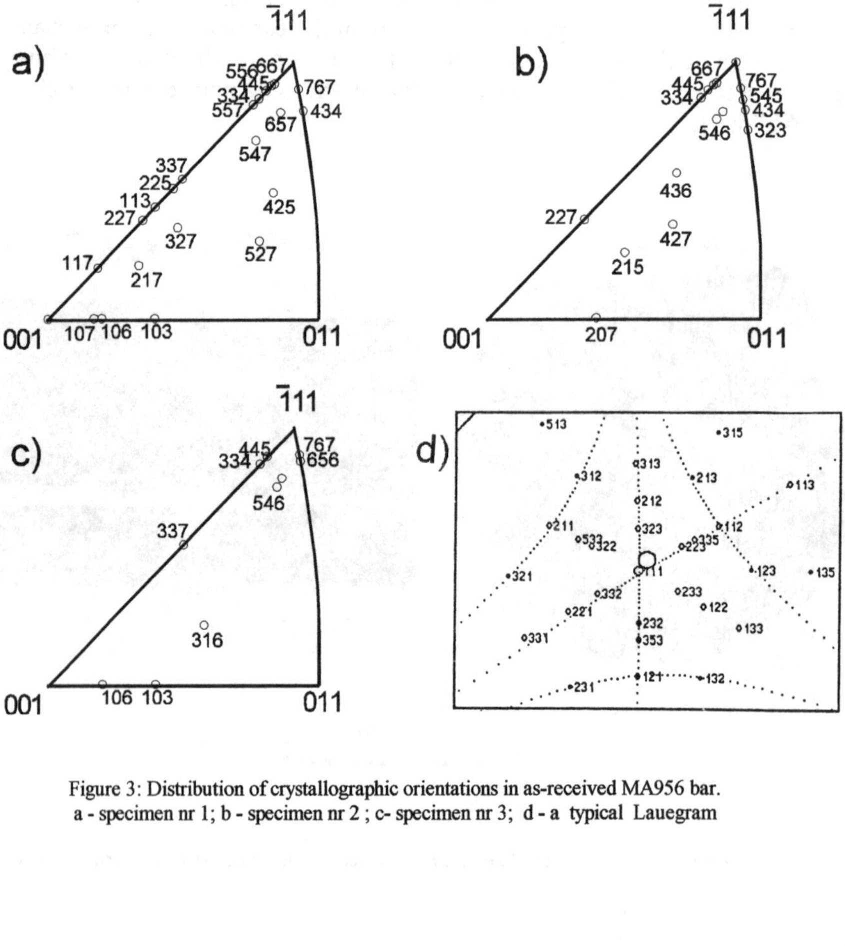

Figure 3: Distribution of crystallographic orientations in as-received MA956 bar.
a - specimen nr 1; b - specimen nr 2 ; c- specimen nr 3; d - a typical Lauegram

For a standard creep specimen of 10mm gauge diameter, the Schmid factor distribution was calculated. Table 1 shows the mean value of a maximum Schmid factor and the dominant slip systems in <111> textured MA956 bar.

Table 1. Quantitative determination of dominant slip planes in <111> textured MA956 bar

specimen nr	mean maximum Schmid factor	dominant slip systems fraction , %		
		{110}<111>	{112}<111>	{123}<111>
1	0.328	0	60	40
2	0,397	4	65	31
3	0,490	7	50	43

The dominant slip plane for the MA956 alloy with <111> texture was established as {112} plane (about 60%). The next preferred slip system was {123}<111> one (about 30%). The slip plane {110} should not be dominant for the material investigated.

4 Conclusions

1. The extruded and recrystallized MA956 bar exhibited crystallographic texture around <111> in the core and a random orientation near the surface of the bar.
2. The results show that the maximum Schmid factor varies on the bar cross section due to the local differences in the grain orientations.
3. Analyses of orientation measurements of the individual grains allow to establish the preferred slip systems for MA956 bar investigated. The {112} slip plane should be the dominant one.

Acknowledgements

The authors appreciate the financial support of State Committee of the Scientific Research (grant nr 18.110.367). The valuable contribution of P. J. Ennis (Research Centre Jülich) to joint studies on the properties of ODS alloys is gratefully acknowledged.

References

1. J. J. Fischer, J. J. de Barbadillo and M. J. Shaw in *Proceedings of an ASM Int. Conference on Structural Application of Mechanical Alloying*, p.79, eds F. H. Froes and J. J. de Barbadillo (Myrtle Beach, South Carolina, 1990)

2. A. Wasilkowska, H. J. Penkalla and A. Czyrska-Filemonowicz in *Proc. of 5th European Conf. on Advanced Materials and Processes and Applications* **vol.1,** p.185, eds L. A. J. L. Sarton and H. B. Zeedijk (Maastricht, Holland, 1997)

3. B. Dubiel, W. Osuch, M. Wróbel, P. J. Ennis, A. Czyrska-Filemonowcz, *J. Mat. Processing Technology* **53**, 121 (1995)

4. B. Dubiel, M. Wróbel, P. J. Ennis, A. Czyrska-Filemonowicz, *Scripta Mat.* (in print)

Phase Transformation

STRUCTURE AND PHASE TRANSFORMATION OF NANOMETER-SIZED PARTICLES OF ALLOYS EXAMINED BY TEM AND ELECTRON DIFFRACTION

T.Tadaki

Division of Physics, Department of Natural Science, Osaka Women's University, 2-1 Daisen-cho, Sakai, Osaka 590, Japan

We prepared nanometer(nm)-sized particles of several alloys undergoing order-disorder or martensitic transformations in their bulk states by vacuum evaporation, and examined their atomic structures and phase transformations therein by transmission electron microscopy and electron diffraction. As a result, remarkable size effects were observed on both the two kinds of phase transformations, *i.e.*, large depressions of their transformation temperatures. A computer calculation of the ground state energy of nm-sized particles of the Cu_3Au alloy by the modified embedded atom method and the Monte Carlo simulation of the order-disorder transformation indicated that the depression of the critical temperature T_c is essentially due to surface effects. Nm-sized austenite particles of Fe-Ni alloys around 25 at.% Ni were observed to be stabilized to a great extent, as previously reported. It was found, however, that the well-stabilized austenite particles transformed into the martensite ones when kept at 77 K for a prolonged period. This result revealed that contrary to previous reports even the nm-sized austenite particles undergo the martensitic transformation at subzero temperatures, although isothermally. The causes giving rise to the drastic stabilization brought about by the reduction of size are discussed.

1 Introduction

When the size of materials decreases down to nanometer(nm) scale, some materials exhibit unique crystal structures and physical properties which are different from those of the corresponding bulk materials[1-4]. Such a particle is often called atom cluster[5], in which are involved atoms ranging from 10^2 to 10^5 and of which the size ranges from 2 nm to 20 nm in diameter. One of the important characteristics of atom clusters is that the proportion of surface atoms to the whole becomes significantly large(80~20%). Because of this, lattice softening takes place typically in atom clusters, manifesting itself as large decreases of melting[6] and characteristic temperatures[7], and force constants[7]. Then, we have been studying size effects on order-disorder and martensitic transformations in nm-sized particles of several alloys by means of transmission electron microscopy(TEM) and electron diffraction(ED), because these phase transformations are considered to be closely related to lattice softening. A computer calculation and the Monte Carlo simulation as to an order-disorder transformation have also been made in order to make clear the essence of size effects on it. Here are summarized some results that we have obtained so far for order-disorder and martensitic transformations in nm-sized particles of Cu-Au[8, 9] and Fe-Ni alloys[10], respectively.

2 Experimental Procedures

Nm-sized particles of Cu-Au alloys with nominal compositions Cu_3Au and CuAu were prepared by simultaneous vacuum evaporation of high purity Cu(99.99%) and Au(99.95%) wires at room temperature in a base pressure of 3 x 10^{-4} Pa onto amorphous carbon films. After evaporation, specimens were heat-treated at 773 K for various periods so that nm-sized particles were formed, and subsequently at 563 K for 1 h for ordering in a JEM-2010 type electron microscope using a specimen heating stage. The base pressure inside the electron microscope was kept at 7×10^{-5} Pa. The order-disorder transformations in the alloy particles were followed by ED at temperature. Lattice imaging was carried out to examine if nm-sized particles are crystallized and ordered.

Nm-sized particles of Fe-Ni alloys with compositions 22.1, 25.2 and 28.6 at.%Ni were also prepared by vacuum evaporation of these master alloys onto amorphous carbon or alumina films at room temperature. The master alloys were made by high frequency induction melting of pure Fe(99.95%) and Ni(99.99%) metals. After evaporation, specimens were annealed in the electron microscope at various temperatures below 873 K for up to 2 h. Chemical compositions of the vacuum-evaporated specimens were examined by energy dispersive X-ray spectroscopy (EDX) using an H-600FE type analytical electron microscope.

3 Order-Disorder Transformations in Nanometer-sized Particles of Cu-Au Alloys

Atomic orderings in submicron- or nm-sized particles of alloys have been studied so far by several workers by TEM[11, 12]. According to their results, the fine particles possessed the same ordered structures as those in bulk. However, the interaction energies between atoms in nm-sized particles are expected to be modified to a great extent by the large proportion of surface atoms. The critical temperature for an order-disorder transformation T_c which is expressed by the interaction energies[13] is thus expected to be significantly affected thereby. Then, we examined atomic structures and the order-disorder transformations of nm-sized particles of Cu-Au alloys with nominal compositions of Cu_3Au[8, 9] and CuAu[14] by TEM and ED.

Well-defined nm-sized particles of the Cu_3Au alloy with average size less than 10 nm in diameter were formed by heat-treating at 773 K for 1 h after evaporation. Lattice fringes with an interval corresponding to interplanar spacing of {111} planes of the fcc lattice were observed not only in the particles as large as about 10 nm, but also in those as small as 2 nm or less. It was thus seen that even nm-sized particles are entirely crystallized, possessing fundamentally the same crystal structure as that of the bulk. The particles were subsequently heat-treated for ordering at 563 K for 1 h, and were explored by TEM and ED whether or not they were ordered as in the bulk. In the upper part of Fig.1 is shown the size of the Cu_3Au alloy particles in which lattice fringes with an interval corresponding to the interplanar spacing of {100} superlattice planes of the $L1_2$ type superlattice were observed, while the lower part of the figure is for the fundamental lattice fringe. It was noted that while the fundamental lattice fringe was observed even in the particles as small as about 2 nm or less, the superlattice fringe was not observed in those smaller than about 4 nm. A similar experiment was also made for the nm-sized particles of the $L1_0$ type CuAu[14] alloy. In this case the superlattice fringes were not observed in the particles smaller than about 6 nm. The TEM observations thus suggested that there seem to be certain critical sizes for the $L1_2$ and $L1_0$ orderings. In fact, as to the nm-sized Cu_3Au particles, Yasuda and Mori[15] recently reported a result consistent with ours by TEM.

Temperature changes of ED patterns taken from the nm-sized Cu_3Au particles during

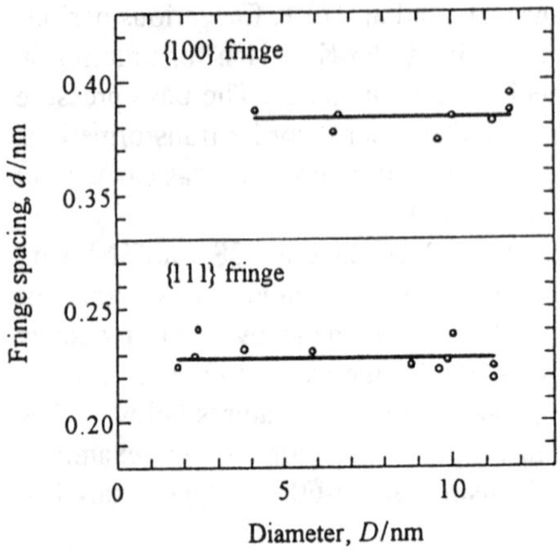

Fig.1: A diagram showing the size of alloy particles in which {100} and {111} lattice fringes were observed.

heating up to 773 K revealed that the critical temperature T_c for the $L1_2$ ordering lies at around 574 K. A similar experiment was also made for the nm-sized particles of the $L1_0$ type CuAu alloy, for which T_c was observed to be located at around 628 K[14]. In order to compare the observed T_c's of the nm-sized Cu-Au particles with those of the corresponding bulk, we estimated average compositions of the particles by measuring average lattice parameters of their assemblies. To do this, a lattice parameter versus Au content curve was used which schematized experimental data found in a literature[16]. Lattice parameters of the assemblies were calculated from the diameters of the Debye rings, referring to a camera constant predetermined by using TlCl. Average Au contents of the nm-sized particles thus obtained were 26.4 at.% and 47.3 at.% for the nominal compositions of Cu_3Au and CuAu, respectively. Then, according to the phase diagram[17], the critical temperatures T_c's for the Cu-26.4 at.%Au and Cu-47.3 at.%Au alloys are roughly 663 K and 683 K, respectively, in their bulk states. Strictly speaking, the latter value of T_c stands for the $A1$ to CuAuII transformation. However, the CuAuII ordered phase with one dimensional long-period ordered structure was not observed in the nm-sized particles. It thus turned out that the differences in T_c between the particles and the corresponding bulk amount to about 90 K and 55K for the Cu_3Au and CuAu alloys, respectively.

In order to understand what the observed size effects originate from, we calculated the ground state energy of the nm-sized Cu_3Au particles by utilizing the modified embedded atom method (MEAM) proposed by Baskes[18], by which many-body interactions among atoms are taken into account semiempirically. The calculations were made for three model crystals of spherical nm-sized particles with 2 and 3 nm in diameter, containing 321 and 1061 atoms, respectively, and of a bulk with a size of 5x5x5 unit cells, on which the periodic boundary condition was imposed. The interatomic interaction was taken up to the second nearest neighbors. The results thus obtained are shown in Table 1.

Table 1: The ground state energies(eV/atom) of nm-sized particles and bulk of the Cu_3Au alloy in perfectly ordered and disordered states, and their differences, calculated by MEAM.

	$D = 2$ nm	$D = 3$ nm	bulk
E_{order}	3.322	3.441	3.712
$E_{disorder}$	3.269	3.384	3.645
$E_{order} - E_{disorder}$	0.053	0.057	0.066

The followings are to be noted: (a)the ground state energy is the largest in the bulk among the three model crystals irrespective to atom configuration, and it decreases with decreasing size; (b)not only in the bulk, but also in the nm-sized particles the ground state energy in ordered state is larger than that in disordered state; (c)the difference in ground state energy between perfectly ordered and disordered states is the largest for the bulk, and it decreases as the size decreases. It thus turned out that the disordered state tends to be stabilized as the particle size decreases. Consequently, it is supposed that the energy difference would vanish eventually at a certain particle size. This trend is consistent with the experimental results that there seem to be certain minimum sizes for the $L1_2$ and $L1_0$ orderings.

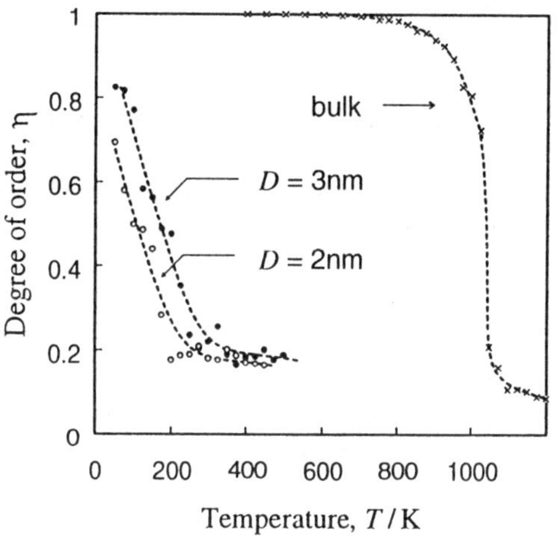

Fig.2: Long-range order parameter versus temperature curves obtained by Monte Carlo simulation for nm-sized particles and bulk of the Cu_3Au alloy.

The Monte Carlo simulation was made for the $L1_2$ ordering by using the many-body potential obtained by MEAM. We prepared simulation cells corresponding to the same model crystals with the stoichiometric composition of Cu_3Au as assumed for the calculation of the ground state energy. For the three model crystals the thermal equilibrium long-range order parameter η's defined by Bragg and Williams were calculated and are shown in Fig.2 as a function of temperature. As seen from the figure, the atomic ordering in bulk is well developed at lower temperature, and upon heating the order parameter decreases sharply at a certain temperature (~1060K). The latter demonstrates that the order-disorder transformation in Cu_3Au is of the first order, as in fact observed in the bulk. However, the corresponding T_c obtained by the Monte Carlo simulation is obviously much higher than that actually observed, 663 K. This overestimation is, as previously pointed out[19], due to the local lattice-distortion effects, which were ignored in the present simulation. Apart from this discrepancy, the followings are to be noted: (i)T_c in the nm-sized particles is much lower than that in the bulk, and it decreases further with decreasing size; (ii)the atomic ordering in the nm-sized particles do not develop sufficiently even near 0 K; (iii)the decrease of η with increasing temperature is much slower in the nm-sized particles. Because of the result (iii), we cannot define the transition point well for the nm-sized particles, and hence can only say that T_c in the nm-sized particles is as a whole depressed to a great extent, compared with that in the bulk.

Now, the depression of T_c may be understood in the following way: Since the energy difference between perfectly ordered and disordered states becomes smaller in nm-sized particles, it would become easier to be overcome by thermal energy; accordingly the disordering is expected to take place at much lower temperatures, as in fact observed. The decrease in total energy difference between the ordered and disordered states with decreasing particle size and the result (ii) mentioned above imply that the perfectly ordered state is not of the lowest energy in the nm-sized particles possibly because of the surface energy. In other words, it appears that in the nm-sized particles the reduction of the surface energy takes preference over the energy gain due to ordering. It thus may be concluded that the depression of T_c in the nm-sized particles is essentially due to surface effects.

4 Martensitic Transformation in Nanometer-sized Particles of Fe-Ni Alloys

The martensitic transformations in ultrafine particles of metals and alloys ranging from 10 nm to 200 nm in size have recently been studied by several workers for Fe-Ni and Co-Fe alloy systems[20-23]. According to them, above room temperature the ultrafine particles undergo martensitic transformations and the reversions at the same temperatures as in bulk materials, while below room temperature the transformations are suppressed even when cooled down to 4 K or when lattice defects are introduced by deformation. Kajiwara *et al.*[21] thus suggested that the stabilization of austenite is not simply due to the lack of favorable nucleating sites, as suggested earlier by Cech and Turnbull[24], but due to the difficulty of the accommodation of plastic deformation in the austenite particles. Considering that the ultrafine particles studied so far may not be so small as that lattice softening become dominant, we have studied the martensitic transformation in nm-sized particles of Fe-Ni alloys in order to clarify size effects on it.

As-deposited specimens of Fe-Ni alloys were island-like, and consisted of fine grains of the bcc martensite smaller than several nm, and in addition $NiFe_2O_4$ oxides. Chemical compositions of the as-deposited specimens examined by EDX were regarded as substantially the same as those of the respective master alloys. When heating the specimens up to 773 K, it was noted that the specimens became entirely a single phase of the fcc austenite, the average size of the austenite particles being several nm. In Fig.3 is shown an example of temperature variations of ED patterns taken from an as-deposited specimen of the Fe-22.1 at.% Ni alloy, indicating such a phase change upon heating. The ED patterns shown in Fig.3 were recorded by a slow scan CCD camera. The reflections as denoted by figures 1, 2 and 3 in Fig.3 correspond to the 311 reflection of the cubic oxide, $NiFe_2O_4$[25], and the 110 and 200 reflections of the bcc martensite of Fe-Ni alloys[26], respectively.

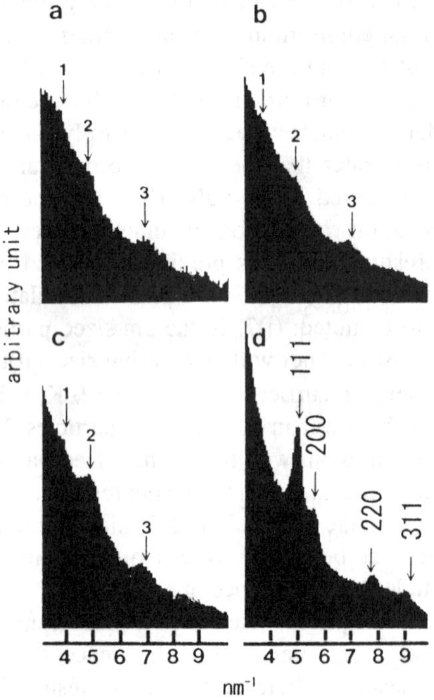

Fig.3: Variation of ED profile upon heating up to 773 K. (a) to (d) are at room temperature, 573 K, 673 K and 773 K, respectively.

Referring to the composition dependence of the martensitic transformation temperatures in the Fe-Ni alloy system[20], the Fe-22.1at.%Ni alloy, for example, should be fully martensitic at 773 K in the bulk state. Nevertheless, the initial fine grains of martensite transformed into the austenite during heating at 773 K. Moreover, it was noted that the austenite particles thus formed were stabilized even when cooled to room temperature. This means that not only the martensite start temperature, M_s, but also the austenite start temperature, A_s, are depressed to a great extent by the reduction of size down to nanometer scale. Therefore, the stabilization of austenite in this case is not simply due to the decrease in M_s, but due to that in T_0, the temperature at which the free energies of austenite and martensite are equal to each other. So far as M_s temperature is concerned, the decrease amounted to at least about 200 K, compared with that of the bulk state for the Fe-22.1at.%Ni alloy. The nm-sized particles thus formed remained fcc even when cooled to 105 K. This drastic stabilization of austenite was in agreement with the previous reports. However, it was observed[27] in the present study that the nm-sized austenite particles undergo the time-dependent martensitic transformation at subzero temperatures: When merely cooled to 105 K, no indication was observed for the martensitic transformation, nor at room temperature after being cooled at 105 K for 12 h; but, when kept at 77 K for 10 d, the austenite reflections disappeared, and instead the 110 and 200 reflections from the martensite appeared. These observations thus indicated that contrary to previous reports even the nm-sized austenite particles undergo the martensitic transformation at subzero temperatures, although isothermally.

There may be no doubt for that the reduction of size results in a drastic stabilization of austenite, because previous reports by several workers are in agreement with one another in this respect. It thus appears that the lattice softening as well as the increase of surface areas as the favorable nucleation sites brought about by the reduction of size is ineffective for enhancing the transformation. Then, the surface energy change upon the transformation is supposed to be a key factor suppressing the transformation. If taking the surface energy change into account, the total free energy change upon the transformation, $\Delta G^{A \to M}$, is given by[28]:

$$\Delta G^{A \to M} = G^T_M - G^T_A = (G^B_M - G^B_A)v + (\gamma_M - \gamma_A)s \quad , \tag{1}$$

where $G^T_{A,M}$ are the total free energies of austenite and martensite, $G^B_{A,M}$ the free energies of their bulk states, $\gamma_{A,M}$ the surface energy densities of the two phases, v the particle volume and s the surface area of a particle. Suppose that the austenite particles are formed under a thermal equilibrium condition. Then, they would possibly be surrounded by surfaces with low surface energy densities such as {111} and {100} planes. Upon the martensitic transformation, however, the {111} planes of austenite are converted into {110} planes of martensite, and the {100} planes of austenite into {110} and {100} planes of martensite. In a bcc crystal, the {110} and {100} planes are of low surface energy densities. The absolute values of the surface energy differences may be small, but, as in fact demonstrated recently for the fcc to hcp transformation in Co particles[28, 29], the surface energy term in the right hand side of Eq.(1) becomes large enough to reduce the chemical driving force when the size is 10 nm. For the Fe-Ni alloy system, however, accurate evaluations of the surface energies of those planes of austenite and martensite have not been made yet. Then, to evaluate them somehow as precisely as possible may be crucial to understand quantitatively why the fcc phase of the Fe-Ni alloys becomes so stable when the size is reduced down to nanometer scale.

It was recently demonstrated[30] that Fe-Ni and other alloys undergo time-dependent martensitic transformations even at temperatures definitely above M_s. This implies that martensitic transformations, which have been believed to be typically athermal in nature, can also be thermally activated even above M_s. Our observation suggests that this is the case for the nm-sized austenite particles. However, the time-dependent transformation in the nm-sized particles is obviously needed to be studied more in detail. Effects of the oxide films actually observed on the transformation remain also to be solved. A study on these points are now in progress.

5 Summary

Size effects on order-disorder and martensitic transformations, which are diffusion-controlled and diffusionless, respectively, have been examined by TEM, ED and in addition a computer calculation and simulation. As a result, it is suggested that the observed size effects, *i.e.*, large depressions of the transformation temperatures for both the two kinds of transformations are essentially due to surface effects. More detailed experiments and calculations are no doubt necessary to fully understand the size effects on these kinds of phase transformations of alloys.

Acknowledgments

The author would like to express his thanks to Prof. Y. Hirotsu, Mr. A. Koreeda, Dr. Y. Nakata, Mr. T. Ohkubo, ISIR, Osaka University, for discussion and encouragement. He is also grateful to graduate students, Messrs. T.Kinoshita and Y.Murai for their collaborations in computer calculation and experiment. The present study was supported by the Grant-in-Aid for Scientific Research(C)(1995-1996) and for Scientific Research on Priority Areas(1997) of the Ministry of Education, Science, Sports and Culture, Japan.

References

1. K. Kimoto and I. Nishida, *J. Phys. Soc. Jpn.* **22**, 744 (1967).
2. H. Poppa, R.D. Moorhead and M. Avalos-Borja, *J. Vac. Sci. Techno.* A **7**, 2882 (1989).
3. H. Mori, M.Komatsu, K.Takeda and H.Fujita, *Philos. Mag. Lett.* **63**, 173 (1991).
4. N. Tanaka, F.Yoshizaki, K.Katzuda and K.Mihama., *Acta Metall.Mater.* **40**, S275 (1992).
5. H. Fujita, *Ultramicroscopy* **39**, 369 (1991).
6. M. Takagi, *J. Phys. Soc. Jpn.* **9**, 359 (1954).
7. K. Oshima and J. Harada, *J. Phys. C: Solid State Phys.* **17**, 1607 (1984).
8. T. Tadaki, A. Koreeda, Y. Nakata,T. Kinoshita, *Surf. Rev. Lett.* **3**, 65 (1996).
9. T. Tadaki, T. Kinoshita, Y. Nakata, T.Ohkubo and Y. Hirotsu, *Z. Phys.* **D40**, 493 (1997).
10. T. Tadaki, Y. Murai, A. Koreeda, Y. Nakata and Y. Hirotsu, *Mater.Sci.Eng.* **A217/218**, 235 (1996).
11. K. Mihama, *J. Phys. Soc. Jpn.* **31**, 1677 (1971).

12. T. Kizuka, N. Mitarai, Y. Ishida and N. Tanaka, *Proc. Int. Symp. on Intermetallic Compounds for High-Temperature Structural Application*, p.1165 (1993).
13. S.C. Moss and P.C. Clapp, *Phys. Rev.* **171**, 764 (1968).
14. T. Tadaki, T. Kinoshita, Y. Nakata and A. Koreeda, unpublished work, Osaka University(1994).
15. H. Yasuda and H. Mori, *Z. Phys.* **D37**, 181 (1996).
16. W.B. Peason, *A Handbook of Lattice Spacings and Structures of Metals and Alloys* (Pergamon, London, 1958), p.411.
17. *Binary Alloy Phase Diagrams* (ASM International Ohio, 1990), ed. T.B. Massalski, p.358.
18. M.I. Baskes, *Phys. Rev. B*, **46**, 2727 (1992).
19. T. Horiuchi, S. Takizawa, T. Suzuki, T. Mohri, *Metall. Trans. A*, **26A**, 11 (1995).
20. Y.H. Zhou, M. Harmelin and J. Bigot, *Mater.Sci.Eng.* **A124**, 241 (1990).
21. S. Kajiwara, S. Ohno and K. Honma, *Philos. Mag. A* **63**, 625 (1991).
22. Y.H. Zhou, M. Harmelin and J. Bigot, *Mater.Sci.Eng.* **A133**, 775 (1991).
23. Y. Chen, G. Deng, H. Lu, J. Wang and G. Li, *Jpn. J. Appl. Phys.* **34**, 113 (1995).
24. R.E. Cech and D. Turnbull, *Trans.AIME* **206**, 124 (1956).
25. W.F. McClune(ed.-in-chief), *Powder Diffraction File*, Set 44, International Center for Diffraction Data, Newtown Square, PA, p.598 (1994).
26. R.P. Reed and R.E. Schramm, *J.Appl.Phys.* **40**, 3453 (1969).
27. T. Tadaki, Y. Murai, Y. Nakata and Y. Hirotsu, unpublished work, Osaka University(1995).
28. M. Alden, S. Mirbt, H.L. Skriver, N.M. Rosengaard and B. Johansson, *Phys. Rev. B* **46**, 6303 (1992).
29. H. Sato, O. Kitakami, T. Sakurai and Y. Shimada, *J. Appl. Phys.* **81**, 1858 (1997).
30. T. Kakeshita, T. Fukuda and T. Saburi, *Scripta Mater.* **34**, 147 (1996).

TEMPERATURE RESOLVED X-RAY DIFFRACTOMETRY

A. BUCHAL
Technical University of Brno, Technická 2, BRNO
CZ 616 69, Czech Republic

A Taylor [1] presumes that the first high-temperature camera was produced by Wilson & Jay in 1941 [2]. My first high-temperature camera was made by *Materials Research Corporation* and was (with vacuum-system PT450 fy. *Laybold-Heraeus*) very robust. My present instrument with the temperature camera *PAAR* from Austria hangs on a goniometer inside the radiation protective cover. But the fundamental advance means standard facilities by the **Position Sensitive detector.** It makes possible to watch also fast phase transfornations, or other structural changes, 'in situ'. Perhaps the most attractive investigations was made on the casting powders in liquid state (exactly in two phase region between liquidus and solidus).

Introduction

An old problem, mamely the replacement of the standard specimen holder with a device capable of heting and cooling of a specimen during the diffraction recording, is being resolved by introduction of commercially available position sensitive detectors. The reproducibility is a problem in all diffraction experiments but it becomes especially pronounced when the measurement takes place over awide range of temperatures in different environments (vacuum, inert gas, etc.). The only apparent advantage of high temperature measurements is the absence of humidity.

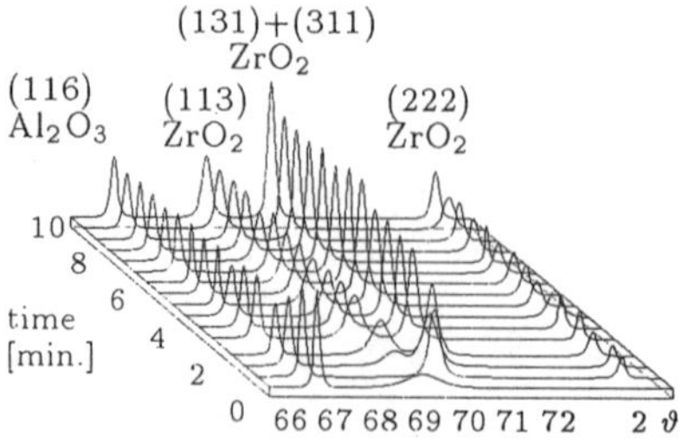

Figure 1: Axonometric projection of zirconia tetragonality development and disolution of corundum

According to A. Taylor [1], the first attempts to devise high-temperature chamer were performed by Wilson & Jay [2] in 1941. Presumably the first Debye-

Scherrer high-temperature camera commercially available is that designed by Goldschmidt & Cunningham [3] for the Central Research Laboratories in 1950. The first high-temperature camera based on the focusing principle that may be adapted for use with Geiger or proportional counters was built by Chiotti [4] in 1954.

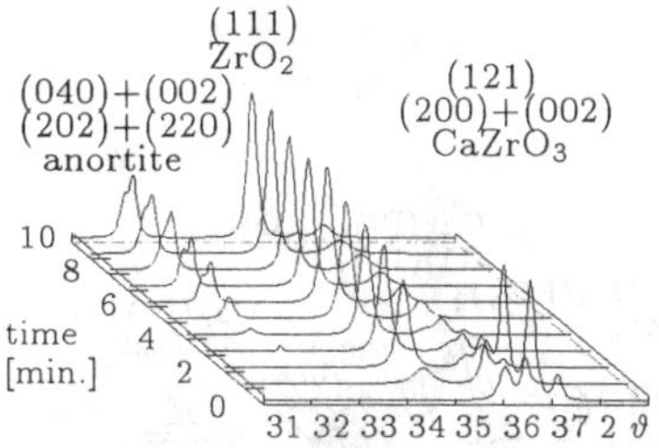

Figure 2: Time dependence of $CaZrO_3$ decomposition and anortite precipitation.

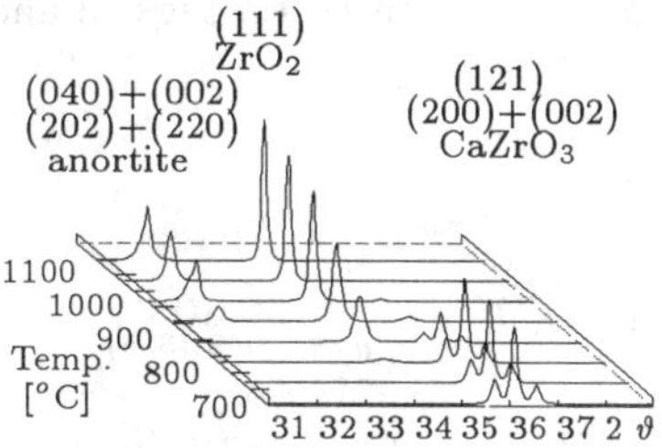

Figure 3: Temperature dependence of $CaZrO_3$ decomposition and anortite precipitation.

1 Instrumental posibilities

The first high-temperature (and also low-temperature) apparatus was produced by American *Materials Research Corporation* and was coupled with vacuum set PT450 fy. *Leybold-Heraeus*. The vacuum set was build in a box of aproximately 1x1x1 m large. My present instrument was supplied by *PAAR* (Austria) also with *Laybold-Heraeus* vacuum set, but a turbomolecular pump is suspended from camera (the size of beer bottle).

Principal progress represents standard equipment with a position sensitive detector. It accomodates a proportional detector *ELPHYSE* that makes possible software adjustment of the number of canals from 128 to 1024 over the angle aprox. $8^o 2\vartheta$. The detector has two operation modes: 1. **Fast scan** and 2. **Stationary**. An exapmle of using stationary detection is shown on Fig. 1. Here the evolution of ZrO_2 tetragonality during the decomposition of $CaZrO_3$ at 850°C (exposition time 40 sec.) was followed.

2 Application possibilities

While in the *MRC* camera it was possible to replace cooling water with liquid nitrogen and achieve in this way temperatures down to -200°C, the *PAAR* camera offers only higher than room temperatures (from RT to +1600°C. Nevertheless all attempts to perform (successful) experiments (with the *MRC* camera) at low temperature where futile. On the other hand, the coupling of high-temperature camera *PAAR* with a small position sensitive detector

ELPHYSE was highly effective. The above mentioned study of phase transformations kinetics in the dielectric paste was on the whole succesful. Stationary position sensitive detector adjusted on $35^o 2\vartheta$ (using CoK_α radiation) excellently recorded temperature and time dependencies of the decomposition of $CaZrO_3$ into tetragonal ZrO_2 (stabilized by CaO) and precipitation of anortite, as demonstrated on Figs. 2 and 3.

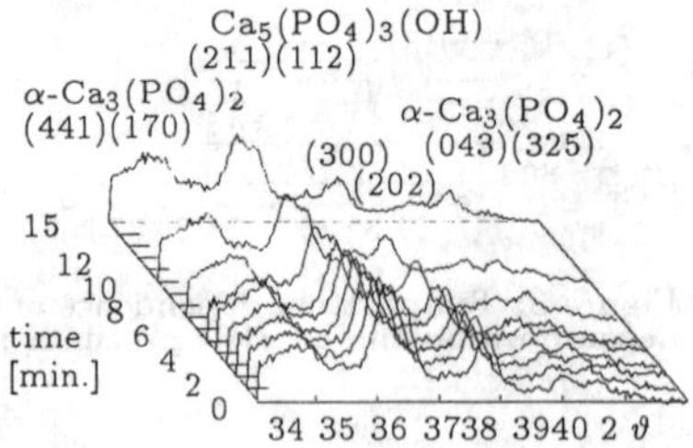

Figure 4: Time dependence of $Ca_5(PO_4)_3(OH)$ decomposition and $Ca_3(PO_4)_2$ precipitation.

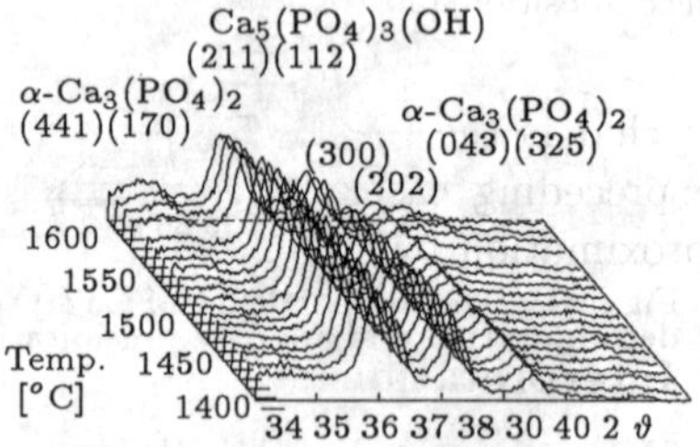

Figure 5: Temperature dependence

The calculations based on these measurements made it possible to determine the activation enthalphy of the $CaZrO_3$ decomposition and also of ZrO_2 tetragonality development. This tetragonality is important for thermal stability of tetragonal zirconia (the transformation of tetragonal zirconia to monoclinic baddeleyite is associated with 4% volume increase and leads to the disintegration of dielectric layer).

This example seems to be exquisitely perfect, nonetheless, the structures with lower symetry do not yield perfect diffraction spectra. During the study of the thermal stability of hydroxyapatite thermal stability used for the heat coating we were not able to simulate thermal conditions existing in the plasma discharge. Nevertheless the decomposition of $Ca_5(PO_4)_3(OH)$ started in air at tepmeratures below 1600^oC. Figure 4 shows the axonometric projection of the interesting parts of diffraction spectra obtained on specimen sullpied by *AMDRY* (labeled A6020) during the time dependence measurement at 1600^oC. Low acquisition time causes considerable Poisson scattering and that intensity changes are, therefore, difficult to resolve. Analogue temperature dependence for temperatures between 1400^oC and 1600^oC is in Fig.5.

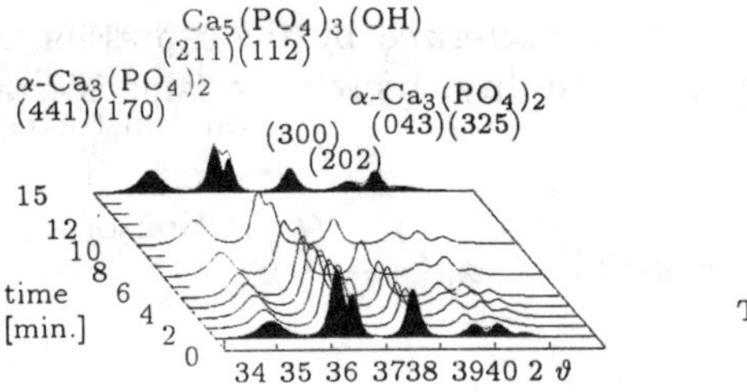

Figure 6: Time dependence of $Ca_5(PO_4)_3(OH)$ decomposition and $Ca_3(PO_4)_2$ precipitation.

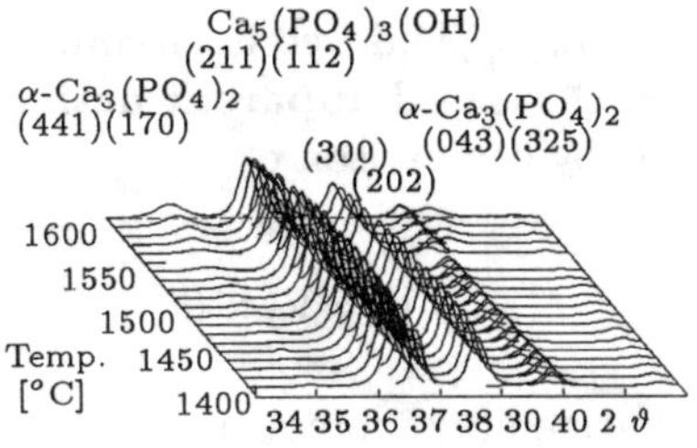

Figure 7: Temperature dependence

That is the reason why eventual intensity variations are far less than in the preceding case. The evaluation of these spectra requires therefore the approximation by suitable analytical functions, as shown in Figs. 6 and 7.

3 Interpretation tools

All common methods of identification could be used for the interpretation of measured spectra.[a]

The intensities for quantitative analysis could be obtained by standard methods. If the number of reflections is relatively small, the individual diffraction profile may be approximated by

$$Intensity_i = \frac{\vec{h}}{\left[1 + Q\left(\frac{i-\vec{p}}{\vec{b}}\right)^2\right]^2}, \tag{1}$$

where $I = \vec{h} \times \vec{b}$ is integral intensity, if $Q = 4 \times (\sqrt{2} - 1)$. However, high-temperature experiment usually consist of series of spectra and it is necessary to find a physical model for intensity variations. For thermal-activated process characterized by parameter κ, it is possible to describe temperature dependence by:

$$\kappa = \kappa_0 \times \exp^{-(\frac{\bar{Q}}{T})} = \exp(\bar{P} - \frac{\bar{Q}}{T})$$

[a]Usually it is necessary to take into account both the positions shift to lower 2θ caused by lattice expansion and, if Pt heating element is perfectly adjusted, the displacement out of focus of the specimen formed by the layer 0.1 mm thick. Fortunately, often the expansion is compensated by the excentricity.

Time dependence is, according to[5], characterized by the expression τ^n. Term $\bar{Q}$ is called **apparent activation enthalpy**. Its relative value is established by comparison of

$$V = V_0 \times (\tau \times \kappa^{\frac{1}{n}})^n = v_0 \times [\tau \times \exp(\frac{\bar{P} - \frac{\bar{Q}}{T}}{n})]^n$$

with canonical expression

$$V = V_0 \times [\tau \times \exp(P + \frac{Q}{T})]^n .$$

Thus we may get $P = \frac{\bar{P}}{n}$ and $Q = \frac{\bar{Q}}{n}$, where Q is called **relative activation enthalpy**. Actual activation enthalpy **H** could be obtained from well known relation

$$\frac{Q}{T} = \frac{\mathbf{H}}{RT}$$

as $\mathbf{H} = R \times Q$, where R is the universal gass constant (R = 0.008314 kJ mol^{-1} K^{-1}).

Conclusion

High-temperature diffraction experiments add a new dimension to the application of X-ray diffraction methods. Further improvement depends on the progress in technology. At present, due to several restrictions spectacular results cannot be expected. Nonetheless, this type of measurement should be available in well equipped X-ray laboratory.

References

1. A. Taylor, *X-Ray Metallography*, (John Wiley & Sons Inc., London, 1961).
2. A.J.C. Wilson, *Proc. Roy. Soc.*, **53** (1941).
3. H.J. Goldschmid and J. Cunningham, *J. Sci. Instru.*, **27**, (1950).
4. P. Chiotti, *Rev. Sci. Instru.*, **25** (1954).
5. H.J. Eckstejn, *Warmebehandlung von Stahl* (VRB-DV für Grundstoffindustrie, Leipzig, 1971).

STRUCTURE OF $Al_{50}Fe_{25}Ti_{25}$ ALLOY FORMED BY MECHANICAL ALLOYING

M. KRASNOWSKI, D. OLESZAK, H. MATYJA
Department of Materials Science and Engineering, Warsaw University of Technology, Narbutta 85, 02-524 Warsaw, Poland

An elemental powder mixture composed of 50at.% aluminium, 25at.% iron and 25at.% titanium added with 1wt.% ethanol was milled in a high energy planetary ball mill. The structural changes and the formation of new phases occurring in the material during mechanical alloying at various milling times were examined using X-ray diffractometry (XRD) and transmission electron microscopy (TEM). The XRD patterns show that, as a milling time increases, the Al peaks disappear gradually and all the Fe peaks become slightly shifted and broadened. These changes can be explained in terms of a b.c.c. Fe(Al) solid solution formation. It was also observed that the intensity of the Ti peaks decreased and new f.c.c. phase peaks appeared. When the milling time was not very long, the f.c.c. phase was identified to be titanium hydrocarbide, whereas after the milling process was completed it was titanium carbide. The final product obtained in this experiment was a composite built of titanium carbide distributed in the Fe(Al) solid solution matrix. TEM observations have shown that, after the longest milling time, the structure of the powder particles becomes nanocrystalline.

1. Introduction

Mechanical alloying (MA), which means the formation of alloys by solid state reaction during ball milling, has been intensively studied over last years. As a result of MA various kinds of structures can be obtained. Among them are amorphous, crystalline or nanocrystalline ones[1]. In the case of nanostructured materials physical and mechanical properties improvement, resulting from grain size reduction to the nanometer scale, makes these materials potentially attractive for many applications.

Al-Fe and Al-Ti alloys possess profitable properties, mainly high specific strength and corrosion resistance at elevated temperatures[2,3], therefore they are very attractive for air and defence industry. The binary Al-Fe and Al-Ti alloys are widely prepared by mechanical alloying (MA)[4,5] while there are not many works related to ternary Al-Fe-Ti alloys obtained by MA. The addition of a third element to binary alloys mentioned above can improve their properties. It is known[6] that the phase composition of the final product of MA depends on the environment in which MA takes place. In order to avoid sticking of milled powders to the walls of a vial or to the balls, or simply to control the environment of MA several agents are used. These agents can be divided into neutral and active. When the active agent in MA process is used the reaction between the agent and one of the milled elements can occur. This kind of MA process is called reactive milling (RM).

The aim of this work was to study structural and phase transformations taking place during ball milling starting from mixture of Al, Fe and Ti powders added with ethanol.

2. Experimental

The samples were prepared by ball milling of pure crystalline Al, Fe and Ti powders in a high energy planetary ball mill Fritsh Pulverisette 5. Initial mixture composed of 50 at% Al, 25 at% Fe and 25 at% Ti was added with 1 wt% ethanol. Ball to powder ratio was 10:1. In order to prevent powder oxidation, the samples were sealed in the vial under argon atmosphere. The milling process was performed at room temperature.

Structural changes in the synthesised material were studied using X-ray diffractometry (XRD) and transmission electron microscopy (TEM). XRD patterns were obtained using Cu K_α radiation in a Philips 1830 diffractometer. TEM observations were performed with a Philips EM300 microscope. Specimens for TEM investigations were prepared by embedding the powders after MA into an aluminium matrix with the procedure described elsewhere[7] and also by suspending the powders in an epoxy resin. Further sample preparation was performed by dimpling and final thinning by Ar ion beam milling using a precision ion polishing system Gatan 691.

3. Results and Discussion

The XRD patterns of powders after various times of ball milling are plotted in Figure 1. The XRD patterns show that, as a milling time increases, the Al peaks disappear gradually and all the Fe peaks become slightly shifted and broadened. These changes can be explained in terms of a b.c.c. Fe(Al) solid solution being formed. It was also observed that the intensity of the Ti peaks decreased and new f.c.c. phase peaks appeared. When the milling time was not very long, the f.c.c. phase was identified to be titanium hydrocarbide, whereas after the milling process was completed it was titanium carbide (see Fig.1). The final product obtained in this experiment was a composite built of titanium carbide distributed in the Fe(Al) solid solution matrix. The XRD results lead to the conclusion that mechanochemical reactions between Ti and dissociation products of ethanol took place during milling. The sequence of the mechanochemical reactions with increasing milling time is as follows: first the titanium hydride is created, than it changes into titanium hydrocarbide and finally it transforms to titanium carbide.

Figure 2 shows TEM micrographs and corresponding selected area diffraction patterns (SADP) of powders milled for 166 h. A particle of powder appears to consist of many crystallites with sizes from tens to hundreds of nm. The TEM images and the corresponding SADPs of powders after 280 h of MA are shown in Figures 3 and 4. It is easy to observe that after the longest milling time the structure of the powder particles becomes nanocrystalline with crystallites size ranging from several to 20 nm.

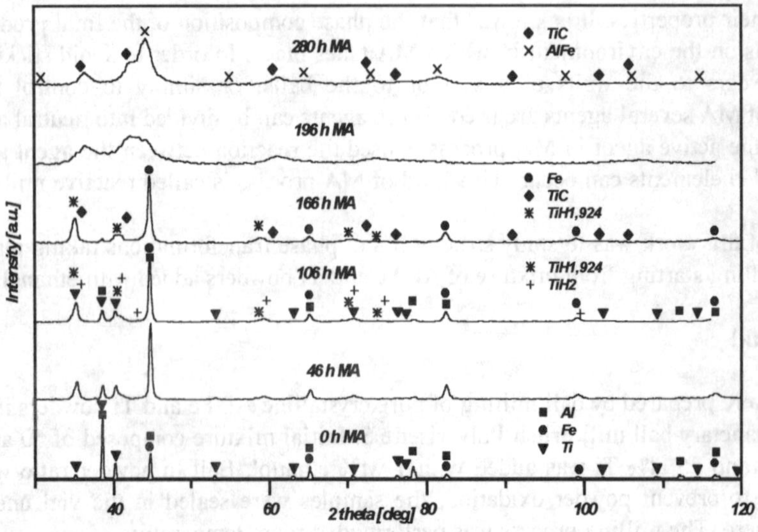

Figure 1: XRD patterns of $Al_{50}Fe_{25}Ti_{25}$ powder mixture added with ethanol as a function of milling time.

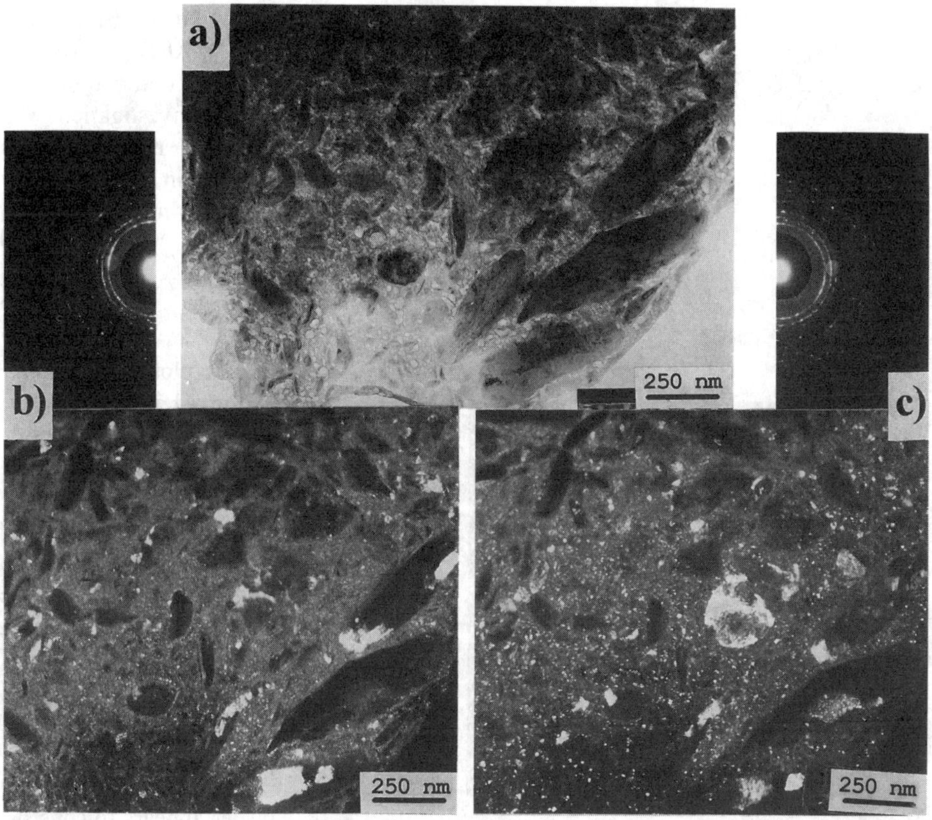

Figure 2: TEM micrographs and diffraction patterns of the sample milled for 166: a) bright field image, b) dark field image taken using a part of the first ring, and corresponding SADP, c) dark field image taken using a part of the second ring, and corresponding SADP. The parts of the rings are indicated by the image of objective aperture.

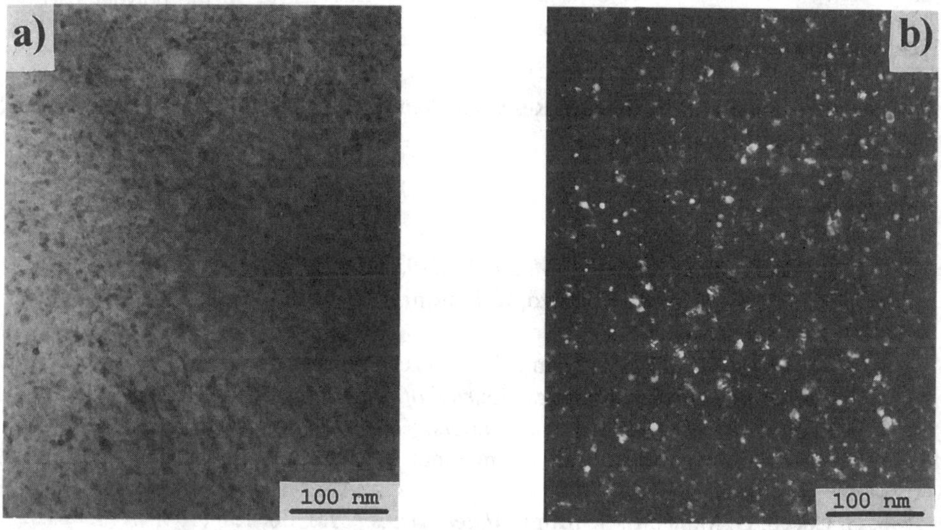

Figure 3: TEM micrographs of the sample milled for 280 h: a) bright field image, b) dark field image taken using a part of the first and the second ring.

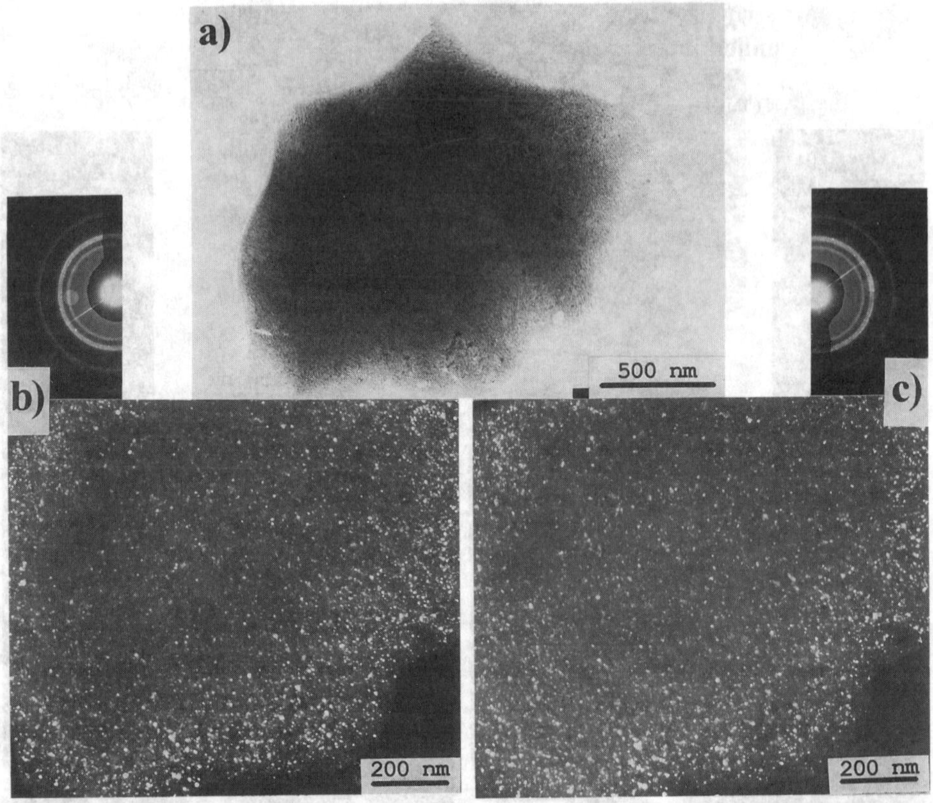

Figure 4: TEM micrographs and diffraction patterns of the sample milled for 280 h: a) bright field image of a powder particle, b) dark field image taken using a part of the first ring [(111) TiC], and corresponding SADP, c) dark field image taken using a part of the second ring [(200) TiC, (100) Fe(Al)], and corresponding SADP. The parts of the rings are indicated by the image of objective aperture.

Acknowledgements

The financial support of the Science Research Committee under the contract No. 7 T08D 043 10 is gratefully acknowledged.

References

1. C. C. Koch, *Mater. Trans., JIM,* **36, 2,** 85-95 (1995).
2. E. Bonetti, G. Scipione, S. Enzo, R. Frattini, L. Schiffini, *NanoStr. Mater.*, **6,** 397, (1995).
3. F. H. Froes, C. Suryanarayama, D. Eliezer, Review*: Synthesis, Properties and Applications of Titanium Alumides, Journal of Mater. Sci,* **27,** 5113, (1992).
4. D. Oleszak, *Ph. D Thesis, Warsaw University of Technology,* (1995).
5. T. Itsukaichi, K. Masuyama, Mumemoto, I. Okane, J. G. Cabanas-Moreno, J. *Mater. Res.,* **8, 8,** 1817-28, (1993).
6. G. Cocco, G. Mulas, L. Schiffini, *Mater. Trans., JIM,* **36,** 150 (1995).
7. Angiolini, F. Cardellini, M. Krasnowski, G. Mazzone, A. Montone, M. Vittori-Antisari, *Microsc, Microanal, Microstruct,* **6, No. 5-6,** 601-609 (1995).

The atomic structure features of rapidly quenched, ion irradiated and ball milled intermetallic compounds in Ni-Nb-V system.

N.P.DJAKONOVA, T.A.SVIRIDOVA, YU.A.SKAKOV, E.V.SHELEKHOV

Moscow State Institute of Steel and Alloys,
Leninsky pr.4, 117936, Moscow, Russia
e-mail: tim@x-ray.misa.ac.ru

Abstract

The features of atomic structure of some Ni-Nb-V intermetallic compounds after rapid quenching, ion irradiation and ball milling were studied by means of X-ray powder diffraction technique and transmission electron microscopy. The behaviour of phases μ (Fe_7W_6-type), λ_1 ($MgZn_2$-type), σ (FeCr-type) and ε (β-Cu_3Ti-type) was examined. The alloys with μ- and λ_1-structures transformed to amorphous state most readily under all the treatments. The stability of crystal structures under study increases in the sequence μ, λ, σ and ε without respect to the treatment used. If treatment intensity was insufficient to reach amorphization the following non-equilibrium crystal states were found: (i) μ, λ_1, ε-phases with stacking faults and polytypes of μ- and ε-phases, including hitherto unknown two-layer polytype of μ-phase (type Cs_6K_7); (ii) change of the type and extent of the chemical long-range order (ε-phase); metastable phases with new coordinations (bcc-phase and phase with structure of Ti_2Ni).

Introduction

The problem of development of new materials with special properties stimulates investigations concerning stability of intermetallic compounds under non-equilibrium conditions[1-4]. Particular attention is given to the study of atomic structure of the produced non-equillibrium phases, both amorphous and crystal, since their comparison with initial structures allows to establish laws of phase transitions and to predict final structural state of materials at their operation under severe conditions. The juxtaposition of states produced by different high energy treatments of the same alloy opens additional opportunities to evolve the theory of phase transformations under non-equilibrium conditions.

It is known, that introduction of defects into the crystal lattice leads to the increase of free energy that, in turn, makes the thermodynamic preconditions for solid-state reactions including amorphization. However which type of defects creates highest free energy is disputed.

Here we concern phase composition change induced by rapid quenching (RQ), ion irradiation and mechanical milling (MM) in various intermetallic compounds (IC) of the same ternary system. Particular attention was given to examination of such crystal lattice defects as stacking faults and chemical disordering.

Experimental details

The alloys compositions are presented in fig. 1. The RQ samples were produced by rapid solidification of melt droplets on the inner surface of a revolving copper cylinder. The range of coolling rates studied was $10^3 \div 10^7$ K/s. He^+ ions with energies 2÷3.25 MeV and fluences $10^{15} \div 5 \cdot 10^{16}$ cm^{-2} were used for ion irradiation. The mechanical milling (MM) of alloys was performed in the planetary ball mill in argon atmosphere. The mechanical energy dose was estimated using computer simulaton of balls movement[5]. The range of doses was 3÷250 kJ/g whereas corresponding treatment time varied 5÷60 min.

The x-ray diffraction was measured by means of a powder diffractometer DRON-4.0 using CoK_α and CuK_α radiations. The special technique was developed to simulate X-ray diffraction patterns of IC (μ-, ε- and Laves phases) containing stacking faults. The quantitative phase analysis was carried out using reduced Rietveld method (with fixed lattice base coordinates and atomic occupancies of the sites[6]). Rapidly quenched samples were also examined by means of transmission electron microscopy (TEM).

Results and discussion

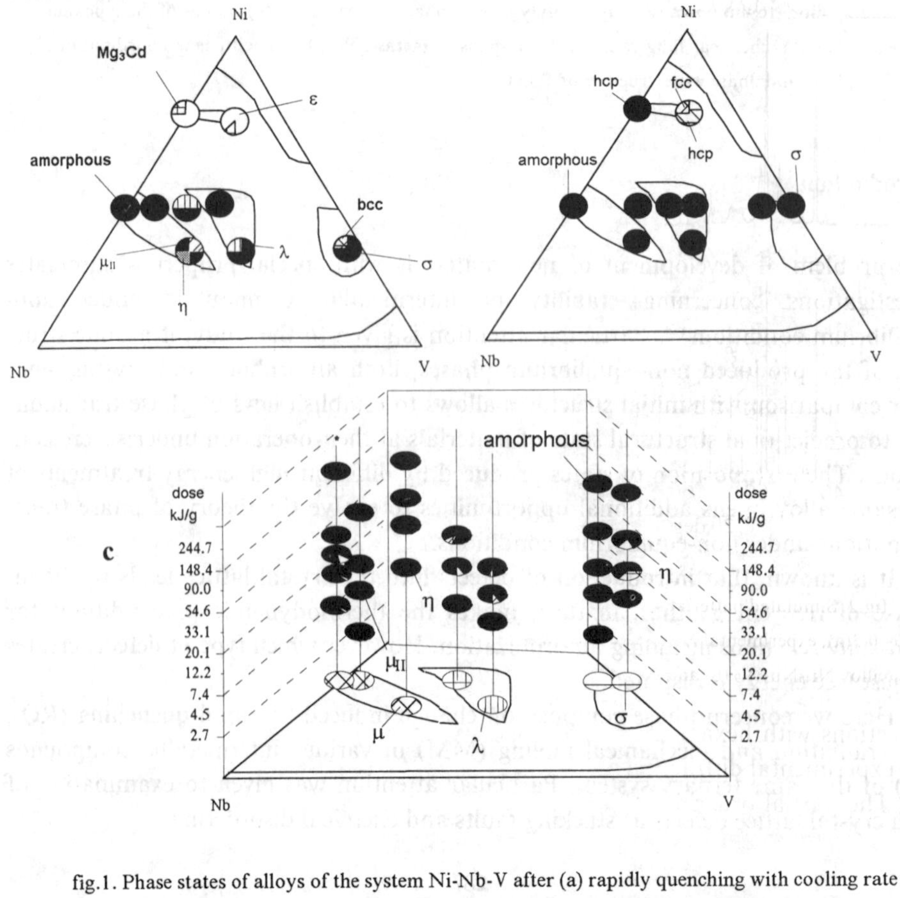

fig.1. Phase states of alloys of the system Ni-Nb-V after (a) rapidly quenching with cooling rate ~10^7 K/s, ion irradiation with the fluence $5 \cdot 10^{16}$ cm^{-2} and (c) mechanical milling.

The crystal structure of IC under study relates to the two main groups of the transition metals compounds: (i) the spherically close-packed structure – ε-Ni_3Nb (β-Cu_3Ti structure type); (ii) tetrahedrally close-packed phases μ-NiNb (Fe_7W_6-type), σ-NiV_2 (FeCr-type) and ternary Laves phase $(Nb,V)(Ni,V)_2$ (λ_1-$MgZn_2$). Since all these phases belong to the same ternary system, it means that distinctions in alloys behaviour are related to the difference in crystal structures.

Complete or partial amorphous state formation was found in the most samples (fig. 1). As regards the tendency to amorphization under RQ and ion irradiation all the phases of the system Ni-Nb-V may be arranged in the following sequence: μ, λ, σ and ε, i.e. the predisposition to amorphous phase formation depends on crystal-chemical factor. The stability of the initial crystal phase subjected to MM also increases in the succession given above (μ, λ, σ, ε), where the measure of stability is the amount of initial phase after treatment. As follows from fig. 1 mechanical milling and ion irradiation are most effective treatments with respect to amorphization.

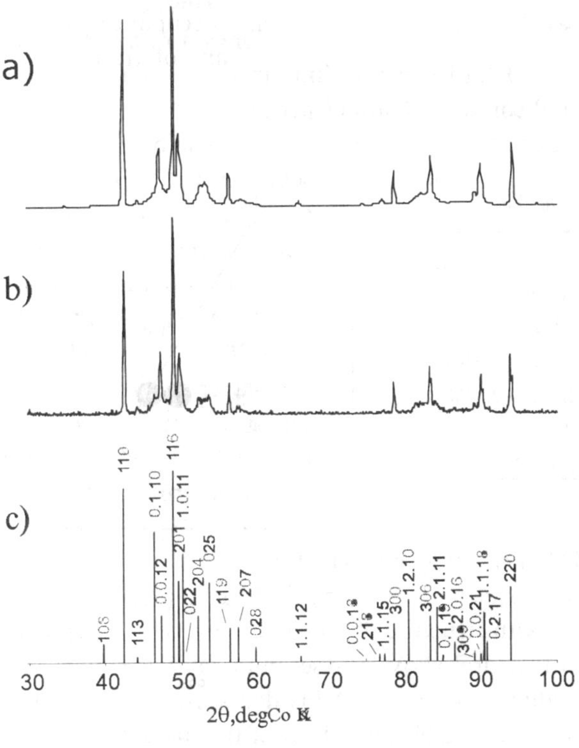

fig.2. Simulated pattern of μ-phase with SF probability α=0.3(a), experimental X-ray diffraction pattern of the RQ alloy $Ni_{45}Nb_{55}V_9$ (b) and stick-diagram of μ-phase (c).

Besides amorphous phase formation some crystal non-equilibrium states preceding amorphization were revealed under all the treatments studied.

The polytypic stacking faults (SF) formation was observed by TEM in the RQ samples within the μ-phase composition range and after cooling rates less than critical one for amorphization (~10^5-10^6 K/s). Recently we have proposed[7] the description of μ-phase in terms of the close-packing as the stacking of complex three-dimensional units in the sequence ABC. It allows to polytypic SF in the planes $(001)_{hex}$ and to simulate X-ray patterns with various SF concentration. These stacking faults are similar to conventional ones in fcc- and hcp-phases and affects the reflections with hexagonal indices H-K≠3n only as shown in fig. 2 b, which shows the experimental diffraction pattern with estimated SF concentration α=0.2÷0.3.

The partial ordering of SF may give rise to the formation in alloy of some domains where the local change of stacking order occurs. For example, high concentration of ordered SF in rapidly quenched alloy $Ni_{35}Nb_{45}V_{20}$ results in predominance of stacking sequence AB, i.e. new two-layer μ-phase polytype forms (called μ_{II}). Fig. 3 shows the experimental diffraction pattern of this alloy and simulated pattern of μ_{II} with SF probability equal to 0.2.

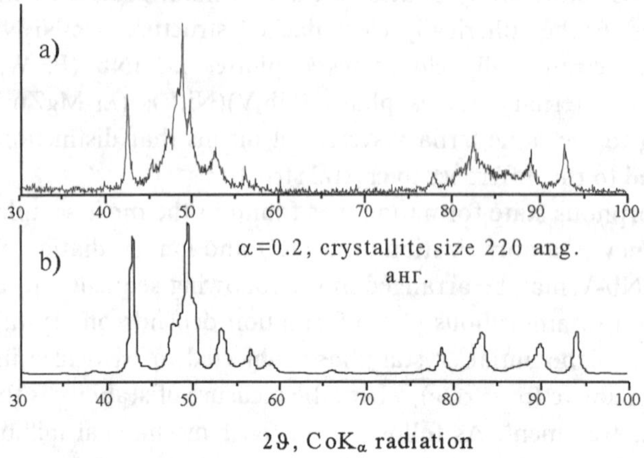

fig.3. X-ray diffraction pattern of RQ alloy $Ni_{35}Nb_{45}V_{20}$ (a) and simulation pattern of μ_{II}-phase with SF (b).

This μ_{II}-phase belongs to the structure type Cs_6K_7 and has not been observed earlier in transition metals systems. The μ_{II}-phase has space group $P6_3/mmc$ and contains 26 atoms per unit cell. Lattice base of the μ_{II}-phase and atomic occupancies of the sites for the alloy $Ni_{35}Nb_{45}V_{20}$ are presented in the table. Atomic positions correspond to ones of the Cs_6K_7 phase, the lattice parameters are equal to $a_{\mu_{II}}=a_{\mu}=4.936$, $c_{\mu_{II}}=2/3\ c_{\mu}=17.90$ Å. Two-layer polytype of μ-phase with high concentration of SF is detected also as an intermediate phase during the transition to amorphous state as a result of MM of the alloy $Ni_{35}Nb_{45}V_{20}$.

Table. Lattice base of the μ_{II}-phase.

Occupation	Site
Nb	2 (**b**) 0, 0, 0.25
Nb	2 (**c**) 1/3, 2/3, 0.25
Nb	4 (**f**) 1/3, 2/3, 0.9769
Nb	4 (**f**) 1/3, 2/3, 0.8208
0.67 Ni+0.33 V	2 (**a**) 0, 0, 0
0.67 Ni+0.33 V	12(**k**) 1/6, 1/3, 0.115

The stacking faults formation was also observed by TEM in rapidly quenched alloys $Ni_{75}Nb_{25}$ and $Ni_{73}Nb_{13.5}V_{13.5}$ with structure of ε-phase. However, the SF concentration is not equal in both isostructural alloys, the second alloy has higher one, i.e. lower energy of SF. The RQ alloy $Ni_{75}Nb_{25}$ contains domains of Mg_3Cd-type phase in addition to stacking faults. The equilibrium phase of the β-Cu_3Ti-type and observed metastable hexagonal (MH) phase of Mg_3Cd-type are related structures based on close-packing of spheres. The both phases are hcp-superstructures, i.e. they may be considered as two-layer sequence AB of closely packed planes but they differ by the type of chemical order in planar layer. The orientation relationship between both phases brings to parallel arrangement of close-packed planes.

Thus if SF-energy is high enough to prevent the stacking faults formation, the domains of MH-phase appear. This phase and equilibrium ε-phase have the same topological (but unlike chemical) order, i.e. stacking sequence of close-packed layers is identical in both phases.

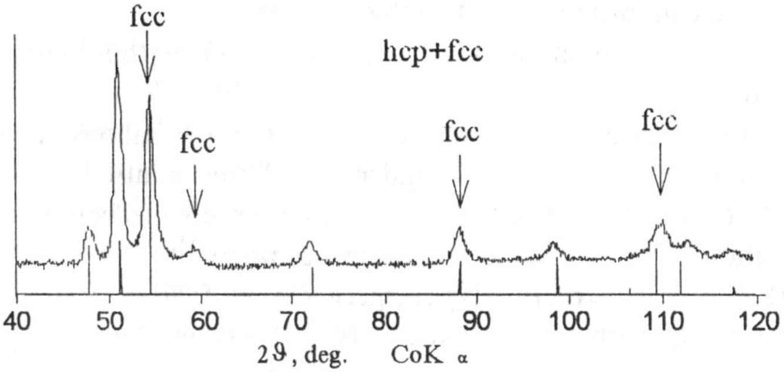

fig.4. X-ray pattern of the alloy $Ni_{73}Nb_{13.5}V_{13.5}$ after ion irradiation.

SF energy affects also the phase transitions in ε-phase under ion irradiation. Apart from amorphization ion irradiation brings to disodering of initial ε-phase and eventually to the formation of hcp-phase as it occurs in the $Ni_{75}Nb_{25}$ alloy. In the case of ternary alloy amorphization doesn't take place but fcc-phase (another close-packed polytype) forms in addition to hcp-phase (see fig. 4) and amount of fcc-phase increases with irradiation dose.

Polytypic SF formation was also found in ε-phase during its mechanical milling. The final phase relates to the structure type $TiAl_3$ and is defined as six-layer «cubic» polytype (fcc-superstructure) of equilibrium ε-phase, provided lattice parameters ratio c/a equals 2.

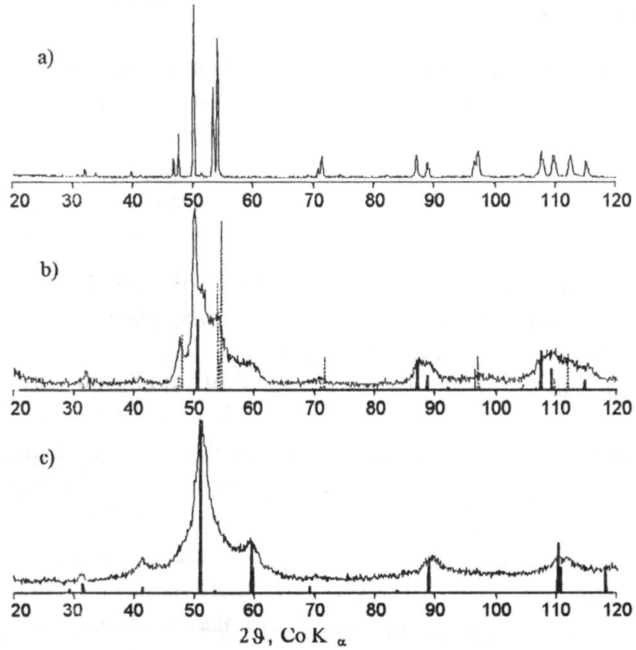

fig.5. Diffraction patterns of the alloy $Ni_{75}Nb_{25}$ before (a) and after ball milling with doses 18 kJ/g (b), 72 kJ/g (c).

The presence of weak superstructure reflections in the diffraction pattern indicates that the chemical ordering in planar layer is partially conserved (fig. 5 c). The diffraction pattern taken from the alloy after intermediate dose may be interprered as one from ε-phase with SF. Stacking faults influence on reflections selectively in dependence on their indices. In the fig.5 b dotted lines mark the reflections affected by SF.

Thus the transformation in the ε-phase during MM can be described as the storage of superstructural SF followed by the transition from hexagonal arrangement of close-packed planes in equlibrium phase to cubic one.

Furthermore, formation of the two «non-diagram» phases with new topological and chemical order was detected.

The metastable bcc-phase arises under RQ and ion irradiation of σ-alloys. The appearance of the phase can be related to the preponderance of coordination number 14 in the structure of both phases: Frank-Kasper coordination polyhedron with 14 vertices in the σ-phase and rhombododecahedron in the bcc-phase.

The metastable η-phase ($E9_3$ or Ti_2Ni structure type) emerges in the following cases:

1) as second phase under RQ with colling rates $\sim 10^5$-10^6 K/s in the μ-alloy $Ni_{35}Nb_{45}V_{20}$ and in the λ-alloy $Ni_{35}Nb_{32.5}V_{32.5}$;
2) as intermediate phase during amorphization of rapidly quenched alloys under irradiation (dose is given in cm^{-2}):
   μ-alloy $Ni_{35}Nb_{45}V_{20}$:

$$\mu_{II}(85\%)+\textbf{amorph.}(10\%)+\eta(5\%)\xrightarrow{10^{16}}\eta(50\%)+\mu_{II}(40\%)+\textbf{amorph.}(10\%)\xrightarrow{5\cdot10^{16}}\textbf{amorph.}$$

and λ-alloy $Ni_{35}Nb_{32.5}V_{32.5}$:

$$\lambda_1(75\%)+\textbf{amorph.}(15\%)+\eta(5\%)\xrightarrow{10^{16}}\lambda(40\%)+\textbf{amorph.}(40\%)+\eta(20\%)\xrightarrow{5\cdot10^{16}}\textbf{amorph.}$$

3) as intermediate phase during amorphization under mechanical milling in all μ- and λ-alloys and binary Ni-V σ-alloys.

Conclusions

All treatments under study resulted in the similar final states for each IC type. Complete or partial amorphous state formation was found in the most samples. Whatever high energy processing had been used, the stability of phases increased in the succession μ–λ–σ–ε, what is an evidence of close connection between alloy stability and its crystal structure.

If cooling rate, fluence or deformation dose are not enough to cause the transition to amorphous phase, the following crystal states are observed:

1) polytypic stacking (growth) faults formation:
   - during rapid solidification of μ, λ, ε-phases and occurrence of the new two-layer μ-phase polytype;
   - under mechanical milling of ε-phase that eventually led to the formation of cubic polytype with $TiAl_3$-structure;
2) change of type and extent of long-range order in the ε-phase:
   - rapid solidification gives rise to formation of Mg_3Cd-type superstructure;
   - ion irradiation causes disordering of ε-phase and its transition to hcp-phase or polytypic transformation to fcc-phase;

3) metastable phases formation:

- bcc phase (W-type) in alloys with initial σ-phase structure;
- $E9_3$ phase (Ti_2Ni-type) in alloys with initial μ- and λ_1-structures.

Similarity of new phase states induced by different treatments of the same phases let one to argue that this states are thermodynamically determined and may exist at low temperature part of metastable phase diagrams.

6. Acknowledgments

This work was supported by Russian Fond for Basic Research (Grant No. 97-02-17915).

References

1. H. Bakker, G.F. Zhou, H. Yang. Materials Science Forum. –Vols. 179-181, 1995, p.47-52.
2. B.C. Giessen. *Rapidly quenched metals.* Proc. II Int. Conf., 1978. p.119-134.
3. D.F. Pedraza. Mater.Science & Engineering. v.90, 1987, p.69-80.
4. C.C. Koch. Mater. Trans, JIM, v.36, 1995, №2, p.85-95.
5. Shelekhov E.V., Salimon A.I. Proc. of Inter. aerosol symp., Moscow, 1996; Aerosol, 1996, v.2, p.61.
6. E.V.Shelekhov. Proc. of III Int. conf. *Applied X-ray analysis of metals*, Moscow, 1994, p.4.
7. E.V. Shelekhov, N.P. Djakonova, T.A. Sviridova, Yu.A. Skakov, E.V. Shtanskaya. Fizika metallov i metallovedenie, 1995, № 3, p.121-125.

THE STRUCTURE CHANGES IN 18/10 STAINLESS STEEL CAUSED BY IRRADIATION WITH AR^+ IONS (40 KEV).

A.N.IVANOV

Moscow State Institute of Steel and Alloys,
Leninsky pr.4, 117936, Moscow, Russia
e-mail: ivanov@x-ray.misa.ac.ru

Abstract.

The powder diffractometry and microhardness measurements revealed the formation of strongly distorted surface layer (~4-5 mkm in thickness). It is found that under that layer the mixture of weakly and strongly distofted regions has been formed. Such a mixture exhibits higher hardness than on the surface. Strongly distorted regions seems to be a Coulomb type spherical defects.

Results

The water-quenched (1100°C) samples of 18/10 steel were irradiated at 400°C (dose $0.8{\cdot}10^{18}cm^{-2}$). X-ray investigations (CoK_α-rad.) revealed that the powder diffraction pattern consisted of two FCC sets of lines (111, 200, 220, 311 and 222). At first such an effect was found by V.S.Khmelevskaya[1]. One of the sets has as narrow lines as before irradiation. The calculation gives the same lattice spacings a=(0.3587±0.0002) nm as before irradiation. The other shows broad bell-shaped maxima with a=(0.3609±0.0005) nm, fig.1, table 1.

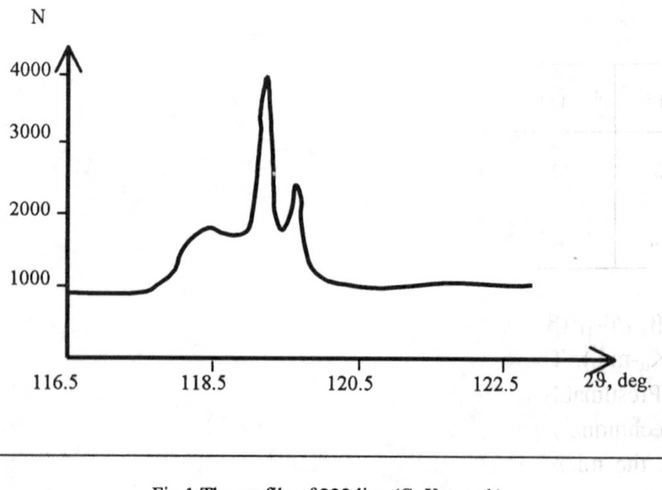

Fig.1 The profile of 222 line (CoK a -rad.)

Table 1. Physical profile breadth of wide lines, deg.

HKL	111	200	220	311
Co-K_α	0.17 ± 0.01	0.25 ± 0.02	0.47 ± 0.01	0.82 ± 0.02
Cu-K_a	0.12 ± 0.01	0.21 ± 0.02	0.37 ± 0.01	0.53 ± 0.03

It is worth noticing that the pattern taken with CuK_α-rad.(X-rays penetration depth is about five times less then for CoK_α-rad., table 2) shows "the line doubling" for 311 reflection only, fig.2, and 111, 200, 222 lines are broad but symmetrical ones.

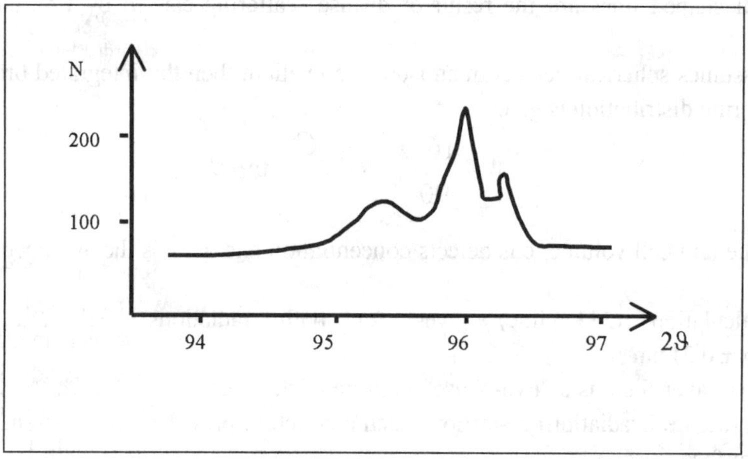

Fig.2 The profile of 311 line (CuK_α-rad.)

Table 2. X-rays penetration depth, microns.

Rad	111	200	220	311	222
Co-K_α	12.6	14.6	20.5	24.2	25.1
Cu-K_α	2.7	3.8	4.5	4.7	5.4

The ratio β_{311}/β_{111} (β – is physical profile breadth) equals to 4.8 ± 0.5 (CoK_α-rad.) and 4.4 ± 0.6 (CuK_α-rad.). Those values are higher than the ratio of tangents (3.05 and 2.50 respectively). Presumably the effect reflects inhomogeneouty of defects distribution as the sliding beam technique[1]) gave for CoK_α-rad. $\beta_{311}/\beta_{111}=2.5 \pm 0.6$.

Moreover, the microhardness of the irradiated surface shows the rise with the load increase, table 3.

[1] For CoK_α-rad. the exit angle 17.8° for 311 line provides the same penetration depth for 311 and 111 lines.

Table 3. Microhardness at different load values, Mpa

Irradiated side		Nonirradiated side	
0.2 n (20 g)	0.5n (50 g)	0.2 n (20 g)	0.5n (50 g)
225 ± 14	361 ± 9	180 ± 20	163 ± 22

Discussion.

The narrow line appears only for 311 reflection(CuK_α). It means that the thin surface layer (about 4.5 microns) is defected so strongly that the normal scattering is supressed. Then the observed bell shaped lines are the result of diffuse scattering caused by Coulomb type defects[2].

If one assumes spherical defects in an isotropic medium then the integrated breadth of diffuse scattering distribution is given[2]:

$$\beta = \frac{16 \cdot \pi^3}{90\sqrt{3}} \cdot \frac{c \cdot |C|}{V} \cdot \tan\theta$$

where V is the unit cell volume, c is defects concentration and $|C|$ is the averaged defects power.

The calculation (311 line) gives for both radiations the same value $c \cdot |C| = (1.5 \pm 0.2)\ nm^2$.

Under that layer there is a "two-phase" mixture of defected volumes with the volumes of the initial (before irradiation) distortion. Such a structure provides higher strenghtening than on the surface.

Conclusion

Thus, the results reveal the strong inhomogeneouty of defects distribution after irradiation. The defects responsible for the formation of the wide bell-shaped maxima are thought to be Coulomb type ones.

Maximal distortion is observed under the surface. The total depth of the defected layer (more than 25 microns) is 1000 times higher than the calculated radiation damage depth.

References.

1. V.S.Khmelevskaya et all. Two-phase decomposition in irradiated FCC alloys. (Rus). Vestnik atomnoj nauki i techniki (Physica radiasionnich povrejdenij), 1985, v.4, No 37, pp.22-25.

2. M.A.Krivoglaz. The theory of X-rays and thermal neutrons scattering by real crystals (Rus). Moscow, Nauka, 1967.

STRUCTURAL STUDIES OF SHAPE MEMORY ALLOYS BY MEANS OF HREM

J. PONS
Departament de Física, Universitat de les Illes Balears, Ctra. de Valldemossa km 7.5, E07071 Palma de Mallorca, Spain

The paper reviews some applications of the High Resolution Electron Microscopy (HREM) technique to the study of particular structural and microstructural features of the shape memory alloys. These alloys present unusual thermomechanical properties, like the shape memory effect and pseudoelasticity, which are related to a thermoelastic martensitic transformation. Several topics are treated, like the intervariant boundary structure of 18R martensite in Cu-based alloys, the substructure of this phase, the twin-boundary structures in Ni-Ti martensites, the structure refinement of Ni_2Al precipitates in Ni-Al alloys, the γ-phase precipitates in Cu-Zn-Al alloys and the premartensitic effects in Ni-Al and Ni-Mn-Ga alloys. At the end, a new method for HREM image processing, based on the phase of the Fourier transform, is commented.

1 Introduction

The so-called shape memory alloys exhibit some non-usual thermomechanical behaviours, like the *shape memory effect* and *pseudoelasticity*, which are closely related to a thermoelastic martensitic transformation. Such a transformation can either be induced by temperature changes or by the application of an external stress. In the first case, the transformation from the parent phase (austenite) to the martensite phase takes place on cooling, starting at the temperature Ms and finishing at Mf. On subsequent heating, the reverse transformation occurs along a different path (it starts at a temperature As, which is higher than Mf, and finishes at Af, higher than Ms), exhibiting hysteresis. At temperatures above Ms, the martensitic transformation can be induced by an externally applied stress. The transition starts at a given critical stress (which depends on the temperature by a Clausius-Clapeyron type equation) and brings about a high deformation (up to values above 10%, for well oriented single crystals). If the temperature is sufficiently higher than Ms (but not too much), the reverse transformation takes place upon releasing the stress (again showing hysteresis) and the deformation is completely recovered. This constitutes the *pseudoelasticity effect*. If the temperature is only slightly higher than Ms, the reverse transformation does not completely occur upon unloading, due to the hysteresis, and an apparent plastic deformation remains at zero stress. Moreover, the application of an external stress at temperatures below Mf (material completely in the thermally induced martensitic phase) also brings about an apparent plastic deformation, by means of the reorientation of the martensitic domains. In the last two cases, the apparent plastic deformation can be removed upon heating above Af, through the thermal reverse transformation to the parent phase, and the material recovers in this way its original shape. This constitutes the *shape memory effect*. These non usual properties are being used in an extensive variety of industrial applications. For a general review of the martensitic transformations and shape memory alloys see the reference 1.

The parent phase usually has an ordered bcc structure, either with nearest neighbour order (B2 structure, like Ni-Ti, Ni-Al and Cu-Zn alloys) or next-nearest neighbour (DO_3

or $L2_1$ structures, like Cu-Al, Cu-Zn-Al, Cu-Al-Ni and Ni-Mn-Ga alloys). The structure of the martensitic phase depends on the alloy system, but it can usually be interpreted as a long period structure obtained by the stacking of atomic planes derived from {110} or {111} planes of the parent phase.

The martensitic transformation parameters, such as the transformation temperatures or stresses, the hysteresis, the deformation attained, ..., are strongly dependent on the previous thermal or thermomechanical treatment performed to the samples, i.e. on the microstructure. In this sense, the electron microscopy, both conventional TEM and High Resolution Electron Microscopy (HREM), becomes very useful for the study and development of this kind of materials.

The present paper will review some examples in which the HREM has entailed a step forward in the knowledge of thermoelastic martensitic transformations and related aspects. Four main topics will be treated: the structure of intermartensitic boundaries, the substructure of martensite (stacking faults), precipitation of other phases and premartensitic effects (precursor phenomena). Additionally, a new method for the HREM image processing will be commented at the end of the paper.

2 Intermartensitic boundaries.

The mobility of the intercrystalline boundaries in the martensitic microstructure plays a key role in the deformation mechanisms of this phase as well as in the shape memory effect. So, this topic has been well studied, either by TEM, electron diffraction and HREM, specially in the alloy systems mostly used for applications: the Ni-Ti-based and Cu-based alloys.

2.1 Twin boundary structures of B19' martensite in Ni-Ti alloys.

A big variety of twinning modes have been observed in the martensitic structure of Ni-Ti alloys. Besides the twin relationships between the martensite plates (variant accommodation twins), internal twins exist in order to provide the lattice invariant shear accompanying the martensitic transformation. The <011> Type II, $\{11\bar{1}\}$ Type I, {011} Type I, (100) compound and (001) compound twins have been observed by CTEM and electron diffraction [2-5], and they have also been studied by high resolution electron microscopy [6-8]. A summary of the main HREM observations is shown next. The boundary of the <011> Type II twin is randomly curved. In a first study [6], some "blurred" regions were interpreted as $(11\bar{1})$ ledges and $(01\bar{1})$ steps, but in a more recent paper [8] the atomic resolution images showed broad strain contrast without ledge or step structure, which suggests that the strain around the boundary is elastically relaxed by gradual atom displacements. Conversely, the boundary of $\{11\bar{1}\}$ Type I twin (which corresponds to a variant self-accommodation twin) is completely straight and coherent, indicating that the variant accommodation takes place up to the atomic scale [8]. The internal {011} Type I twins are of lattice invariant shear type and they show a rather broad boundary in comparison with that of $\{11\bar{1}\}$ Type I twins. Concerning the (100) and (001) compound twins, they have been recognised as deformation twins [3-8], since their characteristics do not give solutions to the phenomenological theory of martensitic transformations. The boundary of the (100) compound twins is straight, although a broad strain contrast is visible in some places, which has been attributed to some local elastic deformation at the interface [8].

2.2 Intervariant boundary structure of 18R martensite in Cu-based alloys.

The thermally induced 18R martensite shows a plate microstructure, each plate corresponding to a variant of martensite. The 24 possible variants can be separated in groups of 4 variants, which are able to form selfaccommodating arrays. The crystallographic relationships between the variants corresponding to neighbour plates, as well as the orientations of their boundaries and twin relationships, have been extensively studied by means of electron diffraction and conventional TEM [9-11]. Three basic plate combinations are commonly observed, i.e. A:C, A:B and A:D couplings, following the notation of Saburi and Wayman[10], were the letters stand for the four variants of a selfaccommodating group (in the notation of Tas *et al* [9], the A:B, A:C and A:D interfaces are referred to as the habit plane interface, the twin type interface of the first kind and the twin type interface of the second kind). The A and B variants are twin related, although their interface does not correspond with the twinning plane, but it corresponds with their common habit plane (interface with the parent phase). The A:C and A:D interfaces correspond with the respective twinning plane of each pair of variants, the $(12\overline{8})$ and (2 0 20), respectively. A detailed high resolution one-dimensional lattice image study of the A:C and A:D interfaces has been reported by Adachi and Perkins[12]. In all the images, a random distribution of basal plane stacking faults were visible in the martensite plates, many of them with hexagonal (2H) structure. This topic will be treated later in more details. The A:C interface appeared straight and coherent at the non faulted regions and it was stepped at the faulted regions. The observations fitted well with the atomistic interface model, in which a coherent atomic matching can be achieved along the A:C interface plane with two perfect 18R crystals at each side of the interface. On the other hand, the A:D interface was very irregular, with an average orientation along the (2 0 20) plane. In this case, at difference with the A:C interface, straight traces are only seen at the junctions of perfect 18R crystal in one side of the boundary with a faulted region in the variant on the other side. The atomistic study again supports the observations: a good atomic matching of two 18R crystals is not possible along the interface plane, but it is indeed possible a good matching between the 18R and 2H structures. As claimed by the authors, these observations are significant in terms of the reverse transformation and shape memory effect, since the regions with a "wrong" matching will require additional energy to move [12]. HREM observations of martensitic boundaries at atomic resolution have also been performed by Cook *et al* [13] and Lovey *et al* [14]. In the latter paper, a study of the A:C interface, which constitutes a $(12\overline{8})$ Type I twin, has been presented [14]. The interface is quite planar and coherent, but the two twin parts are not exactly related by a mirror operation, due to the basal plane stacking faults existing in each crystal. The stacking faults also cause the existence of zones with perfect and imperfect matching between the atom planes alternating along the interface [14]. These observations are relevant in the same terms as those discussed in ref. 12, i.e. the interface mobility. Moreover, as remarked by the authors, the mismatched regions can be more prone to promote the segregation of impurities, also influencing the interface mobility [14].

3 Substructure of faulted 18R martensite.

The structure of 18R martensite, which is obtained in Cu-based alloys with a next-nearest neighbour ordered parent phase, can be regarded as a stacking of close packed planes, derived from {110} planes of parent phase, with the sequence AB'CB'CA'CA'BA'BC'BC'AC'AB'. However, the observed stacking sequence is hardly ever perfect, but a high density of randomly distributed stacking faults is present [15]. Most of them are basal plane stacking faults (which, basically, provide the lattice invariant shear coming with the martensitic transformation [16]) but other planar defects lying out of the martensite basal plane and altering the stacking sequence are also frequent. The study of basal and non-basal stacking faults has been the subject of several studies by electron and x-ray diffraction, TEM and HREM.

Concerning the basal plane faults, the diffuse scattering in diffraction patterns was well modelled by assuming two kinds of stacking faults: the hexagonal faults (β-type) and cubic faults (α-type) [15]. The displacement vector associated to them was analysed by Andrade *et al* [17] and they obtained the vectors -1/18(691) and 1/18(691), for hexagonal and cubic faults, respectively. The HREM images taken at atomic resolution in conveniently oriented zone axes make possible to directly read-off the stacking sequence of the close packed planes (although it is not possible to distinguish the planes with and without prime, which have the same lattice positions but occupied by different atomic species). In such images, the presence of hexagonal and cubic faults has been made evident [12,18,19]. The hexagonal faults are originated by the insertion of two extra planes, which locally creates an hexagonal structure extended along 6 atomic layers. In its turn, the cubic faults are characterised by the absence of two consecutive planes in the perfect structure, giving rise to a 4-layer cubic environment. In addition to these single faults, the combination of several of them to create longer lamellae with hexagonal structure has also been observed [18,12]. Such kind of lamellae were formerly deduced from conventional TEM observations and electron diffraction [20,21]. More recently, Zhu *et al* also observed a "composite" fault by means of HREM, which can be obtained by the extraction of one plane from the perfect structure and the shear of all the following planes (changing from plane A to plane B, from B to C and from C to A) [19]. An interesting result has been obtained in the comparison of Cu-Zn-Al alloys with different electron concentration, e/a, ranging from 1.4 to 1.54 [22]. The density of hexagonal and cubic stacking faults in 18R martensite (obtained from a specific analysis of the diffraction patterns, as proposed in ref. 15, and also from the HREM images) shows a different evolution with e/a : the density of cubic faults decreases with e/a, while the density of hexagonal faults increases [22]. This result is well correlated with the known fact that the hexagonal (2H) martensite is preferred for high values of e/a.

The non basal faults have been firstly observed, without interpretation, by Nishiyama and Kajiwara [23]. Later, they have been reported to be parallel to the $(\bar{1}28)$ or $(12\bar{8})$ planes [24,25]. A detailed study made by Andrade *et al* has shown that there are two kinds of defects, either parallel with the habit plane (HP, close to the $(\bar{1}\,\bar{2}\,8)$ plane) or nearly parallel to a plane crystallographically equivalent to the habit plane (EHP, close to the $(\bar{1}28)$ plane), which were called defects of "Type II" and "Type I", respectively [26]. Lovey *et al* reported a detailed analysis of the Type I defects from HREM images at atomic resolution [18]. The defects are not simply planar, but they have a certain width of a few atomic layers. They constitute the interface between two crystal parts with the same basal plane but which contain different stacking sequences and are translated with respect to each other. Bending of the basal plane rows was observed across the Type I defect traces [18]. In a deep analysis, the authors related these defects with the second variant of bcc. From a single crystal of martensite, two orientation variants of the parent phase are possible: the first one is the orientation of the bcc structure from which the martensite is obtained by a martensitic transformation, and the second one is a differently oriented bcc structure also compatible with the martensite crystal. The second variant has been observed under special experimental conditions [27]. The relationship between the two variants is such that the equivalent habit plane (EHP) of variant 1 is the same as the habit plane (HP) of the second variant. Based on this fact, the authors interpreted the observed bending of the basal plane rows when crossing the Type I defects as a relaxation towards the second variant of bcc [18]. The Type I defects (and also the Type II defects) would then be created at the impingement of two different plates of the same martensite variant nucleated separately [18]. Finally, it has to be mentioned that Zhu *et al* reported that the non-basal plane defects are differently oriented (parallel to the $(\bar{1}\,25)$ planes). These authors proposed another mechanism for the formation of such defects, based on the combinations of two basal plane stacking faults which end inside the same martensite plate and are located at different levels [19].

4 Precipitates of other phases.

4.1 Ni_2Al precipitates in Ni-Al alloys.

The Ni-rich Ni-Al alloys (Ni_x-Al_{100-x}, $50 \leq x \leq 66$) exhibit a number of phase transformations depending on the thermal treatment performed. The ordered cubic NiAl β_2 phase (B2 structure) is only stable at elevated temperatures (above 1270 K for x~63). A rapid quench from the β_2-phase region avoids the decomposition to the stable phases and to retain the β_2 phase until the martensitic transformation takes place (above or below room temperature, depending on the composition). When the quenched alloys are aged at intermediate temperature, the precipitation of the Ni_2Al phase takes place. This is, however, a metastable phase because longer ageing periods result in the decomposition to the Ni_5Al_3 and Ni_3Al phases.

The Ni_2Al phase has a trigonal superstructure and develops into small precipitates. It was first observed independently by Lasalmonie [28] and Reynaud [29,30] in alloys annealed between 673 and 773K. Concerning the crystal structure of this phase, some controversy existed. The structure can be understood as stackings of $(111)_{B2}$ planes, which become the (001) plane of the trigonal Ni_2Al phase. The $<112>_{B2}$ direction becomes the <010> direction of the Ni_2Al phase. Based on electron diffraction patterns along different zone axes, Reynaud proposed a purely replacive transformation from B2 to Ni_2Al, in which, basically, some $(111)_{B2}$ planes occupied by Al atoms are substituted by Ni atoms. On the other hand, based on intensity anomalies of the superlattice spots, Lasalmonie proposed the existence of shuffles of $(111)_{B2}$ planes (following a q=1/3<111> longitudinal displacement wave), in a similar way as the ω transformation. By means of x-ray diffraction experiments, Georgopoulos and Cohen [31] obtained that a ω-type collapse of $\{111\}_{B2}$ planes takes place, with diffusional reactions not being considered. To elucidate from this controversy, a detailed structural analysis by means of HREM, image simulations and microdiffraction was performed by Muto *et al* [32]. This work constitutes a nice example of a crystal structure refinement completely obtained by means of HREM and electron diffraction. The authors considered the Ni_2Al structure as resulting from a displacive-replacive transformation from β2 phase, characterised by the substitution of some formerly Al sites by Ni atoms (in the way proposed by Reynaud), but also with a partial ω-type displacement of the two {111} planes above and below the plane in which the Al atoms have been substituted by Ni (in the way suggested by Lasalmonie) [32]. The displacement of such {111} planes could not be directly seen in the microscope since the spacing of these planes becomes too small to be resolved. Thus, the HREM image simulation was necessary. Several possibilities were analysed, from which the existence of a full ω-type collapse of these {111} planes was undoubtedly ruled out. The upper limit for the partial ω-type displacement of {111} planes was found to be about 50% of full collapse, but no lower limit of such displacement could be obtained from the image simulation study (even it was not possible to strictly prove that such a ω-type shuffle exists). A subsequent careful microdiffraction experiment complemented with dynamical calculations of diffracted intensities showed that the partial ω-type collapse indeed exists and it lies between 20 and 50% of full collapse [32]. In the same work, the as quenched samples were also studied. Diffuse intensity in electron diffraction patterns was observed around the positions of the superlattice reflections of Ni_2Al phase. The HREM images, together with image processing, revealed an underlying pattern of one-dimensional strings along <111> directions, which were interpreted as precursor phenomena of the Ni_2Al precipitation which takes place after annealing [32].

The same authors used the HREM to study the Ni_2Al phase embedded in martensite [33]. Formerly, it was reported that the precipitates undergo the same shear as the parent phase during the martensitic transformation to the 7M martensite, which imply that they do not act as obstacles for the martensitic transformation [34]. The electron diffraction patterns as well as the HREM images taken along different directions clearly verified the latter conclusions [33]. Moreover, it was established that the partial ω-type shuffle existing in the Ni_2Al precipitates embedded in β_2 phase is presumably no longer present in the sheared precipitates embedded in martensite [33].

4.2 *γ precipitates in Cu-Zn-Al*

Ageing the β Cu-Zn-Al alloys at intermediate temperatures (the β phase being retained by a rapid cooling from high temperatures, in the β region of the phase diagrams) promotes the precipitation of γ phase. Nevertheless, γ-type precipitates have been observed just after quenching Cu-Zn-Al single crystals with e/a=1.48 from temperatures in the range 670 - 970 K. The size of these precipitates depends on the quenching temperature, being about 10 nm for quenching temperatures of 770K [35]. In this case, the diffraction patterns show the main characteristics of the γ phase, although the extra reflections corresponding to the γ phase are slightly shifted from 1/3 and 2/3 of the distance between the consecutive β fundamental reflections (as should be in a perfect γ phase), giving rise to incommensurate diffraction patterns in relation to the β phase. Upon subsequent ageing at intermediate temperatures, the precipitates grow and the diffraction patterns show the correct positions of the γ extra reflections [36]. The presence of distributions of such precipitates in the β parent phase has been observed to influence the martensitc transformation [37-41]. The precipitates improve the performances of the material in several topics, like the hardening of the β phase [42], the reproducibility of the transformation temperatures upon thermal cycling [40] and the ease to induce the two way memory effect by mechanical or themomechanical cycling [41,43].

The HREM has been used to interpret the incommensurate diffraction patterns of the γ-type precipitates [44] and to better characterise the accommodation of the precipitates in the parent and in the martensite phases [45]. The HREM images of the precipitates obtained by quenching from 770K showed small microdomains of ~ 5 nm average size inside the precipitates, separated by antiphase-like boundaries [44]. A detailed study comparing the experimental and simulated images has shown that such boundaries are not true antiphase boundaries but planar defetcs which break the translational symmetry of the lattice [45]. The observed shifts of the γ electron diffraction spots have been interpreted as a consequence of interference effects among the coherent scattering from each microdomain [44]. The HREM images have shown that the small precipitates are perfectly coherent with the β matrix, but they lose the coherency when the martensitic transformation takes place [45]. At difference with the Ni_2Al precipitates, the γ precipitates do not follow the shear deformation accompanying the martensitic transformation, which causes problems for the accommodation of the precipitates in martensite. When the precipitates are embedded in a single martensite plate, a nanoplate of another variant has been observed to form at the interface between the main martensite plate and the precipitate. The nanoplate, having a width which takes some tens of atoms, is selfaccommodating with the main plate, in order to reduce the shear strain at the interface with the precipitate. Nevertheless, a fully elastic accommodation of the martensitic shear strain is not possible, as indeed confirmed by the observation of dislocations at the interface with the precipitate [45]. The accommodation problems between martensite and γ precipitates constitutes the most important factor which contribute to the changes in the martensitic transformation parameters observed in samples containing small precipitates [37-41].

5 Premartensitic effects (precursor phenomena).

In most of the alloys that undergo thermoelastic martensitic transformations, different anomalies in the electron and neutron diffraction have been observed at temperatures close but above Ms, which are accompanied by an elastic softening (decrease of the parent phase elastic constants). These anomalies are usually referred to as *premartensitic effects*, or *precursor phenomena*.

The Ni-Al alloy system is a paradigm of the study of such effects. The first observations of diffuse intensity in electron diffraction patterns, together with a "tweed" microstructure in the conventional TEM images as well as softening of the parent phase elastic constants, were reported in the seventies [46,47]. A detailed study of the tweed microstuctures by means of conventional TEM and electron diffraction was presented by Robertson and Wayman [48]. The authors attributed the observed effects to a modulation of the cubic B2 structure by static transverse displacement waves ($\{110\}\langle 1\bar{1}0\rangle$ modes) with a variety of wavelengths higher than 1 nm. The precursor phenomena are also manifested by an incomplete softening in the phonon dispersion curves (obtained by means of neutron diffraction experiments), particularly in the transverse acoustic branch along the $\langle 110\rangle_{B2}$ direction ($\langle\zeta\zeta 0\rangle TA_2$ mode, with displacements along $\langle 1\bar{1}0\rangle$

direction) [49,50]. The phonon anomalies provide strain centres which give rise to the diffuse scattering. As for the strain centres, embryos with a near 14M martensite structure have been proposed, having an average size of 40A at 329K [49]. The phonon anomalies have an electronic origin. Indeed, the experimental phonon dispersion curves were in very good agreement with theoretical calculations in which the softening is related to electron-phonon coupling and other specific properties of the Fermi surface[51,52]. In this field, the HREM has been a crucial tool to further clarify the origin of the "tweed" microstructure and diffuse scattering. The first lattice fringe images of Ni-Al premartensitc state were presented in 1979 by Portier *et al* [53], showing an overall variable intensity modulation and regions perpendicular to <1 $\bar{1}$ 2> and < $\bar{1}$ 12> with variable lattice spacing (clearly detected as diffuse streakings in the optical Fourier transform of the HREM image). These features were interpreted as a result of anharmonic coupling of <110> <1 $\bar{1}$ 0> transverse acoustic phonons [53]. In Ni-Al alloys, the displacement waves in the premartensitic state are mostly static, which allows to detect the variable lattice spacing in the HREM images (better revealed by the streakings in the Fourier transform of the HREM image), at variance with the Cu-Zn-Al alloys, in which the electron scattering effects are due to displacement waves mostly of dynamic type, which impede to record the atom displacements in the HREM images [54]. A detailed HREM study of the Ni-Al premartensitic state was performed by Schryvers and Tanner, who showed that the contrast modulations giving rise to the diffuse streaks in the Fourier space are confined in regions of a few nm size [55]. These regions would correspond to the embryos with nearly 14M structure considered to interpret the neutron diffraction experiments [49]. A model for the atomic configuration corresponding to the modulated regions was build as a superposition of transverse modulation waves with a wavelength very close to the 7-layer periodicity of the 14M martensite structure and an homogeneous shear on {110}<1 $\bar{1}$ 0>, the same as the shear associated to the martensitic transformation [55]. Thus, the structure of such domains (embryos) in the premartensitic state has many common features with the 14M martensite structure which develops upon further cooling to Ms.

In Ni-Mn-Ga alloys with compositions closed to Ni_2MnGa and martensitic transformation temperatures below room temperature, premartensitic effects characterised by softening of <ζζ0> TA_2 phonon modes have been observed, in a similar way as in Ni-Al alloys [56-58]. Nevertheless, at difference with these alloys, a soft-mode condensation occurs in Ni-Mn-Ga alloys at temperatures above Ms, resulting in a weakly first order phase transition to an intermediate phase between austenite and martensite, as observed by internal friction, *in-situ* cooling experiments in the TEM and also by analysing the <ζζ0> TA_2 phonon dispersion curves [58-60]. In this case, the HREM images (taken at room temperature, in the parent phase [58]) again helped to better understand the structure of the intermediate phase [58,60].

6 Recent developments in HREM image processing.

With the appearance of the modern high quality CCD cameras, the computer digitisation and further processing of HREM images has become a standard tool for practically all HREM users. The so-called *Bragg filtering* constitutes the most classical image processing technique. It consists in calculating the Fourier transform of the lattice-fringe image, placing a mask around the intensity maxima of the Fourier transform (which come from the spacial periodicities present in the image) and back Fourier transforming to the direct space. In this way, the noise present in the original image, which gives diffuse intensity in the Fourier space, is filtered. The method can also be used to extract the information coming from a specific portion of the Fourier space. Being the Fourier transform a complex function, it has its amplitude (modulus) and its phase. Recently, a new method based in the phase of the Fourier transform has been developed by Hÿtch [61]. The method uses the same ideas as the holographic reconstruction. A direct relationship is established between the phase and the displacement of lattice fringes and between the gradient of the phase and the local reciprocal lattice vector, which allow to determine the displacement field around a defect as well as local variations of lattice fringe spacing [61]. The method has been applied to accurately determine the translation vector of antiphase boundaries, the displacement fields around dislocations and the changes in the reciprocal lattice vector in carbon nanotubes [61]. In the domain of shape memory alloys, the new processing method has been applied in Ni-Ti-based alloys. The rapidly solidified (melt spun) $Ni_{47}Ti_{50}Co_2$ alloys annealed at 700°C show very small precipitates which, so far, have not been identified. The application of such method has permitted to determine that the lattice planes of the matrix around these precipitates are displaced towards the precipitate by a distance of 0.08 ± 0.01 times the normal matrix lattice spacing, and that the reciprocal lattice vector in the precipitate differs from the matrix by $-17 \pm 3\%$ [62]. These data can be very useful to identify the nature of such precipitates. Considering the topics described in the present paper, other applications of the method in the field of shape memory alloys can be rapidly envisaged, like, for instance, the analysis of the mismatched regions in the martensite boundaries, or precise determinations of the lattice spacing variations in the premartensitic state.

Acknowledgements

The author would like to thank Prof. R. Portier and Dr. D. Schryvers, who provided some images for the presentation of this paper.

References

1. See, for instance: Z. Nishiyama, *Martensitic Transformations* (Academic Press, 1978), and L. Delaey, R.V. Krishnan, H. Tas, H. Warlimont, *J. Mat. Sci* **9**, 1521, 1536 (1974)
2. K. Otsuka, T. Sawamura and K. Shimizu, *phys. stat. sol.* **5**, 457 (1971)
3. S.P. Gupta and A.A. Johnson, *Trans. Japan Inst. of Metals* **14**, 292 (1973)
4. K.M. Knowles and D.A. Smith, *Acta Met.* **29**, 101 (1981)
5. M. Nishida, H. Ohgi, I.Itai, A. Chiba and K. Yamauchi, *Acta Met. Mat.* **43**, 1219 (1995)
6. K.M. Knowles, *Phil. Mag. A* **45**, 357 (1982)
7. T. Onda, Y. Bando, T. Ohba and K. Otsuka, *Mater. Trans. JIM* **33**, 354 (1992)
8. M. Nishida, K. Yamauchi, I. Itai, H. Ohgi and A. Chiba, *Acta Met. Mat.* **43**, 1229 (1995)
9. H. Tas, L. Delaey and A. Deruyttere, *Met Trans.* **4**, 2833 (1973)

10. T. Saburi and C.M. Wayman, *Acta Met.* **27**, 979 (1979)
11. K. Adachi, J. Perkins and C.M. Wayman, *Acta Met.* **34**, 2471 (1986)
12. K. Adachi and J. Perkins, *Met. Trans. A* **16**, 1551 (1985)
13. J.M. Cook, M.A. O'Keefe, D.J. Smith and W.M. Stobbs, *J. of Microscopy* **129**, 295 (1983)
14. F.C. Lovey, G. Van Tendeloo, J. Van Landuyt and S. Amelinckx, *Proc. Int. Conf. on Martensitic Transformations'86*, (The Japan Inst. of Metals, 1987) p. 762
15. Z. Nishiyama, J. Kakinoki and S. Kajiwara, *J. Phys. Soc. Japan* **20**, 1192 (1965)
16. F.C. Lovey, *Acta Met.* **35**, 1103 (1987)
17. M. Andrade, M. Chandrasekaran and L. Delaey, *Acta Met.* **32**, 1809 (1984)
18. F.C. Lovey, G. Van Tendeloo, J. Van Landuyt, L. Delaey and S. Amelinckx, *phys. stat. sol. (a)* **86**, 553 (1984)
19. M. Zhu, D.Z. Yang and C.L. Jia, *Met Trans. A* **20**, 1631 (1989)
20. L. Delaey and H. Warlimont, *Z. Metallk.* **56**, 437 (1965)
21. L. Delaey and J. Lefever, *Trans. Japan Inst. of Metals* **10**, 372 (1969)
22. J. Lelatko, M. Gigla and H. Morawiec, *Proc. IX Conf. on Electron Microscopy of Solids* (Ed. A. Czyrska-Filemonowicz, B. Garbarz, H. Adrian, J. Wojtas and A. Zielinska-Lipiec, Krakow, Poland, 1996) p. 561
23. Z. Nishiyama and S. Kajiwara, *Trans. Japan Inst. of Metals* 3, 127 (1962)
24. K. Otsuka, T. Nakamura and K. Shimizu, *Trans. Japan Inst. of Metals* **15**, 200 (1974)
25. J.W. Brooks, *Electron Microscopy* **1**, 178 (1980)
26. M. Andrade, L. Delaey and M. Chandrasekaran, *J. Physique,* Coll. C4, **43**, C4-673 (1982)
27. F.C. Lovey, A. Amengual and V. Torra, *Phil. Mag. A* **64**, 787 (1991)
28. A. Lasalmonie, *Scripta Metall.* **11**, 527 (1977)
29. F. Reynaud, *J. Appl. Crystall.* **9**, 263 (1976)
30. F. Reynaud, *Scripta Metall.* **11**, 765 (1977)
31. P. Georgopoulos and J.B. Cohen, *Acta Met.* **29**, 1535 (1981)
32. S. Muto, N. Merk, D. Schryvers and L.E. Tanner, *Phil. Mag. B* **67**, 673 (1993)
33. S. Muto, D. Schryvers, N. Merk and L.E. Tanner, *Acta Met. Mat.* **41**, 2377 (1993)
34. I.M. Robertson and C.M. Wayman, *Phil. Mag. A* **48**, 629 (1983)
35. R. Rapacioli, M. Chandrasekaran and F.C. Lovey, *Proc. Int. Conf. on Solid-Solid Phase Transf.*, ed. H.I. Aaronson, D.E. Laughlin, R.F. Sekerka and C.M. Wayman (Metall. Soc. AIME, Warrendale, PA, 1982) p. 739
36. F.C. Lovey, R. Rapacioli and M. Chandrasekaran, *phys. stat. sol (a)* **68**, K105 (1981)
37. R. Rapacioli and M. Chandrasekaran, *Proc. Int. Conf. on Martensitic Transf.' 79*, ed. by The Dept. of Materials Science and Engineering (MIT, Cambridge, MA, 1979) p. 596
38. C. Auguet, E. Cesari, R. Rapacioli and Ll. Mañosa, *Scripta Met.* **23**, 579 (1989)
39. J. Pons and E. Cesari, *Thermoch. Acta* **145**, 237 (1989)
40. J. Pons and E. Cesari, *Acta Met. Mat.* **41**, 2547 (1993)
41. J. Pons, M. Sade, F.C. Lovey and E. Cesari, *Mater. Trans. JIM* **34**, 888 (1993)
42. A. Isalgué, F.C. Lovey and M. Sade, *Scripta Met. Mat.* **28**, 1183 (1993)
43. A. Amengual, E. Cesari and J. Pons, *J. Physique* Coll. C8, **5**, C8-871 (1995)

44. F.C. Lovey, G. Van Tendeloo, J. Van Landuyt, M. Chandrasekaran and S. Amelinckx, *Acta Met.* **32**, 879 (1984)
45. J. Pons and R. Portier, *Acta Mat.* **45**, 2109 (1997)
46. A. Lasalmonie, *C.R. Acad. Sc. Paris* **280**, C1501 (1975)
47. K. Enami, J. Hasunuma, A. Nagasawa and S. Nenno, *Scripta Met.* **10**, 879 (1976)
48. I.M. Robertson and C.M. Wayman, *Phil. Mag.* **48**, 421, 443, 629 (1983)
49. S.M. Shapiro, J.Z. Larese, Y. Noda, S.C. Moss and L.E. Tanner, *Phys. Rev. Lett.* **57**, 3199 (1986)
50. S.M. Shapiro, B.X. Yang, Y. Noda, L.E. Tanner and D. Schryvers, *Phys. Rev. B* **44**, 9301 (1991)
51. G.L. Zhao, T.C. Leung, B.N. Harmon, M. Keil, M. Müllner and W. Weber, *Phys. Rev. B* **40**, 7999 (1989)
52. G.L. Zhao and B.N. Harmon, *Phys. Rev. B* **45**, 2818 (1992)
53. R. Portier, D. Gratias and W. Stobbs, *Proc. Int. Conf. on Martensitic Transf.' 79*, ed. by The Dept. of Materials Science and Engineering (MIT, Cambridge, MA, 1979) p. 541
54. G. Van Tendeloo, M. Chandrasekaran and F.C. Lovey, *Proc. Int. Conf. on Martensitic Transformations'86*, (The Japan Inst. of Metals,1987), p.868
55. D. Schryvers and L.E. Tanner, *Ultramicroscopy* **32**, 241 (1990)
56. G. Fritsch, V.V. Kokorin and J. Kempf, *J. Phys.: Cond. Matt.* **6**, L107 (1994)
57. V.A. Chernenko and V.V. Kokorin, *Proc. Int. Conf. on Martensitic Transformations*, ed. C.M. Wayman and J. Perkins (Monterrey Inst. of Advanced Studies, 1993) p. 1205
58. A. Zheludev, S.M. Shapiro, P. Wochner, A. Schwartz, M. Wall and L.E. Tanner, *Phys. Rev. B* **51**, 11310 (1995)
59. V.V. Kokorin, V.A. Chernenko, E. Cesari, J. Pons and C. Seguí, *J. Phys.: Cond. Matter* **8**, 6457 (1996)
60. E. Cesari, V.A. Chernenko, V.V. Kokorin, J. Pons and C. Seguí, *Acta Mat.* **45**, 999 (1997)
61. M.J. Hÿtch, *Microscopy, Microanalysis and Microstructure* **8**, 41 (1997)
62. M.J. Hÿtch, R. Portier, "Determination of atomic displacement fields from HREM micrographs by geometric phase analysis", Proc. European Symp. on Martensitic Transformations, ESOMAT'97, to be published in *J. de Physique* (1997), in the press

SURFACE AND STRUCTURAL MODIFICATIONS INDUCED BY THE DEZINCIFICATION OF THE α AND β PHASES OF Cu-Zn

H. E. TROIANI*, J. L.. PELEGRINA*, M. AHLERS.
Centro Atómico Bariloche, (8400) San Carlos de Bariloche, Argentina.
*CONICET.

When Cu-Zn alloys are heated in vacuum to temperatures T ≥ 450°C, Zn evaporates. When α phase Cu-25.4at%Zn is the starting material, the homogeneous evaporation is rapidly replaced by the formation of single pores, which on further dezincification grow together to form strings and grooves. When the initial composition is β phase Cu-48 at%Zn, a phase transformation to fcc occurs during Zn loss. By optical, transmission and scanning electron microscopy, the homogeneous nucleation of a nonequilibrium hexagonal structure within the β phase matrix and a heterogeneous nucleation leading to the growth of large equilibrium phase needles is observed.

1. Introduction

Very little information is available on the processes that occur during dezincification of Cu-Zn. It is surprising that the possibility to modify the surface properties of materials by the preferential evaporation of one component has not yet been more fully explored. For this reason, a more systematic study has been initiated, beginning with α Cu-25.4at%Zn and continuing with β Cu-48at%Zn. A detailed description can be found in [1,2]. Here only the main results and principal ideas will be presented. The experimental set up is very simple: Discs of about 5 mm diameter and 0.4 mm thickness of a given surface orientation were kept within a quartz tube, that was dynamically evacuated by a mechanical double stage pump. The tube was inserted into a furnace at a given dezincification temperature and during a certain time, and subsequently the surface was observed by optical, transmission (TEM) and scanning (SEM) electron microscopy. In order to study by TEM regions near the surface, one side of the disc was protected by a lacquer and the other side was thinned electrolytically to the desired thickness.

2. The Dezincification of α Phase FCC Cu-25.4at%Zn

In Fig. 1 is shown the $(100)_\alpha$ surface after 7 min at 450°C, the lowest dezincification temperature at which surface inhomogeneities are observed. The surface is covered by pores whose average diameter is about 0.25 μm and whose separation is about 1 μm. When the dezincification temperatures and times are increased, the pore density and diameter change. But in addition there is a tendency for the pores to align themselves in irregularly shaped rows, which finally collapse to form grooves, as can be seen from Fig. 2, obtained after a dezincification of 15 min at 800°C.

We have here another example of a phase transformation far from equilibrium. The high Zn concentration gradient in the vapor phase close to the surface makes the homogeneous Zn evaporation, described by a one dimensional profile normal to the surface, already unstable at an early stage, and replaces it first by zero dimensional pores, and finally by one dimensional grooves. Very similar patterns have been observed for another transformation far from equilibrium, the solidification during constitutional supercooling [3]. An attempt to simulate the pore formation during dezincification, in analogy to the solidification is now under way [4].

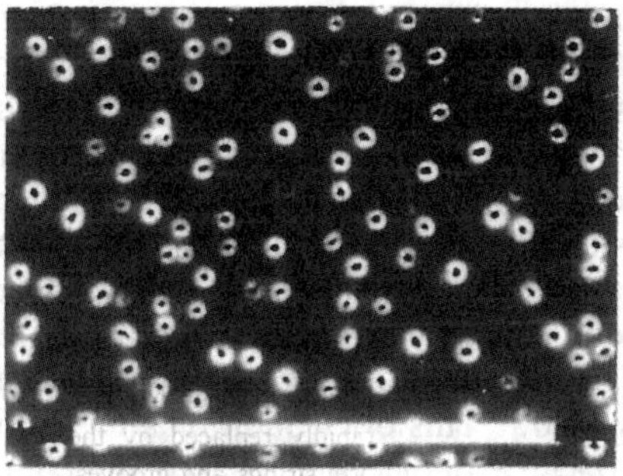

Fig. 1: $(100)_\alpha$ surface dezincified for 7 min at 450°C. The bar corresponds to 10 μm (SEM image).

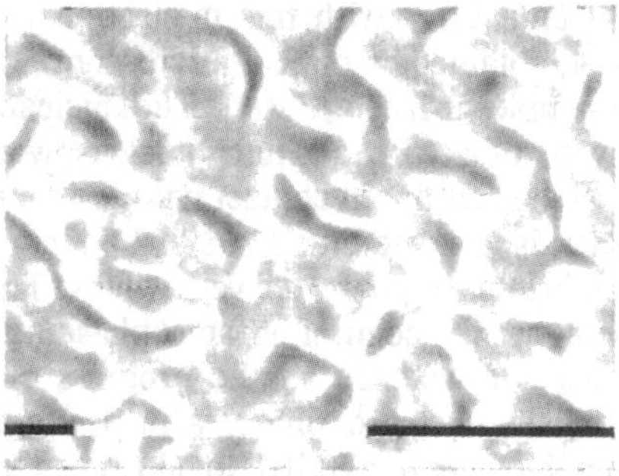

Fig. 2: $(100)_\alpha$ surface dezincified for 7 min at 800°C. The bar corresponds to 10 μm (SEM image).

3. The Dezincification of β Phase BCC Cu-48 at%Zn

When during dezincification the Zn concentration is reduced continuously, at one moment or the other a transformation to the equilibrium fcc α phase has to occur. The expectation that an incoherent α phase forms with a Zn concentration lower than that of the β phase according to the phase diagram [5], does not materialize: The time necessary to get the critical fluctuations is not available, compared to two other competitive nucleation mechanism. In Fig. 3 is shown one of them, consisting of nearly spherical coherent particles, which have their basal plane parallel to a $(110)_\beta$ phase matrix, but which have a stacking corresponding to a hexagonal 2H or 4H structure containing a high density of faults.

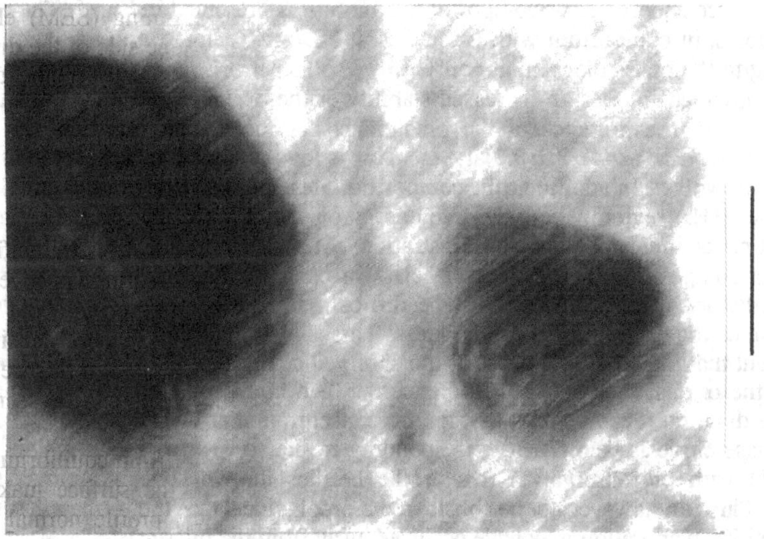

Fig. 3: $(100)_\alpha$ surface dezincified for 12 min at 450°C. The bar corresponds to 50 nm (TEM image).

Competitive with this nucleation is another one, whose consequences are seen in Fig. 4, obtained by optical microscopy: Long straight needles which cross a large part of the sample surface are seen. The nucleation must be extremely difficult compared to the growth, otherwise a single nucleus would not lead to such extended needle growth. Many needles are in contact with the sample border, which may be an indication that the border is a preferential nucleation site. The structure of the needles is that of the equilibrium α phase as verified by TEM. The orientation relationship is the same, whether at the beginning the α phase needles are in contact with retained β phase at the surface, or whether on the progress of the dezincification the surface layer consists completely of the α phase, with β retained only below. This orientation relationship is given by $(112)_\beta$ parallel to $(112)_\alpha$ at the interface, and the close packed directions parallel to each other. From the α and β phase lattice parameters [5] follows that the $(112)_\beta$ plane is not an undistorted interface: The $(112)_\alpha$ is by 2.4 % inflated with respect to $(112)_\beta$, and to make it strain free interface dislocations on each 40th plane or at 10 nm distance have to be incorporated. This is a small density which hardly increases the interface energy. In addition, the nonconservative climb of dislocations with their Burgers vector in the interface should not be an important obstacle to the lateral growth of the interface, since during dezincification sufficient vacancies are introduced.

Thus instead of the expected incoherent homogeneous α phase nucleation, the two nucleation mechanism lead to quasicoherent precipitates. One homogeneous but having a nonequilibrium structure, the other heterogeneous with the equilibrium α phase (neglecting the small amount of misfit dislocations). This competition between various nucleation mechanisms is frequently observed in solid state phase transformations, and apparently the transformation induced by dezincification is no the exception.

Due to the smallness of the round precipitates it has not been possible to analyse with sufficient detail its structure, lattice parameters and orientation relationship, nor could the nucleation sites for the needles be identified. For this reason, only a very approximate picture can be given, to rationalize the results. This can be best done with the help of free energy versus composition diagrams as presented in Fig. 5. The common tangent for α in equilibrium with β touches at 38.7at%Zn for α and 45.0at%Zn for β according to the phase diagram at 450°C [5]. The fact that a nonequilibrium hexagonal phase is observed implies an increased stability of the β phase, and indeed the β phase composition in the presence of the precipitates has been determined by the EDAX method in the SEM as (42.2 ± 0.5)at%Zn. This is lower than the expected 45.0at%Zn in equilibrium with α. In addition, this composition lies approximately in the middle of the two phase (α+β) region, where the free energy of α and β are the same. In Fig. 5 this composition is indicated, just for illustration. At this composition α can be formed without a composition change. Nevertheless, a coherent nucleation with the required $(112)_\beta$ interface orientation can still be difficult, but once formed can lead to rapid growth since the driving force can be large. This is also indicated in Fig. 5, by showing the tangent that touches the β phase at the observed composition, and a parallel one, that touches the α phase curve. The difference between both constitutes the driving force. From the diagram it can be seen that the Zn concentration at which the tangent contacts the α phase curve is below that corresponding to the (α+β) two phase region. This is consistent with the observation that at the needles the Zn concentration is (33.7 ± 0.5)at%. This can however not be considered a proof, since once the α phase has formed, additional Zn evaporation may have occurred away from the interface.

References

[1] H.E. Troiani, J.L. Pelegrina, M. Ahlers, Phys. Stat. Sol. (a) 156, 93 (1996).
[2] H.E. Troiani, J.L. Pelegrina, M. Ahlers, to be submitted to Phil. Mag.
[3] H. Biloni, W.J. Boettinger, in Physical Metallurgy 4th ed., R.W. Cahn, P. Haasen editors, (1995).
[4] H.E. Troiani, M. Kuperman, in preparation.
[5] S. Srinivasa Rao, T.R. Anantharaman, Z. Metallkde. 60, 312 (1969).

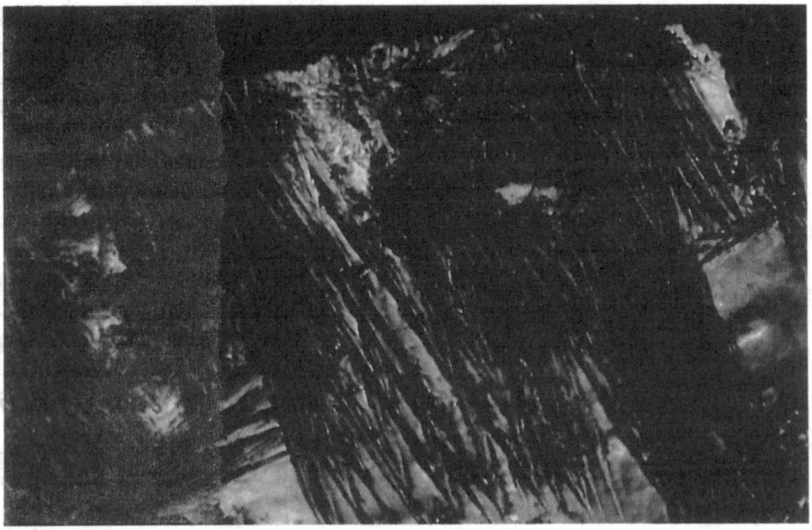

Fig. 4: Needles of α phase embedded in the β matrix. Surface dezincified at 450°C for 12 min. (Optical microscope image).

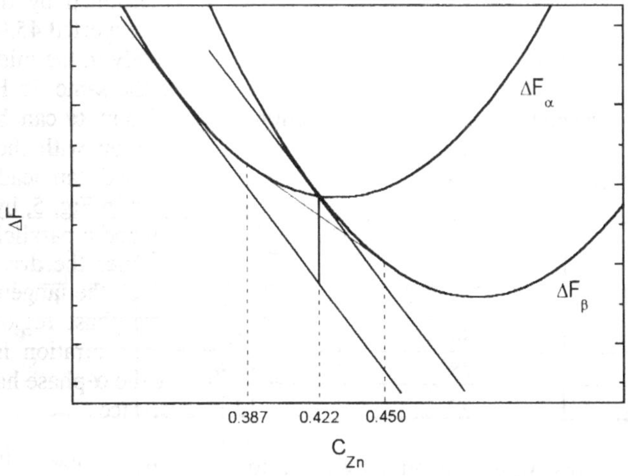

Fig. 5: Schematic phase diagram showing the α / β region.

STRUCTURE OF Cu-Al-Nb SHAPE MEMORY ALLOYS

J.LELĄTKO, H.MORAWIEC, M.GIGLA
Institute of Physics and Chemistry of Metals, University of Silesia, Bankowa 12, 40 007 Katowice, Poland

Yu.N.KOVAL, V.I.KOLOMYTSEV
Institute of Metal Physics, National Academy of Sciences, 36 Vermadsky Blvd., 252142 Kiev, Ukraine

The structure of the Cu-13.4wt.%Al-Xwt.%Nb (X - 0.27, 2.56, 7.86) alloys were investigated. It was found that niobium introduced to the alloys created two phases which differ in both chemical composition and lattice defects - $Nb(Cu,Al)_2$ and Nb(Cu,Al). Niobium addition caused increase the temperature of the reversible martensitic transformation and widening of its hysteresis. The alloy with small niobium addition shows a high degree of the shape memory effect which decreases with niobium increasing in the alloy.

1. Introduction

Two groups of copper - based shape memory alloys found commercial application: Cu-Zn-Al and Cu-Al-Ni [1]. The improved mechanical properties of them is achieved by adding minor elements like: Ti, Zr, B, Co which cause the grain refinement effect [2]. Considerable interest has been expressed in the possibility of producing alloy with a martensitic transformation above 200°C. It is well know that the better mechanical strength and thermal stability as well as higher M_s temperature exhibited by the Cu-Al-Ni alloys. On the other hand these alloys show the presence of brittle γ_2 - precipitates which lowers their deformability. The improving of the deformability may be achieved by lowering of the Al content to 12wt.% but this cause an considerable increase of the M_s temperature. In order to lower this temperature it is necessary to add such alloying elements like manganese and nickel [3-5].

The aim of this paper was to find a new type of alloys for high temperature application. For this purpose the structure and properties of the Cu-Al-Nb alloys have been studied.

2. Materials and methods

Four alloys were produced by melting in an induction furnace (Table 1).

Table 1 Chemical analysis of the tested alloys

Alloys	Compositions, wt.%		
	Cu	Al	Nb
A	86.6	13.4	0.27
B	84.4	13.0	2.56
C	78.6	13.4	7.86
D	86.5	13.5	-

The homogenised ingots were mechanically cut into 1mm thick plates. All specimens were solution-treated at 850°C for 20min., followed by quenching to ice-water.

The martensitic transformation temperatures of the alloys were determined by using DSC method. In order to analyse the alloy structure a transmission electron microscope

(JEM 3010 operating at 300kV with EDS system) was used. Thin foils for TEM examination were prepared initially by jet electro-polishing and finally by ion etching.

3. Experimental results and discussion

Even small amounts of niobium addition (alloy A) in the two - component Cu-Al alloy causes significant increase in the martensitic transformation temperatures (Table 2).

Table 2 Martensitic transformation temperatures [^{0}C] and shape recovery effect η [%]

Alloys	M_s	M_f	$\Delta M.$	A_s	A_f	ΔA	$A_f - M_f$	η
A	313	280	33	341	388	47	108	98
B	305	276	29	345	398	53	122	91
C	302	267	35	341	295	54	128	86
D	250	210	40	250	333	83	123	47

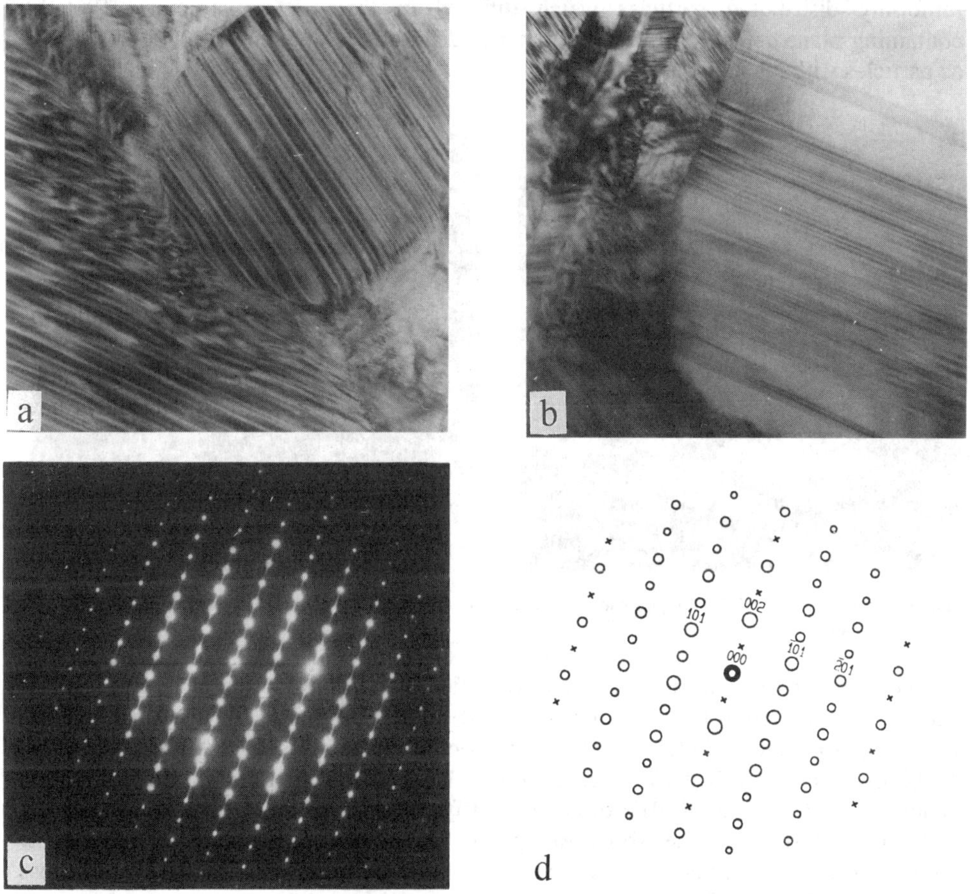

Fig. 1 Particles of the $Nb(Cu,Al)_2$ phase in the alloy B (a, b) and a diffraction pattern with streakes caused by presence of stacking faults (c) and its diagram (d)

In the Cu-Al-Nb alloy, however, niobium addition (alloys A-C) causes small but systematic decrease of the M_s and M_f temperatures while the A_f is increased. This results in a significant widening of the A_f - M_f transformation hysteresis. At a near similar hysteresis of the transformation for these alloys the niobium addition leads to increase of the temperature ranges of the transformations: martensitic - ΔM and reverse - ΔA.

In the Table 2 there are also given values of the η parameter characterising the one-way shape memory effect. It can be seen that the alloys containing niobium exhibit much higher degree of shape recovery than the two-component alloy. The amount of the shape recovery decreases slightly with the increasing niobium content.

To explain the changes of the martensitic transformation parameters and degree of shape memory effect the TEM studies of the alloys structure were carried out. The studies showed that the alloy A has a martensite structure with small spheroidal inclusions. The martensite had a complex structure. Except for plates of the 18R martensite there were also plates being a mixture of 18R/2H structures. In some of them the 2H phase was a dominating one.

Addition of higher niobium content (alloys B-C) introduces large number of uniformly distributed particles which differed in their defect structure. Particles containing plane defects with small number of dislocations were observed (Fig. 1) as well as particles without such defects (Fig. 2).

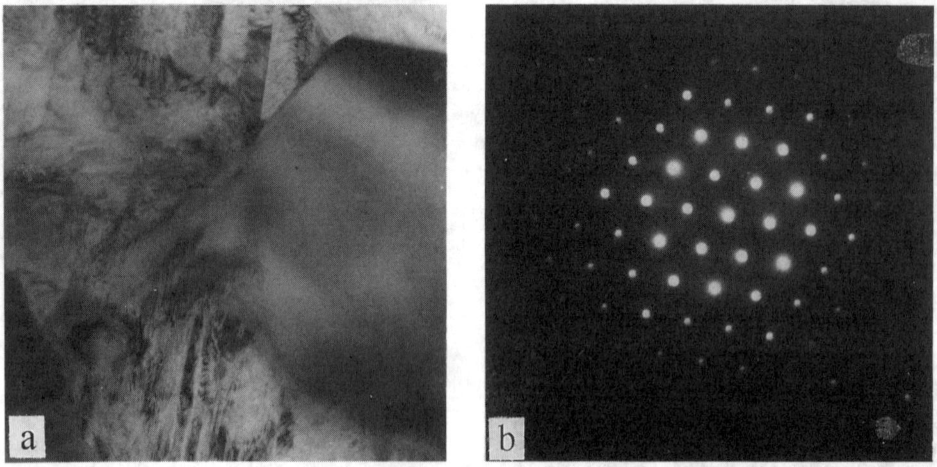

Fig. 2 Particles of the Nb(Cu,Al) phase in the martensite matrix (a) and their diffraction pattern of [001] orientation (b)

The particles visible in Fig. 1 were identified as particles of the $Nb(Cu,Al)_2$ phase whereas particles from Fig. 2 as the Nb(Cu,Al) phase. Basing on the chemical analysis results approximately following stoichiometric formulas were found for the phases: $Nb_2(Cu,Al)_5$ and $Nb_4(CuAl)_5$, respectively. Characteristic streakes in the diffraction patterns (Fig. 1c) in the <001> direction testify presence of large number of stacking faults in the particles of the $Nb(Cu,Al)_2$ phase. The number of the stacking faults differ for particles and depends on their sizes. In smaller particles no areas free from those defects were observed (Fig. 1a)

Increase in the niobium content in the Cu-Al. alloy causes significant difference in the martensite structure. The amount of the 2H martensite increases (Fig. 3) and except for plates containing a mixture of the 18R/2H structures also plates of a single 2H structure were identified.

The obtained results show that niobium addition to the Cu-Al alloy improves the structure and properties of these alloys. Niobium introduced to the alloy to lower its brittleness practically does not dissolve in the matrix and causes increase of the characteristic temperatures and shape memory effect.

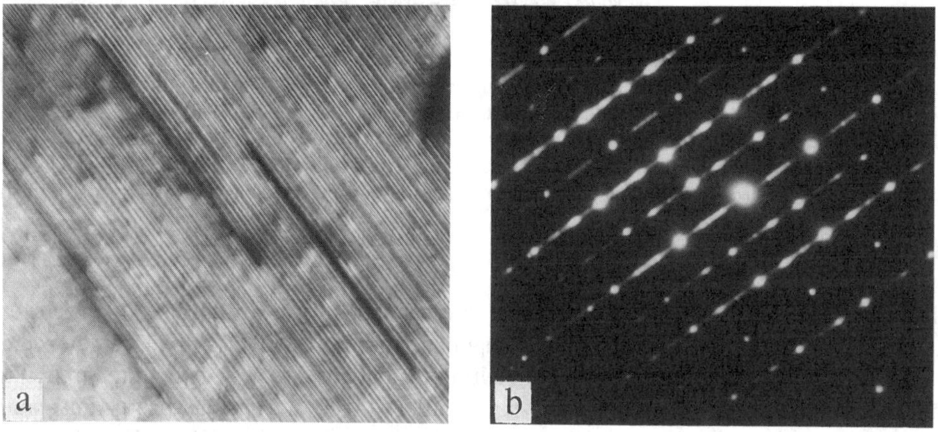

Fig. 3 Mixture of the 18R/2H structures in a martensite plate (a) and its diffraction pattern of $[010]_{18R/2H}$ orientation

4. Summary

Niobium addition in a Cu-Al alloys causes significant increase in the characteristic martensitic transformation temperatures.

Niobium present in Cu-Al-Nb alloys creates two types of phases: $Nb(Cu,Al)_2$ and Nb(Cu,Al). The determined chemical compositions differ considerably from their stoichiometric compositions

The presence of these particles in the studied alloys does not influence significant the shape recovery effect. However, increase in the particle number causes proportional decrease of the recovery effect.

Increasing of niobium content in the studied alloys causes increase of the amount of the 2H martensite.

Acknowledgements

The JEM 3010 transmission electron microscope was presented to the Institute of Physics and Chemistry of Metals, University of Silesia by Polish Foundation for Science.

References

1. N.F.Kennon, D.P.Dunne, L.Middleton, *Metall. Trans.* **A 13**, 551 (1982)
2. G.N.Sure, L.O.Brown, *Metall. Trans.* **A 15**, 1613 (1980)
3. K.Matsushita, K.O.Takashi, o.Taire, *J.Mat. Sc.* **20**, 689 (1985)
4. P.R.Swan, H.Warlimont, *Z. Metallkde.* **69**, 709 (1963)
5. K.Sugimoto, K.Kamei, M.Nakaniwa, Engineering aspects of shape Memory Alloys, ed. by T.W.Duerig, K.N.Melton, D.Stockel, C.M.Wayman, London 1990, p.89

ORDERING EFFECTS IN Ni_3Al POLYCRYSTALS

E. JEZIERSKA
Warsaw University of Technology, Department of Materials Science and Engineering, Narbutta 85, 02-524 Warsaw, Poland

Transmission electron microscopy (TEM) has been employed to examine the structure of ordered intermetallic compound Ni_3Al ($L1_2$ superstructure). TEM investigations were performed on JEM 100B and JEM 3010 microscopes using bright and dark field images, centered superlattice dark field (CSDF) images, selective area diffraction paterns (SAD), convergent beam electron diffraction (CBED) and energy dispersive X-ray spectroscopy (EDS). The following ordering effects in Ni_3Al polycrystals were detected; perfect order $L1_2$, ideal symmetry m3m, antiphase domains, antiphase domain boundaries, ordered grain boundaries, superdislocations.

1 Introduction

The structure and properties of intermetallic compounds form a subject of great current interest. Ni_3Al is an intermetallic compound forming the $L1_2$ long-range-ordered superstructure which is stable up to the melting temperature. Low temperature strength is dependent both on the degree of long-range order (LRO) and on the size of anti-phase domains [1]. For several atom distances away either side of antiphase boundary, the structure is uniformly ordered; yet at the antiphase boundary it is apparent that the order is "out of step" with an irregular wall of unnatural neighbours. Moving dislocations can pass through an existing antiphase boundary only by creating a new APB and, therefore, additional energy must be expended [1]. A unit dislocation in an ideally ordered structure is a double dislocation in the atomic site lattice, an unlikely configuration because of the very high local strain energy. The expected condition is then a dissociation of the unit dislocation of the ordered material into partials separated by stacking fault. An addition of boron has been shown to have a marked effect on the ambient ductility of the polycrystalline Ni_3Al [2]. On the other hand, it has been found that B element exhibits a significant solid solution strengthening effect in the Ni_3Al [3]. The perfect ordering of the constituent atoms of Ni and Al was not established in undoped (pure) Ni_3Al alloy even at just-stoichiometric composition. However, the addition of C and B atoms into the Ni_3Al induced the further ordering of the constituent atoms [4]. That effect is more significant at the off-stoichiometric compositions of Ni_3Al. The strongly attractive interaction between Ni atom and interstitials stabilizes the $L1_2$ structure of the Ni_3Al.

2 Results and Disscussion

Antiphase domain boundaries (APBs), expected in an ordered phase, were observed in the microstructure (Fig.1.). Centered superlattice dark field (CSDF) images of ordered domains shows the size of antiphase domains (in the range of 30-300 nm) and the ordering to the vicinity of the grain boundaries (Fig.2.). Ordered grain boundries enhanced with boron and nickel were found (Fig.3.). The intensity of CSDF image is higher at grain boundary region, so the degree of order is also higher. In CSDF image the disordered region appears black when the ordered matrix appears bright. Grain boundary ledges associated with boundary dislocations are visible on Fig.3 with periodicity of 10 nm. The effect of enhancing with boron and nickel was found by energy dispersive X-ray spectroscopy (EDS) measurement at such boundary layer.

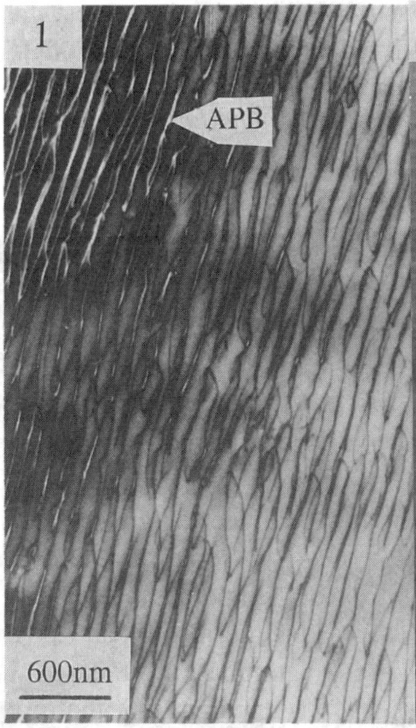

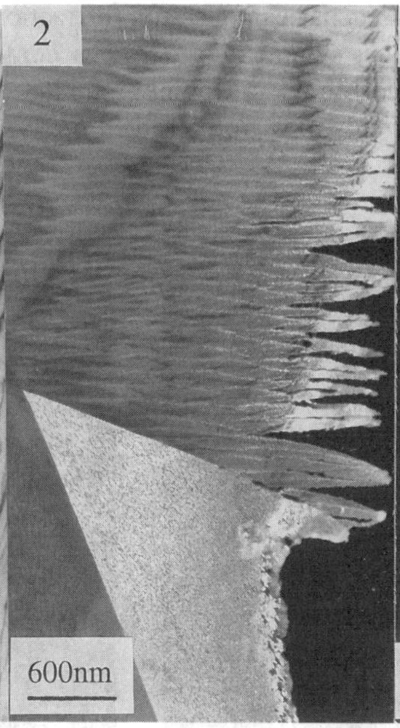

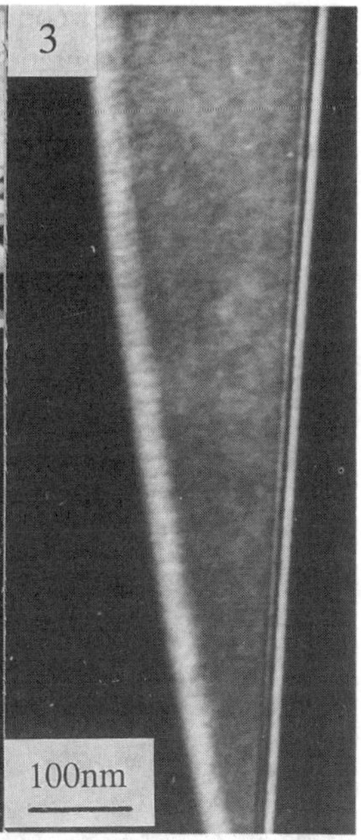

Fig. 1. Antiphase domain boundaries (APBs) in Ni_3Al, zone axis [001]
Fig. 2. Centered superlattice dark field (CSDF) images of ordered domains in Ni_3Al
Fig. 3. Ordered grain boundaries in CSDF image (330 superlattice spot) in Ni_3Al polycrystal

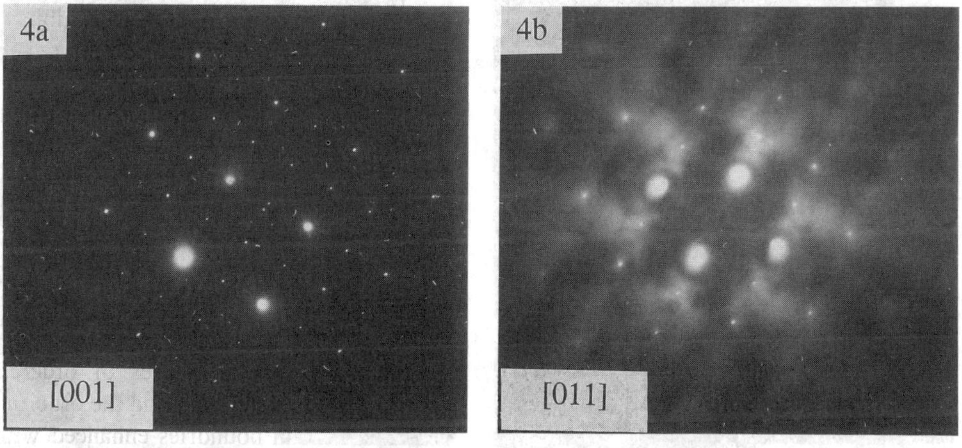

Fig. 4. Long range order (LRO-$L1_2$) at [001] zone axis a/ and short range order (SRO) effects on diffraction patterns at [011] zone axis in Ni_3Al+B b/

Long range order (LRO-$L1_2$) was determined (Fig.4a.) and short range order (SRO) effects on diffraction paterns were noticed in Ni_3Al with boron addition (diffuse scattering effects in SADP on Fig.4b).

From the results of convergent beam electron diffractions (CBED) on Ni_3Al polycrystals, with different orientations, symmetry group was confirmed using Loretto method [5]. Perfect order $L1_2$ and ideal symmetry (point group m3m) are evident from CBED images. Exemples of different orientations are presented on Fig 5.

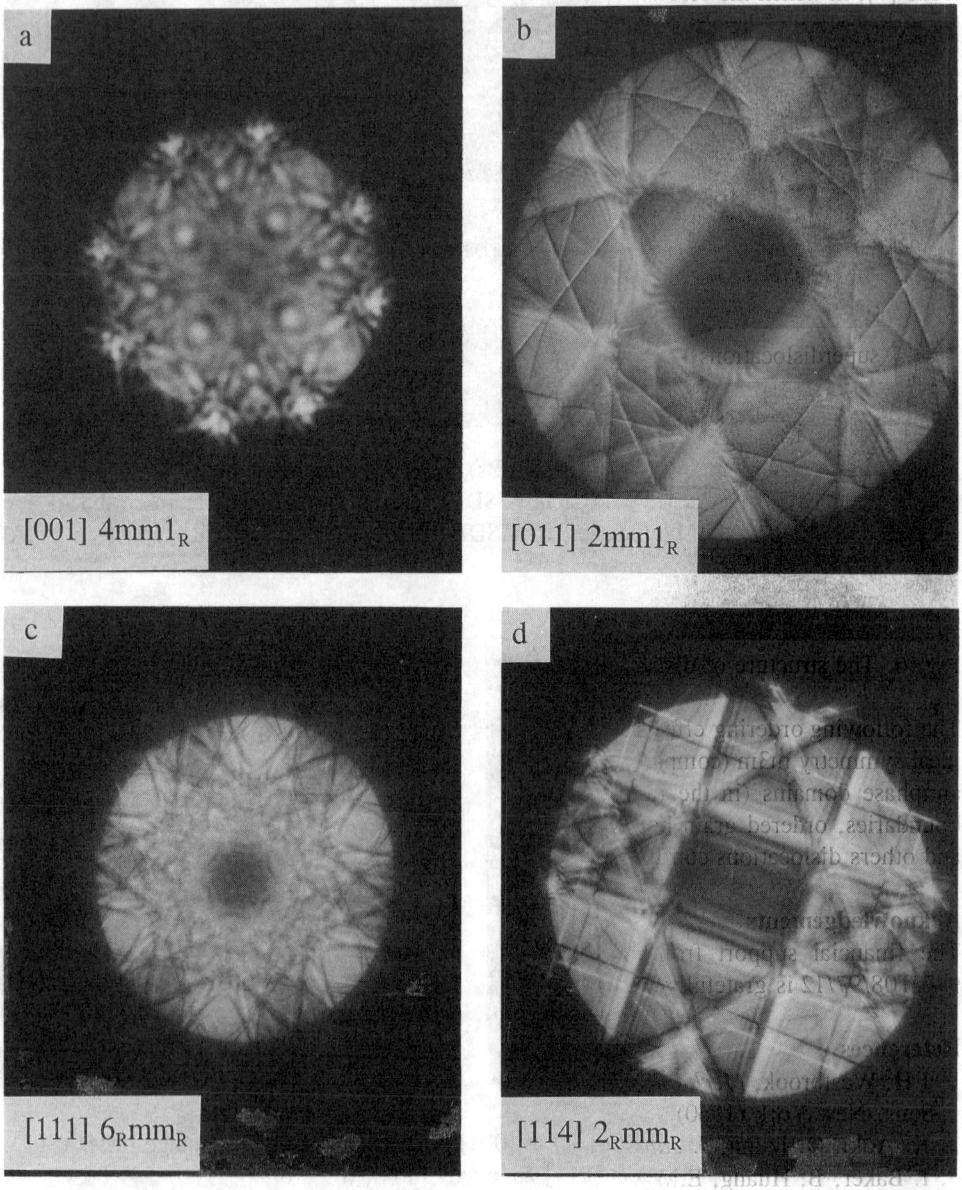

Fig. 5. Convergent beam electron diffractions (CBED) on Ni_3Al polycrystals with different orientation a/ zone axis [001], diffraction group $4mm1_R$, b/ [011], $2mm1_R$, c/ [111], 6_Rmm_R, d/ [114], 2_Rmm_R

The dislocations structure is typical for ordered phases; superdislocations, complex stacking faults (CSF), superlattice intrinsic stacking faults (SISF), dipoles and pairs of dislocations. The superdislocations dissociated into superpartials are likely to control the plastic deformation in this material [6]. Observations of the deformed structure ($\epsilon=2\%$) show the presence of screw superdislocations, most of which are in form of Kear Wilsdorf (KW) locks [7], in which the screw partials dislocations bound an APB on (010) plane (Fig.6.).

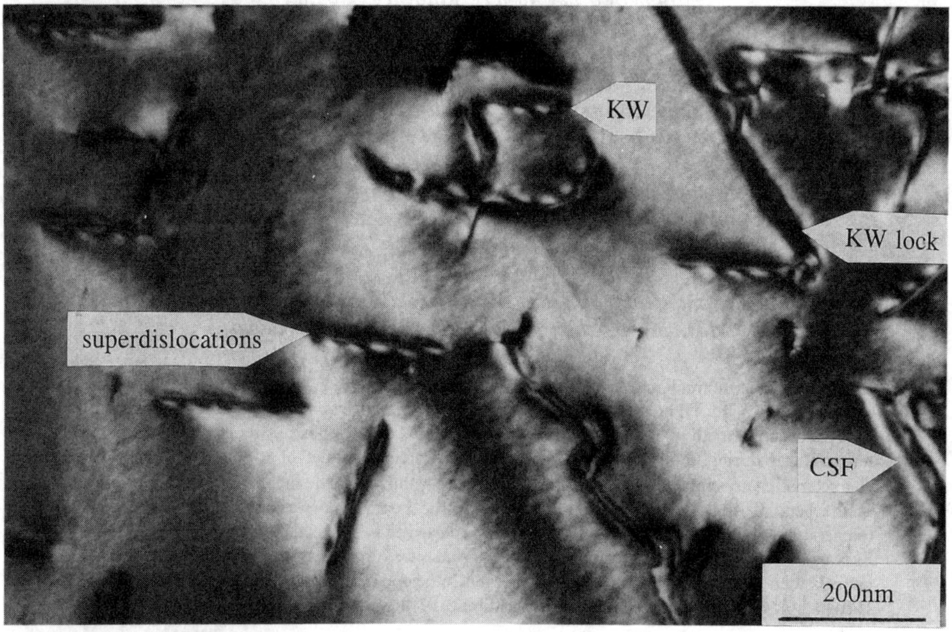

Fig. 6. The structure of dislocations in Ni_3Al polycrystals

The following ordering effects in Ni_3Al polycrystals were detected; perfect order $L1_2$ and ideal symmetry m3m (compared by conventional electron diffraction and CBED patterns), antiphase domains (in the range 30-300 nm), the ordering to the vicinity of the grain boundaries, ordered grain boundaries enhanced with boron and nickel, superdislocations and others dislocations configurations typical for ordered phases.

Acknowledgements
The financial support from the Committee for Scientific Research under grant PB 977/T08/97/12 is gratefully acknowledged.

References

1. J.H. Westbrook, *Mechanical Properties of Intermetallic Compounds,* ed. John Wiley & Sons, New York (1960)
2. K. Aoki, O. Izumi, *J. Japan Inst. Metals* **45**, 1190 (1979)
3. I. Baker, B. Huang, E.M. Schulson, *Acta metall.* Vol. **36,** No. 3, 493 (1988)
4. N. Masahashi, T. Takasugi, O. Izumi, *Acta metall.* Vol. **36**, No. 7, 1815 (1988)
5. M.H. Loretto, *Electron beam analysis of materials,* Chapman & Hall, New York, (1984)
6. V. Vitek, *Progress in Materials Science,* Vol. **36**, 1 (1992)
7. P.B. Hirsch, *Progress in Materials Science,* Vol. **36**, 63 (1992)

X-RAY STUDIES OF THE REVERSIBLE B2 ⇔ R TRANSITION IN TiNiCo SHAPE MEMORY ALLOY FOR MAXILLOFACIAL AND ORTHOPAEDIC SURGERY

Z. LEKSTON
Institute of Physics and Chemistry of Metals, University of Silesia, Bankowa 12, 40-007 Katowice, Poland

J. DRUGACZ
II Faculty and Clinic of Maxillofacial Surgery, Silesian Academy of Medicine, Francuska 20/24, 40-027 Katowice, Poland

K. JANUSZEWSKI
II Faculty and Department of Pathomorphology, Silesian Academy of Medicine, Medyków 18, 40-752 Katowice, Poland

B. WÓJCIK
Faculty and Clinic of Orthopaedic Surgery, Silesian Academy of Medicine, p.o. box 168, 41-200 Sosnowiec, Poland

The influence of thermal and thermomechanical treatment on the B2 to R-phase transformation temperatures of the TiNiCo shape memory alloy was examined by high and low temperature X-ray diffraction method. For sample after aged at. 400°C the R-phase exists in a wide range of temperatures during cooling. During heating the diffraction doublet of R-phase is small. The characteristic temperatures of B2⇔R transitions were stablized by the incomplete thermocycling. In the bending tests during heating and cooling the best memory effects and the recovery temperature close to human body temperature was observed in material after treatment at 400°C. This material is suitable to manufacture shape memory implants. The clinical treatment of patients with mandibular fractures by use of TiNiCo shape memory implants was satisfactory. Histopathological tests showed good biocompatibility of these memory implants.

1 Introduction

In recent years the unique characteristics of shape memory, superelasticity, acceptable resistance to corrosion and biocompatibility exibited by titanium nickel memory alloys have been exploited in the manufacture of biomedical devices [1-3]. One simple clinical application of shape memory implants is for bone clamps for maxillofacial and orthopaedic surgery.

In the II Faculty and Clinic of Maxillofacial Surgery of Silesian Academy of Medicine in Katowice seventy seven patients with mandibular fractures were treated using the TiNiCo shape memory clamps. The treatment progressed satisfactory. The application of the shape memory clamps facilitates treatment due to ensuring stable fixation of the bone fragments. No pathologic tissue reaction to the implants was observed [4].

When using shape memory clamps in bone surgery, it is essential that the temperature at which shape is regained is approximately equal to the body temperature. For an alloy with cobalt addition the recovery properties can be changed by thermal treatment, cyclic repeating of the tranformation or thermomechanical cycling. In this alloy the R-phase transition occurs.

In this paper the fragment of studies of phase transitions during cooling and heating, the stability of T_R temperature after thermal cycles and the best heat treatment for obtain medical implants with recovery temperature adapted to the body temperature are presented.

2 Material

The alloy with nominal composition Ti - 48.7At.%Ni - 1.3At.%Co was obtained by vacuum induction melting technique. The material was melted in a graphite crucible and casted into a graphite mould. The ingot was homogenized under a vacuum /10^{-5} Torr/ at 900°C for 48 h and hot-rolled to rods of diameter about 4 mm. Subsequently the rods were hot rolled to plates of thickness about 1 mm and hot drawn to wires of diameters 2-1 mm. The specimens for X-ray experiments and shape memory tests after hot rolling and hot drawing were quenched from 700°C/1h and aged at 300 to 600°C for 1 to 9h. Some of plates and wires were cold rolled and cold drawn with total deformation of about 10% and subsequently annealed from 200 to 600°C for 1h and cooled in water with ice.

3 Experimental procedures

A phase analysis of the specimens was performed at room temperature using X-ray Philips diffractometer equipped with a graphite monochromator. CuK_{α} radiation was used.

The studies of phase transitions during cooling and heating were carried out using Philips X-ray diffractometer equipped with a temperature camera with a temperature control system. The samples were heated by an electric heater and cooled by liquid nitrogen vapours environment. The diffraction patterns in the 37-47° /2Θ/ angular region were observed during cooling from the room temperature to about -100°C and during heating to about +100°C. Part of specimens were cyclicly cooled and heated in the temperature range of B2⇔R transition.

The shape memory effects were examined by a bending test. The wires after thermal treatments were placed into ethanol cooled by dry ice to about -60°C and deformed to angle of about 80 degrees and then heated up to about +80°C. Two way memory effect was obtained by thermomechanical cycling.

4 Results and discussion

4.1 X-ray Phase Analysis

At room temperature in the specimens after quenching from 700°C/1h and after quenching and ageing at 200-300°C /1h the parent phase B2 was observed [5]. After ageing at 400°C/9h the rhombohedral R-phase with small quantity of parent phase B2 coexists at room temperature /Fig.1/.

4.2 Phase Changes measured by X-ray temperature diffraction

During cooling and heating of the specimens after various thermal treatment the B2⇔R+B19'⇔B19' sequence of phase transitions was observed. During cooling of sample after ageing at 400^{C} the R-phase exists in a wide range of temperatures. During heating the doublet of R-phase was hardly visible. Ageing at 400°C considerably raised the temperature range of the phase transitions /Fig.1/. The temperatures of start and finish of the B2⇔R transition was stable after incomplete thermal cycles without transition to the monoclinic martensite B19' /Fig.2/.

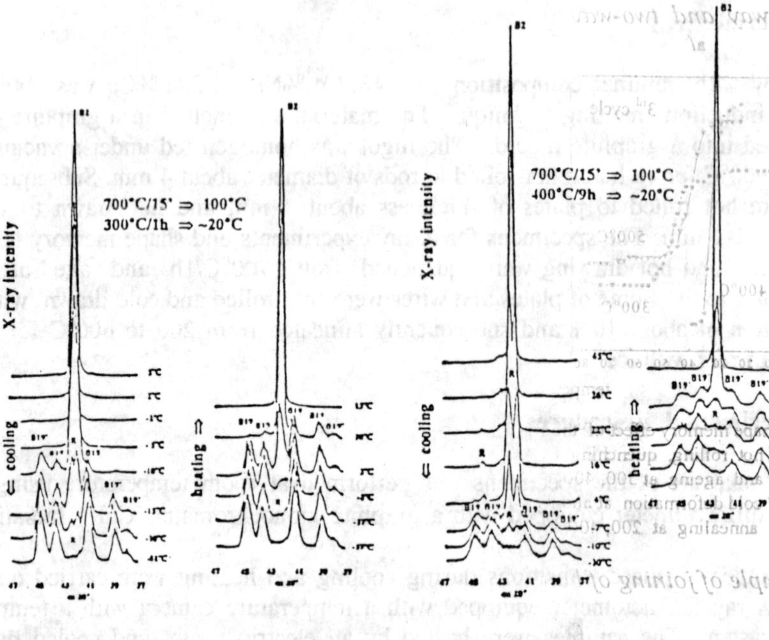

Figure 1: Phase transitions in the TiNiCo alloy after hot-rolling, quenching from 700°C/1h and ageing at 300/1h and 400°C/9h.

4.3 Thermal Cycling in the Temperature Range of the B2 ⇔ R Transition

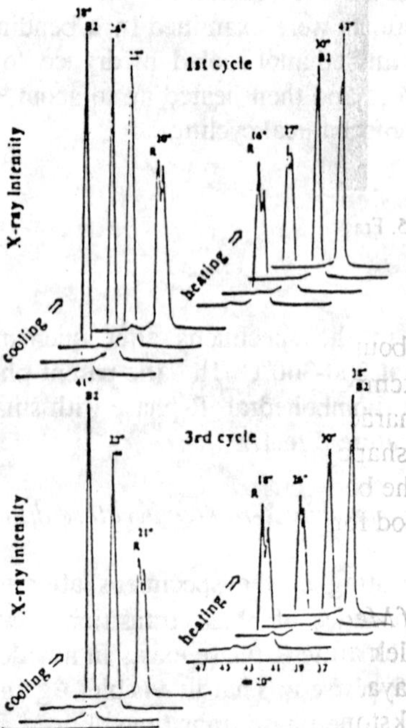

Figure 2: Cyclic cooling and heating in the temperature range of the B2 ⇔ R transition for the TiNiCo sample after quenching from 700°C/1h and ageing at 400°C/9h.

4.4 One-way and two-way shape memory

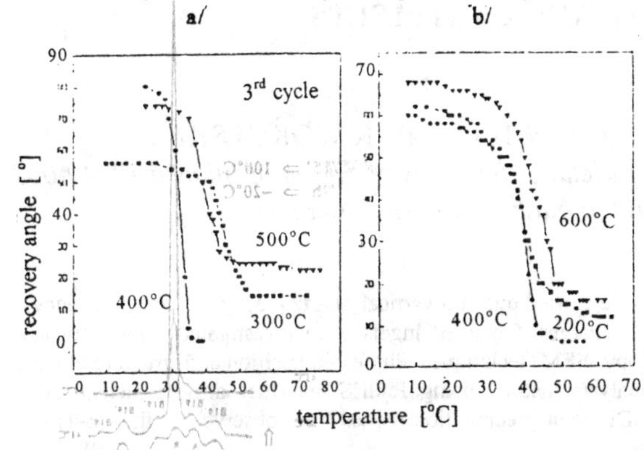

Figure 3: Shape memory effect of the TiNiCo samples after:
a/ hot rolling, quenching from 700°C/1h and ageing at 300, 400 and 500°C/1h
b/ cold deformation at about 10% with subsequent annealing at 200, 400 and 600°C/1h

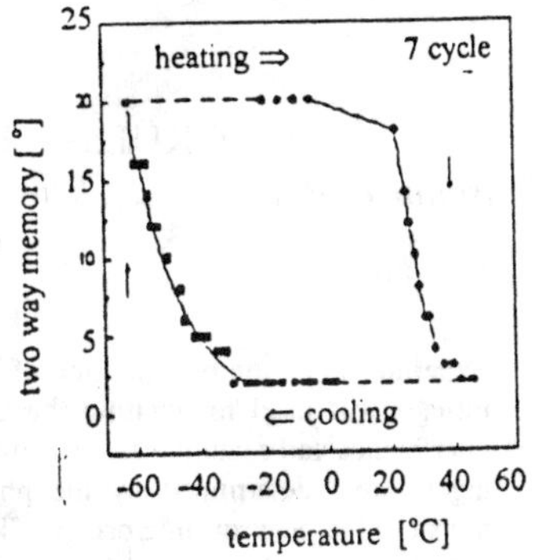

Figure 4: Two - way memory effect obtained by cylic repeating of one -way memory effect for the TiNiCo sample after hot rolling, quenching from 700°C/1h and ageing at 400°C for 1 hour.

4.5 Example of joining of bones with the shape memory clamps

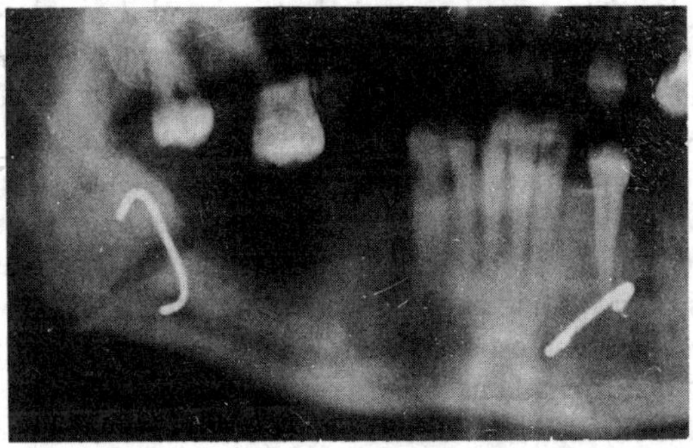

Figure 5. Fracture and joining with shape memory clamps of mandibular bones

5 Summary

In this alloy aged at about 400°C the R-phase dominates at room temperature. After thermal cycles in the temperature range of B2⇔R transition the phase transition is reversible and their characteristic temperatures are stable. In the samples with the R-phase transition the shape memory properties are better. The temperature of shape recovery is adapted to the body temperature in the wires after annealing at about 400°C. This material is very good for maxillofacial and orthopaedic surgery.

References

1. C. Wayman, *J. of Met.* **6**, 129 (1980)
2. P. Filip, J. Musialek and K.Mazanec, *J.Phys IV,* **5,** C8-1211 (1995)
3. S. A. Shabalovskaya, *ibid,* C8-1199
4. J. Drugacz, Z. Lekston, K. Januszewski, H. Morawiec, *J. Oral Maxillofac.Surgery* **53,** 665 (1995)
5. Z. Lekston, *Proc. of XVth Conf. Appl. Cryst. (World Scientific, Singapore 1995)*

ON PRECIPITATION OF TiC_X CRYSTALS BY REACTION BETWEEN GRAPHITE AND Co-Ti MELTS

Z.WOKULSKI, D.STRZĘCIWILK AND K.WOKULSKA
Institute of Physics and Chemistry of Metals, University of Silesia, Bankowa 12, 40-007 Katowice, Poland; e-mail: wokulski@us.edu.pl

Experiments on the precipitation of TiC_x were carried out in a vertical quartz reactor. A high frequency furnace was used for melting the Co-Ti-C charges. Obtained ingots were investigated using optical, interference and scanning electron microscopy (SEM) techniques. Phase composition and structure of the ingots were determined by the phase analysis method using Philips and Inel diffractometers and transmission electron microscopy (TEM). Different precipitation modes were observed in the Co-Ti-C system examined.

1 Introduction

Titanium carbide (TiC) belongs to interstitial compounds that crystallise in B1 structure, Fm3m space group. This is a material that distinguishes itself with a very high melting point (3067°C), a high hardness and a good electrical and thermal conductivity and also with a low chemical reactivity. This makes that it has been widely used in numerous, strategically important, areas of modern technology [1-2]. It is used in production of MMC type (metal matrix composite) composites as an addition to the matrix, e.g. aluminium dispersion hardened with TiC particles [3-4]. TiC-Co cermets have huge practical importance as cutting tools or gas turbine blades are manufactured on their base. Structural and tool materials based on such composites can operate at high temperatures, in chemically aggressive environments and at high mechanical loads.

One of the simplest methods of MMC type composites production is obtaining *in-situ* composites strengthened with refractory carbides by means of the HTS (high temperature solution) method. So far TiC_X crystals were obtained by a number of authors by means of HTS method [5-7] at the temperature of high-temperature solution homogenisation of 1200÷1600°C, using Al, Fe and Ni as high-temperature solvents.

This paper presents results of TiC_X crystals obtaining by the HTS method, using Co as the high-temperature solvent.

2 Principles of TiC crystallisation from high-temperature solutions

Various crystals of high-melting compounds and phases were obtained by the HTS method, especially numerous high-melting carbides, including TiC. Specific crystallisation conditions must be met to obtain crystals of interstitial phases by this method. This results from mutual relations between the high-temperature solvent (S) and the crystallised compound (MNm). To obtain TiC crystals it is necessary that in the ternary system M-Nm-S (Metal-Nonmetal-Solvent) there is an area of equilibrium between the solvent S (Co) and the MNm compound (TiC) as in Fig. 1. That is because the lack of solubility of the crystallised compound (TiC) in the solvent (Co) is crucial to obtain the stoichiometric TiC. To achieve such a state the solvent S (Co) shall be chosen for the binary system M-Nm

(TiC) in such a way, that it is in equilibrium with the $MNm_{stoich.}$ ($MNm_{stoich.}$-S). The selected composition C_x is heated up to temperature T_1 and exposed to homogenisation at this temperature, at which M-Nm components (Ti-C) are solved in the solvent S (Co). Primary MNm (TiC) crystals will appear after lowering temperature to T_2. MNm crystallisation ends after reaching the solidus curve that corresponds with the eutectic temperature T_E, and the entire crystallisation process ends with crystallisation of a ternary eutectic.

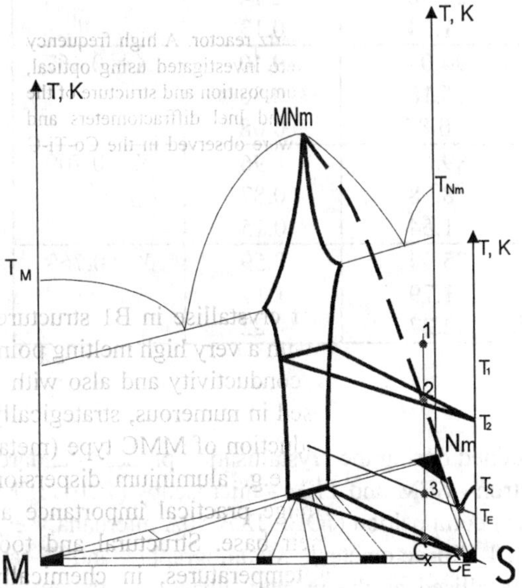

Figure 1: Diagram of the M-Nm-S system, with areas of existing equilibrium between solvent S and MNm compound.

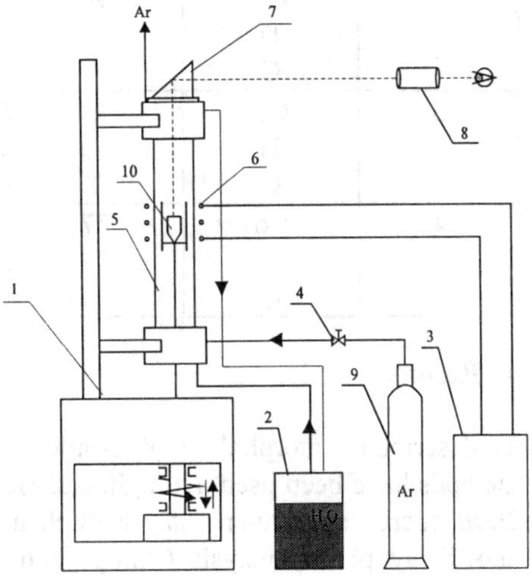

Figure 2: Diagram of the system for obtaining TiC crystals.
1 - control system; 2 - cooling system; 3 - induction heating generator; 4 - electronic regulator of mass flow of gas; 5 - quartz tube; 6 - induction heating coil; 7 - prism; 8 - pyrometer; 9 - cylinder with argon; 10 - graphite heating element with an alumina crucible.

3 Experiment

TiC_x crystallisation processes were conducted using a Metals Research BCG-265 stand. The arrangement of components in the experiment is presented in Fig. 2. Input materials: 99.0% pure titanium, 99.5% pure cobalt and specpure graphite were placed in an alumina crucible and then in a graphite heating element, that was subjected to induction heating up to about 1500 - 1700°C. Then the melt solution was heated for 1-4 hours, until its total homogenisation. Temperature was lowered during another 2-4 hours with a rate of 2.5-5° C/min. The temperature was measured by means of an optical pyrometer. All alloys were conducted in a helium or argon protective atmosphere at a pressure of about 1 atmosphere in the quartz tube. Weighed amounts of input materials were chosen in such a way as to achieve assumed values by the X = C/Ti atomic ratios. Percentage compositions for individual melts are presented in Table 1.

Table 1: Compositions of melts using cobalt as the solvent.

Growth No	Element	at %	wt %	mass [g]	atomic ratio X = C/Ti
1	Co	70	77.16	7.71	X = 0.250
	Ti	24	21.50	2.14	
	C	6	1.34	0.13	
2	Co	90	94.05	9.40	X = 0.667
	Ti	6	5.10	0.50	
	C	4	0.85	0.08	
3	Co	83	89.68	8.96	X = 0.700
	Ti	10	8.78	0.87	
	C	7	1.54	0.15	
4	Co	77	85.94	8.59	X = 0.769
	Ti	13	11.79	1.17	
	C	10	2.27	0.22	

4 Results

To describe the morphology of samples obtained during the crystallisation process standard methods have been used, i.e. light and electron - SEM and TEM - microscopy (JOEL JEM 200B microscope), microanalysis of chemical composition (JOEL JXA 50A microanalyser) and X-ray phase analysis (Philips and Inel diffractometer). Metallographic and X-ray observations have confirmed that TiC_x crystallised as the primary phase and the whole process was finished with crystallisation according to the reaction: *liquid* $\rightarrow Co + TiC_X$. Local chemical composition of the samples was examined using electron-probe microanalysis. Backscattered electron images in the atomic number contrast mode (COMPO) were used for observations of TiC_x precipitates, Fig. 3. As a result of metallographic observations the TiC_x occurrence was found mainly in upper parts of melts, that resulted from various densities ρCo and ρTiC_x. Obtained microcrystals had shapes of cubes or irregular shapes in Co matrix. The performed TEM observations aimed at determination of existing phases and finding possible defects of structure. Thin foil samples were obtained by ionic thinning. Pictures of examined areas and corresponding electron diffraction patterns were performed by means of a JEM 200B microscope with 200kV accelerating voltage. The existence of two phases has been noted in samples during observations. As it can be seen from Fig. 4b TiC_x occurs in the form of long single fibres, while the Co matrix (α-Co) had a martensitic structure, Fig. 4a, that has been confirmed by the electron diffraction pattern, Fig. 4c,d. The martensitic transformation occurring here consists in changing the β-Co structure into α-Co by creation of a stacking fault in every second {111} plane of fcc lattice. Visible in Fig. 4b single fibres situated in martensitic cobalt have been identified as TiC_X, that has been confirmed by an electron diffraction pattern performed for this area, which shows reflexes originating from cobalt apart from TiC phase reflexes. As a result of observations carried out no other precipitations have been found in the volume of examined samples.

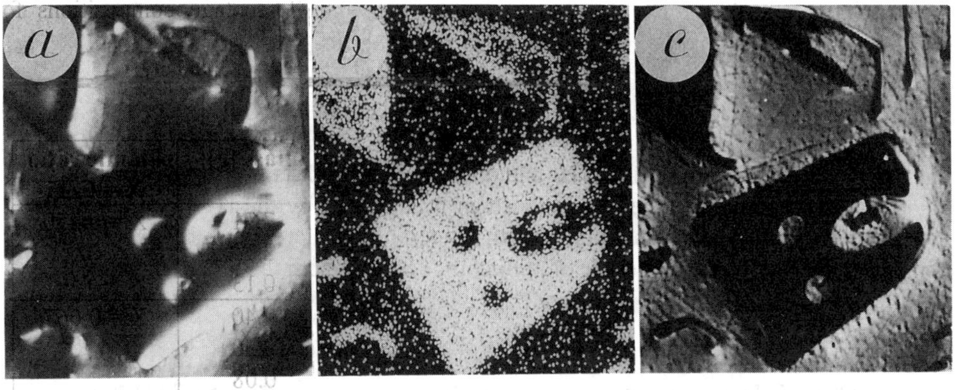

Figure 3: TiC precipitates found in the 4 sample. (a) - SEM image, (b) - distribution of Ti, (c) - COMPO (atomic number contrast) image (magnification: 1000x).

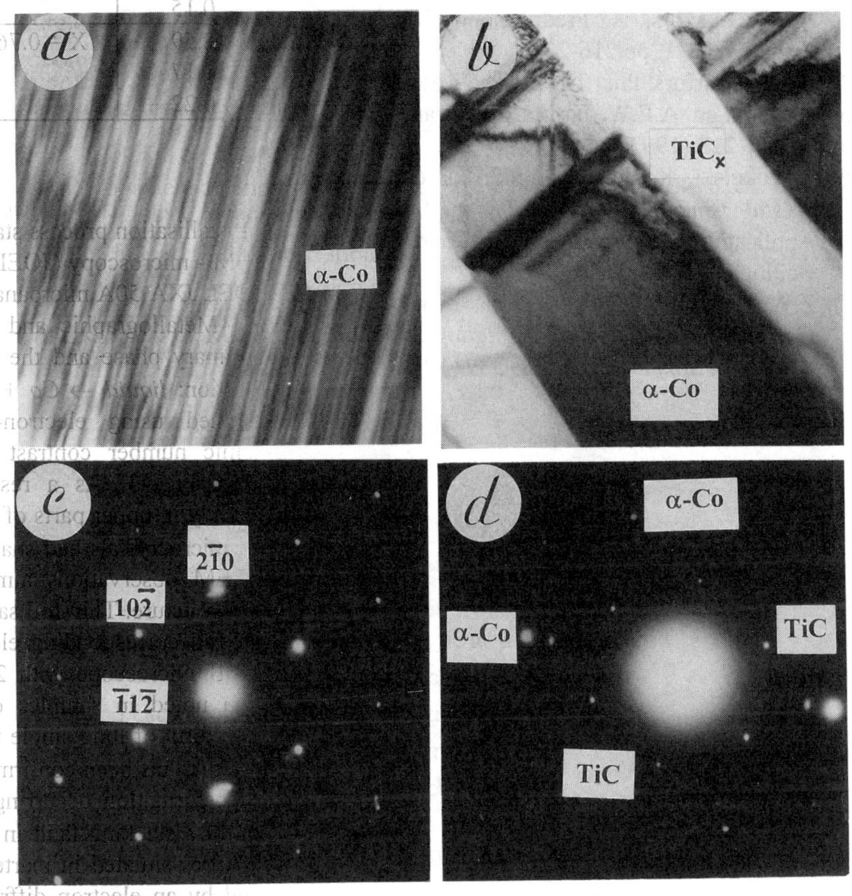

Figure 4: (a, b) -TEM pictures (magn. 40 000x and 15 000x, respectively); (c, d) - electron diffraction patterns for these areas.

5 Conclusion

The performed investigations have confirmed that cobalt can be used as a high-temperature solvent for TiC_x crystallisation and also as a material for MMC type composites production, strengthened with titanium carbide particles. Crystals obtained in the conducted melts had a shape of cubes or irregular and their dimensions did not exceed 0.5 mm.

TiC_x crystals of large dimensions can be obtained by the HTS method, however that requires modification of the growth system, i.e. increasing the mass of input components and decreasing the rate of pulling the crucible from the high-temperature zone.

References

1. S.Otani, Y.Ishizawa, Prog. Cryst. Growth Charact. Mater. **23,** 153-77 (1992).
2. Ch.Taschner, A.Leonhardt, M.Schonherr, E.Wolf and J.Henke, Mater. Sci. Engng. **A139,** 67 (1991).
3. A.E.W. Jarfors, H. Fredriksson, Microgravity Science and Technology, III/4, February 1991, pp.216-221.
4. A.E.W. Jarfors, PhD Thesis, The Royal Institute of Technology, Stockholm 1992.
5. A. Bergman, A.E.W. Jarfors, Z. Liu and H. Fredriksson, Key Engineering Materials **79-80,** 213 (1993).
6. Z.Wokulski, K.Wokulska, S.Suski, *On the possibility of preparation TiC single crystal from molten solution* in ***Solid State Crystal:*** Materials Science and Application, J.Zmija, A.Rogalski and J.Zieliński, Editors, Proc. **SPIE 2373,** p.91 Bellingham, USA (1995).
7. Z.Wokulski, K.Wokulska, Preparation of TiC_x crystals from Ni melts, *Proc. of XVI Conference on Applied Crystallography,* Editors: H.Morawiec, D.Stróż, World Scientific, Singapore, pp.336-339 (1995).

THE INFLUENCE OF SUBSTITUTIONAL CARBON AND OF HYDROSTATIC PRESSURE ON THE THERMAL STABILITY OF SiGe/Si HETEROSTRUCTURES

P. ZAUMSEIL
Institute for Semiconductor Physics, Walter-Korsing-Str.2, D-15230 Frankfurt(Oder), Germany

A. MISIUK
Institute of Electron Technology, Al.Lotnikow 32/46, PL-02-668 Warsaw, Poland

We have studied the annealing behavior of pseudomorphic $Si_{0.603}Ge_{0.397}/Si$ and $Si_{0.597}Ge_{0.391}C_{0.012}/Si$ superlattice structures in the temperature range between 750 °C and 900 °C. Carbon incorporation of 1.2% changes the thermal stability of SiGe structures significantly. The plastic relaxation is suppressed due to an effective dislocation pinning. The interdiffusion of Ge and Si strongly increases. The activation energy of this diffusion process for a Ge content of 40 % decreases from about 4.8 eV for the pure SiGe to about 2.0 eV with an additional C content of 1.2 % assuming a negligible diffusion of the carbon. Anneals of SiGe layer structures for 3 hours under hydrostatic pressure of 10 kbar show an enhanced relaxation and a Si/Ge interdiffusion with strong reduction of the activation energy.

1 Introduction

Heteroepitaxial structures based on group IV elements and grown on Si substrates are of increasing interest because of new possibilities in device application. Pseudomorphically grown $Si_{1-x}Ge_x$ layers on a Si substrate are under a strong compressive stress depending on the Ge content. This leads to metastability as soon as the layer exceeds an upper equilibrium critical thickness.[1] Relaxation by the introduction of misfit dislocations is possible during post growth annealing. One possibility for reducing the inherent strain of $Si_{1-x}Ge_x$ layers is the addition of carbon.[2]

For technological application of such new heteroepitaxial layer systems, their thermal stability (including relaxation and Si/Ge interdiffusion) is an important point. We have studied this in detail by annealing experiments with $Si_{1-x}Ge_x/Si$ and $Si_{1-x-y}Ge_xC_y/Si$ superlattice structures especially in the technologically interesting temperature range between 750 and 950°C. Annealing experiments under high hydrostatic pressure can give further information about diffusion mechanisms. We have found that adding of carbon as well as high hydrostatic pressure changes the thermal behavior of SiGe structures significantly.

2 Experiment

The investigated SiGe/Si and SiGeC/Si superlattice (SL) structures were grown by molecular beam epitaxy (MBE) on (100) oriented Si substrate at the Max-Planck-Institute für Festkörperforschung, Stuttgart.[3] The nominal Ge concentration was 40 %. The superlattice consisted of 15 periods with 10 nm thick $Si_{0.60}Ge_{0.40}C_y$ / 40 nm thick Si (y = 0 and 0.012, respectively). With a total thickness of 750 nm, both heterostructures were in a metastable state. A set of $Si_{1-x}Ge_x$ single layer samples for high pressure anneals with different Ge content (x = 0.25, 0.30, 0.35) and layer thickness (d = 30, 60, 120 nm) was grown by chemical vapor deposition (CVD).

X-ray investigations were carried out using CuK_α radiation and the symmetrical (400) reflection. For the double (DCD) and triple (TCD) crystal diffractometer measurements symmetrical (n,-n) and (n,-n,n) arrangements were used. Additional measurements were carried out with synchrotron radiation of 1.05 Å wavelength. A high temperature chamber mounted on a conventional single crystal diffractometer was used for annealing experiments in a N_2 atmosphere. To simulate the DCD rocking curves (RC), an algorithm was used, which is based on a semi-kinematical model, but modified to take into account the increased diffraction power due to the Ge content.[4,5] High temperature anneals under high hydrostatic pressure (10 kbar, 3 hours) were carried out in a pressure furnace filled with purified argon gas at the High Pressure Research Center, Warsaw.

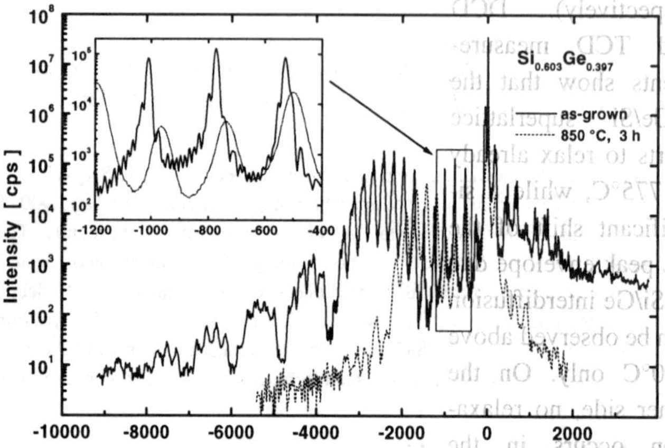

Fig. 1: Comparison of DCD rocking curves of the $Si_{0.603}Ge_{0.397}/Si$ superlattice structure in the as-grown state and after 3 h annealing at 850 °C. Synchrotron measurements with 1.05 Å wavelength. The insert shows a detail of the curve with the 0th order superlattice peak at $\Delta\Theta = -520$ arcsec.

3 Results and discussion

3.1 Carbon incorporation

Figure 1 shows as an example two double crystal diffractometer rocking curves of the $Si_{0.603}Ge_{0.397}/Si$ superlattice structure measured with synchrotron radiation (SR) of 1.05 Å wavelength in the as-grown state and after 3 h annealing at 850 °C. The main advantage of the SR measurements is the wider dynamical range of intensity, that allows to see the superlattice peaks up to the -40th order in this case. But, the angular range between $\Delta\Theta$ = -3500 and +1000 arcsec is sufficient for a simulation of the strain profile of the structures, and this can be also measured with a conventional X-ray source and CuK_α radiation.

The most important features of the DCD rocking curves can be discussed by means of figure 1. The angular distance of the individual SL peaks is determined by the period of the superlattice. The width and position of the envelope of the SL peaks is related to the thickness and the perpendicular lattice strain in the SiGe(C) layers of the structure. The insert in figure 1 shows the fine structure of the DCD curves in the as-grown state. This fine structure disappeares after annealing and the width of the SL peaks increases due to diffuse scattering generated by misfit dislocations. This effect of relaxation can be observed by triple crystal diffractometer measurements [3], or by ω-scans in a DCD arrangement as it is shown in figure 2, or by a shift of the 0th SL peak in the DCD curve towards the substrate peak.

A simulation and fitting of the rocking curves in the as-grown state with the semi-kinematical model [4] leads to a correct determination of the layer parameters of both samples ($x = 0.397 \pm 0.007$, $d_{SL} = (52.5 \pm 0.5)$ nm; $x = 0.391 \pm 0.007$, $y = 0.012 \pm 0.001$, $d_{SL} = (52.5 \pm 0.5)$ nm for the SiGe and the SiGeC superlattice, respectively). DCD and TCD measurements show that the SiGe/Si superlattice starts to relax already at 775°C, while a significant shift of the SL peak envelope due to Si/Ge interdiffusion can be observed above 850°C only. On the other side, no relaxation occurs in the SiGeC/Si superlattice up to 875°C, but a Si/Ge interdiffusion can be observed above 750 °C. The absence of diffuse scattering and consequently of misfit dislocations in the SiGeC/Si SL sample cannot be simply explained by the strain reduction due to the carbon incorporation. The same result was obtained by annealing of $Si_{0.58}Ge_{0.42}$ and $Si_{0.50}Ge_{0.49}C_{0.01}$ single layer structures with equal strain.[3] We suppose that this effect of structural stabilization is due to dislocation pinning in the strongly disturbed surrounding of the substitutional carbon atoms. [6]

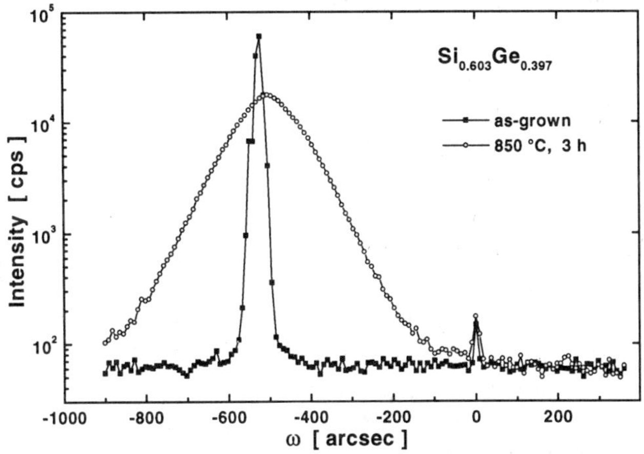

Fig. 2: ω-scan through the 0th SL peak (the detector is placed to a $2\Delta\Theta$ value corresponding to the 0th SL peak, and the sample is rotated). Synchrotron measurement with 1.05 Å wavelength. The broadening of the peak after annealing is due to strain relaxation by misfit dislocations.

A simulation of the DCD curves of the annealed samples allows the determination of the depth profile of strain. With increasing temperature the maximum strain in the individual SiGe(C) layers decreases and the width of the layers increases due to diffusion. Figure 3 shows the difference in the behavior of both samples. The addition of 1.2 % carbon reduces the activation energy of the Si/Ge interdiffusion from 4.8 eV to 2.0 eV. This strong reduction may be related to a significant modification of the point defect spectrum that changes the diffusion mechanism.

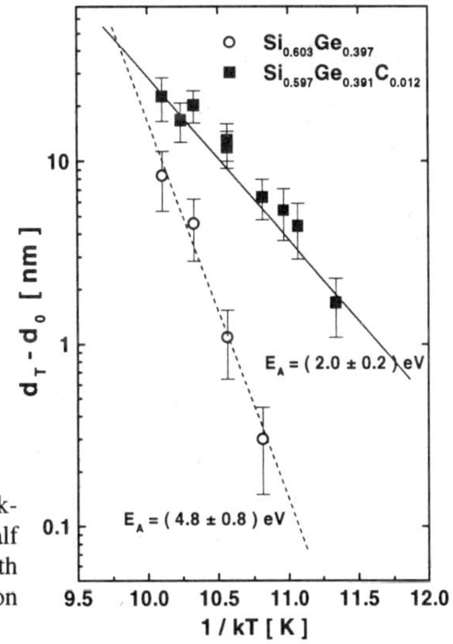

Fig. 3: Arrhenius plot of the increase of the individual layer thickness (d_0 - SiGe(C) layer thickness in the as-grown state, d_T - half width of the strain profile after annealing) due to diffusion for both SL samples. The straight lines are the best fits, giving activation energies of 4.8 eV (SiGe) and 2.0 eV (SiGeC), respectively.

3.2 *High hydrostatic pressure*

The SiGe single layer structures annealed under hydrostatic pressure (10 kbar, 3 hours) show relative to treatments under atmospheric pressure an increase of relaxation and diffusion. [7] The aim of this investigation is to estimate the activation energy of the Si/Ge interdiffusion under these conditions. The best results are obtained for 60 nm thick SiGe layers (x = 0.25, 0.30, 0.35, samples 1-3) with a 60 nm thick Si cap layer to avoid any reaction of the Ge with the atmosphere. The diffusion coefficient is roughly estimated from the measured maximum Ge concentration of the SiGe layer. The Arrhenius plot in figure 4 shows that the diffusion coefficients obtained for all samples and temperatures T = 800 - 950°C lie on a straight line with an activation energy of (2.3±0.4) eV. The higher values for 1000°C are probably due to the limited thickness of the cap layer.

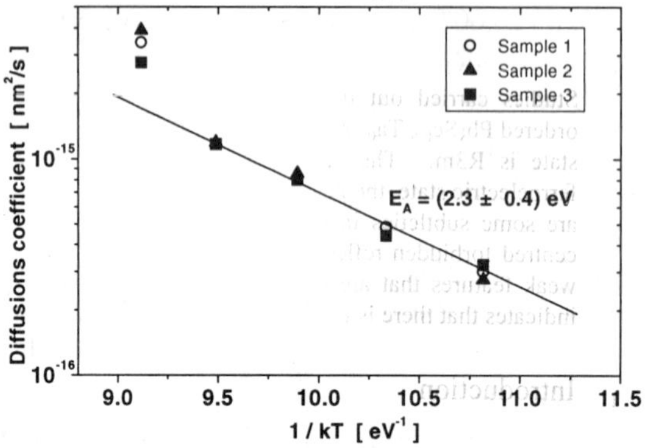

Fig. 4: Arrhenius plot of the diffusion coefficient of Si/Ge intediffusion measured at samples with x = 0.25, 0.30, 0.35 (samples 1-3) after annealing under 10 kbar hydrostatic pressure for 3 hours.

4 Summary

The investigations have shown that the incorporation of 1.2 % carbon significantly changes the thermal stability of SiGe structures. The plastic relaxation is suppressed, or at least delayed. This is not only due to the strain reduction, but probably also due to dislocation pinning. Presence of carbon in the lattice strongly increases the interdiffusion of Ge and Si. The activation energy for a Ge content of 40% decreases from about 4.8 eV for the pure SiGe to about 2.0 eV with an additional C content of 1.2 %. Under hydrostatic pressure we also find a strong reduction of the activation energy of the Si/Ge interdiffusion.

Acknowledgement

The authors would like to thank ESRF, Grenoble, for the possibility of synchrotron measurements at beamline 10.

References

1. J.W. Matthews and A.E. Blakeslee, *J.Crystal Growth* **32**, 265 (1976).
2. S.C. Jain, H.J. Osten, B. Dietrich and H. Rücker, *Semicond. Sci. Technol.* **10**, 1289 (1995).
3. P. Zaumseil, G.G. Fischer, K. Brunner and K. Eberl, *J.Appl.Phys.* **81**, 6134 (1997).
4. R.N. Kyutt, P.V. Petrashen and L.M. Sorokin, *phys.stat.sol.(a)* **60**, 381 (1980).
5. P. Zaumseil, submitted to phys.stat.sol.(a).
6. B. Dietrich, H.J. Osten, H. Rücker, M. Methfessel and P. Zaumseil, *Phys.Rev. B* **49**, 17185 (1994).
7. P. Zaumseil, G.G.Fischer, Ch.Quick and A.Misiuk, *phys.stat.sol.(a)* **153**, 401 (1996).

STRUCTURAL CHARACTERISTICS OF THE PSEUDO-PEROVSKITE RELAXOR FERROELECTRIC $Pb(Sc_{0.5},Ta_{0.5})O_3$

K.Z. BABA-KISHI
The Hong Kong Polytechnic University, Department of Applied Physics, Hung Hom, Hong Kong, PR China.

Studies carried out using high resolution powder diffractometery (HRPD) on highly ordered $Pb(Sc_{0.5}\,Ta_{0.5})O_3$, indicate that the paraelectric phase is Fm3m and the ferroelectric state is R3m. The calculated atomic displacements show that on passing into the ferroelectric state, the polarization arises mainly from distortions in the lead cations. There are some subtleties in this system observed in the TEM, where extremely weak face-centred forbidden reflections are observed. In the neutron data, there are also extremely weak features that are consistent with these primitive reflections. Structure refinement indicates that there is a major movement of the Pb-cations along the polar axis.

1 Introduction

Over several decades there has been considerable interest in ferroelectric materials for use in a range of applications. In the field of infra-red sensors, pyroelectric materials offer advantages over conventional photo-effect sensors such as low power and ambient temperature operations. Among the large number of potential materials, $Pb(Sc_{0.5},Ta_{0.5})O_3$, (PST) for-short, displays interesting pyroelectric and dielectric properties. In common with a large number of ferroelectric oxides, PST has a structure similar to that of the mineral perovskite, $CaTiO_3$. Adopting the general formula ABO_3, these materials can be viewed as consisting of BO_6 octahedra centred on the corners of a simple cube, with the A-cations occupying the positions between octahedra. In the case of PST, Pb-cations occupy the A-sites and Sc and Ta cations occupy the B-sites. The degree of structural ordering of the Sc and Ta ions on the B-sites, affects the dielectric properties of PST[1].

From the knowledge of the structural studies carried out on crystals and ceramics of PST, it is evident that PST undergoes paraelectric (PE) to ferroelectric (FE) and to antiferroelectric (AFE) phase transitions [2&3]. It has been shown that the antiferroelectric transition is weak, depends on the degree of structural order and could co-exist with both the PE and FE states [3]. The structural changes that lead to these transitions are not fully understood, primarily because of the insensitivity of the techniques used so far to study the subtle changes in the structure upon transition. Most of the X-ray and electron diffraction studies carried out indicate that further advances on the structural understanding of PST, using these techniques, are not possible [4].

Although the PE state is known to be cubic, the lattice constant at the PE to FE transition, and the Bravais lattice in the PE state have been the subject of controversy. The space group of the FE state is also the subject of controversy, particularly concerning the presence of oxygen octahedral tilting. To evaluate some of these structural problems in PST, electron and neutron diffraction studies were carried out. The results from these studies are discussed in some detail.

2 Transmission electron microscopy (TEM) of $Pb(Sc_{0.5}Ta_{0.5})O_3$

2.1 Structural order and the forbidden reflections

Heat treatment of PST can produce various degrees of structural order of the B-site cations. The ordering of the Sc^{3+} and Ta^{5+} ions which occupy alternate B-sites along the <111> directions, creates a face-centered superstructure on the B-site sub-lattice and doubles the basic perovskite unit cell. The basic unit cell has space group Pm3m with lattice parameter a=4.07Å, and the doubled superstructure unit cell may be described by the space group Fm3m with lattice parameter 8.14Å. Dark-field imaging in the TEM with one of 1/2 1/2 1/2 superstructure reflections (referred to the Pm3m unit cell), reveals ordered domains, Fig. 1a. The state and the degree of structural order can be assessed qualitatively from the configuration and density of antiphase boundaries.

The configurations of the structural ordered domains in PST ceramics is more complex than that for crystals, and are often seen to be influenced by grain boundaries and dislocations. The mechanisms of ordering in not clear in PST, but it is possible that large ordered domains adjacent to grain boundaries are a result of the greater mobility and numbers of point defects in such regions. Lattice vacancy of the B-sites can also have some influence, since a vacancy makes additional space available to promote atomic migration.

Electron diffraction patterns recorded from both the disordered and ordered PST, consistently show both 1/2 1/2 0 and 1/2 0 0 types of reflections, Fig. 1b. These reflections have not been observed in X-ray diffraction studies, except using X-ray precession technique, which shows the presence of the 1/2 1/2 0 reflection[4]. Studies carried out shows that these reflections do not arise from double diffraction[3]. The forbidden reflections occur both at room temperature and low temperatures and that they are not related to the PE and FE phase transitions.

It was also observed that the intensities of the 1/2 1/2 0 and 1/2 0 0 reflections vary markedly, depending on the locality from which they originate. This indicates that local fluctuations in composition have significant effects on the intensities of these reflections. Such fluctuations in local stoichiometery in PST are inherent to all the relaxor materials and are believed to cause the spread in the Curie temperature. In small regions where the distribution of a species of B-site cation, e.g. Sc, is dominant over the other species, e.g., Ta, then the intensities of the 1/2 1/2 0 reflection arising from them are expected to change. The fact that there is a strong possibility of mixed ordering of Ta and Sc, the existence of these forbidden reflections can be attributed tosuch mixed ordering. There is also the possibility of oxygen octahedral tilting together with the operation of a different space group, as a potential explanation for the existence of these reflections.

2 Neutron diffraction studies of $Pb(Sc_{0.5}Ta_{0.5})O_3$

High-resolution powder diffractometry (HRPD) was carried out on a highly ordered PST ceramic. Neutron powder diffraction data were collected on the HRPD, at the ISIS spallation source. Data were collected in time-of-flight, from 30000ms to 230000ms corresponding to a d-spacing range of 0.6 to 4.6Å at 200K and 300K in thePE and FE phases, respectively. In the PE phase, the diffraction pattern was consistent withthe space group Fm3m and a unit cell constant of 8.14866(1)Å, Fig. 2. The diffraction data showed asymmetric line broadening where the weak reflections hkl; h+k+l = 2n+1 were 50% broader than the strong reflections. Structure refinement was carried out using a modified version of the Rietveld method to take into account the hkl dependent line broadening.

The face-centred space group permits ordering of the Sc and Ta and also allows the B-site cation-oxygen bond length to be different between the ordered sites. At convergence, the B-sites were found to be 90(1)% ordered.

At about 200 K the sample underwent a FE transition to space group R3m with lattice constants a = 8.15156(6)Å, α = 89.8815(7). Two space groups, R3 and R3m, consistent with the lack of systematic absences, observed in the neutron pattern, were tested, and in both cases the fit to the data was superior in R3m than R3. Figure 2 shows the line splitting that is a characteristic of rhombohedral symmetry; 444 cubic reflection splitting into two peaks with ratio 3:1. Identical hkl dependent line broadening was observed in the FE phase. In addition, very weak superstructure reflections indexable on a primitive 8Å unit cell were observed. The reflections appeared to have non-systematic absences with hkl, two odd one even. The presence of these superstructure reflections is consistent with the TEM observations.

Structure refinement showed that not only did the B-site cations become displaced from the centres of their regular oxygen octahedra, but also there was significant movement of the lead ions along the polar axis. The major displacement occurred for the lead(s) and was in the order of 0.25Å. Figure 3 illustrates the displacements in the lead(s) and the B-sites octahedra. A large asymmetry was found in the 3.0Å co-ordination sphere of Pb, which changes its co-ordination number from 12 to 9 on FE transition [5]. The co-ordination of the oxygen around the Pb suggests that the phase transition may be related to the delocalisation of a stereochemically active lone-pair when passing from the PE to FE state, Fig. 4.

3 Conclusions

The onset of the PE to FE transition in PST is marked by a change from Fm3m to R3m symmetry as observed in the neutron patterns. In the FE state, PST is non-centrosymmetric as a result of the combined effects of the Pb displacements and off-centered B-sites cations. TEM investigations enabled us to observe the configurations of the ordered domains and verify the existence of the very weak forbidden reflections. Various forbidden reflections were also seen in the neutron diffraction patterns of the ferroelectric phase. Although the refinement of the ordered phase of PST is satisfactory, it is clear that the structure derived is still only an approximation to the true structure.

Acknowledgments

The assistance of staff at Rutherford Appleton Laboratory, where the neutron studies were carried out, is gratefully acknowledged.

References

1) CGF Stenger and AJ Burgraaf, Phys.Status Solidi A, 61, 275 (1980)
2) L E Cross, Ferroelectrics, 76, 241 (1987)
3) KZ Baba-Kishi and DJ Barber, J. Appl. Cryst., 23, 43 (1990)
4) KZ Baba-Kishi, G Cressey and J Cernik, J. Appl. Cryst.. 25, 477 (1992)
5) KS Knight and KZ Baba-Kishi, Ferroelectrics, 173, 341 (1995)

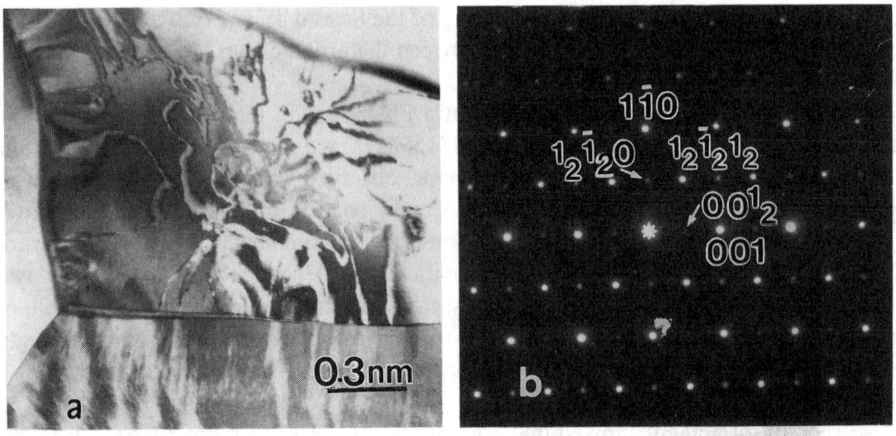

Figures 1a and b: a) Structural ordered domains present within and close to a grain boundary in a ceramics of PST. B) Electron diffraction pattern showing the very weak forbidden superstructure reflections.

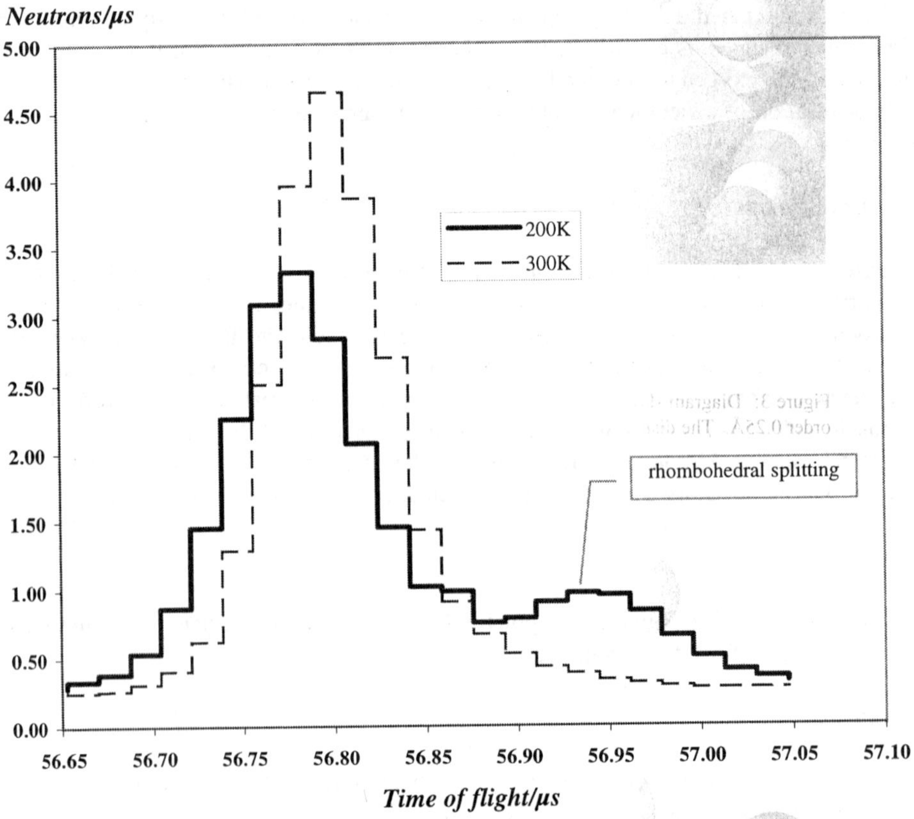

Figure 2: Neutron diffraction pattern from the cubic phase (dashed line) and the rhombohedral phase (solid line) of PST, clearly showing the peak splitting originating from the PE to FE transition.

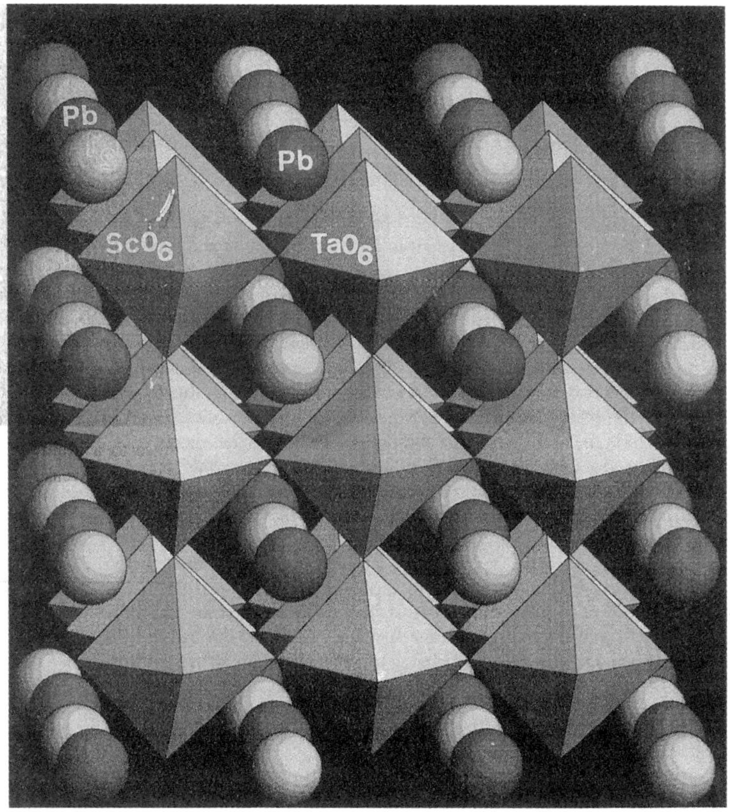

Figure 3: Diagram showing the major displacement associated with the Pb atoms of the order 0.25Å. The displacements in the B-sites octahedra are much smaller.

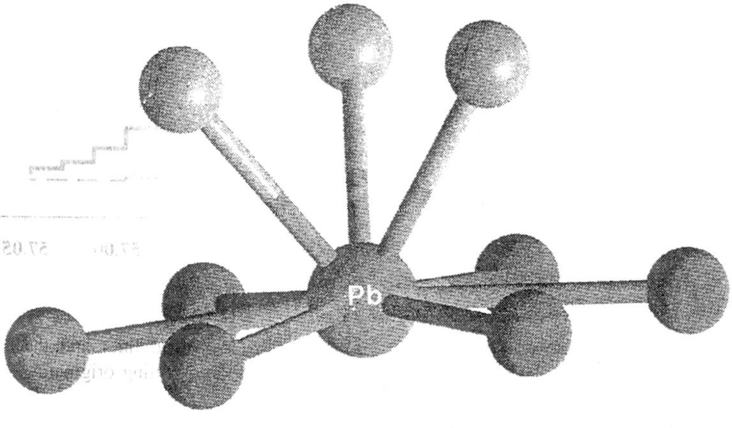

Figure 4: 3.0Å co-ordination shell surrounding the Pb in the FE state of PST. The asymmetry of the bonding is indicative of an electron lone-pair.

DIFFUSE SCATTERING BY WEAKLY AND STRONGLY DISTORTED CRYSTALS CONTAINING PRECIPITATES

R.I.BARABASH
National Technical University of Ukraine, Metal Physics Department, pr.Pobedy 37, Kiev 252056, Ukraine

The presence of new phase particles causes local lattice distortions due to the existence of lattice misfit between matrix and forming phase. It also changes structure amplitude and sometimes structure type. All the above factors influence diffraction conditions under X-ray, neutron or electron scattering. Precipitates of new phase in the crystal cause the appearence of diffuse scattering besides the usual regular reflections. The positions of both diffuse and regular intensity maxima are displaced relatively to the ideal positions of intensity maxima corresponding to crystal without defects. Presence of precipitates also cause the decrease the intensity of regular reflections. The analysis of diffuse scattering is different for two limit cases of weakly and strongly distorted crystals. These cases can be distinguished by the value 2M of the exponent of Debye-Waller factor exp (-2M). If $2M\leq 1$ crystals are considered to be weakly distorted, if $2M>>1$ they are strongly distorted. The value of the parameter 2M depends on the displacement fields caused by the defect in the matrix and on the diffraction vector. In weakly distorted crystals we have to distinguish two following cases. The first and most simple is the case when a precipitate with a diffuse boundary is formed in a uniform matrix. This case can be described by a simple one parameter function and corresponds to continuous concentration distribution. To analyse the influence of a depletion layer existence on diffuse scattering a simple model of sphere particles with the same structure as that of matrix and with diffuse boundaries was discussed. In this case intensity distribution tends to zero in reciprocal lattice points and often is asymmetric and concentrates on one side of the point being crescent-shaped. In contrast to the case of homogeneous matrix in the last case isodiffuse surfaces do not pass through reciprocal lattice points but surround their neighborhood. For strongly distorted crystals in the case of **2M>1** the regular I_0 as well as diffuse I_1 peaks of the matrix have approximately the same intensity and are present in the intensity distribution simultaneously.

1 Introduction

Ageing of supersaturated solid solutions results in the formation of relatively large inhomogeneities occuping a volume much more larger than the volume of a unit cell. The presence of new phase particles causes local lattice distortions due to the existence of lattice misfit between the matrix and the forming phase. These distortions are described by the law: $|\vec{U}(\vec{r})|\sim Cr^{-2}$ (C - constant which characterizes the strength of defects, its value being of the order of absolute meaning of volume change $|\Delta V|$ caused by a defect). They cause the appearance of intensive X-ray diffuse scattering in the vicinity of reciprocal lattice pionts. These intensity distributions may differ essentially for different materials after different treatment. The type of intensity distribution depends on mutual effects caused as by the difference of structure amplitudes between precipitate and matrix as by the distortions around precipitates and inside them. The picture may become still more complicated due to the correlation between the precipitate distribution. If there are no distortions there appear symmetric intensity distributions surrounding reciprocal lattice points. The presence of distortions makes them asymmetric and leads to the hyperbolic intensity increase with the approach to the reciprocal lattice point. In previous papers [1-4] the intensity distributions were analysed for the case when diffuse scattering is mainly influenced by distortions.

Often it is very important to take into account the interference of scattering caused by the difference of structure amplitudes and distortions. Below we have analysed such effects for some elastically anisotropic crystals. We have also considered the influence of the depletion layer surrounding precipitates.

2 Weakly Distorted Crystals with precipitates

In weakly distorted crystals the Debye-Waller factor exp (-2M) connected with distortions fields around the new particles is of the order of unit. For simplicity let us consider the case when similar precipitates are formed in the cubic matrix and correlation between their positions is not essential. Even in this simple model we can get a lot of intensity distributions.

2.1 Diffuse scattering intensity for weakly distorted crystals

Let us analyse diffuse scattering $\mathbf{I_1}$ by the crystal containing Np equal particles of a new phase with the same orientation being chaotically distributed in the matrix. If the particles have a sharp boundary with the matrix and their structure differs from that of the matrix then according to [5,6,7,13] the intensity $\mathbf{I_1}$ in the vicinity of matrix reciprocal lattice points is defined by the equation:

$$I_1 = N_p v^{-2} \left|\Delta f - f\vec{Q}\vec{A}_q\right|^2 \left|s(\vec{q})\right|^2. \tag{1}$$

Here V- is a unit cell volume; f- the matrix structure amplitude; Δ f- the difference between new phase and matrix structure amplitudes (if precipitate and matrix have the same structure and are coherent); or Δf=-f if their structures are different; $\vec{Q}$- diffraction vector ($\vec{Q}=\vec{k}-\vec{k}_0$), $\vec{k}$ and $\vec{k}_0$ are the wave vectors of the scattered and initial waves; $\vec{q}=\vec{Q}-\vec{G}$,where $\vec{G}$ is the reciprocal lattice vector nearest to $\vec{Q}$, $s(\vec{q})$ - Fourier component of a shape function of the precipitate; $\vec{A}_q$ determines the Fourier component of the displacement field $\vec{U}(\vec{r})$ in the matrix around the new phase particle. At small q they may be found from the equations of macroscopic elasticity theory. The solution of corresponding equation system for cubic particles in cubic matrix looks as follows ([13])

$$\vec{A}_q = \frac{3L\vec{\alpha}_q}{q}; \quad \alpha_{qx} = \frac{c_{11}+2c_{12}}{3D}\left(1+\xi n_y^2\right)\left(1+\xi n_z^2\right)n_x, \tag{2}$$

where Lij is the tensor of deformation due to the formation of precipitates; c_{11}, c_{12}, c_{44} - elastic modulus and

$$D = c_{11} + \xi\left(c_{11}+c_{12}\right)\left(n_x^2 n_y^2 + n_x^2 n_z^2 + n_y^2 n_z^2\right) + \xi^2\left(c_{11}+2c_{12}+c_{44}\right)n_x^2 n_y^2 n_z^2;$$

$$\xi = \frac{c_{11}-c_{12}-2c_{44}}{c_{44}}; \qquad \vec{n} = \vec{q}/q. \tag{3}$$

In (1) we suppose that local distortions are small and $\left|\vec{Q}\vec{U}\right| \langle\langle 1$ or $q \ll (LR^3Q)^{-1/2}$, (R- effective particle size). If precipitates have the same structure as matrix does and are formed as a result of contineous diffusion change $\delta c(\vec{r})$ of concentration c of atoms A in a binary solution A-B then they do not form sharp boundary with the matrix. In this case I_1 can be determined with formula (1) . Suppose that for this case

$$\Delta f = f_A - f_B; \quad S(\vec{q}) = \int \delta c(\vec{r}) \exp(i\vec{q}\vec{r}) d\vec{r}; \quad L = \frac{1}{3v}\frac{\partial v}{\partial c} \tag{4}$$

From (1)-(4) the intensity distributions around reciprocal lattice points are influenced by the dependence of $\vec{q}$ on two factors: $\left|\Delta f - f\vec{Q}\vec{A}_{\vec{q}}\right|^2$ and $\left|S(\vec{q})\right|^2$. The first of them is connected with the displacement field in elastically anisotropic matrix and the second describes the influence of concentration distribution on the diffuse scatttering. To analyse the influence of the last factor let us consider as in [6,10] two simple functions $\delta c(\vec{r})$ for precipitates with diffuse boundary.

2.2 Precipitates in a homogeneous matrix

Concentration distributions are the most simple after relatively large annealing time when the layers around the particles depleted with atoms A unite and create almost constant concentration level. In this case concentration distribution around separate spherical particles can be described for example by a simple Gauss function

$$\delta c(\vec{r}) = \delta c_m \exp\left(-{r^2}/{2R^2}\right) \tag{5}$$

where δc_m- is maximal concentration change, R- characteristic radio of the distribution. According to (4) $S(\vec{q})$ "shape function" for distribution (5) looks like

$$S(\vec{q}) = nv\exp\left(-\frac{1}{2}R^2q^2\right); \qquad n = \frac{(2\pi)^{3/2}R^3}{v}. \tag{6}$$

here n is the effective number of atoms in the region where the composition is changed. According to (1),(6) in this case

$$I_1 = K\left|C - \frac{\vec{m}\vec{\alpha}_{\vec{q}}}{qR}\right|^2 \exp\left(-R^2q^2\right); \; K = N_p n^2 f^2\left(\frac{1}{v}\frac{\partial v}{\partial c}QR\right)^2; \tag{7}$$

$$\vec{m} = \vec{Q}/Q, \quad C = \frac{\Delta f}{f}\frac{1}{QR}\left(\frac{1}{v}\frac{\partial v}{\partial c}\right)^{-1}. \tag{8}$$

The parameter C describes the relative influence of the effects connected with the difference of atomic scattering factors of the alloy and with distortions. Different sections of isodiffuse surfaces by certain planes in the reciprocal latice space (isodiffuse curves) for different meanings of parameter C were calculated by computer simulations by mcans of formulas (2),(3),(7) for elastic modulus corresponding to Cu being the example of the crystal with an average anisotropy. We have analysed sections by planes (100) near the points (00h) (Fig.1). The numbers near the curves show the values I_1/K for the corresponding curves and on the rest of figures they show α I_1/K (α is shown at each figure). At relatively small value of C=0,4 the intensity distribution is mainly influenced by distortions . The isodiffuse curves asymmetry is very weak and can be observed only for the curves corresponding to very small intensity level. When C=0 scattering takes place only on the distortions and isodiffuse curves are strictly symmetric and zero intensity curve coincides with the line [010]. If C differs from zero this curve differs from straight line and is shown in dash. With the C increase (for C=1,C=2) the curves become essentially asymmetric. But at further C increase the curves asymmetry decreases again although the shape of the curves changes. At $C\langle 0$ the isodiffuse curves have the same shape , but are inverse to the corresponding curves with positive C relatively the reciprocal lattice point. It means that the asymmetry sign is determined by the signs of Δf and $\partial v / \partial c$.

2.3 Influence of the Depletion Layer.

At early stages of ageing the effective diffusion length is essentially less than the distance between the particles. Around each precipitate there will apprear the depletion layer with less concentration of one of the components. To analyse the influence of the depletion layer we shall consider a simple model of spherical particles having the same structure as the matrix. We shall suppose that both the concentration increase inside the particle and its decrease in the depletion layer can be described by simple distribution

$$\delta c(\vec{r}) = \delta c_m \left[\exp\left(-\frac{r^2}{2R^2} \right) - \left(\frac{R}{R'} \right)^{3/2} \exp\left(-\frac{r^2}{2R'^2} \right) \right]; \qquad (9)$$

$$\int \delta c(\vec{r}) d\vec{r} = 0$$

At the same time the total atom concentration near each type of precipitates is constant. If we determine $S(\vec{q}\,)$ using formulas (4), (9) and substituting this expression into (1) we find that intensity distribution I_1 can be written as following

$$I_1 = K \left| C - \frac{\vec{m}\,\vec{\alpha}_q}{qR} \right|^2 \left[\exp\left(-R^2 q^2 \right) - \exp\left(-R'^2 q^2 \right) \right]^2 . \qquad (10)$$

In this case $S(\vec{q}\,)\rightarrow 0$ and $I_1(\vec{Q}\,)\rightarrow 0$ at $\vec{q}\rightarrow 0$. It means that in the reciprocal lattice points intensity I_1 tends to zero. That's why intensity distributions have minima in the reciprocal lattice points and are symmetric in the limit cases when one of the items in (7) is much bigger than another. It results in spherically symmetrical halo when $|C|>>1$, $qR\sim 1$ and in the appearance of two almost equal intensity maxima at $|C|<<1$. If both items are of the same order the resulting intensity distribution is essentially asymmetric and is concentrated mainly on one side of the reciprocal lattice point having crescent shape.Opposite to the case of homogeneous matrix in latter case isodiffuse surfaces do not pass through the reciprocal lattice points but surround them in their vicinity.

Such crescent shape of isodiffuse surfaces is a result of precipitate induced displacement fields disappearence due to the compensation of different sign effects from the particle and the depletion layer. Such compensation takes place upon linear dependence v on c (Vegard law) and if the condition $\int \delta c(\vec{r})d\vec{r} = 0$ is true. The deviations from Vegard law will result in the absence of full displacement fields compensation at large distance from the particle and intensity $I_1(q \to 0)$ will not tend to zero. But still the value of the coefficient near q^{-2} in the expression for I_1 will be much smaller comparing with the case of a homogeneous matrix. Its value will characterize the extent of distortions penetration through the depletion layer into the matrix. Such transformation of isodiffuse curves for concentration distribution (9) was illustrated by computer simulations (Fig.1,2) upon the linear dependence of v on c for different values of the parameter C .Calculations were fulfilled for the case R'=2R for crystals with elastic modulus of Cu and NaCl (signs of asymmetry parameter ξ for Cu and NaCl are opposite). Isodiffuse curves in the plane (100) around the reciprocal lattice points (00h) for Cu (Fig.1) in this case differ from the case of a homogeneous matrix. Presence of the depletion layer makes them essentially asymmetric even upon the value C=0.4 (in the case C=0 which was cosidered earlier [8,9] asymmetry disappears). They consist of different intensity maxima displaced from their reciprocal lattice points at approximately the distance equal to the reverse particles size. The positions of the displaced peaks are indicated by the cross. Besides such peaks displaced along the direction $\vec{Q} \approx \vec{G}$ there exist also weak isolated maxima in the direction being situated at certain angle to $\vec{G}$. When C=1.0 and 2.0 intensity distribution becomes crescent like. In the region of large q isodiffuse curves tend to the circle. For AgCl precipitates in NaCl crystals isodiffuse curves were simulated in two different sections (100) and $(1\bar{1}0)$ near the point (hhh) (Fig.2). Section of the point (hhh) by the plane (100) has axial symmetry. But its section by the plane $(1\bar{1}0)$ has no axial symmetry. In this crystal with $\xi > 0$ crescent shape of isodiffuse curves at C=0.5 and C=2.0 is much more distinct than for Cu crystals. At Fig.2 for C=2.0 the edges of isodiffuse curves with lower intensity value unite almost into a circle and form the asymmetric halo. Computer simulations were compared with interesting experiments [11,12] for decomposition in the system NaCl-AgCl. In these experiments the AgCl concentration (17 mol%) was chosen in such a way that for odd h_i structure amplitude of the matrix f=cfAg+(1-c)fNa-fCl was equal zero (for even h_i $f \neq 0$). According to (10) for odd h_i (f=0, C=∞) there was no scattering on the distortions and isodiffuse curves formed symmetric halo surrounding reciprocal lattice points. Their intensity decreased with Δf for reflections with higher Miller indices. For even h_i distortions were essential and caused the asymmetry of the intensity distributions. The value of their intensity increased with Q (for distant reciprocal lattice points). The shape of isodiffuse curves for the section of the reciprocal lattice point (hhh) by the plane $(1\bar{1}0)$ for precipitates of AgCl in NaCl corresponds to negative C. That's why we must take inverse curves relatively to the reciprocal lattice point. Similarly as experimental one [11,12] they are turned by their crescent side to the zero point. Moreover around reflections (222) or (220) intensity distributions are crescent while near (444) or (442) (for them Q is twice larger and according to (8) C is twice smaller) the isodiffuse curves consist of two separate peaks with different intensity. In the alloy with 30 mol (AgCl for odd reflections $f \neq 0$ and asymmetric intensity distributions appear around all reciprocal lattice points. According to (8) and taking into account that Δ f/f does not depend on Q we obtain for reflections (hhh) the following parameters ratios $C_{444}:C_{222}:C_{333}:C_{111}=1/16\div 1/8\div 2/3\div 2$. The increase of $|C|$ with the transition from (444) to (111) reflections results in the same change of intensity distribution as for values of C parameters equal 0 : 0.5 : 2.

Thus the interference of scattering connected with distortions and with depletion layer gives the same intensity picture as observed in [11,12] even without correlation between precipitates. Some outer part of the isodiffuse curve can be lost experimentally due to existence of some initial intensity level.

2.4 Correlation between Precipitates

The influence of particles correlation in the case 2M<<1 and small volume fraction of the particles for equal and eqiaxial precipitates results in the additional factor in the expression for intensity I_1

$$I_1 = I_1^{(0)}\frac{N}{c}\langle\left|C_{\vec q}\right|^2\rangle = I_1^0\left[1+\frac{1}{c}\sum_{\vec\rho\neq 0}\varepsilon(\vec\rho)\cos\vec q\vec\rho\right] \tag{11}$$

$$\varepsilon(\vec\rho)=\left\langle\left(c(\vec R_t)-c\right)\left(c(\vec R_t-\vec\rho)-c\right)\right\rangle,$$

where I^0_1 is defined by the formulas (1), (7) or (10) and the radio vecor $\vec R_t$ determines possible positions of precipitates centers in different crystals cells t, N- number of cells. $C(\vec R_t)=1$ if precipitate center is in the point t and $C(\vec R_t)=0$ in the opposite case. $\varepsilon(\vec\rho)$ - correlation parameter of precipitates divided by vector $\vec\rho$; <...> means averaging over statistical ensemble. So the correlation results in the appearance of the additional factor $\left\langle\left|C_{\vec q}\right|^2\right\rangle$ in the expression for intensity. The above results correspond to diffuse scattering intensity:

$$I_1^{(0)} = I_1\frac{c}{N}\left[\left\langle\left|C_{\vec q}\right|^2\right\rangle\right]^{-1} \tag{12}$$

They can be used if the dependence $\left\langle\left|C_{\vec q}\right|^2\right\rangle$ on $\vec q$ is not too strong. If it is possible to calculate function $\left\langle\left|C_{\vec q}\right|^2\right\rangle$ using a certain model for decomposition kinetics these additional correlation effects may be simply taken into account by formula (11).

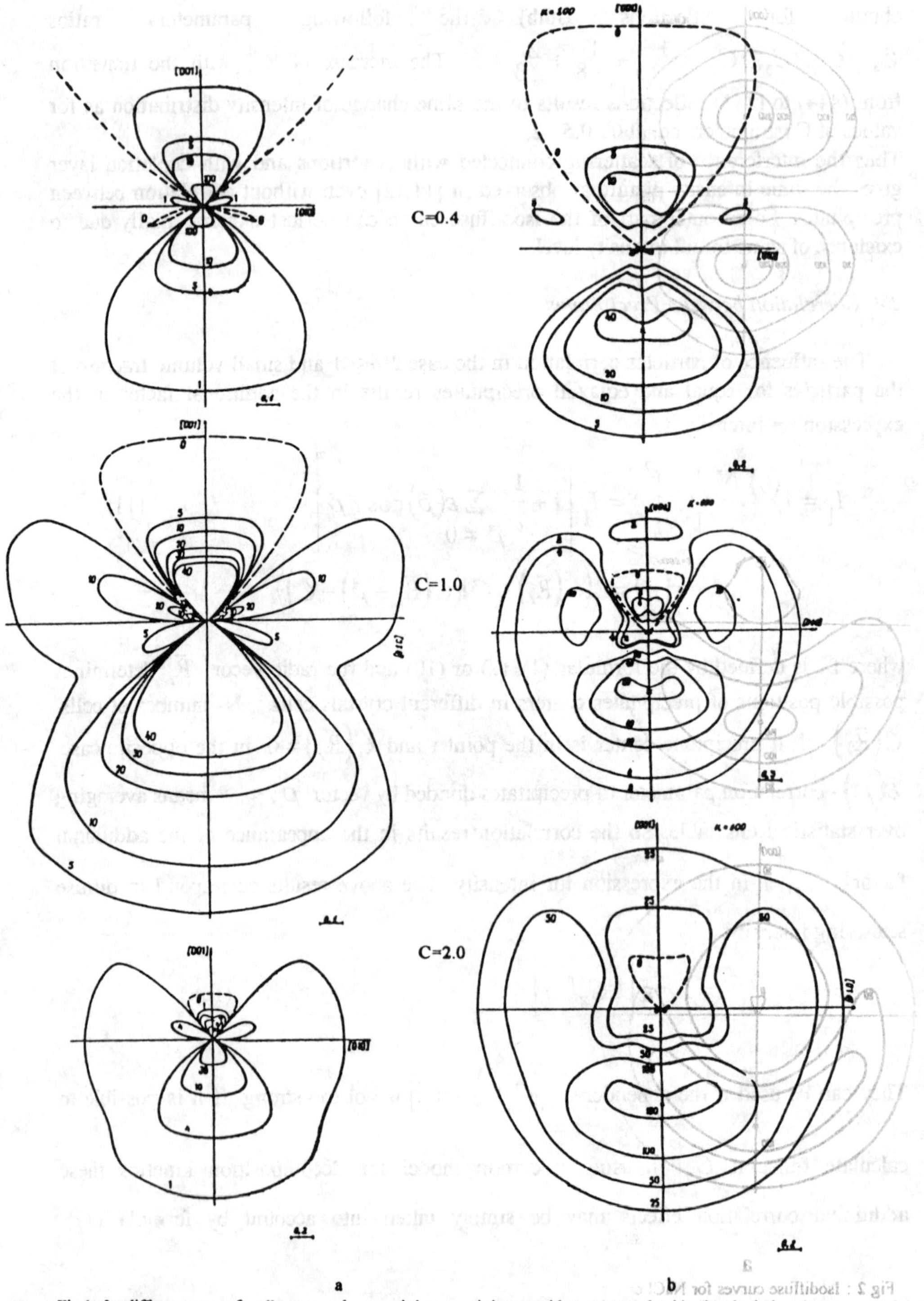

Fig.1: Isodiffuse curves for Cu crystals containing precipitates without (a) and with (b) depletion layer around reflection (ooh) in the plane (100) for different values of parameter C

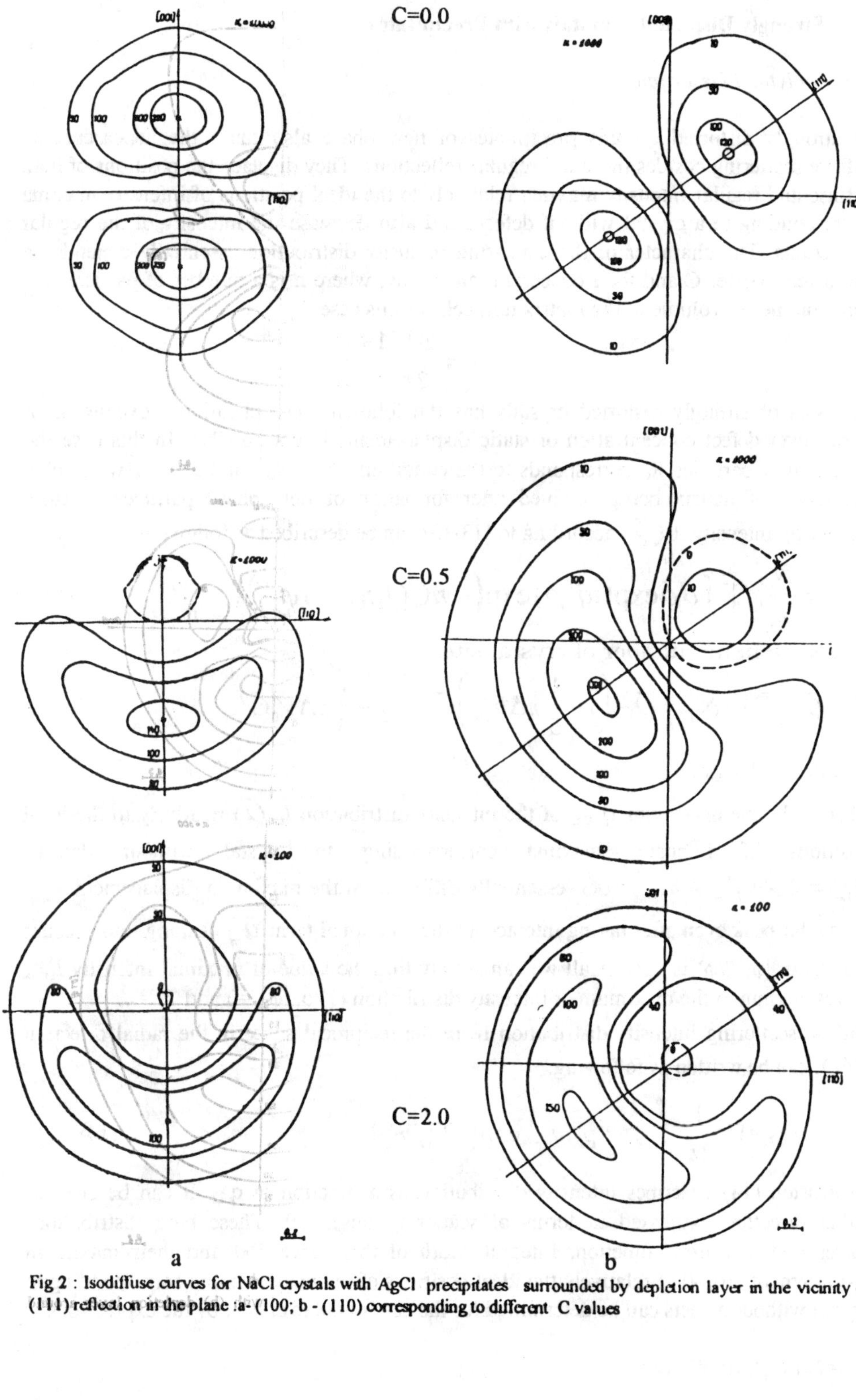

Fig 2 : Isodiffuse curves for NaCl crystals with AgCl precipitates surrounded by depletion layer in the vicinity of (111) reflection in the plane :a- (100; b - (110) corresponding to different C values

3 Strongly Distorted Crystals with Precipitates

3.1 Analytical Description

In strongly distorted crystals precipitates of new phase also cause the appearence of diffuse scattering besides the usual regular reflections. They displace the positions of both diffuse and regular intensity maxima relatively to the ideal positions of intensity maxima corresponding to a crystal without defects and also decrease the intensity of the regular reflections. The character of the appearing intensity distribution essentially depends on the defects power C and their concentration c = nv, where n is a number of particles per unit volume, v- volume of the matrix unit cell. In this case

$$c = \frac{\Delta V}{2\pi}\frac{1+\sigma}{1-\sigma}$$

Analysis of strongly distorted crystals has the following complication: expansion in series over defect concentration or static displacements is not possible. In this case the size of new particles R_0 corresponds to the condition $QC>>R_0^2$ and $2M>>1$ with cubic symmetry of matrix being retained after formation of new phase particles diffuse scattering intensity $I_1(\vec{Q})$ according to [13-15] can be described as following:

$$I_1 = \frac{N}{v}|f|^2 \int d\vec{\rho}\exp(i\vec{q}'\vec{\rho})\exp\left(-nCQ\rho\left(\varphi'+i\varphi_1''\right)\right), \tag{13}$$

here N - general number of crystal sites,

$$\vec{q}' = \vec{Q}-\vec{G}^0-\delta\vec{G}' = \vec{Q}-\left(1-\frac{1}{3}n\Delta V_{im}\right)\vec{G}^0 = \vec{q}-\frac{1}{3}n\Delta V_0\vec{G}^0$$

$$\vec{q} = \vec{Q}-\vec{G}^0-\delta\vec{G},$$

That's why the maximum $\vec{q}_{0m}$ of the intensity distribution $I_1(\vec{Q})$ relatively to the ideal positions of intensity maxima corresponding to crystal without defects, $\vec{q}_{0m} = \delta\vec{G}'+\vec{q}'_m \neq \vec{q}^{\,0}_{0m} = \delta\vec{G}$ essentially differs from the maximum displacement $\vec{q}^{\,0}_{0m}$ of regular peak even after taking into accout the additional term $\vec{q}'_m$. Taking into account that ratio $|\varphi_1''|/\varphi'$ is very small we can simply find the value of maximal intensity I_{1max} corresponding to the maximum of intensity distribution (1) at $\varphi_1''=0$ and $\vec{q}'=0$. Diffuse scattering intensity distribution I_{1r} in the reciprocal space in the radial direction ($\|\vec{Q}$), can be written as following:

$$I_{1r}(q_{0r}) = Ir_i \frac{1}{2\pi}\int_{-\infty}^{+\infty} d\rho\exp(iq_{0r})\exp[-T(\vec{\rho}\vec{m})]. \tag{14}$$

Expression (14) describes intensity distribution as a function of q_{0D}. It can be also for radial direction expressed in terms of scattering angle 2θ. These both distributions correspond to Lorenz function. Integral width of this curve 2δθ and their maximum displacement $2(\theta_m-\theta_0^0)$ relatively the ideal positions of intensity maxima corresponding to crystal without defects can be determined for the case of cubic crystal by the expressions

$2\delta\theta = 2\pi nC\varphi'(\vec{m},\vec{m})\mathrm{tg}\theta,$

$$\theta_m - \theta_0^0 = nC\left[\frac{\vec{m}\,\delta\vec{G}}{nCQ} + \varphi''(\vec{m},\vec{m})\right]tg\,\theta = \left(-\frac{\Delta V_{im}}{3} - C\varphi_1''\right)ntg\,\theta = -\frac{\Delta d}{d}tg\,\theta$$

Here $\Delta d/d$ - relative change of lattice parameter corresponding to diffuse maximum displacement. Displacement of regular maximum $2(\theta_m{}^0 - \theta_0{}^0)$ (or maximum in weakly distorted crystal) is quite different from $(\theta_m - \theta_0{}^0)$ and is equal

$$\theta_m{}^0 - \theta_0{}^0 = \frac{\Delta d}{d}tg\,\theta = \frac{1}{G}\delta\vec{G}\vec{m}\,tg\,\theta = -\frac{1}{3}n\Delta Vtg\,\theta. \qquad (15)$$

In the simple case of cubic defects in elastically isotropic medium halfwidth and maximum displacement in the radial direction can be determined by simple equations:

$$2\delta\theta = \frac{4\pi^2}{27\sqrt{3}}\frac{1+\sigma}{1-\sigma}n|\Delta V|tg\,\theta \approx 0.844\frac{1+\sigma}{1-\sigma}n|\Delta V|tg\,\theta,$$

$$2\left(\theta_m - \theta_0^0\right) = -\frac{2}{9}\frac{1+\sigma}{1-\sigma}\left[\frac{2(1-2\sigma)}{1+\sigma} + 0.160\right]n\Delta Vtg\,\theta, \qquad (16)$$

$$2\left(\theta_m^0 - \theta_0^0\right) = -\frac{2}{9}n\Delta Vtg\,\theta = -\frac{2}{9}\frac{1+\sigma}{1-\sigma}\left[\frac{2(1-2\sigma)}{1+\sigma} + 1\right]n\Delta Vtg\,\theta.$$

It is essential that both halfwidth and maxima displacement depend on $n\Delta Vtg\theta$. Comparing formulas for $\theta_m - \theta_0{}^0$ and $\theta_m{}^0 - \theta_0{}^0$ we notice that both regular and diffuse maxima are displaced relatively the ideal positions of intensity maxima corresponding to crystal without defects but diffuse maxima displacement is approximately six times lower.

Such analytical expressions for halfwidth and maxima displacement can be used only if 2M>>1, but much more often experimental data correspond to the the condition 2M~1. In this case we can observe simultaneous existence of both diffuse and regular peaks of intensity. In the region 2M≤1 analitical analysis of intensity distribution is impossible and computer simulation was used to obtain diffuse scattering intensity distributions.

At Fig.3a there are shown the results of numerical calculations for real T' and imaginary T'' parts of the parameter T which is necessary to calculate diffuse scattering intensity. The results of computer simulations of diffuse scattering intensity distributions for small 2M values are presented at Fig.3d for the case $\Delta V>0$. It is displaced relatively to the ideal position of intensity maxima corresponding to a crystal without defects (it is shown by straight line). The value of this displacement is dependent on the value of 2M. If $V<0$ these intensity distributions will be situated on the other side of ideal peak position.

From the Fig.3 it can be seen, that at 2M~1 intensity distribution is essentially asymmetric and does not correspond to Lorenz shape. With 2M increase curves asymmetry gradually decreases and their shape tends to the Lorenz one. Halfwidth and maximum displacement also decrease. The positions of maxima of quasiline in radial direction q'_{Dm} and their integral widths $qr_i=4\pi\lambda^{-1}\cos\theta\delta\theta$, halfwidths $q_{r1/2}$, being defined as the distance between points $q'_{D\frac{1}{2}} + \delta^+_{\frac{1}{2}}$ and $q'_{D\frac{1}{2}} - \delta^-_{\frac{1}{2}}$, in which intensity is equal to half of maximal values

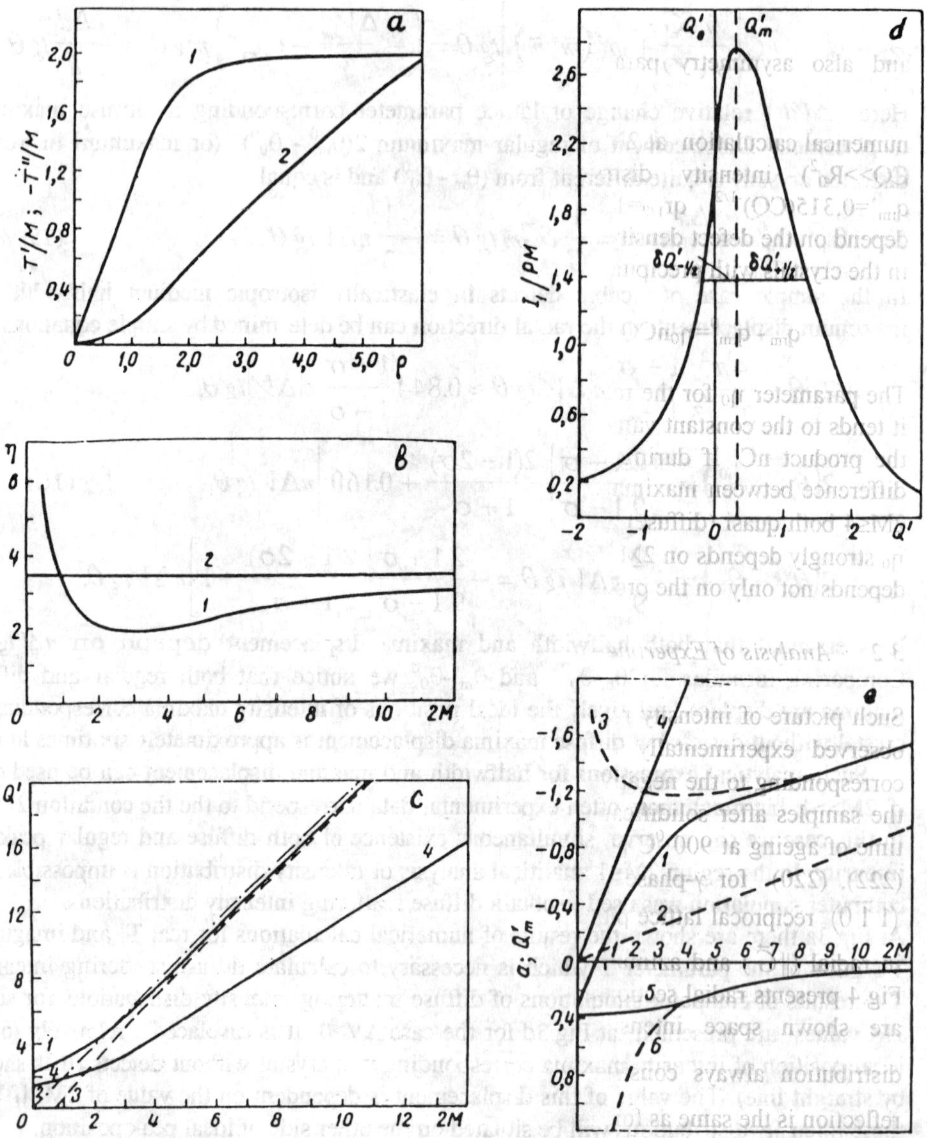

Fig.3: Simulation of diffuse scattering for strongly distorted crystals with precipitates:
a - real (T'/M), and imaginary (T''/M) parts of function T($\vec{\rho}$); b - computer simulation of η dependence on 2M; c - halfwidth intensity dependence in radial direction on 2M: 1-asymptotic value $1{,}89 * 2M(CQ)^{-1/2}$,2-computer simulation, 3-asymptotic value q_{Dr} with additional terms, 4-numerically simulated halfwidth; d - diffuse scattering intensity distribution for small 2M; e - dependence of maxima positions q_{Dm}' (1-computer simulation, 2-asymptotic value 0,08 $2M(CQ)^{-1/2}$, 3-asymptotic value with certain additional terms,4- regular reflection q_{D0}', 5- asymmetry parameter,6- asymptotic asymmetry parameter

and also asymmetry parameter $a = \frac{\delta^+_{1/2} - \delta^-_{1/2}}{\delta^+_{1/2} + \delta^-_{1/2}}$ were determined. The results of numerical calculation at 2M=2...4 differ essentially from analytical one. When $2M \to 0$ ($CQ >> R_0^2$) intensity distribution $I_{1r}(qr)$ becomes symmetric with parameters $q_{rm}' = 0{,}315(CQ)^{-1/2}$; $qr_{1/2} = 1{,}525(CQ)^{-1/2}$; $q_{ri} = 2{,}27(CQ)^{-1/2}$; $Q = 0{,}38(2M << 1)$ which do not depend on the defect density but only on their power. Distance between maxima I_1 and I_0 in the crystals with precipitate is equal

$$q_{rm} - q_{rm}^0 = \eta_0 n C Q \frac{\Delta V}{|\Delta V|}, \tag{17}$$

The parameter η_0 for the region 2M~1 was calculated numerically (Fig.3f). At large 2M it tends to the constant value 3.515. That's why this distance $q_{Dm} - q_{Dm}^0$ depends only on the product nC. If during the ageing this product retains the same then the relative difference between maxima $q_{Dm} - q_{Dm}^0$ will not change. In the region of not very large $2M \le 4$ both quasi (diffuse) and regular maximums can be observed simultaneousely and η_0 strongly depends on 2M. So when 2M is not large the displacement of quasireflections depends not only on the product nC, but also seperately on n and C.

3.2 *Analysis of Experimental Diffuse Scattering*

Such picture of intensity redistribution changes with the increase of defect power was observed experimentally in the region 2M~1 for Ni-based ageing single crystals corresponding to the negative sign of ΔV. X-Ray intensity distributions were received for the samples after solidification, quenching from the temperature 1240 ^{0}C and different time of ageing at 900 ^{0}C [16]. Both structure intensity distributions (002), (004), (111), (222), (220) for γ-phase and superstructure -(001), (003), (110) for γ'-phase in the $(1\bar{1}0)$ reciprocal lattice plane were restored. Intensity distributions were examined both in radial ($\| \vec{G}$) and azimuth (perpendicular to $\vec{G}$) directions around (222) reflection. Fig.4 presents radial section of (111) and (222) for different ageing time. At Fig.5 there are shown space intensity distributions around (222) reflections. This intensity distribution always consists of two maxima. The azimuth section ($\perp \vec{G}$) of the reflection is the same as for I_0 as for I_1 (Fig.5). Intensity distribution is stretched along $\vec{G}$ and has a complicated shape in the radial direction. In the azimuth direction it is relatively narrow and does not change along the whole reflection. Parameter 2M depends on $(h_i^2)^{3/4}$ (Miller indices). It means that for one sample different reflections may correspond to different 2M values. It will result in different contribution of I_0 and I_1 to the resulting intensity distribution. That's why the resulting intensity distributions for the first (111) and the second (222) orders of the same diffraction maximum (Fig.4) are not similar to each other. The angle position of I_0 and I_1 depends on the sign of ΔV being the characteristics of volume change between matrix and precipitates. In our experiments I_1 was always situated at the smaller $\vec{Q}$ values relatively to I_0 . It corresponds to negative ΔV<0.

The parameter η_0 in the region $2M \approx 1$ was determined using the results of numerical calculations (Fig.3f). For this 2M region it was equal $\eta_0 \approx 2 \ldots 2{,}2$. This gives us possibility

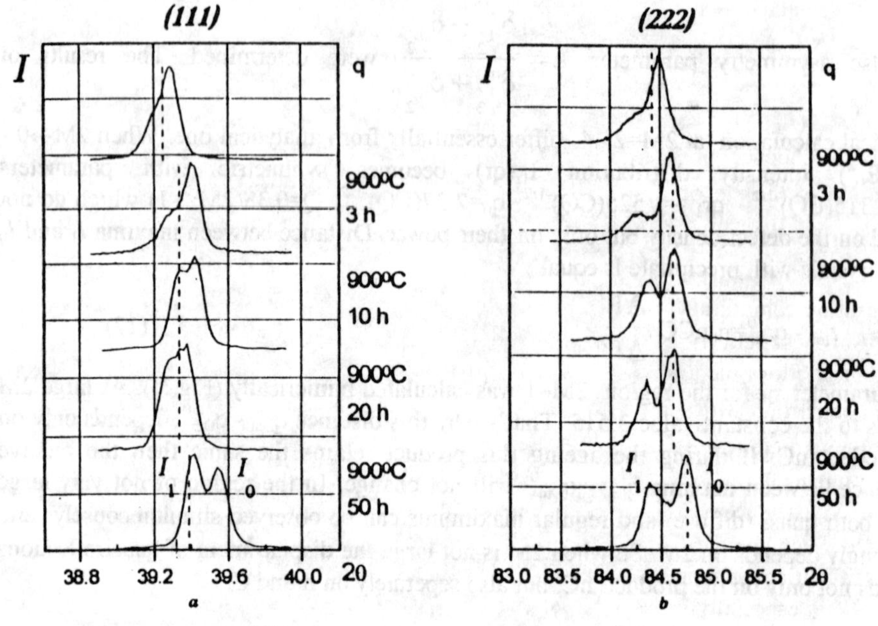

Fig. 4: X-ray intensity distribution in radial direction for reflections (111) and (222) after different ageing time

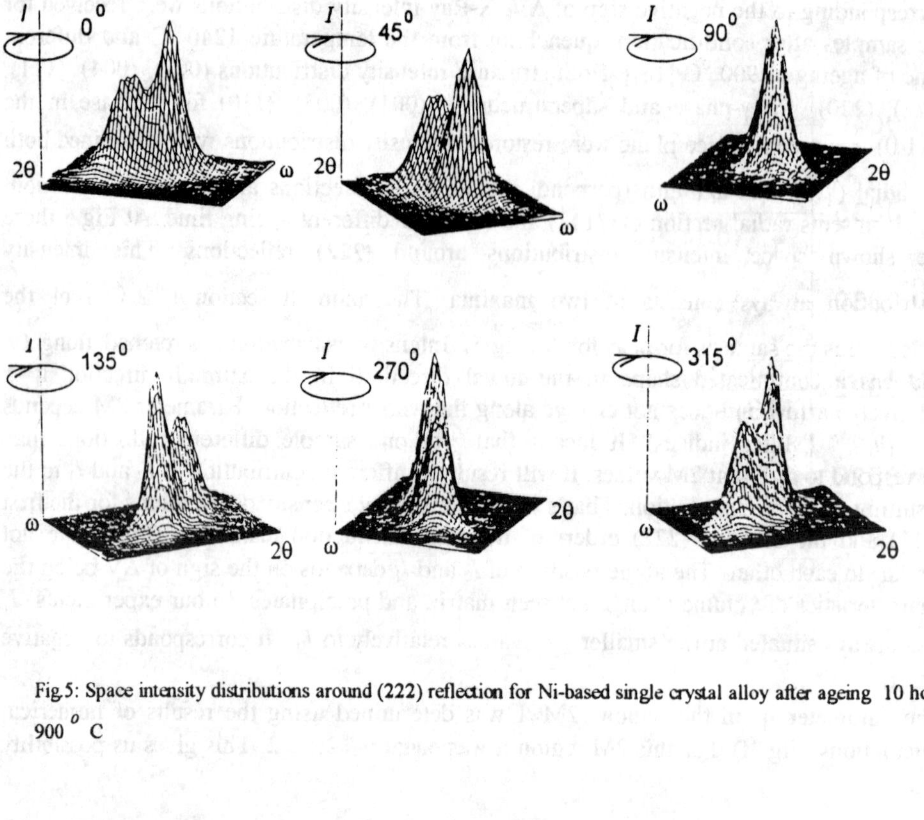

Fig.5: Space intensity distributions around (222) reflection for Ni-based single crystal alloy after ageing 10 hours at 900 ^{o}C

to determine the product nC . Taking into account the dependence of M on ΔV we determined the main parameter for dispersively hardened alloys

$$n\Delta l' = p\frac{\Delta \upsilon}{\upsilon} = -(2-2.3)\cdot 10^{-2}$$

So the dependence of width and displacement of reflections on the modulus and orientation of the vector $\vec{Q}$ are in good agreement with theoretical dependences. The ratio $2\delta\theta/\mathrm{tg}\theta$ also depends on the direction of $\vec{Q}$ and is equal for (420), (331) and (222) reflections: 1.92:1.66:1.22:1. If we use elastic modulus renormed according to Feigt c_{pq}, for the alloy containing 70% Ni the same as for pure Ni it will give the same ratio $2\delta\theta/\mathrm{tg}\theta$ of these reflections 1.8:1.6:1.2:1 (the slight difference can be connected with the deviation of alloys modulus from pure Ni ones and also with the larger modulus of new phase comparing with matrix). The dependence of reflection integral width on $\vec{Q}$ orientation is mainly a result of anisotropy of matrix and precipitates. Ageing results first of all in the displacement of the ideal peak θ_0^0 position. This influences the positions of I_1 and I_0 . That's why during ageing both regular and diffuse maxima positions are displaced. Such changes are explained by formulas (4)-(6). Correspondingly the efffective values of lattice parameters calculated for regular and diffuse peaks will also differ (Fig.6). It is especially essential when these both maxima are observed simalteneously. The same type of effective lattice parameters dependences were observed by other authors [17,18].

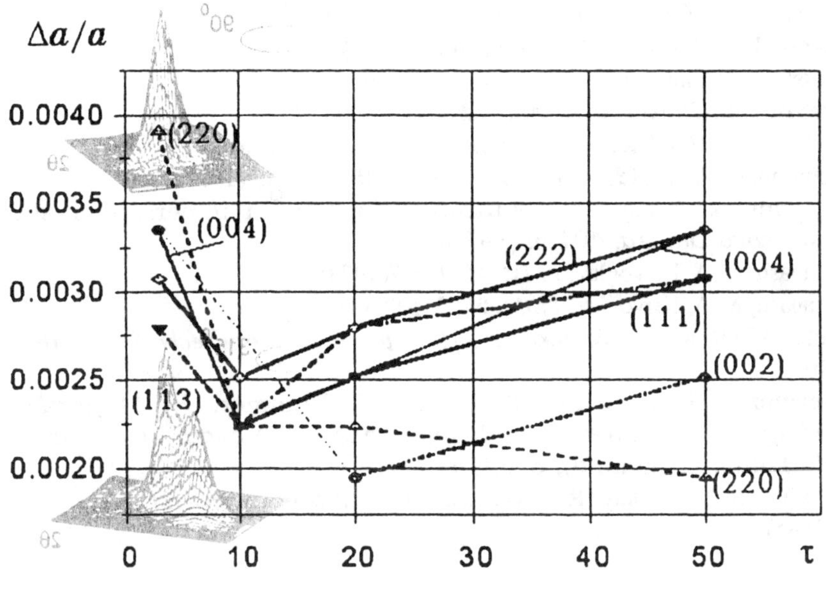

Figure 6: The change of effective lattice parameters difference for different reflections with ageing time

Conclusions

- Precipitate distorts lattice and changes the structure amplitude. Resulting intensity distribution depends on these mutual effects.
- Weakly distorted crystals with precipitates: the scattering interference for anisotropic crystals results in the asymmetry of isodiffuse surfaces. Reciprocal lattice points are no longer inversion centers for intensity distribution. The extent of asymmetry depends on value and sign of defect power. Depletion layers make isodiffuse curves crescent and turn scattering intensity into zero at reciprocal lattice points.
- Strongly distorted crystals with precipitates: there are both regular and diffuse maxima, when $2M \geq 1$. Shift of regular and diffuse maxima depends on sign and value of ΔV. Diffuse scattering intensity is stretched along the diffraction vector.
- On experiment: reciprocal space was restored around structure matrix and superstructure γ'-phase points. Diffuse and regular reflections were observed. The average volume change caused by precipitates was obtained.

References

1. K.V.Chuistov, *Modulated Ctructures in Ageing Alloys*,ed.Naukova Dunka,Kiev,(1975)
2. K.M.Yamaleev,*X-Ray Diffuse Scattering by Ageing Crystals*,Nauka,Moscow,(1973)
3. A.M.Yelistratov,*Reports of Academy of Science USSR*,**87**,4,p.581 (1952)
4. Yu.A.Bagaryatsky,*Crystallography*,**3**,5p.578 (1958)
5. M.A.Krivoglaz,*FMM*,**9**,5, p.641 (1960)
6. M.A.Krivoglaz,*FMM*,**13**,p.17 (1962)
7. A.G.Khachaturyan, *Crystallography*,**10**,4,p.459 (1965)
8. V.V.Kokorin,K.V.Chuistov,*FMM*,**27**,6,p.1067 (1969)
9. V.V.Kokorin,K.V.Chuistov,*FMM*,**27**,5,p.804 (1969)
10. R.I.Barabash, M.A.Krivoglaz,Metallofizika,**5**,1,p.100 (1983)
11. A.V.Dobromislov,V.V.Glebov,N.N.Buinov,*Phys.Stst.Sol.*,**A,201**p.87 (1973)
12. A.V.Dobromislov,V.V.Glebov,*Sov.Phys.Stat.Sol.*,**15**,4,p.1245 (1973)
13. M.A. Krivoglaz. X-ray and neutron diffraction in imperfect crystals. *Kiev, Naukova Dumka*, 407 p. (1983)
14. R.I. Barabash, M.A. Krivoglaz, *FMM*, **45**, 1, p.7(1978)
15. R.I. Barabash, M.A. Krivoglaz. *FMM*, **51**, 5, p.903 (1981)
16. Barabash, A.D.Rud, S.I.Sidorenko etc., *Metal physics and new technology*, , **18**, 11, p.1(1996)
17. H. Biermann, H.A. Kuhn, T. Ungar, J. Hammer, H. Mughrabi,*in Proceedings of "9th International conference on Strength of Metals and Alloys"*.,(London, England), p. 421 (1991)
18. M.V. Nathal, R.A. Mackay, R.G. Garlick, *Material Science and Engineering*, 75, p.195 (1985)

X-RAY ANALYSIS OF AGEING KINETICS IN NICKEL -BASED SINGLE CRYSTALS

BARABASH R.I., NIKITINA I.O., SIDORENKO S.I., TURTSEVICH A.G.

Metal Physics Department, National Technical University of Ukraine, pr. Pobedy, 37,Kiev, 252056, Ukraine.

New method for studying kinetics of ageing process in nickel based single crystals was worked out. It is based on experimental analysis and computer modelling of X-ray diffuse scattering by such alloys. During ageing the precipitation of γ'-phase particles takes place in nickel based single crystals. This influences the strength properties of such materials. Experimentally intensity distribution were restored around different structure reciprocal lattice points of matrix and γ'- phase in the plane $(1\bar{1}0)$. Superalloys were aged at 900^0C during 3, 10, 20, 50 hours. After ageing almost all intensity distributions consisted of two maxima. These distributions were stretched along reciprocal lattice vector having rather complicated shape in this direction, while in azimuth direction it is relatively narrow and analogous for different azimuth cross sections. An appearance of intensive diffuse scattering near regular reflections was established. The ratio between these peaks positions and integral intensities depended on the orientation and value of a diffraction vector. Analysing the character of intensity distribution around reciprocal lattice points, the sign of misfit between γ'- phase and matrix and other parameters of γ' - phase were determined at different ageing time.

1 Introduction

The heat-resistant properties of Ni-based single crystals result from the corresponding heat treatment (in our case, after annealing at 900^0C during different time intervals). During this treatment particles of $(NiMe)_3Al$ - based γ'-phase particles precipitate from supersaturated solid solution. Their size, quantity and distribution influence strengthening of such alloys. The γ'-phase particles formation causes significant matrix distortions and is accompanied by the appearance of intensive diffuse scattering around matrix reciprocal lattice points [1-7]. The theory of the diffuse scattering by such crystals was worked out in [9-11].

2 Experimental Procedure

X-ray diffuse scattering in the reciprocal lattice space was analysed by single crystal diffraction method [2,8]. The data were collected for γ'-phase superstructure reflections using Cu K_α-radiation and for matrix structure reflections using Cu K_β - radiation. The reciprocal lattice point section has been obtained by $\theta-2\theta$ method in the radial direction (along reciprocal lattice vector) and by ω method in the azimuth direction (normal to reciprocal lattice vector). Space intensity distribution has been obtained by turning out around (222) reflection of γ- and γ'-phase in the plane $(1\bar{1}0)$ by $\omega-2\theta$ method. X-ray intensity distribution has been analysed around both structure reciprocal lattice points

(111), (002), (220), (113), (222), (004) of γ and γ'-phase and for superstructure points (001), (110), (003) of γ'-phase in the $(1\bar{1}0)$ plane. Single crystal specimens of GS-40 alloy have been studied after quenching at 1240^{0}C and consequent ageing for 3, 10, 20, 50 hours at 900^{0}C.

3 Results and Their Discussion

Single asymmetric reflection, with slower intensity decrease from small angle side, has been observed for the quenched state around all reciprocal lattice points. Intensity distribution for (002), (004) reflections in radial direction is presented on Fig. 1. During ageing almost all reflections are the superposition of the regular and diffuse peaks displaced along reciprocal lattice vector. Resulting intensity distribution for each reflection is elongated in the direction of reciprocal lattice vector $\mathbf{G}_{hkl}$.

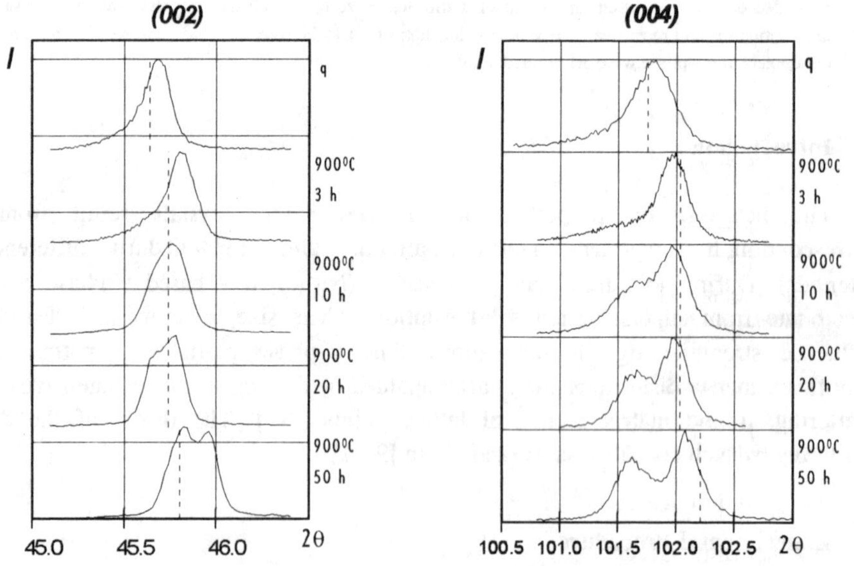

Fig.1 - X-ray intensity distribution in radial direction for reflections (002) and (004) after different aging time

In order to find out the γ'-phase contribution, superstructure reflections (001), (110), (003) have been recorded for all cases and the lattice parameter of γ'-phase has been determined. Similarly to [2], the positions of γ'-phase structure reflection around structure matrix reflections have been calculated with (003) reflection. These positions do not coincide with appearing diffuse peak position and is indicated with dash on Fig. 1. If reflections of γ- and γ'-phases were simply summarised in one resulting intensity distribution then different peak shapes would have to be similar for the same reflection type, for example (111) and (222) or (002) and (004) reflection pairs. But this has not been observed experimentally. On the contrary, maximal intensity ratio decreases while diffraction vector modulus $\left(\sum h_i^2\right)$ or scattering angle θ increases (fig 2 a, b). At the same time this ratio goes up while ageing time increases and then reaches the saturated level for all reflection types. Even if the common full half-width dependence on $tg\theta$ is taken into account and the ratio I_1/I_0 is normalised over $tg\theta$, the qualitative behaviour of this dependence does not change. Because of this the (002) and (004) reflection shapes differ. Such behaviour is connected with the increase of static Debye-Waller factor 2.8 times at the transition from (002) to (004) reflection. According to [9], the latter results in the decrease of diffuse peak maximal intensity against the regular one.

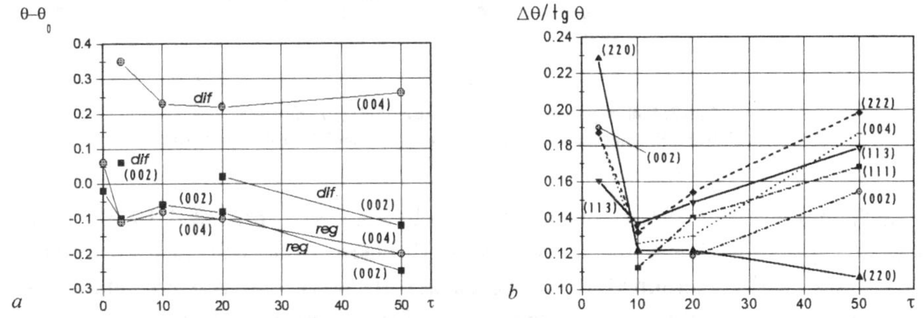

Fig. 2: Displacements of diffuse and regular maximums relatively maximums of standard samples; *b* - normalised difference of angle positions of regular and diffuse maximums depending of ageing time

Fig. 3 presents the regular and diffuse peak displacements for (002) and (004) reflections versus ideal crystal position during ageing. According to [9-11], the displacement value depends on the parameter $2M_{hkl}$. This parameter is proportional to $\left(\sum h_i^2\right)^{3/4}$ and increases with the θ increase. Therefore the reflection (004) corresponds to the $2M_{hkl}$ value larger than the reflection (002) does. The magnitude of displacement absolute value $\theta - \theta_0$ and angle position difference between regular and diffuse peaks $\Delta\theta = \theta_{reg} - \theta_{dif}$ depend not only on the parameter $2M_{hkl}$ (i. e. on the diffraction vector modulus $\left(\sum h_i^2\right)$), but also on its orientation. However, simple proportional dependence $\Delta\theta$ on $tg\theta$ has not been found. This fact becomes more clear if the $\Delta\theta$ is normalised over $tg\theta$ (Fig. 3 b). At ageing time up to 10 hours the $\Delta\theta/tg\theta$ decreases for all reflections. Then it increases for all reflections except (220). It is essential that although diffraction vector modulus of (004) reflection is maximal, but its relative change of ($\Delta\theta / tg\theta$) between peaks of doublet is less than for (222) reflection.

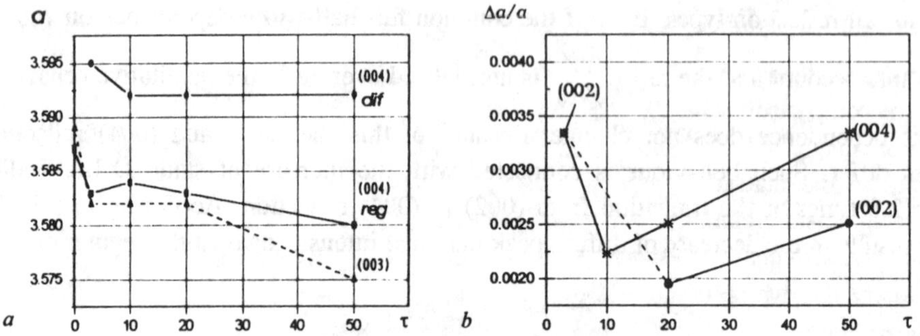

Fig. 3: Lattice parameters for structure (004) (a) reflections
and relative difference $\Delta a/a$ calculated from regular and diffuse maxima (b)

Such enormous reflection displacement behaviour was predicted by theoretical works [9-11]. It was shown that reflections are displaced from their ideal positions for cubic crystal with determined lattice parameter if anisotropy between matrix and precipitation is taken into account. Magnitudes of these displacements differ for (h00), (hh0), (hhh) reflections. In the framework of this conception, reflection positions depend on the orientation $\mathbf{G}_{hkl}$ and X-ray data on lattice parameters do not coincide with those measured by dilatometric method for such ageing solutions. Lattice parameter values for (002) and (004) reflections (Fig. 3 a, b) differ much, which was observed also by other authors [4, 6]. The γ'-phase lattice parameters determined with the superstructure reflection (001), (110), (003) do not coincide with the same parameters calculated for the (002) and (004) reflections with regular and diffuse maxima for different ageing time. Lattice parameter determined with the (003) reflection is indicated on figures with dash (Fig. 3 a).

The relative change of the lattice parameter for these two maxima ($\Delta a / a$) decreases up to 10, 20 hours of ageing and then increases for 50 hours of annealing (Fig. 3 b). These relative differences for (002) and (004) reflections do not coincide with each other which should take place for the same type of reflections. This indicates that dependencies of regular and diffuse maximums positions on diffraction vector modulus is different. The theory of the dependence of regular and diffuse maxima positions on diffraction vector modulus and orientation and also on the parameters and distribution of γ'-phase precipitates was worked out in [9-11].

The composition of solid solution changes during ageing. Correspondingly ideal reflection position as a starting point changes too. That is why both maxima displace during this process. Such behaviour corresponds to different effective values of lattice parameter determined by the regular and diffuse peaks (for example, by (002) and (004) reflections on Fig. 3 a, b).

According to [9-11], these regular and diffuse peaks displacements might lead to the supposed lattice parameter change ($\Delta a/a$) by about (2.0-3.5)$\cdot 10^{-3}$ (Fig. 3 b). The ageing time increase leads to a displacement of regular reflections in the direction of bigger angle in comparison with the quenched state, i.e. correspond to the smaller value of lattice parameter value. This means that ΔV, which is due to the formation of γ'-phase particles in the matrix, is negative. The diffuse maximum is always situated on the smaller angle side relative to the regular peak and corresponds to bigger effective value a, which is also characteristic for $\Delta V < 0$. Using the dependence from [9-11] the total volume change, which is due to all the γ'-phase precipitates per unit volume, might be estimated and is equal $n|\Delta V| \sim (3,7\ldots5,0)\cdot 10^{-2}$. These particles influence strengthening of nickel-based single crystal superalloys during ageing.

References

1. M. Feller-Kniepmeier, T. Link, I. Poschmann, G. Scheunemann-Frerker and C. Schulze, *Acta mater.* **44**, №6, 2397 (1996).
2. Barabash, A.D.Rud, S.I.Sidorenko etc, *Metal physics and new technology*, **18**, №11, 1 (1996).
3. F. Jiao, D. Bettge, W. Osterle, J. Ziebs, *Acta Mater.* **44**, № 10, 3933(1996).
4. H. Biermann, H.A. Kuhn, T. Ungar, J. Hammer, H. Mughrabi, *Proceedings of "9th International conference on Strength of Metals and Alloys", London, England*, 421(1991).
5. M.B.Henderson, J.N.Martin, *Acta Mater.* **44**, №1, 111 (1996).
6. M.V. Nathal, R.A. Mackay, R.G. Garlick, *Material Science and Engineering*, **75**, 195 (1985).
7. M. Verson, Y. Brechet, F. Louchet, *Acta Mater.* **44**, № 9, 3633 (1996).
8. Ya. D. Korol, A.D. Rud, A.I. Ustinov, *Inst. of Metal Physics, Kiev* **IMF 27**, 89 (1989).
9. R.I. Barabash, M.A. Krivoglaz, *FMM*, **45**, №1, 7 (1978).
10. R.I. Barabash, M.A. Krivoglaz, *FMM*, **51**, №5, 903 (1981).
11. M.A. Krivoglaz. X-ray and neutron diffraction in imperfect crystals (Kiev, Naukova Dumka, 1983) 407 p.

PRESSURE EFFECT ON OXYGEN PRECIPITATE NUCLEATION IN CZ-SI, ITS STRUCTURAL EVOLUTION AND PHASE TRANSFORMATION

I.V.ANTONOVA, V.P.POPOV, A.E.PLOTNIKOV,

Institute of Semiconductor Physics RAS 630090,
Lavrentieva 13, Novosibirsk, Russia

A.MISIUK,

Institute of Electron Technology, Al. Lotnikow 32/46,
02-668 Warsaw, Poland

B.SURMA

Institute of Electronic Materials Technology, Wolczynska 133
01-919 Warsaw, Poland

Oxygen precipitate (OP) formation proceeds through the oxygen-rich local areas (OP nuclei) creation in silicon matrix. The accumulation of oxygen atoms in these areas with annealing time leads to phase transformation. Utilization of enhanced pressure displaces the time of phase transformation towards the shorter annealing time and gives rise to creation of OP with smaller sizes and higher OP concentration.

1. Introduction

Oxygen is an important impurity in Czochralski-grown silicon (Cz-Si). The solid solution of oxygen in silicon remains highly supersaturated in a very large temperature range (up to 1250°C). Heat treatments at temperatures when oxygen atoms become mobile (T > 350°C) lead to different forms of oxygen atom agglomerates. Amongst the various factors that have a strong effect on the OP formation process (such as heat treatment regimes, oxygen concentration, impurities, etc.) the impact of mechanical stress is not well understood. The strongest stress effects are obtained to occur for stress magnitude higher than 0.1 GPa. The OP formation and evolution in Cz-Si at T ≥ 600°C and the impact of enhanced pressure on this process are discussed in the present paper.

2. Experimental

The initial material was n-type Czochralski-grown silicon phosphorous-doped with free carrier concentration $1.2 \cdot 10^{15} cm^{-3}$. Oxygen concentration was equal to $(6.5 - 9.1) \cdot 10^{17} cm^{-3}$. Samples were subjected to annealing at 600-1130°C . Enhanced created in argon atmosphere pressure has magnitude up to 1.2 GPa. Infrared (FTIR) spectroscopy, deep level transition spectroscopy (DLTS), scanning electron microscopy (SEM) in regime of secondary electron emission and selective etching were used in the investigation.

For DLTS studies, the samples after all heat treatments were irradiated by electrons with the energy 2.5 MeV and dose 10^{15} cm^{-2} causing an electrical active oxygen - vacancy complex (A-center) creation. The inhomogeneity in A-center distribution due to inhomogeneity in oxygen distribution was studied. The method of oxygen agglomerate investigation allows to obtain the parameter N_t. The N_t value shows the rise of the total oxygen concentration in the agglomerates in comparison to the oxygen concentration in the crystal matrix. If $N_t = 1$, the oxygen agglomerates are not formed in the samples. Oxygen incorporated in precipitates which are amorphous SiO_x phase does not take part in oxygen-vacancy complex introduction and drops out of field of vision of this method. Thus for the crystal with OP the N_t is again equal to 1.

3. Results and discussion

The dependences of N_t parameter on annealing time t for 600 and 730°C at ambient pressure (10^{-4} GPa) are presented in Fig.1. $N_t(t)$ dependences are seen to have maxima for annealing time t = 1 h at T = 730°C and for t = 10 h at T = 600°C. The increase in $N_t(t)$ at an early stage of oxygen clustering can be considered as the proof of the oxygen atom accumulation in some local areas. We define such areas as the nuclei for further oxygen precipitation. At the beginning of the clustering the oxygen atoms occupy practically the same position in the nuclei as that in the silicon crystal lattice. But after more prolonged annealing some kind of phase transformation occurs for some part of the nuclei. It follows from the decrease in N_t value for annealing time t > 1 h and t > 10 h for 600 and 730°C correspondingly. The oxygen atoms incorporated in SiO_x phase are not detected by DLTS. After treatment at t > 2 h for T = 730°C and t > 20 h for T = 600°C, the N_t value became to be almost equal to 1, thus the OP formation process can be considered as already finished in all nuclei. More prolonged treatments did not result in an increase in the N_t concentration (N_t = 1). Thus OP growth occurs for prolonged annealing times without new nuclei formation.

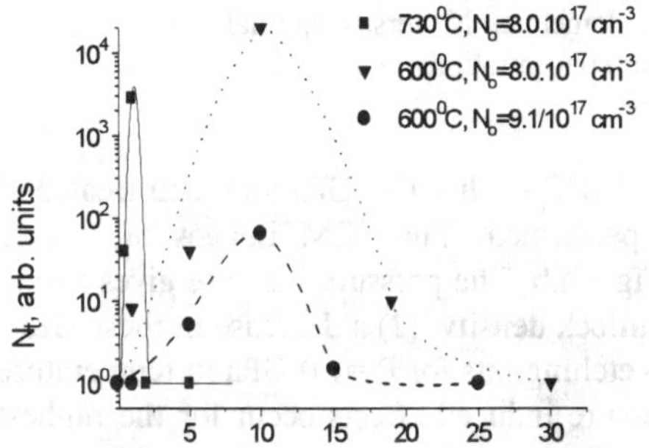

Fig.1 N_t as a function of annealing time for temperature T = 600°C and 730°C for different silicon crystals annealed at atmosphere pressure.

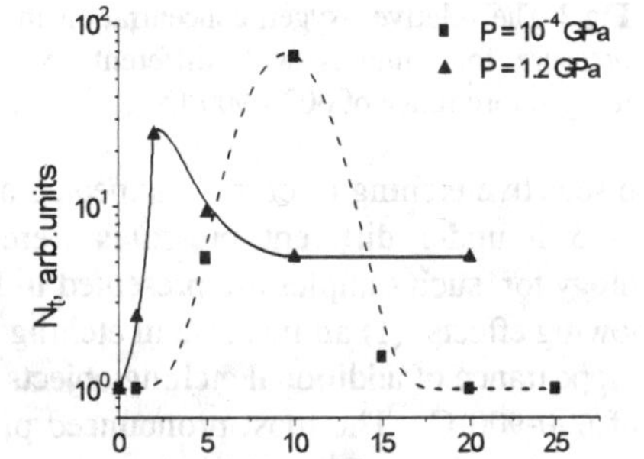

Fig.2 N_t as a function of annealing time for temperature T = 600°C for samples annealed at atmosphere pressure 10^{-4} GPa and at P=1.2 GPa. Initial oxygen concentration is equal to $9.1 \cdot 10^{17}$ cm^{-3}.

The increase in annealing temperature up to 900°C causes a decrease in OP formation time below 5 min. Nucleation processes in different crystals have the same time parameters (the onset and completion of phase transformation) but the oxygen concentration in agglomerates, N_t, is lower for crystals with a higher oxygen concentration (Fig.1). It may be determined by various number of OP nuclei or by structural particularity of nuclei due to variation in oxygen concentration as well as due to different crystal perfection.

Utilization of enhanced pressure changes kinetics of the OP formation. The dependence of $N_t(t)$ for samples annealed at 600°C upon pressure P = 1.2 GPa is presented in Fig.2. The time of phase transformation onset is displaced to t = 2 h. The prolonged treatments do not lead to $N_t = 1$ as it was for atmospheric pressure. The curve $N_t(t)$ for P = 1.2 GPa flattens out at $N_t = 5.2$ for the used crystal. N_t dependence on external hydrostatic pressure P for the samples annealed at different temperatures is presented in Fig.3. One can see that the enhanced stress at the nucleation stage gives rise to decrease in N_t, but for the pressure value P~1.0 GPa the oxygen agglomerates are appearing in the crystal ($N_t > 1$) again. The interstitial oxygen concentration does not depend on enhanced hydrostatic pressure value for used treatments, according to the FTIR data.

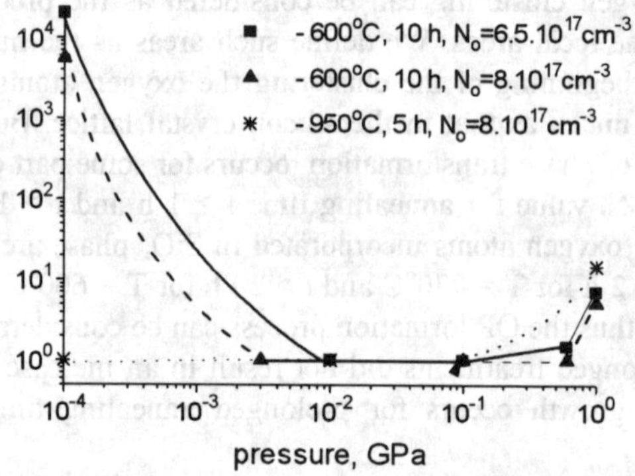

Fig.3 The relative oxygen concentration in agglomerates N_t versus internal pressure for samples with different oxygen concentrations annealed in temperature range of 600 - 900°C.

The selective etching of crystals annealed at 750°C - 8 h - 10^{-4} GPa and then treated at 900°C - 5 h under different pressures were performed. The SEM images of surface morphology for such samples are presented in Fig.4 a,b. The pressure increase gives rise to the following effects: (1) an increase in etching hillock density, (2) a decrease in these sizes, (3) the appearance of additional etching objects - etching pits for P =1.0 GPa in temperature range of 600-900°C. The most pronounced pressure induced effects occur for the highest pressures (≥1.0 GPa). The SEM data for samples subjected to selective etching after annealing at 1130°C at pressures 10^{-2} and 1.0 GPa are presented in Fig 5 a,b. For all pressures used annealing at this temperature caused creation of the etching hillocks only. The pressure effect is the same as that observed at lower temperature: etching hillock density is increased and hillock sizes are decreased for P = 1.0 GPa.

For the samples annealed in different conditions (T = 900-1130°C, different annealing times and pressure values) etching pits were observed only for samples for which $N_t > 1$. In our opinion the etching pits are mostly connected with OP nuclei whereas etching hillocks correspond to the SiO_x particles.

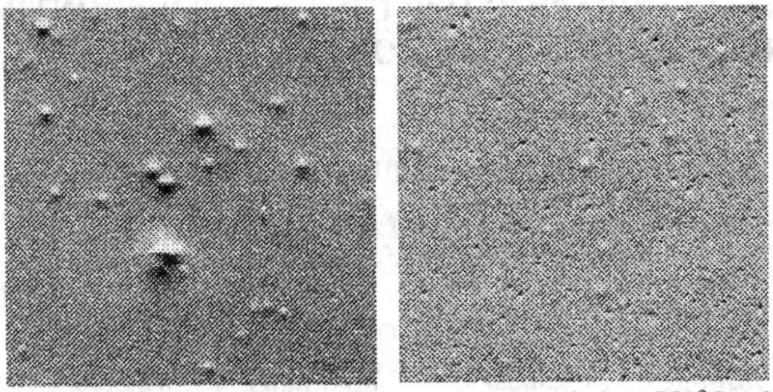

Fig.4 The SEM images of surface morphology for samples annealed at 750°C - 8 h, - 10^{-4} GPa + 900°C, 5 h at different pressure and subjected to selective etching during 10 min. Pressure value P, GPa: a - 10^{-2} , b - 1.0.

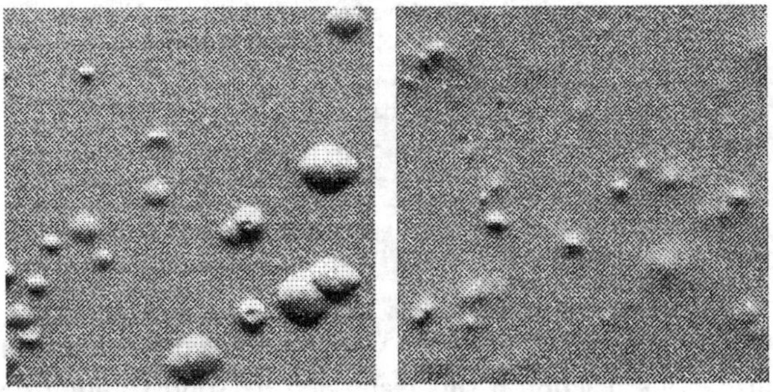

Fig.5 The SEM images of surface morphology for samples annealed at 1130°C - 5 h, at different pressure and subjected to selective etching during 3 min. Pressure value P, GPa: a - 10^{-2}, b - 1.0.

4. Conclusions

It was stated that oxygen precipitate formation proceeds through two stages at 600-750°C treatment. At the first stage the accumulation of oxygen atoms in some local areas (OP nuclei) occurs. OP nuclei are not the particles of silica phase yet. The accumulation of oxygen atoms in these areas (OP nuclei) with annealing time leads to structural transformation and OP as silica particles are formed. It was obtained that the main parameters of nucleation process (onset and completion of phase transformation) are determined first of all by the temperature. The variation in oxygen concentration or structural perfection of the crystal change only the N_t value.

The OP formation under enhanced pressure causes a decrease in the time of onset and completion of phase transformation at an early stage. The enhanced pressure gives rise to creation of oxygen precipitates with smaller sizes and higher concentration in comparison to the case of annealing at atmospheric pressure. Stabilization of OP nuclei occurs at pressure about 1.0 GPa in the temperature range of 600-900°C. The selective etching and SEM make it possible to observe OP nuclei (etching pits) as well as OP (etching hillocks) for samples annealed in temperature range of 600-900°C. These effects are more pronounced for pressure above 0.1 GPa.

CRYSTALLOGRAPHIC ASPECTS OF THE DISCONTINUOUS PRECIPITATION REACTION IN A Ni-4at.% Sn ALLOY

L. LITYŃSKA, P. ZIĘBA, A. PAWŁOWSKI
Institute of Metallurgy and Materials Science, Polish Academy of Sciences, Reymonta Street 25, 30-059 Cracow, Poland

W. GUST
Institute für Metallkunde der Universität Stuttgart, Seestr. 75, D-70174 Stuttgart, Germany

The investigation of the crystallographic orientation relationship between the α and β lamellae formed due to the discontinuous precipitation reaction have been performed in a Ni-4 at.% Sn alloy. The examination has shown the existence of a definite orientation relationship: $[\bar{1}11]_\alpha || [0001]_\beta$ and $(220)_\alpha || (11\bar{2}0)_\beta$. The proposed orientation relationship is preserved during the steady state period of the growth of the discontinuous precipitates which is characterized by a constant thickness of the α lamellae over a large growth distance.

1 Introduction

Discontinuous precipitation (DP) is a very well recognized phenomenon in Ni-rich Ni-Sn alloys[1-7]. During this reaction a supersaturated α° solid solution decomposes at an advancing cell boundary leading to the formation of a lamellar structure: solute depleted α solid solution and $Ni_3Sn(\beta)$ lamellae.

Most of the work which has been done so far was focused on the kinetic aspects of DP like growth rate, diffusivity at the moving cell boundary, activation energy and driving forces of the growth. Only Kampik *et al.*[6] using a photoemission electron microscope (PEEM) showed the initial stages of the process and the crystallographic relationship. They questioned, based on differences in the PEEM contrast, the orientation relationships found by Smith[8], i.e. that the depleted solid solution has the same orientation as the matrix solid solution from which the cell grew.

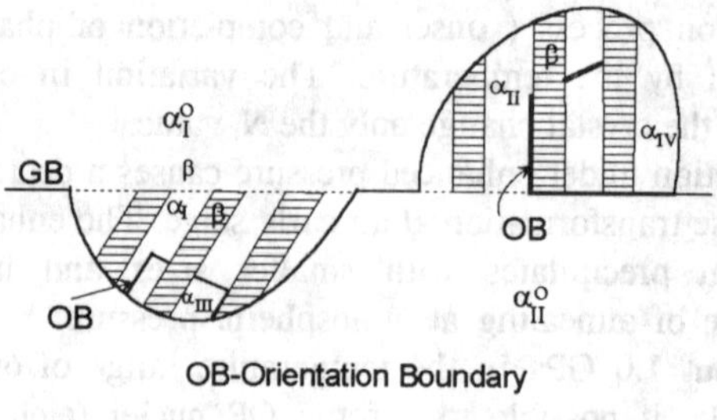

Figure 1: Schematic view of the crystallographic relationship during the initiation of DP. After Kampik *et al*[1].

The new scheme proposed by Kampik *et al.*[6] admits three different possibilities (Fig. 1). The first one is a very well known Smith's orientation, i.e. $\alpha_I^\circ \rightarrow \alpha_I$ which can be changed during the growth to $\alpha_I^\circ \rightarrow \alpha_I \rightarrow \alpha_{III}$. Another possibility is the formation of tin depleted lamellae with a quite different orientation just at the start of DP reaction according to scheme: $\alpha_{II}^\circ \rightarrow \alpha_{IV}$ instead of $\alpha_{II}^\circ \rightarrow \alpha_{II}$.

However, detailed crystallographic features within the colony of the discontinuous precipitates have not been characterized, yet. Therefore, it was the aim of the present paper to identify the crystallographic relationship between the α and β lamellae formed due to the DP reaction. The study was carried out on a Ni-4 at.% Sn alloy where the DP was observed previously using optical microscopy and scanning electron microscopy[9].

2 Experimental

A nickel alloy containing 4 at.% of tin was prepared by a melting together appropriate amounts of Ni and Sn (both of 99.99% purity) in an Al_2O_3 crucible and casting into steel moulds as rods 8.3 mm in diameter. Then the castings were homogenized at 1495 K for 840 hours to remove any microscopic segregation. The resultant grain size (1 mm) was toolarge to facilitate transmission electron microscopy (TEM). Therefore, the rods were cold worked by hammering for 30% reduction in diameter and heat-treated at 1025 K for 2 hours which provided a grain refinement down to 50-200 μm. Then, the rods were cut into slices 1.5 mm thick by the spark erosion method and aged in an evacuated silica capsule for 620 hours at 775 K.

Thin foils for the TEM study were prepared by dimpling and ion beam thinning using a GATAN instrument. The observation of the microstructure was carried out using a transmission electron microscope Philips CM 20 Twin operating at 200 kV and equipped with an energy dispersive X-ray spectrometer (EDX) Link exL, enabling a high spatial resolution analysis of local changes in the solute content.

3 Results

Figure 2 presents a TEM micrograph of a discontinuous precipitation colony formed in Ni-4 at.% Sn alloy after ageing at 775 K for 620 hours. It consists of alternating α and β lamellae separated from the α° matrix by the moving reaction front (RF). The electron diffraction analysis confirmed that the α lamellae are the face-centred cubic (fcc) Ni-rich solid solution with the lattice parameter a=0.352 nm. The β lamellae are a hexagonal phase with the lattice parameters a= 0.528 nm and c= 0.424 nm, which corresponds very well with the Ni_3Sn compound belonging to the space group $P6_3$ /mmc[10].

The chemical composition microanalysis revealed (Fig. 2) a characteristic parabolic shape of the Sn distribution within the α lamellae. The average Sn content taken at the distance $0.3\lambda_\alpha$ (λ_α-the thickness of the α lamella) is much higher than the equilibrium value (x_e=1.8 at.% Sn). On the other hand, the same test performed within the β lamellae showed Sn content around 24 at.%, which is in excellent agreement with the published data on the Sn concentration in the Ni_3Sn compound.

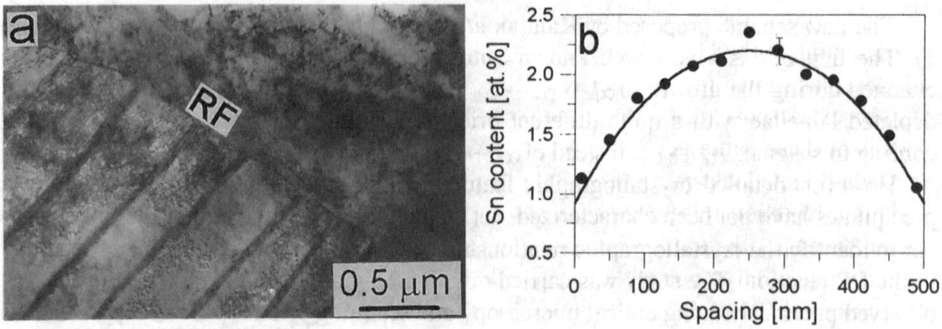

Fig. 2. (a) TEM micrograph showing the reaction front (RF) of the discontinuous precipitation reaction in Ni-4 at.% Sn aged at 775 K for 620 h. (b) A typical Sn concentration profile across the α lamella. The error bar is indicated.

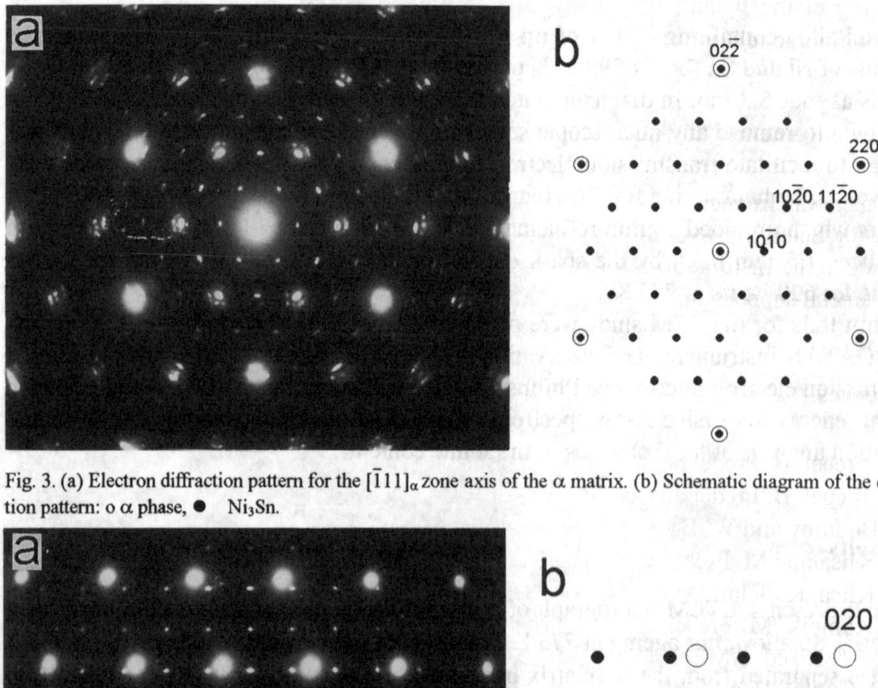

Fig. 3. (a) Electron diffraction pattern for the $[\bar{1}11]_\alpha$ zone axis of the α matrix. (b) Schematic diagram of the diffraction pattern: o α phase, ● Ni_3Sn.

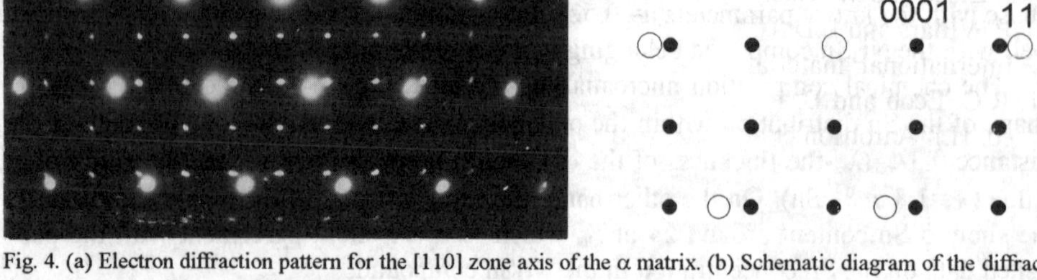

Fig. 4. (a) Electron diffraction pattern for the [110] zone axis of the α matrix. (b) Schematic diagram of the diffraction pattern: ○ α phase, ● Ni_3Sn.

Figure 3a shows the selected area diffraction pattern taken from a lamellar region and corresponding to the $[\bar{1}11]_\alpha$ and $[0001]_\beta$ zone axes in the electron beam direction. It may be indexed as shown in Fig. 3b using the lattice parameters of the α and β phase lamellae. The extra spots in the diffraction pattern arise from double diffraction. From the analysis of these patterns the existence of a definite orientation relationship of between the α and β lamellae can be established as: $[\bar{1}11]_\alpha \,||\, [0001]_\beta$ and $(220)_\alpha \,||\, (110)_\beta$.

This orientation relationship was confirmed by another diffraction pattern taken in the zone axes $[110]_\alpha$ and $(11\bar{2}0)_\beta$ (Fig. 4). Other crystallographic relationships were not found in spite of large number of experiments. This is a prove that the (0001) basal plane of the hexagonal Ni_3Sn phase is parallel to one of the {111} planes of the fcc α phase.

The proposed orientation relationship is preserved during the steady state period of the growth of lamellar colony which is characterized by a constant thickness of the α lamellae over a large growth distance. However, any change of the growth direction and branching of the β lamellae at the reaction front results in some modification of the crystallographic relationship between the α and β lamellae.

Acknowledgements

The paper was financially supported by Grant No 7 TO8A 01210 from the State Committee for Scientific Research in Poland. One of the authors (P.Z.) expresses also his gratitude to the Institute für Metallkunde of the University of Stuttgart for the hospitality and financial support during the stay as a research associate.

References

1. M. Frebel, B. Predel and U. Klisa, *Z. Metallkde.* **65**, 311 (1974).
2. M. Frebel, B. Predel and U. Klisa, *Z. Metallkde.* **65**, 469 (1974).
3. S. Budurov and W. Boshinov, *Z Metallkde.* **71**, 646 (1980).
4. U. Klisa and M. Frebel, *Z. Metallkde.* **73**, 426 (1982).
5. U. Klisa, K. Schur, M. Frebel and B. Predel, *Z. Metallkde.* **73**, 627 (1982).
6. T. Kampik, M. Frebel, K. Schur and G.E. Pfeffekorn in *Scanning Electron Microscopy 1983*, ed. AMF O'Hare (Chicago 1983), p. 633.
7. S.P. Gupta, *Acta Metall.* **35**, 747 (1987).
8. C.S. Smith, *Trans. ASM* **45**, 533 (1953).
9. P. Zięba and W. Gust, *Z. Metallkde.* **88,** 270 (1997), in press.
10. P. Villars and L.D. Calvert, in *Pearson's Handbook of Crystallographic Data*, (ASM International, materials Park, OH, 1991).
11. R.C. Ecob and B. Ralph in *Proc. Int. Conf. on Solid-Solid Phase Transformations*, ed. H.I. Aaronson et al. (The Metal Soc. AIME, NewYork 1982), p. 939.

GRAIN BOUNDARY MIGRATION IN Ni-26 wt.% Mo ALLOY

P. ZIĘBA, A. PAWŁOWSKI

Institute of Metallurgy and Materials Science, Polish Academy of Sciences, Reymonta Street 25, 30-059 Cracow, Poland

The grain boundary migration during ageing of an Ni-26 wt.% Mo alloy has been investigated using an analytical electron microscopy. The detected solute content step across the migration zone and the change of crystal orientation have been attributed to the de-alloying case of diffusion-induced grain boundary migration. No such changes have been related with the dissolution of small precipitates existing in the matrix.

1 Introduction

A class of solid state phase transformations characterized by a formation of a new phase at a moving interface called a reaction front, and showing a chemical step across it, attracts an increasing attention of scientists. This began in 1970s after Hillert and Purdy[1] described in details a diffusion-induced grain boundary migration (DIGM). The processes related to DIGM like: surface corrugation, liquid film migration (LFM) were subsequently discovered[2-5] and widely investigated. They supplemented the very well known discontinuous precipitation widely investigated in the past[6-8] with the new approach.

However, there is still controversy concerning early stages of these reactions; especially a driving force which makes the boundary to migrate. In this paper, two various processes of the grain boundary (GB) migration will be presented for the first time in an Ni-26 wt.% Mo alloy. The Ni-Mo system is well known for DIGM and LFM[5,9,10] but in the Mo-rich part. The occurrence of discontinuous precipitation in Ni-26 wt.% Mo alloy was also reported by Gust *et al*[11]. They showed a dominant role of grain boundary diffusion in the formation of a plate-like structure of the Mo_3Ni compound in the α_{Ni} matrix solid solution.

2. Experimental

A nickel alloy containing 26 wt.% of molybdenum was prepared by melting the elements of 99.95 % purity in an induction furnace. The alloy was cast into a steel mould and annealed at 1315 K for 6 weeks in order to remove segregation of components. The annealing was followed by water-quenching and ageing at 925 K.

An energy dispersive X-ray (EDX) spectrometer Link exL attached to the Philips CM 20 was used to reveal the changes in solute content occurring during ageing. For the foil thickness used it allowed a routine chemical analysis in the area ≈20 nm in diameter. The percentages of Mo at the analysed points were obtained from the observed peak integrals using Cliff-Lorimer[12] "ratio technique", which related compositions to intensity ratios of characteristic spectra. In the present study the Cliff-Lorimer factor was determined with respect to a single element i.e. Mo and Ni. In such a case they are called "elemental sensitivity factors" and they were found to be k_{Mo}=1.61 and k_{Ni} = 1.41.

The observation of structure evolution was performed using a transmission electron microscope Philips CM 20 Twin. Foils for transmission electron microscopy were prepared from the samples mechanically sliced to 0.25 mm and discs 3 mm in diameter were punched. Finally, the discs were electropolished in an electrolyte of 10% H_2SO_4 in methanol at temperature -30 °C and a potential of 40 V.

3 Results and Discussion

Figure 1 presents a typical micrograph of the Ni-26 wt.% Mo alloy aged at 925 K. Allotriomorphs exist at the original grain boundaries (GBs). The presence of these single precipitates gives an input for the grain boundary movement in one directory or two opposite directions by forming a characteristic "S" shape (Fig. 1a). The GB segments migrate away from the centre of curvature.

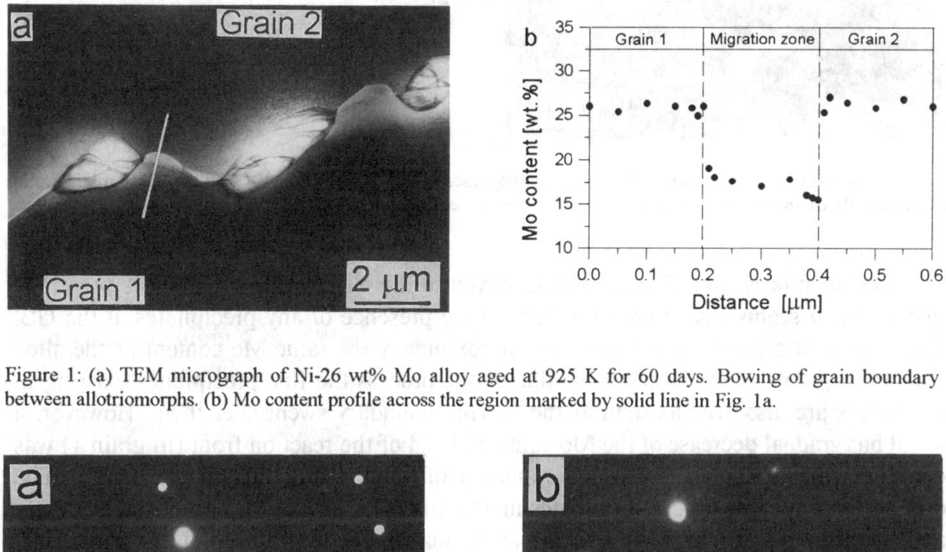

Figure 1: (a) TEM micrograph of Ni-26 wt% Mo alloy aged at 925 K for 60 days. Bowing of grain boundary between allotriomorphs. (b) Mo content profile across the region marked by solid line in Fig. 1a.

Figure 2: Selected area diffraction patterns taken from (a) the depleted region and (b) the matrix ahead of the moving boundary.

The area swept by the moving GBs is depleted in solute as revealed by the EDX analysis (Fig. 1b). A characteristic step in molybdenum content at the GB and an abrupt change in crystal orientation between the depleted region and the grain interior (Fig. 2) indicate that the process is governed by a diffusion along the GB. With time, the boundaries continue to bow, supplying solute atoms to the allotriomorph, whose size consequently increases. It is expected that the migration will be stopped as soon as the equilibrium Mo content is obtained within the allotriomorphs. The described process can be considered as a de-alloying type of diffusion-induced grain boundary migration in which the matrix serves as a source of the Mo atoms.

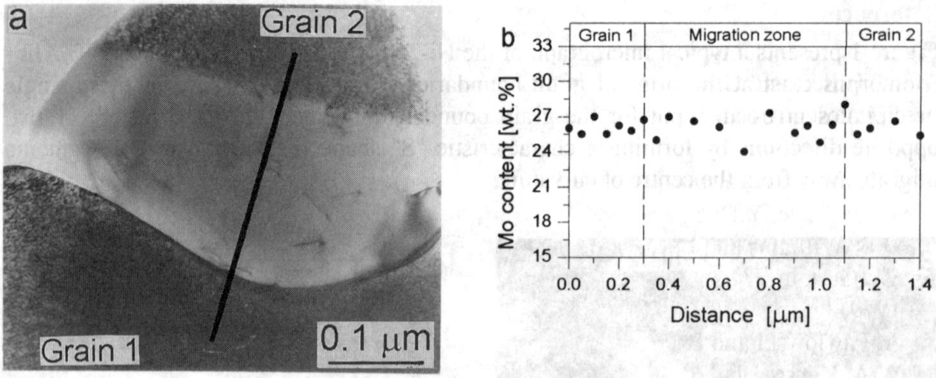

Figure 3: (a) TEM micrograph of Ni-26 wt.% Mo alloy aged at 925 K for 60 days. Grain boundary migration in absence of allotriomorphs. (b) Mo content profile across the region marked by solid line in Fig. 3a.

Another process of GB migration observed in the Ni-26 wt% Mo alloy is presented in Fig. 3a. It seems that it does not require the presence of any precipitates at the GB. The migration area is characterized by approximately the same Mo content as the alloy content (Fig. 3b). This suggests that during the process the fine precipitates existing in the matrix are dissolved as soon as the moving boundary swept over them. However, a small but gradual decrease of the Mo content ahead of the reaction front (in grain 1) was detected which is attributed to some amount of volume diffusion in the matrix. The driving force for this process originates mostly from the surface tension of the boundary and also from a strain gradient existing in the matrix due to presence of the fine metastable precipitates.

4 Conclusions

The investigation revealed two various processes of grain boundary migration in an aged Ni-26 wt.% Mo alloy. The first one is characterized by the formation of the solute depleted zone between two allotriomorphs. The driving force of the process comes from the chemical potential gradient across the reaction front and the curvature of the boundary. Another process is related with the dissolution of small precipitates at the reaction front with the main driving force resulting from the curvature of the boundary and the strain field existing in the matrix.

Acknowledgements

The authors wish to acknowledge the financial support of the State Committee for Scientific Research in Poland in the frame of the project of IMIM PAN No 11/96.

References

1. M. Hillert and G.R. Purdy, *Acta Metall.* **26**, 333 (1978).
2. F.J.A. den Broeder, *Thin Solid Films* **124**, 135 (1985).
3. A.H. King, *Int. Mater. Rev.* **32**, 173 (1987).
4. D.N. Yoon, *Int. Mater. Rev.* **40**, 149 (1995).
5. W. H. Rhee, Y.D. Song and D.N. Yoon, *Acta Metall.* **35**, 57 (1987).
6. D.B. Williams and E.P. Butler, *Int. Met. Rev.* **3**, 153 (1981).
7. W. Gust in *Phase Transformations* 1, II-27, (Institution of Metallurgists, London 1979).
8. A. Pawłowski and P. Zięba, *Archives of Metallurgy* **35**, 401 (1990).
9. W.A. Kaysser and S. Pejovik, *Z. Metallkde* **71**, 649 (1980).
10. Y. J. Baik and D.N. Yoon, *Acta Metall.* **33**, 1911 (1985).
11. W. Gust, T. Nguyen-Tat and B. Predel, *Z. Metallkde* **70**, 241 (1979).
12. G. Cliff, G.W. Lorimer, *J. Microsc.* **103**, 203 (1975).

COMPUTER SIMULATION OF "ORDER-ORDER" KINETICS IN $L1_2$ SUPERSTRUCTURE

P.ORAMUS, R.KOZUBSKI
Institute of Physics, Jagellonian University, Reymonta 4, 30-048 Krakow, Poland

M.C.CADEVILLE, V.PIERRON-BOHNES
Institut de Physique et Chimie des Matériaux de Strasbourg, 23, rue du Loess, 67037 Strasbourg, France

W.PFEILER
Institut für Materialphysik, Universität Wien, Strudlhofgasse 4, 1090 Wien, Austria

Monte Carlo simulations of long-range order (LRO) relaxation in A_3B system with $L1_2$ superstructure have been performed within a model based on a vacancy jump mechanism between nearest neighbour sites. Systematic studies involved different vacancy concentrations, different sets of pair interaction potentials, as well as different activation barriers for the migration of A- or B-atoms. The dependence of resulting relaxation times on vacancy concentration has been checked.
Previously proposed model of the origin of the experimentally observed two (fast and slow) parallel relaxation processes has been supported. It was found that the fast relaxation is damped when high activation barriers are assumed to act between the majority A-atoms. Contribution of particular kinds of atomic jumps to the overall LRO relaxation was monitored.

1. Introduction

The activation energy for "order-order" kinetics in Ni_3Al resulting from recent resistometric measurements [1,2] is equal to 4.6 eV - i.e. substantially higher than the one for Ni diffusion in Ni_3Al (about 3 eV). It is predominantly contributed by the activation energy for vacancy *formation* amounting to 2.9 eV. The monitored "order-order" relaxations involve two parallel processes strongly differing in rates, but showing close activation energies.

The initiated Monte Carlo simulations aimed at the reproduction of experimentally observed LRO relaxation curves and the explanation of the origin of two parallel relaxation processes.

2. Principles of the model

Perfectly $L1_2$ ordered A_3B binary system built-up of 40^3 superlattice cells and containing a fixed number of 10 vacancies was studied. Its energetics was described by means of pair interaction energies of atoms in two co-ordination zones:
nearest neighbours: $V_{AA} = 0.072$ eV; $V_{BB} = 0.036$ eV; $V_{AB} = -0.036$ eV; next-nearest neighbours: $V_{AA} = -0.012$ eV; $V_{BB} = -0.006$ eV; $V_{AB} = 0.008$ eV.

The energies were evaluated arbitrarily so that: (a) $L1_2$ superstructure is stable within possibly wide temperature range and the ordering energies in both co-ordination zones approximate the values yielded by experiments [3], (b) the majority A-atoms interact stronger than the minority B-ones, (c) the vacancy-atom and vacancy-vacancy interactions are equal to zero.

The degree of LRO in the system was determined by means of the Bragg-Williams-type LRO parameter η proportional to the number of antisites present in the system.

Vacancy mechanism for atomic migration was assumed. The probability of a jump of an A(B)-type atom to a nearest-neighbour (nn) vacancy is given by a modified Metropolis term:

$$P(A(B)) = \frac{\exp\left[\frac{-\Delta V}{k_B T}\right]}{1+\exp\left[\frac{-\Delta V}{k_B T}\right]} \times \exp\left[\frac{-E^{+}_{A(B)}}{k_B T}\right] \qquad (1)$$

where: $P(A(B))$ - probability of A(B)-atom jump; ΔV - system energy change caused by the jump, T - absolute temperature; k_B - Boltzmann constant; $E^{+}_{A(B)}$ - saddle-point energy attributed to the jumping atom.

In the present approximation, the saddle-point energy scales only the mobility of particular atoms and meaningful is only a difference $\delta^{+} = E^{+}_{A} - E^{+}_{B}$.

A particular MC step consists of a random choice of a nn lattice site of each vacancy and possible execution of a vacancy jump to this lattice site according to the probability (1). LRO relaxations simulated within the reported study started always at full LRO (η =1, T=0 K) and led to an equilibrium state with η <1 at T > 0 K.

3. Results

3.1 Effect of fixed vacancy concentration

Even if only 1 single vacancy is introduced into the simulated system, the effective vacancy concentration exceeds the realistic one by 3 orders of magnitude. Two parasitic effects are due to high vacancy concentration: (a) formation of divacancies leading to a loss of potential atomic jumps (a jump of a vacancy to a vacancy is meaningless), (b) decrease of the number of atoms due to a fixed number of lattice sites in the simulated sample. It has been shown, however, that the two above effects mutually cancel so that the rate of the relaxation results always proportional to the vacancy concentration. Therefore, the number of vacancies means only a MC time scaling.

3.2 Order-order relaxations

An example of an order-order isotherm of η corresponding to $\delta^{+} = 0$ is shown in Fig. 1a. All the simulated curves fitted weighted sums of two single exponentials:

$$\frac{\eta - \eta_{EQ}}{1 - \eta_{EQ}} = C \times \exp\left[-\frac{t}{\tau_s}\right] + (1-C) \times \exp\left[-\frac{t}{\tau_l}\right] \qquad (2)$$

The relaxation times τ_s and τ_l differed by about an order of magnitude. Their Arrhenius plots (Fig.1b) indicate that only the long one τ_l yielded a definitely non-zero activation energy E_A connected with the applied pair-interaction energies.

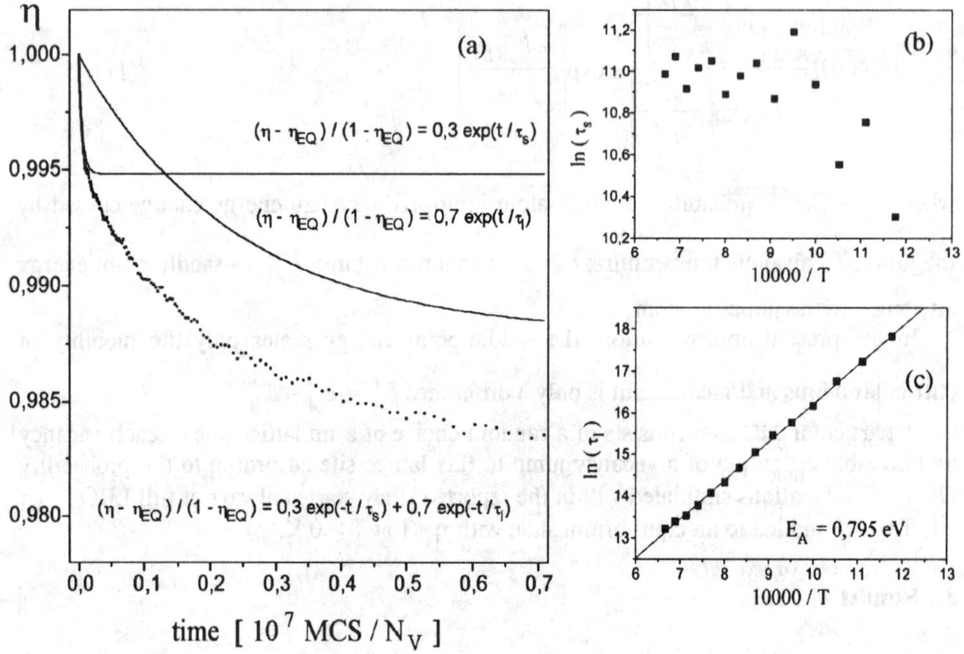

Fig.1. (a) Simulated isotherm of η corresponding to T=1200 K and to $\delta^+ = 0$. Solid lines represent the fast and slow relaxation components. Arrhenius plots of the relaxation times τ_s (b) and τ_l (c).

The influence of non-zero δ^+ on the simulated LRO kinetics has up to now been studied at only one selected temperature by imposing increasing barriers upon either A or B atoms.

It was found that the weight factor C in eq.2 reaches its maximum at $\delta^+ = 0$ and decreases with increasing $|\delta^+|$. The sign of δ^+ controls, however the behaviour of the ratio τ_l/τ_s, which either increases ($\delta^+ < 0$) or remains constant ($\delta^+ > 0$) with decreasing C.

3.3 Contribution of particular kinds of jumps to the simulated LRO kinetics

Figs.2a,b display the time dependence of the number of *executed* jumps of particular kinds per 320000 MC steps at T=1200 K and $\delta^+ = 0$. It is visible that the initial stage of the relaxation was dominated by the jumps of the minority B-atoms between B-type (cube corners) and A-type (face centres) sublattices. During the process the number of such jumps gradually relaxed to a lower level. At the same time the A-atoms jumped predominantly within the A-type sublattice. The number of their jumps between B- and A-type sublattices, initially definitely lower, showed an increasing relaxation throughout the process.

4. Discussion and conclusions

The B-atom jumps in Fig.2a represent both the B-sublattice → A-sublattice and the highly probable reversed (backwards) ones. The efficiency of the B-atom dynamics in decreasing

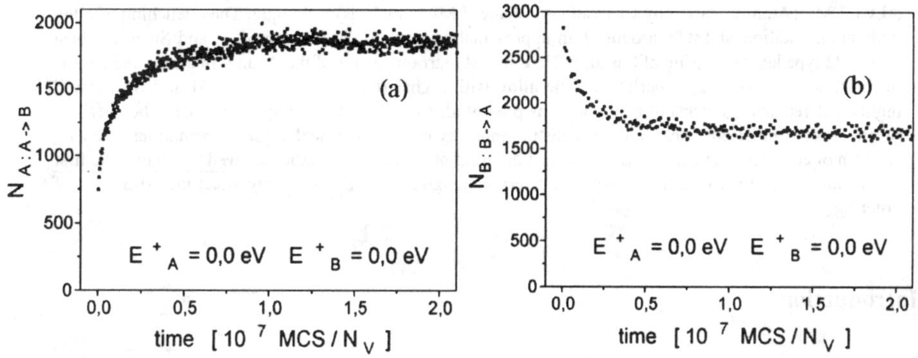

Fig.2. Number of *executed* A-atom (a) and B-atom (b) jumps between A and B-sublattices per 320000 MC steps against MC time

the degree of LRO increases with a decreasing probability of the reversed jumps. This is attained if a jump of an B-atom to a nn vacancy on the A-sublattice is immediately followed by a jump of a nn A-atom to the just created B-vacancy. This, in turn, becomes less probable when A-atoms become less mobile than the B-ones (increasing positive δ^+).

On the basis of the presented results, it is claimed that the above process is the mechanism of the fast relaxation component observed in LRO kinetics in A_3B systems, which is basically in agreement with the model proposed previously [1,2]. In order to make the things clear, it should be remarked that when the mobility of A-atoms is decreased (decreasing negative δ^+) the process is damped but still shows up due to the increasing ratio τ_l/τ_S.

The slow component of the LRO relaxation in $L1_2$ superstructure is a complex co-operative phenomenon involving all kinds of atomic jumps, whose contributions do not substantially change in time.

References:

1. R.Kozubski, W.Pfeiler, *Acta Mater.* **44**,1573 (1996).
2. R.Kozubski, M.Migschitz, W.Pfeiler, in *Stability of Materials*, ed. A.Gonis, P.E.A.Turchi, J.Kudrnovsky (NATO ASI Series B, Plenum Press, New York 1996), p.699.
3. P.Cedenese, A.Marty, Y. Calvayrac, J.Physique (France) **50**, 2193 (1989).

ORDERING PROCESSES IN CuZnAlSn ALLOYS ANALYSED WITH ALCHEMI

J. MORGIEL

Institute of Metallurgy and Materials Science, Polish Academy of Sciences, 25 Reymonta Street, 30-059 Cracow, Poland

The changes in occupation of the β - CuZn lattice sites by Al and Sn additions during ageing of the Cu -48at.%Zn -1at.%Al - 1at.%Sn and Cu -15.4at.% Zn -6at.%Al -6at.%Sn has been investigated using ALCHEMI (Atom Location by Channeling Enhanced Microanalysis) technique. The quenching of alloy with small addition of 1at.% produced an approximately random distribution of Al and Sn in the host CuZn B2 type lattice. Ageing of this alloy leads to fast rearrangements of the Al and Sn atoms and results in formation of $D0_3$ /$L2_1$ superlattice. The alloy with higher (6at.%) additions of Al and Sn showed highly ordered arrangements of both additions present already after quenching. Comparing the obtained results with simulations including such disturbances as non symmetrical crystal orientations, varying strength of electron channeling and non perfect ordering of the reference lattice showed that while the first two conditions just increase the measurements error though the last can seriously affect the estimation of ordering.

1 Introduction

ALCHEMI (**A**tom **L**ocation by **Ch**annelling **E**nhanced **M**icroanalysis) technique was originally derived to describe distribution of substitutional additions within otherwise defect free host lattice [1]. The binary β brasses undergo homogenous second order transition from A2 to B2 at 465^{0}C, what is securing fully ordered reference lattice even during fast quenching. Additions of Sn or Al results in further occupancy rearrangements causing formation of $D0_3$ or $L2_1$ sublattices but at much lower temperatures and slower pace [2,3]. The ordering of brasses was considered as an prerequisite of shape memory effect of these materials but its role is still not fully explained.

The measurements performed on quenched CuZnAl alloys by Nakata at all [4] indicate that Al atoms occupies predominantly positions at the zinc rich {100} planes within the B2 ordered CuZn lattice and do not change their positions in any significant way during ageing. On the other hand, the classical TEM observations indicate both the evolution of $D0_3$ superlattice { 1/2 1/2 1/2 }spots from SRO (Short Range Order} type to LRO (Long Range Order) type and increase in antiphase domain size [3]. The above changes in character of order correspond with the shift of the start of martensite transformation M_S toward higher temperatures.

The present project was aimed at evaluation of the distribution of additions of both Al and Sn into the CuZn host lattice with ALCHEMI technique. The experimental results were discussed against the computer simulations including such disturbances as non symmetrical crystal orientations, varying strength of electron channelling and non perfect ordering of the reference lattice.

2. Experimental

Alloys of nominal composition (I) Cu - 48at.% Zn -1at.% Al - 1at.% Sn and (II) Cu - 15.4at.% Zn -6at.% Al -6at.% Sn were cast, rolled and homogenised as described in [5]. Thin foils were prepared through standard double jet electropolishing. ALCHEMI measurements were carried out in Philips CM20 TWIN (200 kV) microscope equipped with Link EDS system in high resolution mode under planar channelling conditions for {100} planes [4]. The calculations of the occupancy factors were performed through least square minimisation method described in [6].

3 Results and Discussion

The results of simulation of channelling experiments presuming certain channelling strength, statistical disturbances or non symmetrical orientation within host lattice that are showed in Table 1 to 3, pointed out that while non symmetrical orientation is of least importance though the statistical disturbances has to be minimised in real experiment. The one practical way of solving this problem used throughout this work was to take the mean value of several experiment for further calculations. The disordering would just magnify the influence of statistical error on calculated occupancy factors.

Table 1: Data calculated for k=0.5 (Al) and m=0.5 (Sn) without statistical error.

file	I_α	I_β	R1	R2	k~dk	m.~dm
czas$t0	0.95	1.05	0.9	1.1	0.5~0.08	0.5~0.09
czas$t1	0.9	1.1	0.8	1.2	0.5~0.05	0.5~0.05
czas$t2	0.8	1.2	0.7	1.5	0.5~0.02	0.5~0.02
czas$t3	0.5	1.5	0.3	3.0	0.5~0.01	0.5~0.01

Table 2: Data calculated for k=0.5 (Al) and m=0.5 (Sn) with ~ $N^{-1/2}$ as statistical error at non symmetrical orientations.

File	I_α	I_β	R1	R2	k~dk	m.~dm
czas$e0	0.95	1.05	0.9	1.1	0.24~0.08	0.74~0.08
czas$e1	0.9	1.1	0.8	1.2	0.37~0.05	0.62~0.04
czas$e2	0.8	1.2	0.7	1.5	0.44~0.02	0.56~0.02
czas$e3	0.5	1.5	0.3	3.0	0.48~0.01	0.52~0.01

Table 3: Data calculated for k=0.5 (Al) and m=0.5 (Sn) with ~ $N^{-1/2}$ as statistical error and at non symmetrical orientation.

file	I_α	I_β	R1	R2	k~dk	m~dm
czas$e02	0.95/ 0.8	1.05/ 1.2	0.9	1.5	0.42~0.04	0.58~0.04
czas$e03	0.95/ 0.5	1.05/ 1.5	0.9	3	0.46~0.02	0.54~0.02
czas$e12	0.9/ 0.8	1.1/ 1.2	0.8	1.5	0.31~0.02	0.69~0.02
czas$e13	0.9/ 0.5	1.1/ 1.5	0.7	3	0.43~0.01	0.58~0.01

Table 4:Data calculated for k=0.6 (Al) and m=0.9 (Sn) with ~ $N^{-1/2}$ as statistical error .

file	I_α	I_β	R1	R2	k~dk	m~dm
czas$$e0	0.95	1.05	0.9	1.1	0.34~0.08	1.10~0.08
czas$$e1	0.9	1.1	0.8	1.2	0.53~0.05	1.01~0.04
czas$$e2	0.8	1.2	0.7	1.5	0.55~0.02	0.96~0.02
czas$$e3	0.5	1.5	0.3	3.0	0.57~0.01	0.92~0.01

Table 5: Data calculated for k=0.6 (Al) and m=0.9 (Sn) with ~ $N^{-1/2}$ as statistical error at 80% Cu:Zn ordering of the host lattice.

File	I_α	I_β	R1	R2	k~dk	m~dm
czas&&e0	0.95	1.05	0.9	1.1	0.22~0.14	1.54~0.08
czas&&e1	0.9	1.1	0.8	1.2	0.44~0.05	1.36~0.04
czas&&e2	0.8	1.2	0.7	1.5	0.57~0.02	1.22~0.02
czas&&e3	0.5	1.5	0.3	3.0	0.63~0.01	1.21~0.01

For the alloy with small alloying additions (files czas77**) and for particular heat treatment the calculated occupancy factors for Al and Sn are the same within the measurement error increasing from 1/2 for quenched state what denotes random distribution up to 0.9 - 1 after ageing indicating practically complete transfer of this additions to zinc rich planes as shown in Table 6. The increase of these occupancy factors is accompanied by significant growth of antiphase boundaries from 0.1 after quenching to more than 10 after ageing at 200^{0}C for 4 hours, as shown in Fig.1. The strength of channelling was found to be comparable with that obtained for binary CuZn alloy.

Table 6: Occupancy factors for the Al /k/ and Sn /m/ at zinc rich {100} planes together with the estimated strength of channelling at opposing orientations /R1, R2/ and antiphase domain mean size Φ for alloys I and II after quenching and ageing at 200^{0}C.

file	ageing 200^{0}C	R1	R2	Φ - APB [μm]	k~dk	m~dm
czas77q1	-	1.09	0.92	0.1 - 0.2	0.49~0.12	0.53~0.07
czas77s1	10 min.	1.06	0.93	0.2 - 0.5	0.86~0.09	0.84~0.06
czas77s2	4 hrs.	1.07	0.83	>10	0.93~0.09	1.00~0.06
czas78q1	-	1.05	0.9	0.05 - 0.1	0.99~0.06	0.96~0.07

The measurements performed on alloy with higher content of aluminium and tin (file czas 78**) produced occupancy actor near or exceeding 1 indicating either that the additions are located on zinc rich planes already after quenching even as the antiphase domain size was small or that the ordering is not finished but the high alloying interfered with the ordering processes within the host lattice. The ageing of this alloy, which in first case should leave occupancy factors unchanged and in second case shift them toward non physical values above 1 was not possible due to very fast precipitation processes of the γ phase at the antiphase boundaries.

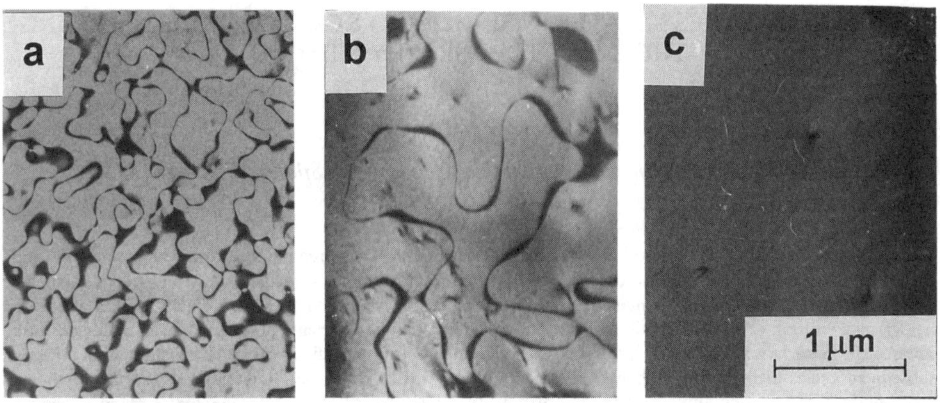

Fig. 1. The antiphase domains in alloy I after a) quenching, and ageing at 200^{0}C for b) 10 min. and c) and 4 hrs..

4 Conclusions

The results of the performed experiments proved that the growth of antiphase domains during ageing of brasses with small Al and Sn alloying additions are correlated with rearrangements of ternary and quaternary additions toward the {100 } rich planes. The consistent high occupancy of alloying additions of Al and Sn of the zinc rich {100} planes obtain for alloy II might indicate either high ordering within this alloys present already after quenching or non perfect ordering of the Cu - Zn B2 type reference host lattice. Additionally, the channelling at this planes for all brasses are weak to medium and therefore the calculated occupancy factors are encumbered with quite large statistical error and therefore should be treated as a semiquantitative ones.

Acknowledgements

The author wish to acknowledge the financial support of State Committee for Scientific Research in Poland (grant no. 3P407 032 07).

References

1. J.C. Spence and J.Tafto, Journal of Microscopy, **130**, 147 (1983).
2. R. Rapacioli and M. Ahlers, Scripta Metallurgica, **11**, 1147 (1977).
3. J.Dutkiewicz and J.Morgiel, Journal of Materials Science, **21,** 429, 173 (1986).
4. Y.Nakata, O.Yamamoto, K.Shimizu, Materials Transactions, **34**, 429(1993).
5. J.Morgiel, J.Dutkiewicz, W.Baliga, Proc. XVI Conf. on Applied Crystallography, Cieszyn 1994, 242.
6. J.Morgiel, Proc. IX Conf. on Electron Microscopy of Solids, Zakopane 1996, 133.

STUDIES OF STRESS STATES IN SINGLE CRYSTALS OF β_1 -CuZnAl USING THE X-RAY TOPOGRAPHY METHOD

W. BOGDANOWICZ

Institute of Physics and Chemistry of Metals, University of Silesia, Bankowa 12, 40-007 Katowice, Poland

The work presents results of studies of a reversible martensitic transformation induced by a tensile stress in single crystals of β_1 - CuZnAl. The studies were carried out with the use of the Auleytner X-ray topography method. It was found that at one-variant martensitic transformation in the specimen there are small stresses visible on topograms in a form of parallel bands. When at the transformation two variants of the martensite plates occur the stresses are much larger. Especially large stresses are created during a cyclic martensitic transformation where in subsequent cycles there occurs a change in variants of the martensite plates. It was found that the relation of components of the rotation angle between the martensite and the parent phase has some hysteresis in a loading - unloading cycle.

1 Introduction

Stresses occurring in shape memory alloys significantly influence stability of the martensitic transformation. The X-ray topography allows determination of stress distribution in a whole single-crystalline specimen by a diffraction image from the specimen surface. The Auleytner method applied in these studies is very „sensitive" for small changes in lattice parameters and positions of crystal planes giving rise to the diffraction [1]. This makes possible determination of changes of angles between crystal planes of order of a few dozen of angular seconds. It is well known that after the martensitic transformation induced by tensile stress in the β_1 phase there are created bands of parallel dislocations [2] which are not related to the plastic deformation. These bands are the source of internal stresses [3]. In [4] it was found that after several cycles of the martensitic transformation induced by stress in the specimens there are created immovable dislocation tangles which cause creation of large local stress and specimen cracks. In [5] it was found with the use of the X-ray topography methods that self-accommodating, thermally induced martensite plates cause large mechanical stresses registered at the topograms in form of reflex bending of order of a few degrees.

The aim of this work was to study with the use of the Auleytner wide beam method the stresses occurring in single crystals of the β_1 - CuZnAl alloy during the martensitic transformation induced by tensile stress.

2 Experimental

The studies were carried out with the use of a tensile machine co-operating with the Auleytner camera. The camera takes X-ray topograms reflexes from both the parent phase part of the specimen and from the martensite part.

From the β_1 - CuZnAl single crystals plate-like specimens were cut in such an orientation that the surface was of the (110) type. The dislocation density was in the range 5.5×10^4 - 8.7×10^4 cm^{-2}. The chemical composition determined by the X-ray microanalysis method was Cu - 74 wt.%, Zn - 18.9 wt. % and Al. - 7.1 wt.%. The characteristic temperature of the alloy were: $M_s = +5 \pm 5°C$, $M_f = -15 \pm 6°C$, $A_s = -7 \pm 6°C$ and $A_f = +16 \pm 5°C$.

The single crystals of the parent phase were oriented by the Laue method. Also the martensite single crystals obtained by tensile deformation of the parent phase were oriented by the Laue method by comparison of distribution of reflexes on the lauegrams with the theoretical crystallographic projections for the 2H and 18R martensite. It was found that the lauegrams are more consistent with the projections for the 2H martensite than for the 18R one.

3. Results and discussion

From the topograms the dependence of components of the rotation angle of the martensite part (γ_1') against the parent phase part (β_1) versus strain (ε) was determined. Fig. 1 shows the characteristic dependence of such a component α_x on

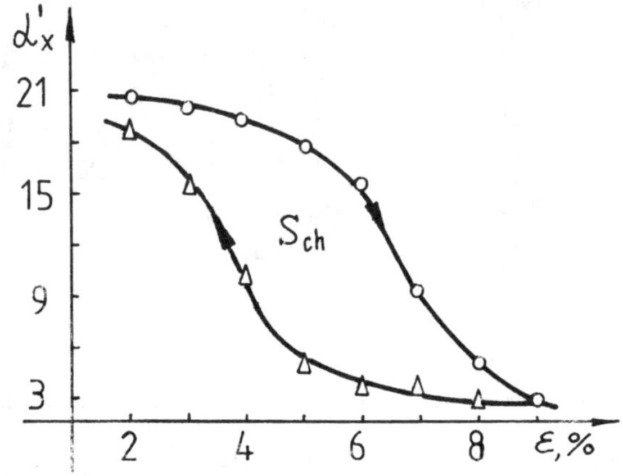

Fig. 1 Characteristic dependence of the α_x component of the angle of rotation of the γ_1' phase against the β_1 phase on strain ε obtained during one martensite transformation cycle, o - loading of the specimen, Δ - unloading. S_{ch} - field of hysteresis.

strain ε obtained during the martensite transformation cycle, where α_x - the component of rotation against the x axis perpendicular to the direction of deformation. Fig. 2 shows characteristic topograms obtained at increased loading. This increase caused increase of the single crystal area of the γ_1' phase which occurred by annexing newly formed plates arranged in the [101] direction. The increase of strain causes deformation of reflexes of the both phases in the area close to their interface (D area in Fig. 2c). Further increase of strain causes an increase of contrast which becomes less sharp and the image of block structure vanishes (D_1, D_2, D_3 areas in Fig. 2f). These changes in the diffraction patterns prove that near the interface large stresses are created already at the strain of 5.2%. Additionally, on the topograms of the γ_1' phase stress bands are created parallel to the direction of the martensite plates which form the single crystal (Fig. 2e).

In each cycle of the martensitic transformation only one variant of martensite plates parallel to [101] or $\left[\bar{1}01\right]$ direction was induced in the samples. Sometimes, after some

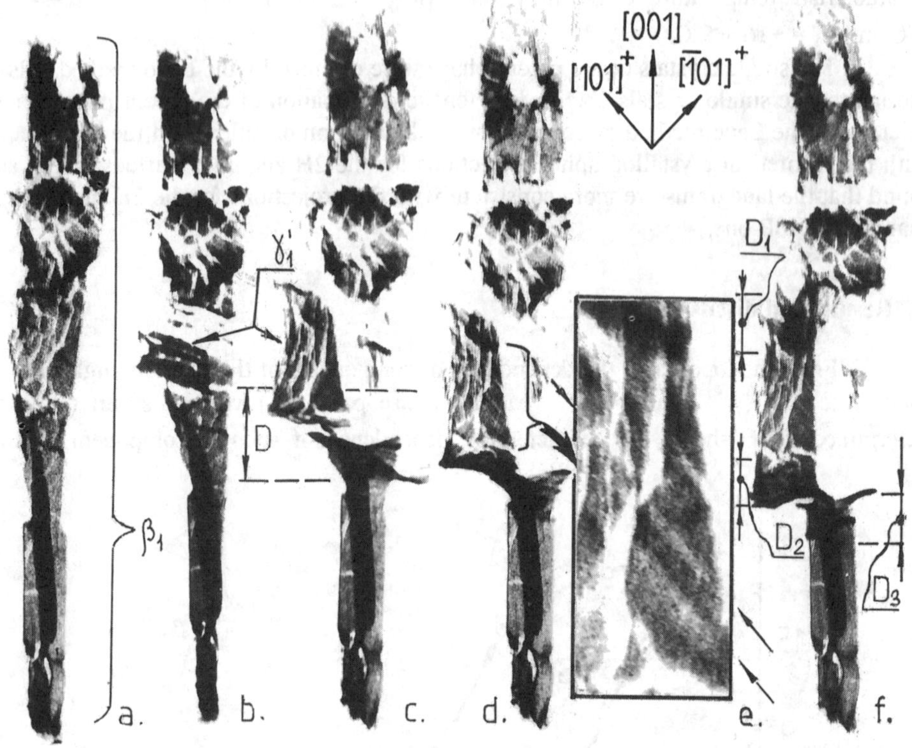

Fig. 2. Topograms obtained at different strain: **a)** $\varepsilon = 0.00\%$, **b)** $\varepsilon = 3.50\%$, **c)** $\varepsilon = 5.2\%$, **d)** $\varepsilon = 7.2\%$, **f)** $\varepsilon = 7.9\%$, **e)** an enlarged part of the topogram **d**. β_1 phase - reflex 220, γ_1' phase - 002. Vectors $[101]^+$ and $\left[\bar{1}01\right]^+$ are projections on the (110) plane of [101] and $\left[\bar{1}01\right]$ directions. D, D_1, D_2, D_3 areas of stresses.

number of cycles a change of the plates variant from [101] to $\left[\bar{1}01\right]$ or opposite occurred.

Fig. 3 shows changes on topograms occurring during the change of the $\left[\bar{1}01\right]$ variant to [101]. It can be seen that directly after the change of the martensite variant starts, there are large stresses in the sample (bending of the reflexes, Fig. 3b). After 4 hours these stresses disappear and the γ_1' area consists of two parts: a single crystal formed from plates of the [101] orientation (m_1 area in Fig. 3c) and a single crystal formed from plates of the $\left[\bar{1}01\right]$ orientation (m_2 area in Fig. 3c). Further deformation causes increase of this area which was formed from plates of [101] orientation and disappearing of the area formed from the other plates (Fig. 3d). As it is seen from Fig. 1 the $\alpha_1(\varepsilon)$ relation shows a hysteresis. For stress - strain curves the hysteresis gives value of energy dissipated during the transformation cycle. The S_{ch} field in Fig. 1 is also connected with energy dissipation.

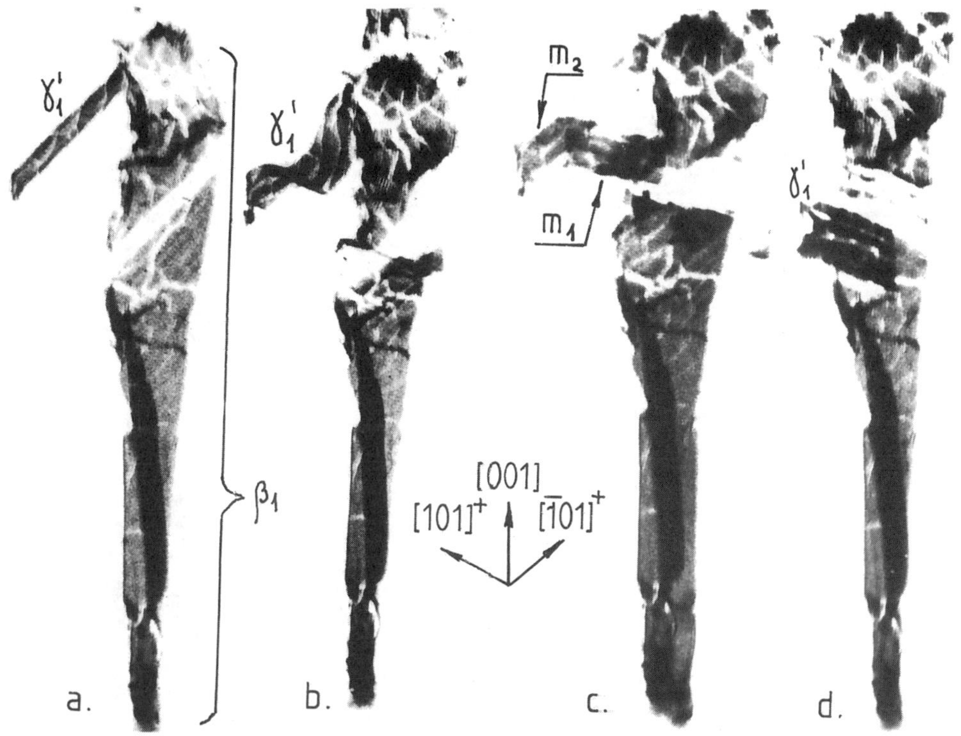

Fig. 3 Changes of X-ray topograms caused by change of the martensite plates variant. **a**, **b**, **c** - obtained at deformation ε =3.5%, **d** - obtained at ε = 4.%. **a** - obtained at the second cycle of the transformation; **b**, **c**, **d** - obtained at the third cycle of the transformation. **b** - obtained directly after the deformation was applied, **c** - obtained after loading the specimens for 4 hours with the deformation degree 3.5%

4 Conclusions

1. Inducing in single crystals of one variant of the martensite plates causes occurring small stresses indicated at the topograms in form of bands parallel to the martensite plates. When two variants of plates are induced the stresses are much bigger which can be seen in form of bent diffraction patterns. Especially large stresses occur when a change of the martensite plate variant takes place at repeated cycles of the transformation.

2. The obtained relation of α_x component of the angle of γ_1' and β_1 phases mutual rotation shows hysteresis during a full cycle of the transformation. The hysteresis is connected with dissipation of energy during the transformation..

References

1. W.Bogdanowicz and A.Lehr, *Arch.Nauki o Mat.*, **3**, 78 (1989).
2. J. Perkins, *Mater. Sci. Eng.*, **51**, 181 (1981).
3. K. Takalzawa, K. Adachi, and S. Sato, *J. Jap. Inst. Met.*, **44**, 846 (1980).
4. M. Sade, R. Rapacioli and M. Ahlers, *Acta. Metal. Mater.*, **33**, n3, 487 (1985).
5. C. Jourdan, J. Gastaldi, G. Grange, S. Belkahla and C. Guénin, *Journal De Physique IV*, **5**, C2-257 (1995).

Surface and Layers

X- RAY REFLECTIVITY IN THIN FILM STUDIES

G. A. STERGIOUDIS, S. LOGOTHETIDIS AND P.A. PATSALAS

Department of Physics, Aristotle University of Thessaloniki

54006 Thessaloniki, Greece.

Specular reflectivity measurements (XRR) were used to study the changes of density and surface roughness of amorphous carbon (a-C) thin and ultra-thin films, with regards to thickness. Also oxidation process of TiN_x films, was followed. Density, thickness and surface roughness of the films were deduced using x-ray reflectivity measurements in a series of films within a range of nominal thickness from 27 Å to 326 Å for amorphous carbon and of about 700 Å for TiN_x thin films. Both kinds of films were deposited by sputtering. Density and roughness of films vary as a function of thickness and it was found that amorphous carbon films deposited with positive substrate bias voltage exhibit a reduction in density and an increase in surface roughness. In the case of titanium nitride (TiN_x) films, XRR measurements showed a constant increase in density and thickness and a decrease in surface roughness after exposure to air. Only the films deposited at low bias voltage are prone to oxidation.

1 Introduction

X-ray reflectivity is recently widely used in thin films characterization[1-6] to measure the density of films through changes in the critical angle of the total external reflection. It is also used to determine directly the thickness and microscopic roughness (in the scale of several Å) of thin films.

The density, thickness and microscopic surface roughness are important parameters since they influence the functional properties of thin and ultrathin films. X-ray reflectivity technique fulfills requirements of direct and nondestructive measurements.

In our case amorphous carbon and fcc titanium nitride (TiN_x) thin and ultrathin films with variable thickness were deposited by RF magnetron sputtering and DC reactive magnetron sputtering, respectively. Precise thin-film thickness (d) and density (ρ) have been derived from X-ray specular reflectivity data collected with a conventional powder diffractometer equipped with a Göbel mirror and reflectivity sample stage (RSS). Surface roughness (σ) values were also determined.

In the present work the evolution of density and surface roughness of amorphous carbon thin and ultrathin films, in function of their thickness and the oxidation process of thin fcc titanium nitride (TiN_x) films, were observed by X-ray reflectivity (XRR) measurements.

2 Description of the Method and Information on the Films Parameters

The method of x-ray reflectivity based on the total external x-ray reflection offers much larger possibilities than the common techniques used for determination of the parameters of thin and ultrathin films. The information on parameters such as density, thickness and surface roughness can be deduced from the dependence of the reflection coefficient on the angle of incidence. The measured intensity in a specular scan exhibits many orders of reflections, except the main specular reflection. This comes from fulfilling the interference condition, namely the phase difference. The peaks outside the region of total reflection are related to the thickness of the film by the modified Bragg equation:

$$\sin^2\theta_i = \theta_c^{\,2} + (n_i + \Delta n)^2\lambda^2/4d^2 \quad (1)$$

where θ_i is the position of maximum or minimum of the i_{th} interference fringe, θ_c is the critical angle for total reflection, n_i is an integer, Δn is ½ and 0 for maximum and minimum respectively, and λ = the x-rays wavelength.

To the first approximation and for very small angles, the angular position of an extremum is given by the following equation:

$$i\lambda = 2d(\theta^2_i - \theta_c)^{1/2} \quad (2)$$

The angle of the total external reflection is related to the refraction index of x-rays $n=1-\delta-j\beta$ via the optical constant δ:

$$\theta^2_c = \delta = 2N_o(e^2/2\pi mc^2)(Z\tilde{n}/A)\lambda^2 \quad (3)$$

Where $\beta=\lambda\mu/4\pi$, N_o is Avogadro's number, Z is the average atomic number, ñ is the mass density of the film, A is the atomic mass and λ the x-ray wavelength.

It is known that surface roughness is related to the overall reduction in the reflected intensity and the interface roughness reduces specular reflectance due to the scattering in the non-specular directions.[7,8] Usually the distribution of widths of one or the other material at each interface is assumed to be a Gaussian one and taking into account approximation of the average interfacial roughness we can write the specular reflectivity I_s as:

$$I_s = I_o \, ex(-q^2_z \sigma^2) \tag{4}$$

where I_0 is the perfect reflectivity, q_z the perpendicular component of the momentum transfer vector and σ is the roughness.

Since the value I_o can not be obtained through measurements because no perfect surface exist, a dynamical calculation based on Fresnel diffraction theory[9] or on equivalent Ewald-von Laue dynamical diffraction is used. The exponential term in the (4) has the effect of attenuating the reflectivity. This approach is formally analogous to the treatment of the dynamic Debye-Waller factor[10] resulting from thermal vibration, except that here the disorder is frozen in. Consequently the exponential in eq. (4) has been called the static Debye-Waller factor. Equation (4) is independent on how the roughness is correlated from layer to layer.

3 Experimental

A D-5000 Siemens diffractometer equipped with Reflectivity Sample Stage (RSS) and with a Göbel mirror was used. The Göbel mirror is a new device which transforms a divergent x-ray beam into a parallel, highly brilliant beam.[11,12] The RSS consists of a translation stage on which an edge diaphragm and a vacuum sucker are mounted. The absolute position of the edge diaphragm is displayed by a dial indicator. The zero point is calibrated to ensure reproducible results for the reflectivity measurements. The set up enables large samples to be fully illuminated during reflectivity measurements for up to several degrees of the incidence angle. For example a 1cm wide parallel beam can bath a 10mm long sample up to 5.70°. Thus a conventional x-ray diffractometer equipped with a special RSS and a Göbel mirror is a useful tool to determine thin-film density, thickness and roughness with high accuracy and good resolution.

XRR measurements were performed in θ-2θ scaning mode (2.2kW Cu tube 12mmx0.04mm line focus). In order to fix the experiment parameters as optimization of the scan ranges, step size, power of generator and time per step some preliminary measurements were performed. Subsequently the qualitative reflectivity measurements were done under the conditions showed in Table I.

Table I. Data for reflectivity measurements

range *	2θ (start in °)	2θ (stop in °)	Generator Voltage (kV)	Generator Current (mA)	Time per step (sec)
1	0	0.66	20	5	0.4
2	0.66	1.20	40	30	5
3	1.20	2.46	45	40	10

* A delay time between two ranges is possible, giving the opportunity to change the parameters.

The step size was 0.006^0 for all the ranges. The specimens were located on the RSS. Depending on the film flatness, the film surface was purposely tilted in order to make he incident beam for $\theta=0^o$ to coincide with the direction parallel to the film surface. Thus the critical angle for total reflection, which determines the density of the film can be measured accurately. The rocking curves for the carbon films measured for $\Delta\theta$ between 0^0 and 0.2^0 were almost ideal with a single peak due to the good flatness of the films surface. It means that no effect of misalignment was observed.

A series of amorphous carbon a-C thin films were deposited by RF magnetron sputtering. Their thickness was gradually increased from a minimum nominal value of 27 Å to a maximum of 326 Å. Details on the films preparation are given elsewhere.[13] The TiN_x films were deposited on silicon (100) wafers by DC reactive magnetron sputtering at various negative bias voltage (V_b) and then exposed to air at room temperature. Oxidation of TiN_x films was monitored by real time ellipsometry.

4 Results and Discussion

All measured parameters are constant within one range of measurements but one or more parameters are different if different ranges are compared. The individual ranges are scaled in respect to each other and they are included to a single range to make them readable for a fitting procedure. Moreover as the intensity differences within one range of a measurement were of several orders of magnitude a dead time correction was necessary. The relation used for the dead time correction was :

$$I = I_o \cdot \exp\left(\frac{I_o \cdot \tau / t}{1 - I_o \cdot \tau / t}\right) \tag{5}$$

with the t = time per step and τ = dead time.

From the XRR spectra film thickness, density and roughness as a function of time were calculated using the REFSIM fitting procedure and assuming a single layer (Si - substrate, followed by a-C film, plus an interface of silicon-carbon layer) model for a-C films. For TiN_x films the same procedure was used assuming a composite titanium nitride and oxide film plus the substrate. Simulation of the assumed model showed very good agreement with the experimental curves, especially for thicker films, as it is shown in figure 1.

Figure 2a shows XRR from the thinner (27 Å) and from the thicker (326 Å) of the examined a-C films and the curves shown in figure 2b were obtained immediately after deposition and after a 172 h exposure to air. The difference in the spectra of the two different films is obvious as well as is their density. The calculated ρ is 2.15 gr/cm^3 for the thin and 1.68 gr/cm^3 for the thick a-C films respectively, while ρ is 4.06 gr/cm^3 for as-deposited and 4.25 gr/cm^3 for oxidized TiN_x films. From figure 3, it is clearly seen that the density of the film decreases as its thickness increases. This is indicative of the decrease of the sp^3 to sp^2 bonding ratio as new material is deposited to the previous one.

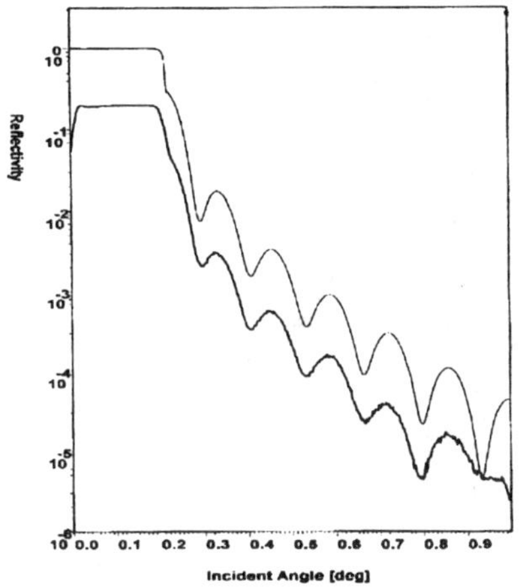

Figure 1: XRR for a a-C film of a nominal thickness 326 Å, with simulation (upper curve) assuming a single layer (Si - substrate, followed by a-C film, plus an interface of silicon-carbon layer), model. (Curves are shifted).

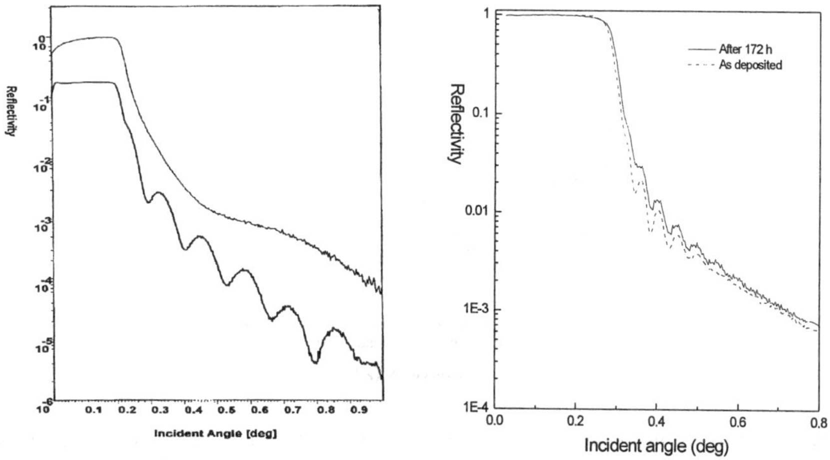

(a) (b)

Figure 2: (a) XRR spectra for the thinner (upper curve) and the thicker film of the whole series of the examined films. (b) XRR spectra for a TiN_x film immediately after deposition (upper curve) and for the same film after a 172 h exposure.

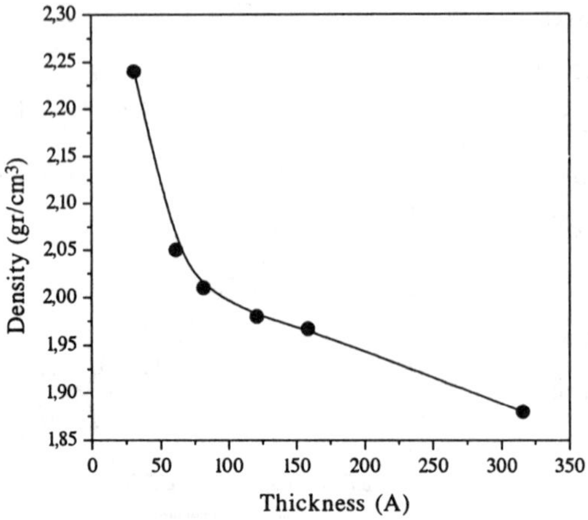

Figure 3: Variation of the density versus thickness of the examined a-C films, calculated from the XRR measurements.

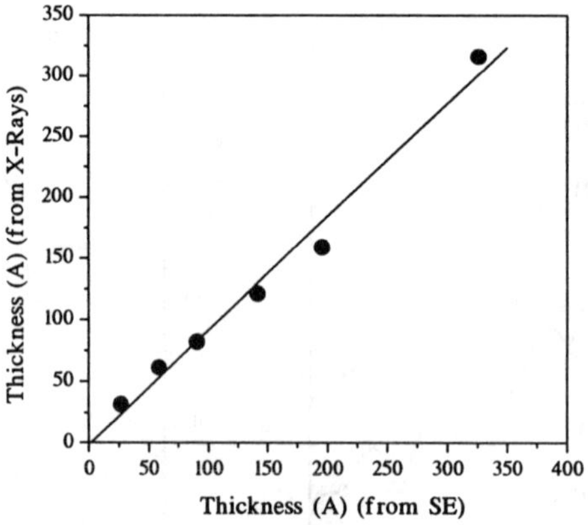

Figure 4: Comparison of the thickness of the films deduced from XRR measurements to those calculated by spectroscopic ellipsometry. It was found that these parameters are dependent on preparation condition, so their determination is particularly important. The as measured parameters, agree well with those taken by other techniques, wherever a comparison is possible.

Figure 4 compares values of d calculated by in situ spectroscopic ellipsometry in the energy range 1.5 eV to 5.5 eV[6] during sputtering and those resulted from XRR measurements. The XRR results are in a good agreement with those from the in-situ spectroscopy ellipsometry. Thus the possibility of this method to calculate with high accuracy thickness of films as thin as 20 Å is demonstrated.

Surface roughness which is related to the overall decrease in reflected intensity varies linearly from 6 Å for the thinner film to 10 Å for the thicker one, stressing the fact that roughness is related to the way and to the mechanism of growth of the films.

Table II. XRR calculations of density (ρ), thickness (d) and roughness (σ) of as-deposited film at -25V bias and after 172 h exposure to air.

Film Property	V_b=-25 V	
	As-deposited	After oxidation
ρ (g/cm³)	4.06	4.25
d(nm)	66.95	70.44
σ(nm)	3.91	3.49

The results of measurements on TiN_x films are summarized in Table 2 and show an increase in density and in film thickness along with a decrease in surface roughness after exposure to air. It has been established that upon exposure to air, the vertical columnar boundaries of TiN_x films act as fast channels for oxygen to enter the film and saturate the intercolumnar spacing.[14,15] After this spacing is filled, the oxidation process slows down due to the barrier layer of titanium oxide. As the time passes a slow bulk oxidation mechanism is dominated, leading to a saturation of oxidation.[16] It is remarkable that although oxygen is absorbed by all TiN_x films, the oxidation occurs only in those deposited at low bias voltage (V_b) values (overstoichiometric). The XRR results supporting this suggestion are in a good agreement with results of different techniques, which also suggest that the TiN films deposited at low (V_b) are prone to oxidation.[15]

5 Conclusions

Taking into account the advantages of the XRR method it is noted that this relatively new, non destructive technique provides an excellent method for accurate determination of density, thickness and roughness of thin and ultrathin films.

The accuracy of the measurements of the method allows calculations of the parameters of the films as well as observations of changes of these parameters in function of the thickness of the films. Depending on the conditions of preparation of the a-C films, a continuous decrease of the density with an the increase of the thickness was observed. This rationalized in terms of the ratio of sp^3 to sp^2 bonds . A slight linear change increment of the surface roughness from the thinner to the thicker film was observed depending on the mechanism of growth. Some other unexpected phenomena such as creation of new interface layers was elucidated and found to be compatible with the analysis of the experiments curves.

For the TiN_x films the results suggest an initial fast oxygen diffusion stage along the intercolumnar space, followed by a slow stage of oxidation.

Acknowledgements

This work was supported, in part, by EU EPET II 333 project.

References

1. M. F. Toney and S. Brennan, J. Appl. Phys. 66, 1861(1989)
2. T.P. Russell, Mater. Sci. Rept. 5,171 (1990)
3. C.A. Lucas, T. D. Nguyen and J. B. Kortright, Appl. Phys. Lett. 59, 2100(1991).
4. T.C. Huang, R. Gilles and G. Will, Adv. X-Ray Anal. 37, 183(1991)
5. T.C. Huang,, J.P. Nozieres, V. S. Speriosu, H. Lefakis, and B. A. Gurney, Appl. Phys Lett. 60, 1573 (1992).
6. Y. Huai, R. W. Cochrane and M. Sutton, Phys. Rev. B 48, 2568(1993).
7. D.E. Savage, J. Kleiner, N. Schimke, Y. H. Phang., T. Jankowsky, J. Jacobs, R. Darious and M.G. Lagally, J. Appl. Phys. 69, 1411(1991).
8. J.B. Kotright, J. Appl. Phys. 70, 4286(1991).

9. M. Schuster, and H. Göbel, Adv. X-Ray Anal. 39,1 (1996)

10. G. Chiari, Sci. and Technology for Cult. Heritage, to be published.

11. M. Born and E.Wolf, Principles of Optics (Pergamon, New York, 1983)

12. James, The Optical Principles of the Diffraction of X-rays (Oxbow, Woodridge,CO,1982)

13. S. Logothetidis, Appl. Phys. Lett. 69(2), 158 (1996).

14. S. Logothetidis and A. Barborica, Microelectronic Enginnering 33, 309 (1997).

15. S. Logothetidis, A.A. Adjaottor, G. Stergioudis and E.I. Meletis, submitted for publication in Thin Solid Films.

16. S. Logothetidis, G. Stergioudis and P. A. Patsalas, in press in Microelectronic Engineering.

CHARACTERIZATION OF THIN LAYERS BY X-RAY REFLECTOMETRY

D. ŻYMIERSKA, E. SOBCZAK, K. GODWOD, S. MIOTKOWSKA

Institute of Physics, Polish Academy of Sciences, Al. Lotników 32/46, 02-668 Warsaw, Poland

The paper presents an application of the grazing incidence X-ray reflectivity method for rapid nondestructive characterization of thin films and interfaces, independently on the amorphous or crystalline structure. The total thickness of multilayer can be determined from the period of interference fringes (Kiessig's fringes) and superlattice period from the positions of Bragg peaks.

1 Introduction

Modern microelectronic devices will require the growth of ultra-thin and uniform layers, therefore it is very important to understand phenomena at crystal surfaces at the nanometer scale and an increase of interest in research of the crystal surface is recently observed. Grazing incidence X-ray methods turned out to be appropriate tools with atomic resolution in space and a depth controlled sensitivity. They are also nondestructive. In addition, due to a weak interaction of X-rays with condensed matter an interpretation of experimental data is relatively easy in comparison to high resolution electron microscopy.

Among many surface specific techniques the grazing incidence X-ray reflectivity method under the total external reflection condition is very useful. The measurement of the X-ray reflectivity around the critical angle for total external reflection makes possible a determination of thin film thickness as well as surface and interface roughnesses, irrespectively of the crystalline structure. The critical angle and the shape of the measured reflectivity are determined essentially by the electron density changes and are otherwise independent of the amorphous or crystalline structure, or of the orientation of crystallites on the surface.

The aim of the paper is to present influence of several physical parameters (electron density of layer, film thickness, surface and interface roughnesses, periodicity and a number of periods in multilayers) on grazing incidence X-ray reflectivity applying the computer simulations.

2 Grazing incidence X-ray reflectivity (GIXR)

In condensed matter the index of refraction for hard X-rays is slightly smaller than unity and can be written as $n = 1 - \delta - i\beta$, with δ and β being positive and of the order 10^{-5} to 10^{-7}; β is directly related to absorption of X-rays in a sample and δ is the unit decrement of the index of refraction given by $\delta \approx (e^2/2\pi m)N_e\lambda^2$, where N_e is the number of electrons in volume unit, e and m are the electron charge and the electron mass, respectively, and λ is the wavelength of incident radiation.

From the refraction law it follows, that X-rays are refracted away from the normal when entering an investigated sample. In the case of grazing incidence the angle of refraction $\theta_2 = \sqrt{\theta_1^2 - 2\delta_2 - 2i\beta_2}$, so θ_2 is complex and becomes zero at a finite value of the incidence angle θ_1 called the critical angle θ_c, given by formula $\theta_c = \sqrt{2\delta}$ when the absorption is neglected ($\beta_2 = 0$). The critical angle for X-rays is below 0.5°. If the grazing incidence angle θ_1 has a value less than that of critical angle θ_c, the total external

reflection occurs.[1] The critical angle depends mainly on the electron density change on the surface of the investigated sample and on the wavelength of incident radiation.
In the region just above the critical angle θ_c, the X-ray wave penetrates succesively into the sample as the angle is increased. If a thin film of different electron density with respect to the substrate is present, interference oscillations (Kiessig's fringes) are observed in the reflectivity profile, containg information on film thickness.[2]

Simple measurement of fringe spacing $\Delta\theta$ for fringes occuring far above the critical angle suffices to estimate the film thickness t, if a single periodicity can clearly be extracted, from the formula $\Delta\theta = \lambda / (2t)$ with λ being the wavelength of incident radiation. The angular position θ_m of the mth maximum is given by

$$\theta_m = \sqrt{[(m+0.5)\lambda / (2t)]^2 + \theta_c^{\,2}} \tag{1}$$

if the refractive index of the substrate is larger than that of the layer, or

$$\theta_m = \sqrt{[m\lambda / (2t)]^2 + \theta_c^{\,2}} \tag{2}$$

if the refractive index of the substrate is smaller than that of the layer.

The considerations lead to the Fresnel equations for the reflected and transmitted amplitudes and intensities of X-rays.[3] If the angle of incidence is below the critical angle θ_c, the reflected intensity is, even in the absence of absorption, no longer proportional to the volume of the sample but only to the illuminated surface area. In this case X-rays are strictly surface specific probes. In the ideal case of non-absorbing solid with perfectly flat and smooth surface the grazing incidence X-ray reflectivity becomes equal to unity below the critical angle θ_c . The decay of the reflectivity curve for θ_1 above θ_c is approximately proportional to θ_1^{-4} according to $R = \delta^2 / 4\theta_1^{\,4}$. The surface roughness, which drastically reduces the reflectivity above θ_c, has to be taken into account in the Fresnel equations.[4,5,6]

3 Results and discussion

The intensity of the grazing incidence X-ray reflectivity was calculated for several semiconductor II-VI compounds using the program GIXS[7], which applies the recursion relation formalism of the Fresnel equations developed by Parratt[3] in distorted wave Born approximation (DWBA) modified by Sinha et al. [4] and Bowen and coworkers [5].

We started with the investigation of a single thin layer on a substrate. In Fig. 1 an influence of electron densities of the layer and substrate on the X-ray reflectivity for three examples is shown, for which the surface and interface roughnesses are identical and equal to 0.5 nm. In the case, when both electron densities of the layer and substrate are slightly different, e.g. CdTe layer on ZnTe, the obtained curve is almost smooth. If the difference in electron densities is large the Kiessig fringes evidently occur, e.g. for CdTe on Si.

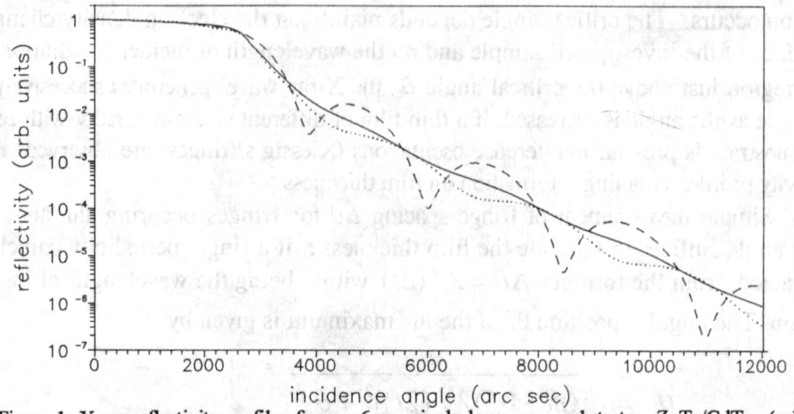

Figure 1: X-ray reflectivity profiles from a 6 nm single layer on a substrate: ZnTe/CdTe (solid line), CdTe/Si (dotted line) and Si/CdTe (dashed line).

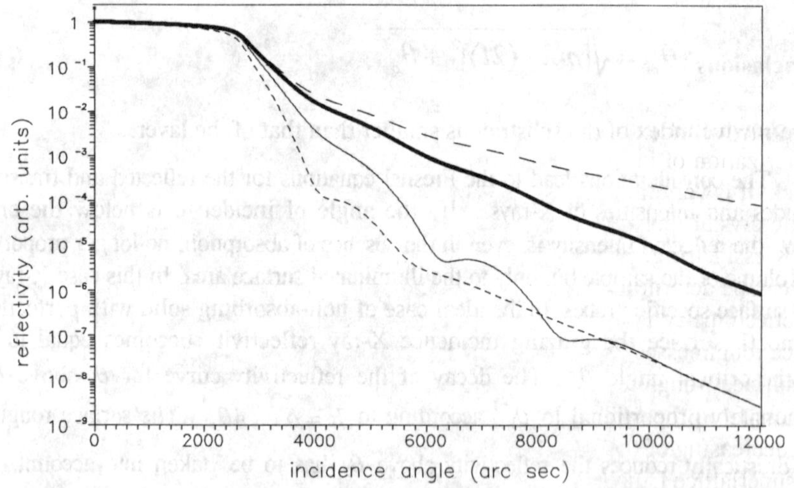

Figure 2: X-ray reflectivity profiles from a 6 nm single CdTe layer on the ZnTe substrate with the constant interface roughness equal to 0.5 nm and varying surface roughness equal to: 0.2 nm (long dashes), 0.5 nm (thick solid line), 1 nm (solid line) and 1.5 nm (short dashes).

The four examples in Fig. 2 present a dependence of the X-ray reflectivity on the surface roughness of the 6 nm CdTe layer on the ZnTe substrate. The curve denoted by a thick solid line is identical with the curve denoted by a solid line in Fig. 1. The substrate roughness is assumed to be constant and equal to 0.5 nm. The surface roughness of the layer is varing from 0.2 nm to 1.5 nm. It is evident that a slope of the reflectivity curve is a probe of the surface roughness of the layer.

In Fig. 3 the simulated X-ray reflectivity curves from CdTe/CdMnTe superlattices are presented. The number of periods, thickness of layers as well as the surface roughness were varied. Multilayers give multiple systems of interference fringes in reflectivity curves. The total thickness of the multilayer is determined by the Kiessig fringe spacing. The waves reflected from individual layers interfere and, as a result, the sharp Bragg peaks are observed for a multilayer. From the positions of Bragg's peaks the superlattice period can be estimated.

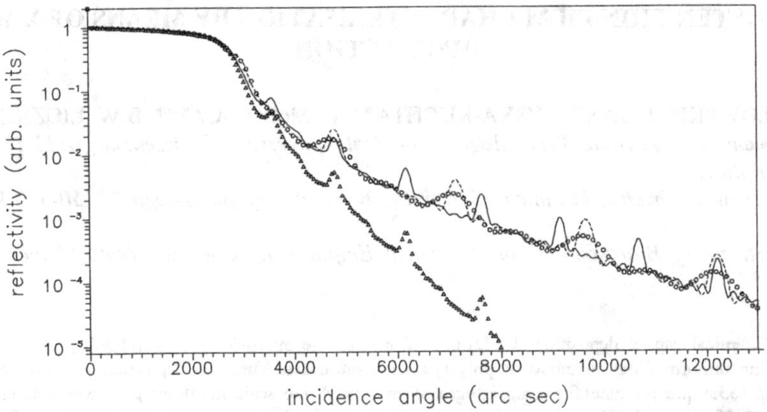

Figure 3: X-ray reflectivity profiles from a CdTe/CdMnTe superlattice on GaAs/ZnTe: 2 periods (6 nm + 6 nm) with 0.2 nm roughness denoted by circles, 4 periods (6 nm + 6 nm) with 0.2 nm roughness denoted by a dashed line, 4 periods (10 nm + 10 nm) with 0.2 nm roughness denoted by a solid line, 4 periods (10 nm + 10 nm) with 0.7 nm roughness denoted by triangles.

4 Conclusions

It was shown that grazing incidence X-ray reflectometry is a powerful technique for characterization of layers and thin films, applicable to films from 1 nm to at least 100 nm thick. It is very important that there are no restrictions on the nature of the sample (a single crystal, polycrystal, amorphous or even liquid sample). Simple measurement of the Kiessig fringe spacing suffices to estimate the film thickness. The superlattice period can be determined from the Bragg peaks positions.

Reflectometry is also a very useful method for investigations of the surface and interface roughnesses of thin films. Whereas microscopic methods allow to determine a local surface roughness, X-ray reflectivity methods give information on average roughness from the large surfaces and interfaces. The surface roughness controls the rate of decreasing of the reflectivity signal, and the interface roughness - the amplitude of the superlattice fringes.

Acknowledgments

This study was partly supported by the KBN project No. 2PO3B 021 09.

References

1. A. H. Compton, *Phil. Mag.* **45**, 1121 (1923).
2. H. Kiessig, *Ann. Phys.* **10**, 715 (1931), *ibid.* 769.
3. L. G. Parratt, *Phys. Rev.* **95**, 359 (1954).
4. S. K. Sinha, E.B. Sirota, S. Garoff and H. B. Stanley, *Phys. Rev.* B **38**, 2297 (1988).
5. D. K. Bowen and B. K. Tanner, *Nanotechnology* **4**, 175 (1993).
6. O. Sakata, A. Y. Nikulin and H. Hashizume, *Jpn. J. Appl. Phys.* **32**, L616 (1993).
7. M. Wormington, *GIXS (Grazing Incidence X-ray Scattering) software*, Bede Scientific Instruments Ltd. and University of Warwick, UK.

TUNGSTEN THIN FILM CHARACTERISATION BY MEANS OF X-RAY DIFFRACTION

J. KOZŁOWSKI*, E. JANKOWSKA-KUCHTA**, C. McCONICA***, B.W. LICZNERSKI*
** Department of Electronic Technology, Technical University, Janiszewskiego 11/17, 50-372 Wrocław, Poland*
*** Department of Physics, Technical University, Wybrzeże Wyspiańskiego 27, 50-370 Wrocław, Poland*
**** Department of Bioresources and Chemical Engineering, Colorado State University, Fort Collins, USA*

Chemical vapour deposition (CVD) is used in both the microelectronics and solar cell industries for depositing thin uniform films on mono or polycrystalline substrates. Tungsten deposition is a commonly used metallization technique for manufacturing integrated circuits. A lab scale nonflowing reactor was build at the Colorado State University (**CSU**) to study chemical vapour deposition reactions and for the purpose of minimising the waste of expensive high purity gases.

This paper presents the results of X-ray diffraction measurements performed on tungsten thin films, deposited on monocrystalline silicon substrate and silicon substrate covered with AlN layer in nonflowing reactor. The obtained structure characterisation was compared with the same parameters measured on the films deposited in commercial CVD flowing reactor (Hewlett Packard Laboratory - **HP**).

X-ray diffractometric data were collected on the Philips MRD* (Materials Research Diffractometer) at the Technical University in Wrocław.

1. Introduction

The traditional method for implementing CVD utilises a flowing reactor which is semibatch. The substrates are loaded and unloaded in batch mode, while the gases flow in and out of the reactor in a steady state fashion. At Colorado State University a lab scale nonflowing reactor was build for the purpose of minimising the waste of expensive high purity gases. This reactor can be filled with exactly the amount of gas which is necessary for the desired film thickness, and the deposition can be run until the reactants are 100% utilised. There is no unnecessary waste. The crystallographic structure of both kinds of tungsten layers (deposited in flowing reactor - HP and nonflowing reactor - CSU) was examined.

2. Sample preparation

The nonflowing reactor was used to deposit tungsten on silicon wafers from tungsten-hexafluoride (WF_6) in hydrogen reduction process [1].

The pieces of silicon substrate and silicon substrate covered with AlN were placed separately in the reactors. Depositions were conducted under the same parameters (reactants, pressure, temperature, time) and controlled via mass-spectrometer.

The following samples were examined:

- the tungsten thin film deposited on Si (commercial, semibatch reactor, HP lab.)
- the tungsten thin film deposited on Si (nonflowing, experimental reactor CSU lab.)
- the tungsten thin film deposited on AlN on Si (nonflowing, experimental reactor CSU lab.)
- the tungsten plate (commercial production)

---

* The purchase of Philips MRD facility has been sponsored by the Foundation for Polish Science.

3. The X-ray diffractometric method

Diffractometric data were all collected using a Philips Materials Research Diffractometer. The parallel beam optics was supplied with CuK_{α} radiation. The optical path is schematically presented in Fig. 1, whereas the basic information about the incident and diffracted beam in Table 1.

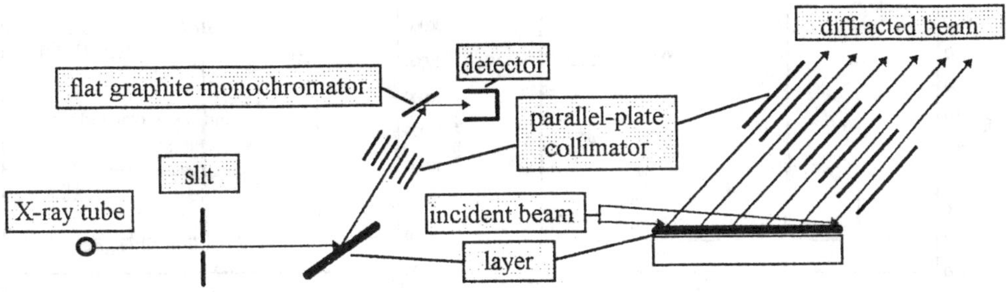

Fig. 1. Schematic diagram of parallel-beam optics.

Table 1. Incident and diffracted beam configurations.

OPTICS	CONFIGURATION	ANGULAR DIVERGENCE
Incident beam	Powder collimator	0.3°
Diffracted beam	Parallel plate collimator Graphite monochromator	0.3° or 0.1°

Two kinds of one-dimensional scans were used:
- the standard theta/2theta
- 2theta scan for the fixed, small angle between the incident beam and the layer's surface.

Additionally the reciprocal space maps were recorded using two-dimensional scan omega, 2theta/omega.
In the first procedure, the omega scan is used to increase the angle between the incident beam and the reflecting planes while leaving the detector in a fixed position. In the second procedure, the 2theta/omega scan, the omega drive (rotating the sample) is used to increase the angle between the incident beam and the reflecting planes while the 2theta drive is used to drive the detector around at the double speed.

4. Results

The diffraction pattern for the reference tungsten-alpha commercial plate is presented in Fig. 2a. It is the typical pattern for polycrystallites with randomly oriented crystallites, which average size is equal to 340A. The stress measured for this plate is small and negative (-250 MPa).
Six commercial HP samples were analysed The results for all of them were very similar and the typical diffraction pattern obtained for one of them using the theta/2theta scan is presented in Fig. 2b. In this sample also the tungsten-alpha phase was identified (card 04-0806) and strong (011) preferred orientation of tungsten crystallites can be visible.

The residual stress inside the tungsten crystallites is positive and rather small (360 MPa). The average crystallite size measured on the basis of the diffraction line broadening was equal to 350A (comparable with the average size of crystallites inside the tungsten plate).

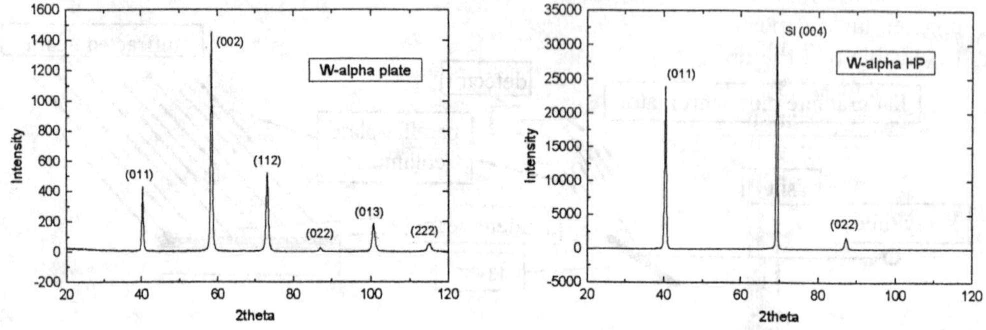

a) b)
Fig. 2 Diffraction pattern of commercial tungsten-alpha plate (a) and tungsten-alpha thin film (HP) (b)

The tungsten thin films deposited directly on the silicon substrate in the nonflowing reactor do not present such a good crystallographic structure as the commercial samples. The diffraction peaks are weaker and more broadened (Fig. 3a). The crystallites are about twice smaller and the stress occurring inside the layers about twice greater.

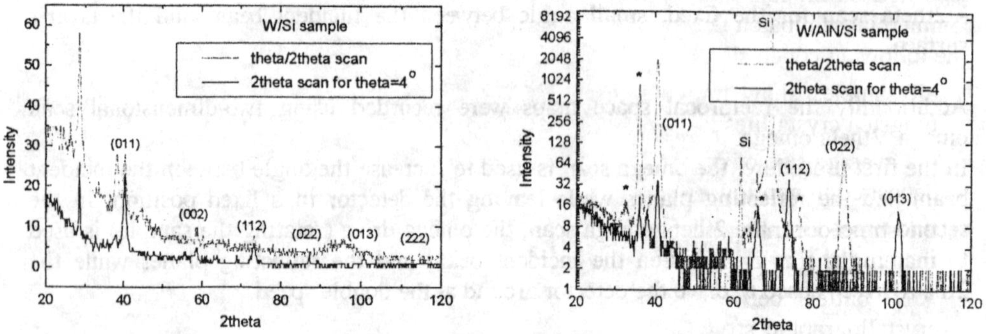

a) b)
Fig. 3 Diffraction pattern of tungsten thin film deposited on silicon substrate (a) and tungsten thin film deposited on AlN on silicon substrate (b) (* AlN peaks)

The tungsten layer deposited on silicon substrate with the AlN buffer shows much better crystallographic structure than the one described before. This layer is highly oriented and (011) direction is dominant The average size of crystallites are slightly greater than for layer deposited directly on Si substrate and stress inside tungsten grains is small and not uniform. That means, that some grains are compressed, and some have extended stresses.

A diffraction space map of the (011) reflection reveals more about the crystallinity (Fig. 4). The horizontal axis (2theta/omega) is the direction of the typical scan and represents variations in „d" spacing for a particular set of lattice planes of defined orientation. The vertical axis (omega) represents a rotation of the sample for a fixed 2theta value. It presents variations in orientation of a particular set of lattice planes of a defined „d" value and is known as „mosaic spread" or „texture".
Fig. 4 shows this map for two kinds of samples: tungsten commercial layer HP and tungsten layer deposited on silicon substrate with AlN buffer in the nonflowing reactor. It is visible that the disorientation of the tungsten crystallites in both samples is similar and equal to +/– 5°.

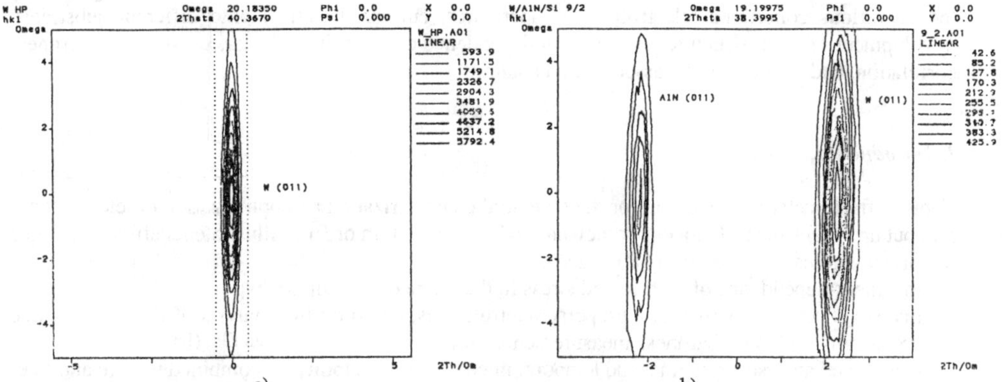

a) b)

Fig. 4 The reciprocal space map at the (011) reflection of the HP layer (a) and W/AlN/Si layer (b).

5. Conclusions

This paper presents the crystallographic analysis of the tungsten thin layers deposited on silicon substrate. Results were compared with the samples produced in commercial semibatch reactor (HP lab) and experimental nonflowing reactor (CSU lab). The following conclusions can be drawn:

- all samples show the tungsten-alpha phase and the preferred orientation of the tungsten crystallites is (011),
- the average size of crystallites is about twice greater in HP samples than in samples deposited in nonflowing reactor,
- the stress occurring inside tungsten grains is about twice smaller in the HP samples than in the similar samples deposited in the nonflowing reactor,
- AlN buffer between the substrate and the tungsten layer improves the crystallographic structure of the layer and reduces its stress making them rather heterogeneous,
- the disorientation of the crystallites in the highly oriented layers (HP and W/AlN/Si) is similar and equal to +/– 5°.

Acknowledgements

This work was supported by the grant from M.Skłodowska-Curie Found II, number MEN/NSF-96-273

Reference

1. D.Bell, K.Baker, E.Kuchta, C.McConica „Batch reactor kinetics, study of tungsten LPCVD from silane and WF_6", J.Electrochem. Soc., Vol 143, No1, 1995, p.269

X-RAY DIFFRACTION ANALYSIS OF MICROSTRUCTURE FORMATION IN ELECTRODEPOSITS

I. HANDREG, P. KLIMANEK, G. LANGE
Freiberg University of Mining and Technology, Institute of Physical Metallurgy,
D-09596 Freiberg, Germany

Advantages of X-ray diffraction methods concerning the structural characterization of electrodeposits are demonstrated in connection with three problems occurring at investigations of Cu-coatings deposited at various conditions (electrolytes, temperatures, current densities) onto different substrates: development of microstructure and stress in dependence on deposit thickness, substructure-hardness correlation, and stability of the as-deposited coating.

1. Introduction

X-ray diffractometry is very useful for the structural characterization and optimization of electrodeposits, but up to now the technique is somewhat underestimated. In order to illustrate its efficiency, three important applications are briefly considered:

- thickness-dependence of texture and stress in the range of 1µm up to 50 µm,
- investigation of substructure - property interrelations, based on the analysis of diffraction - line broadening and microhardness measurements, and
- control and analysis of thermal and temporal microstructure stability by combined texture and line-broadening analysis.

2. Experimental Procedures

Cu-coatings, used in the present work, were deposited from several electrolytes (cyanidic, acid, pyrophosphatic) at various deposition conditions (temperature, current density, additives) onto different substrates (copper, carbon, α-iron and austenitic steel). Detailed information about the deposition conditions can be found in previous papers Handreg et.al.[1,2]. The microstructure characterization of the deposits was carried out by means of optical microscopy (cross section imaging) and conventional (Bragg-Brentano) X-ray powder diffractometry. Quantitative texture analysis was mainly based on three-dimensional orientation distribution functions (Odf´s) calculated from corrected (volume and/or defocussing correction) incomplete pole figures by means of the Bunge [3] formalism. Substructure parameters as, especially, the mean total dislocation densities of the coatings were estimated from the Stokes-corrected Fourier coefficients of the line shapes of selected diffraction via a procedure according to Krivoglaz and Wilkens, proposed by Klimanek [4,5]. Stress measurements where carried out by the $\sin^2\psi$-methode or in the case of sharp (fibre) textures by the utilization of the ideal positions of texture components Lange et.al.[6].

3. Experimental Results and Discussion

3.1 Development of Texture and Microstructure

The microstructure formation depends on the competition of influences due to the substrate (material, surface state and cleanness, texture, grain size) and the deposition conditions (electrolyte composition, temperature, current density) [1,2]. Whereas the texture formation of the initial layers is often determined by the substrat (epitaxial growth), the influence of the electrolyte becomes predominating with increasing thickness (formation of an intrinsic fibre texture) Hnadreg et.al.[1,2,7]. In the same way stress development near the interface is caused by the orientation dependant misfit of substrat and growing deposit. During the electrodeposition process the coating becomes more and more influenced by the deposition conditions [6]. Figure 1 shows the estimated stress development vs. the coating thickness in the case of deposits from a cyanidic electrolyte. The Cu-coatings were electrodeposited within a thickness range from 1μm up to 50μm onto polished and ground surfaces of recrystallized α-iron-sheets. Additionally the (111)-pole figures and micrographs of 3μm and 20μm of the polished surface will be presented in Figure 1 (more about the influence of surface state in [2,7])

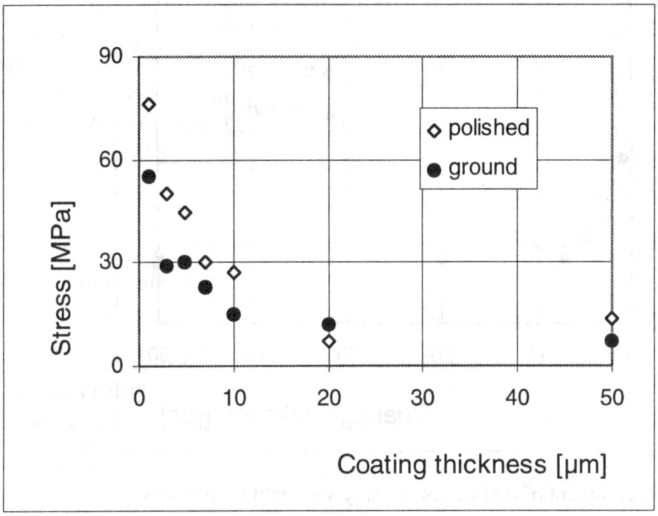

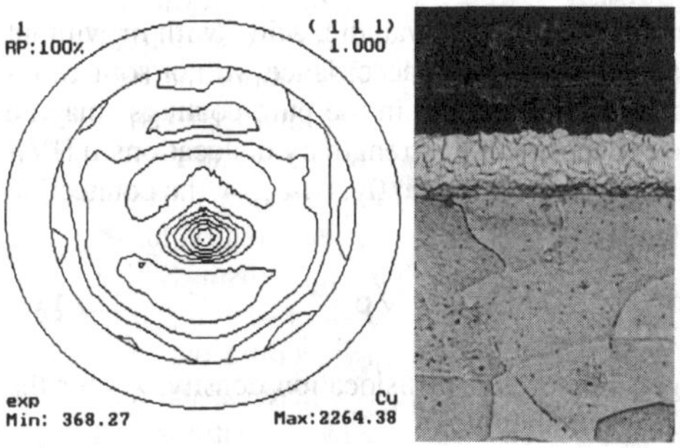

a) d = 3μm (1000x)

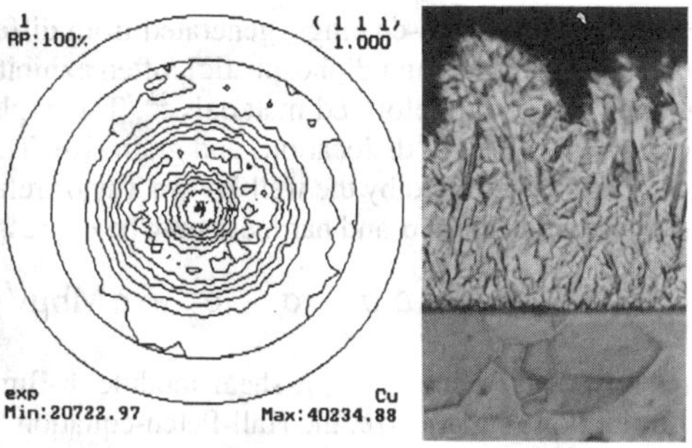

b) d = 20μm (1000x)

Fig. 1: Stress, texture and microstructure development in dependence of deposit thickness

As shown in Figure 1, as well stress, texture as microstructure alter significantly with the coating thickness. The formation of the highest stresses near the substrate surface may be explained in connection with epitaxial growth ({122}<012> component) by the orientation dependant misfit between iron and copper [7]. An enhanced thickness results in an orientation change into a <110>-fibre texture accompanied by a weakness in stress.
To interpret the origin of stresswithout the substrat influence, that means only by the deposition conditions (above d ≈ 20μm), it is usefully to look at the development of the dislocation density (Fig.

2). As it becomes visible from Figure 2 there is a similarity between stress and dislocation dependence on thickness, which refers to common origins of stress und dislocations. Therefore, the stress induced by the deposition process may be explained by the theory of crystallite coalescence, that means at certain distances between two nucleates very high adhesion forces are appear and will cause a spontaneous joining of them. According to nucleate distances the accompanied lattice distortion may be elastic or plastic and result in the formation of stress and/or dislocations, respectively [6].

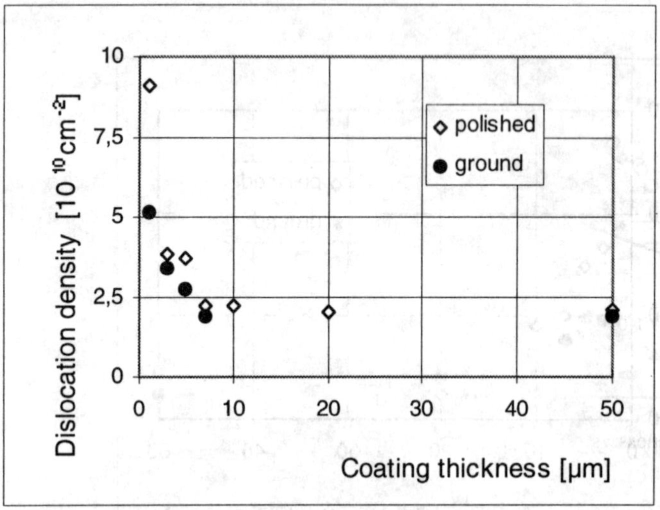

Fig. 2: Development of dislocation density vs. coating thickness

3.2 Correlation between Hardness and Substructure

Electrodeposited Cu-coatings, generated from different electrolytes (cyanidic, acid - with or without hardening additives and diphosphatic), often exhibits microhardness in accordance with or sometimes above heavy cold-deformed materials [2,8]. The explanation of hardness in the pure coatings may be found in the effect of dislocations and grain size. The enhancement of strength by dislocations $\Delta HV\rho$ could be characterized by the well-known Taylor-relation [8] and consequently, based on the connection between yield stress σ and hardness may be expressed by (1):

$$\Delta\sigma = \sigma - \sigma_0 = \alpha M b \mu \sqrt{\rho} \qquad \Delta HV_\rho \sim \sqrt{\rho} \tag{1}$$

α=const., M-Taylor factor, μ-shear module, b-Burgers-vector and ρ-dislocation density. As for the contribution of grain size, the Hall-Petch-equation (2) is commonly:

$$\sigma = \sigma_0 + \frac{k_y}{\sqrt{D}} \tag{2}$$

k_y = const., D - grain size.Figure 3 shows the correlation of microhardness and dislocation density according to (1) for electrodeposited Cu-Coatings and cold-rolled copper as well as the theoretical Taylor curve.

Whereas the values of the cold-rolled material with a grain size of D ≈ 40µm can be approximate by the Taylor curve, the microhardness of the electrodeposits always exceeds the theoretical ones. The regression curve exhibits a comparable loop, but is shifted about $\Delta H \approx 80$. This deviation is caused by the influence of grain size. As already shown, hard coatings posses in addition to the high dislocation density very small grain sizes often in the order of D≈ 0.1μm up to 1μm Schläfer et.al.[9]. In accordance with the literature data concerning the hardening by small grain sizes, a mean contribution of $\Delta H \approx 80$ by grain size effect appears justifiable in view of the scattering of the measured values in Figure 3.

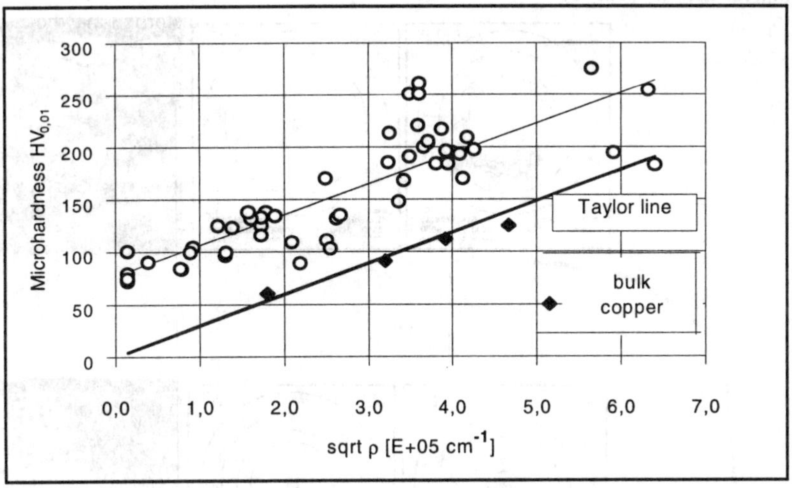

Fig. 3: Microhardness vs. square root of dislocation density

3.3 Thermal and Temporal Stability

Conventional techniques of microstructure investigation (e.g. optical microscopy or microhardness measurements) are unsuitable concerning the examination of the phenomenon for various reasons (small layer thickness, microcrystallinity, preparation requirements etc.), but by X - ray powder diffractometry it can be studied nondestructively and quickly. Considering the microstructure as a mixture of an as-deposited (d) and a recrystallized (r) phase and taking into account that the integrated intensity of an X-ray reflection I*(hkl) is related to the volume fraction Δv of the scattering crystallites by the expression:

$$\Delta v \sim \frac{I^{*}(hkl)}{K(hkl)}; \qquad K(hkl) = W(\theta)\ |F(hkl)|^2\ A(\theta)\ p(hkl) \tag{3}$$

($W(\theta)$ - angle-dependent part of the polarization and Lorentz factor, $|F(hkl)|^2$ - structure factor, p_{hkl} - multiplicity of the lattice planes {hkl}, $A(\theta)$) - absorption (volume) correction) the recrystallized volume fraction X of the coating can be determined in the simple case of a two-component fibre texture as a function of the time t and/or the temperature T via the equation (d - as-deposited, r - recrystallized material)

$$X = X(T,t) = \frac{I^{*r}(hkl)}{I^{*d}(hkl) + I^{*r}(hkl)} = \frac{I^{*r}(200)}{I^{*d}(111) + I^{*r}(200)} \tag{4}$$

Fig. 4 shows an example of a recrystallization curve estimated by X-ray as well as the corresponding texture and microstructure development for a diphosphatic Cu-coating (deposited at 60°C) onto iron, which were altered drastically by recrystallization even at room temperature [1,10,11].
Independent on the annealing temperature (investigated up to 180°C) the process is accompanied by a texture transition from a very strong <111> - fibre texture of the as-deposited state to an <001> recrystallization texture (4). As well the microstructure changes from the fine as-deposited columnar

grains into recrystallized globular ones without twins. The evaluation of such curves contributes to the explanation of recrystallization processes in electrodeposits. Due to the special microstructure characteristics (dislocation content, grain size, texture) coatings deposited at various conditions show different thermal and temporal stability (Tab. 1).

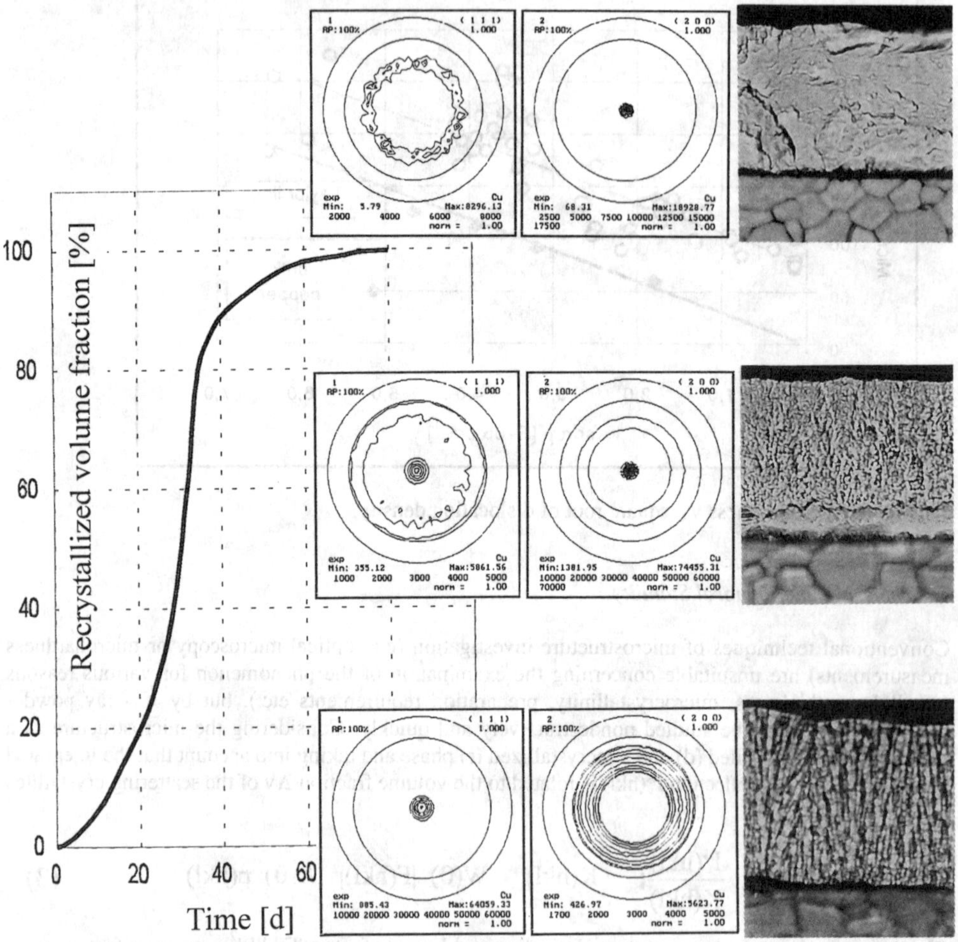

Fig. 4: X-ray recrystallization curve and texture-microstructure development

Tab. 1: Characteristics of electrodeposits from several electrolytes

Electrolyte	T[°C]/ i[A/dm^2]	dislocation density ϱ_d [10^{10}cm^{-2}]	Texture (↑ strong ↓ weak)	Microhardness $HV_{0,01}$
cyanidic	60/ 2	≈3	<110>↓	125
acid	25/12	≈8	<110>↑	160
acid with hardening additive (0.6ml/l)	25/12	≈21	<111>↓ + <311>↓	230
diphosphatic	60/3	≈5 - 30	<111>↑↑	120-280

Contrary to diphosphatic deposits, coatings obtained from acid or cyanidic electrolytes at room temperature, no microstructure changes could be detected by optical microscopy and X-ray texture analysis after a storage of four years or after annealing for several hours up to 180°C. First, there were found a critical dislocation density required for the start of recrystallization processes of about $\varrho_d^c \approx 2*10^{10}cm^{-2}$ and a strong <111>-fibre texture in the as deposited state. Secondly, if the dislocation density is located below $\varrho_d \approx 2*10^{11}cm^{-2}$, the recrystallization process becomes slower and more incomplete (X<<1) [1,10].

Acknowledgments

This work has been supported by the Arbeitsgemeinschaft Industrieller Forschungsvereinigungen e.V. (AIF) of Germany.

References

1. Handreg, I. and Klimanek, P.: Jahrbuch Oberflächentechnik Bd. 51 (1995), Edited by A. Zielonka and R. Schwan (Metall Verlag, Heidelberg) 243-278
2. Handreg, I.: Dissertation, Freiberg University of Mining and Technology (1997)
3. Bunge, H.J.:Theoretical Methods of Texture Analysis, DGM Informationsgesellschaft, Oberursel (19987)
4. Klimanek, P.: J. de Phys. IV, Colloque C7, supp. au J. de Phys., Vol.3(1993), 2149-2154
5. Klimanek, P. : Habilitation thesis, Freiberg University of Mining and Technology (1992)
6. Lange, G., Klimanek, P., Handreg, I.: Metalloberfläche, 51(1997)2, pp. 127 - 132
7. Handreg, I., Klimanek, P., Lange, G., Schneider, M.: Materials Science Forum, Vols. 228-231 (1996), pp. 463 - 468
8. Lange, G., Klimanek, P., Barthel, Th.: Proc. of Dreiländertagung 9-11/Okt.1996, Frierichshaven, Germany, pp. 72-75
9. Schläfer, S., Klimanek, P., Handreg, I., Heinzel, G., Lange, G. : Mat. Science Forum, Vols. 228-231 (1996), pp. 457 - 462
10. Handreg, I., Klimanek P. and Lange, G. (1996): Jahrbuch Oberflächentechnik (1996) Bd. 52, Edited by A. Zielonka (Hüthig Verlag, Heidelberg) 283-310
11. Handreg, I., Klimanek, P., Lange, G., Weiland, E.: Proc. of the 4th EUROMAT 95 (Symposium D), 25-28 September (1995), Padua/Venice, Italy, pp. 471-474

COMPUTER SIMULATION OF X-RAY SCATTERING BY SURFACE LAYERS WITH CONCENTRATION GRADIENT

R.I. BARABASH, S.I. SIDORENKO, A.M. LAKHNIK

Metal Physics Department, National Technical University of Ukraine, pr. Pobedy, 37, Kiev, 252056, Ukraine.

In the framework of kinematic theory of scattering the intensity distribution formed by surface layers with concentration gradient of some elements was developed. We have worked out a computer program to restore X-ray intensity distribution by such surface layers in radial direction (parallel to reciprocal lattice vector). Computer simulation allows to find out diffusion coefficient even in this case. Procedure was applied to analyse different surface layers as to nitriding of iron alloys as to oxidation of HTSC layers.

1. Introduction

During different types of treatment in the surface layers of the material there is formed concentration gradient of one or several elements due to diffusion processes at the surface. It occurs also during oxygen evaporation from superconductive ceramics $YBa_2Cu_3O_{7-x}$ and can cause phase transformation of superconductive to nonsuperconductive phase. Another example is heat-chemical treatment of materials with a formation of protecting layers on its surface. Concentration gradient usually occurs in thin films. Peculiarities of X-ray or neutron scattering by such layers with concentration gradient were analyzed in the number of works [1-6]. For a simple case when the existence of concentration gradient only changes interatomic distance in the solid solution of the surface layer or thin films and thus displaces the positions of reciprocal lattice points depending on the depth of the layer main expression describing X-ray scattering were obtained in [7-10]. General kinematic theory of X-ray scattering by thin films and surface layers taking into account appearing distortions fields was worked out in [11].

Below we use the approach developed in [7-11] for modelling X-ray scattering intensity distribution under different conditions of diffusion in the surface layers. We have worked out the procedure of diffusion coefficient determination by means of numerical methods based on comparison of modelled and experimental X-ray scattering distributions.

2. Theoretical Description of X-Ray Scattering by Solid Solution with Concentration Gradient

Following the approach developed in [7-9] let us describe the diffusion process in the surface layer by well known diffusion equation:

$$\frac{\partial C}{\partial \tau} = D\frac{\partial^2 C}{\partial x^2} \tag{1}$$

Let us also consider, that:

• In the initial moment concentration of interstitial element in the whole sample is equal zero and that at the surface of the sample ($x=0$) concentration C_0 all the time is supported to be constant that means, that:

$$C|_{\tau=0}=0;\ C|_{x=0}=C_0 \tag{2}$$

• Thickness of the sample is considered to be such, that there always exist layers in which the concentration of diffusion element is all the time equal zero. It means that depth of the layer x can be supposed to be changing in the interval $(0,\infty)$ and:

$$\lim_{x\to\infty} C(x,\tau)=0,\ \tau>0 \tag{3}$$

• Vegard law is valid $d=d_0+\gamma C$, (d_0 - initial interatomic distance, γ - proportionally coefficient). The solution of such task is described by the equation:

$$C(x,\tau)=\frac{2}{\sqrt{\pi}}C_0\{1-\operatorname{erf}(\frac{x}{2\sqrt{D\tau}})\} \tag{4}$$

X-ray scattering intensity from different parts of the layer with changing concentration gradient is weakened according to the law:

$$I=I_0e^{-\mu l} \tag{5}$$

μ - is linear weakening coefficient which is supposed to be independent of the concentration of diffusing element in solid solution. Taking into account the geometry of X-ray scattering last expression can be written as following:

$$I=I_0exp(-2\mu x/sin\ \theta)=I_0exp(-\mu' x), \tag{6}$$

where $\mu'=4\mu d/n\lambda$. We considered the case when monochromatic radiation was used, so that λ=const. The resulting intensity distribution can be found by integrating over the depth of the sample:

$$I(\theta)=A\int_0^\infty \delta(\theta-\theta(x))e^{-\mu' x}dx, \tag{7}$$

(A - proportionally coefficient). As the change of interatomic distance in the matrix is usually small in comparing with its value, than: $d\theta=-\alpha dC$, $\alpha=\gamma tg\theta_0/dC$. Taking into account properties of δ-function we may find the expression for the intensity of X-ray scattering by surface layer with concentration gradient:

$$I(\theta)=\frac{A\sqrt{\pi D\tau}}{\alpha C_0}e^{-\mu'^2D\tau}\exp\left\{\left(\frac{x(\varphi)}{2\sqrt{D\tau}}-\mu'\sqrt{D\tau}\right)^2\right\}\Bigg|_{\varphi=\theta_0-\theta} \tag{8}$$

Last expression shows the dependence of X-ray line profile on surface diffusion coefficient and can be used to work out the procedure for determination of diffusion coefficient in the surface layers and thin films. As was shown first in [7] such intensity distribution has single minimum point corresponding to:

$$x_{min}=2\mu D\tau. \tag{9}$$

This regularity results in the appearance of two maximum on both sides of this minimal point. The increase in time of the process causes displacement of scattering angle corresponding to this minimal intensity point θ_{min} to the angle θ_0:

$$\theta_{min} = \theta_0 - \alpha C_0 \frac{2}{\sqrt{\pi}} \int_{\mu'\sqrt{D\tau}}^{\theta_0} e^{-\tau^2} d\tau \xrightarrow[\tau \to \infty]{} \theta_0 \tag{10}$$

3. Modelling of Real X-Ray Intensity Distribution

Real experimental curve of intensity distribution I^* is a result of convolution of standard intensity distribution $f(\theta)$ and theoretical one $I(\theta)$ which was described above by equation (8):

$$I(\theta) = \int_{-\infty}^{\infty} f(z-\theta) I(z) dz \tag{11}$$

Taking into account that $I(z)$ is not equal zero only in the interval $[\theta_0, \theta_0 + \alpha C_0]$ we obtain:

$$I(\theta) = \int_{0}^{\alpha C_0} f(z-\theta) I(z) dz \tag{12}$$

For Gauss function describing standard intensity distribution:

$$f_\sigma(\theta) = \frac{1}{\sigma\sqrt{2\pi}} \exp\left\{ \frac{-(\theta - \theta_0)^2}{2\sigma^2} \right\} \tag{13}$$

According to [8] the resulting intensity distribution is folllowing

$$I_\sigma(\theta) = \int_0^{\alpha C_0} \frac{1}{\sigma\sqrt{2\pi}} \exp\left\{ -\frac{(z-\theta)^2}{2\sigma^2} \right\} \frac{A\sqrt{\pi D\tau}}{\alpha C_0} \exp\left\{ \frac{x^2(z)}{4D\tau} - \mu' x(z) \right\} dz \tag{14}$$

Using these formulas we have modelled X-ray intensity distribution for several values of diffusion parameters (Fig. 1). The broadening parameters of a standard line profile influences the resulting intensity distribution. From simulated line profiles for different values of diffusion parameters it follows that distribution always consists of two maxima. Such X-ray profiles were experimentally observed for surface layers and thin films with concentration gradient in a great number of papers [1-8]. The higher is αC_0 value the more distincdy these two maxima can be observed (Fig. 1c). When the diffusion proceeds the point of minimal intensity dividing the resulting intensity distribution into two parts is displaced to the θ_0 position (Fig. 1). It is in agreement with the analytical expression for intensity (10). With the diffusion time increase the integral intensity (I_1) of the part of the line profile situated near the initial Bragg peak position gradually decreases while the integral intensity (I_2) of the peak localised around αCo position increases (Fig. 1a, b). Analogous effect is caused by the increase of the diffusion coefficient value (Fig. 1d). Such intensity redistribution was used as a basis for diffusion coefficient determination.

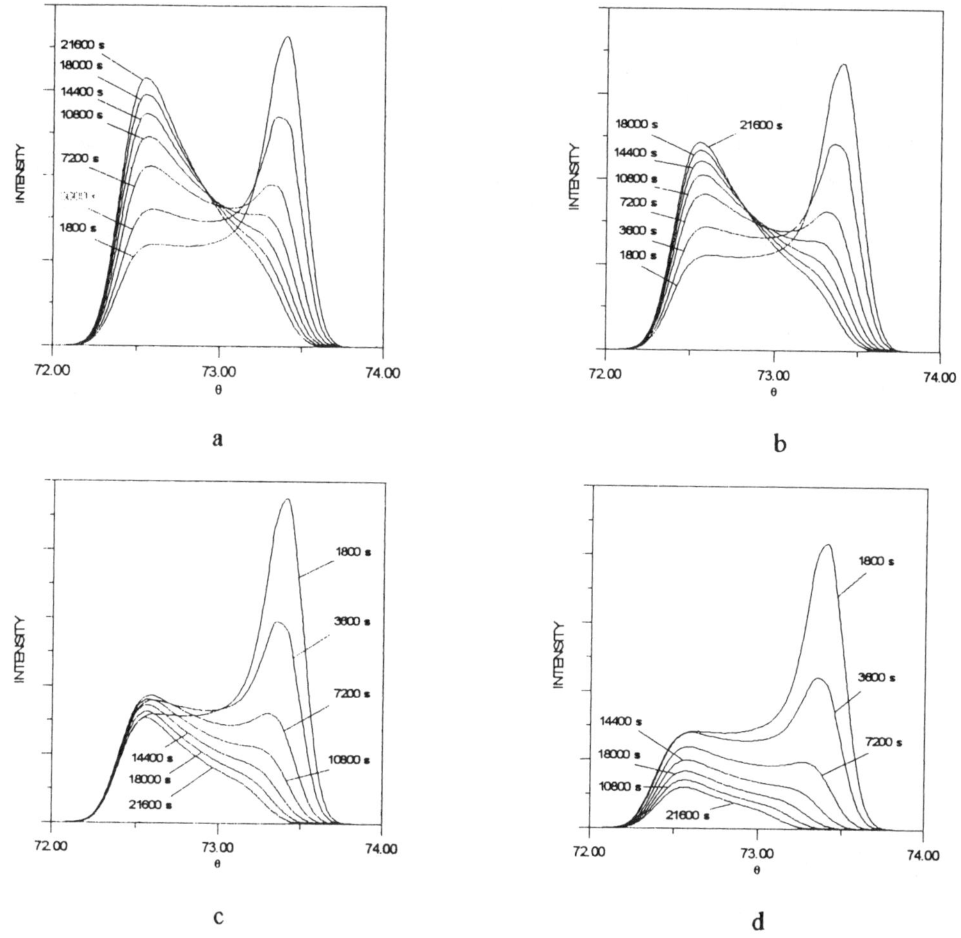

Figure 1: Computer simulation of intensity distributions corresponding to different β ($D=0.5\cdot10^{-10}$ cm^2/s; $\sigma=0.1^0$; $X_{max}=0.3\cdot10^{-2}$ cm.):

a) $\beta=0$; $\alpha C_0=1.00^0$; b) $\beta=0.5\cdot10^{-7}$; $\alpha C_0=1.00^0$;
c) $\beta=0.275\cdot10^{-6}$; $\alpha C_0=1.00^0$; d) $\beta=0.5\cdot10^{-6}$; $\alpha C_0=1.00^0$.

Sometimes as a result of heterodiffusion several layers corresponding to different phases according to equilibrium diagrams are formed at the surface of the sample. In that case a diffraction line is often a result of the interference of diffraction peaks belonging to different phases. In that case resulting intensity distribution becomes still more complicated and the point of minimal intensity will be difficult to observe

4. Numerical Calculating of Diffusion Coefficient

The described approach was used for determination of diffusion coefficients by means of comparison of experimental X-ray intensity distribution and a simulated one. We have worked out an algorithm and a computer program for a model described by espressions (1-9) and supposing that standard line profile corresponded to the Gauss function. In this algorithm we used numerical methods for determination of the minimum of the functional dependence $\sigma = f(D)$, (σ -mean square deviation of the simulated intensity distribution from the experimental one. Calculations were finished after value σ_i corresponding to i-th iteretion for D_i diffusion coefficient became less or equal certain value ε. Then diffusion coefficient value D will be equal to the value D_i determined after i iterations. Program was used to determine diffusion coefficient by analysing X-ray intensity distribution simulated for known diffusion coefficient $D=0.5 \cdot 10^{-10}$ cm^2/s. The calculated value D_i coincided with the real one with the good accuracy.

5. Conclusions

1. X-ray intensity profiles were simulated for different values of surface diffusion conditions parameters in the framework of kinematic theory of scattering.
2. The resulting intensity distribution formed by surface layer with concentration gradient always consists of two maximum.
3. Increase of surface diffusion time of diffusion coefficient value results in the redistribution of X-ray intensity of the resulting line. Minimal intensity point being displaced to the side of initial Bragg reflection. Dependence of X-ray intensity distribution on surface diffusion coefficient value is analysed.
4. An algorithm and a computer program to determine surface diffusion coefficients was worked out and the value of diffusion coefficient value is analysed.

References

1. Fisher B., Rudman P.S., *J.Appl.Phys.* 32 (1961) 1604.
2. Pines B. Y., Chaikovskiy E.F., *DAN SSSR* 111 (1956) 1234.
3. Fogelson R.L., *Fiz. Met. Matallov.* 25 (1968) 492.
4. Houska C.R., *J.Appl.Phys.* 41 (1970) 69.
5. Tenney D.R., Carpenter J.A., Houska C.R., *J.Appl.Phys.* 41 (1970) 4485.
6. Delhez R., Keijser Th. H. de, Mittemeijer E.J., *Surf. Eng.* 2 (1987) 331.
7. Barabash R.I. and etc., *Fiz. Met. Metallov.* 30 (1970) 379.
8. Barabash R.I., Belotskii A.V., Permyakov V.G., *Metallofizika* 45 (1973) 72.
9. Barabash R.I. *Doklady of Ukrainian Academy of Science* (1980) 88.
10. Barabash R.I., Belotskii A.V., Karasevskaya O.P., *Metallofizika* 4 (1982) 103.
11. Barabash R.I. Krivoglaz M.A., *In "X-ray and Neutron Structure Analysis in Materials Science"* (Plenum Press, NY-London, 1989) p.161.

OPTIMISING HEXAPHENYL THIN FILM QUALITY FOR OPTO-ELECTRONIC APPLICATIONS BY HEAT TREATMENT

R.Resel, W.Graupner, C.Hochfilzer, N.Koch, F.Meghdadi, S.Tasch, M.Wohlgenannt, G.Leising
Institut für Festkörperphysik, Technische Universität Graz, Austria

K.Reichmann
Institut für Chemische Technologie anorganischer Stoffe, Technische Universität Graz, Austria

The conjugated organic material hexaphenyl ($C_{36}H_{26}$) is used for various opto-electronic applications; e.g. as blue light emitter in electroluminescent devices. Hexaphenyl films are prepared by vacuum deposition on ITO substrates. Heat treatment in vacuum and in Argon atmosphere does not influence the preferred orientation of the crystallites, only a small fraction of amorphous deposited hexaphenyl changes to a crystalline state and the FWHM of the X-ray reflections decreases. The heat treatment decreases the photoluminescence quantum yield and the life time of long living photoexcited states. Both is assigned to higher order of the hexaphenyl chains. Heat treatment increases the roughness of the hexaphenyl film surface.

Introduction

Due to their semiconducting properties conjugated organic molecules are potential candidates for opto-electronic applications. Hexaphenyl is an interesting material as active layer in future flat panel displays[1]. Thin films of hexaphenyl are prepared by physical vapour deposition. Preferred growth of the crystallites is observed. The direction of the preferred growth depends on the sample preparation conditions: at low substrate temperatures (T_S) and high deposition rates (DR) the long axes of the hexaphenyl molecules are parallel to the substrate; at high T_S and/or low DR the hexaphenyl chains are perpendicular to the surface of the substrate[2].

It has been shown that heat treatment causes a change from the amorphous to a crystalline state in phenyl-amine films[3] and that thermal annealing provides a better performance of a thin film transistor based on hexathiophene[4,5]. Within this paper we study the change of the film properties due to heat treatment in order to improve device performance.

Experimental Details

The samples were prepared by evaporation of hexaphenyl onto ITO coated glass substrates kept at room temperature. The DR and the average thickness (d) was chosen to be different: DR=30Å/sec., d=7000Å for the film heat treated in vacuum (sample VAC) and DR=40Å/sec., d=4000Å for the film treated in an argon atmosphere (sample ARG). The first heat treatment of both samples was performed at 50°C for 4h, then in steps of 20°C for VAC and in steps of 50°C for ARG, time for each step was 4h. The heat treatment was performed up to the limit of full desorption of the hexaphenyl. X-ray experiments were performed in focussing Bragg-Brentano geometry, using Ni-filtered CuK_α radiation. The diffraction peaks

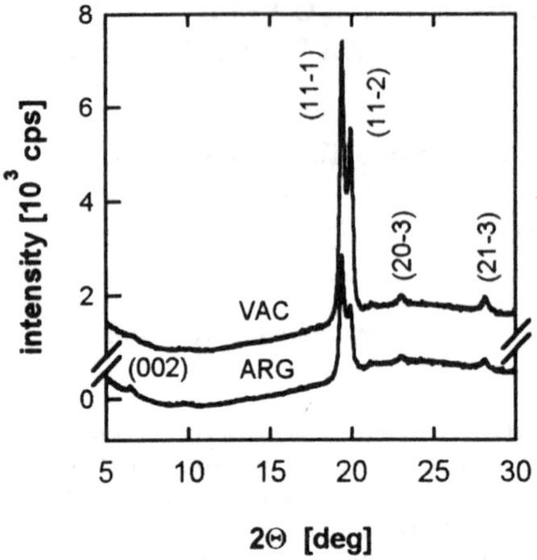

Fig.1: Θ/2Θ scans of the hexaphenyl thin films in the as prepared state.

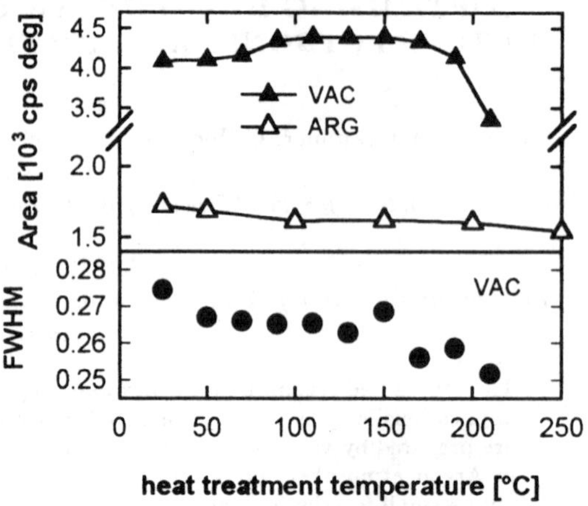

Fig.2: Area of the (11-1), (11-2) peaks (above) and FWHM of the sample VAC (below).

were fitted, determining FWHM and the area under the peaks. A Nanoscope III was used for the atomic force measurement. The experimental setup of photo induced absorption and photo luminescence quantum yield (PLQY) measurements are described elsewere [6].

Experimental Results

The samples were investigated by X-ray diffraction as prepared; Θ/2Θ scans of the samples VAC and ARG are depicted in Figure 1, for clarity the curves are vertically shifted. Based on previous results [2], we conclude that the observation of the dominant (11-1) and (11-2) reflections reveals a preferred growth of hexaphenyl crystallites with the (11-2) plane parallel to the substrate. In comparison with the calculated powder pattern too low intensity is observed at the (20-3), (21-3), (002) reflections. Therefore, only 6% (VAC) and 8% (ARG) of the crystallites are randomly oriented.

Due to heat treatment no additional peaks develop. No change of the orientation of the hexaphenyl chains from parallel to the substrate to a perpendicular direction occurs. Such a reorientation was observed in thin films of molecules consisting of alkyl chains [7,8]. These molecules are flexible in contrast to rigid rod molecules like hexaphenyl or hexathiophene where reorientation of the molecules may be hindered due to the rigid rod nature of the molecules. However, some influence in the arrangement of hexathiophene molecules relative to the substrate was observed by rapid thermal annealing close to the melting point [5].

A more detailed analysis of the Θ/2Θ-scans was done by fitting the (11-1) and (11-2) peaks. Additional to the (11-1) and (11-2) peaks a further peak is detected for both measurement series located between (11-1) and (11-2) at 2Θ=19.70°. This peak cannot be attributed to the crystal structure observed by Baker [9]. However, a polymorph would have its main peak at 2Θ=19.60° [10]. In order to detect a change

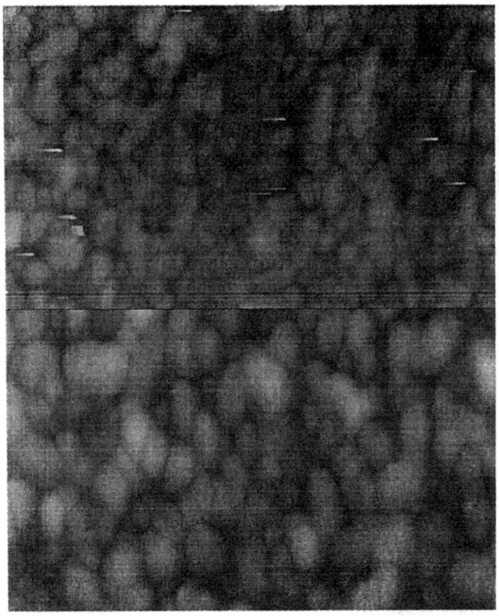

Fig.3: AFM analysis of the sample VAC as prepared (above) and after treatment at 210°C (below); 2x1μm are depicted.

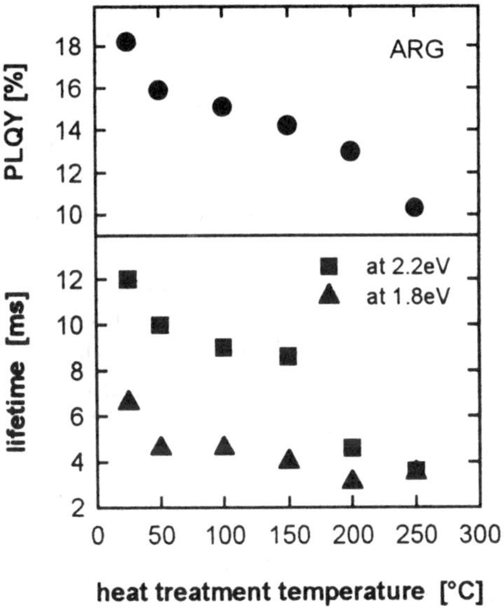

Fig.4: photoluminescence quantum yield of ARG (above) and lifetime of the long living photoexcited states at 2.2, 1.8eV.

of the hexaphenyl material from the amorphous state to a crystalline state by heat treatment, the area under the peaks is analysed as a function of the heat treatment temperature; the result is depicted in Figure 2. In case of the sample VAC the area under the peaks first slightly increases and then rapidly decreases, for the sample ARG it decreases over the whole temperature range. For both samples the total decrease is about 10%. This behaviour of both samples is referred to loss of hexaphenyl material by desorption from the substrate. The samples VAC and ARG show linebroadening of the (11-1) and (11-2) reflections. The heat treatment causes a decreasing FWHM like it is depicted for the sample VAC in Fig.2.

AFM measurements were taken from the sample VAC after each heat treatment. The results for the film in the as prepared state (top) and the film after the last annealing step (bottom) are shown in Figure 3, in both cases a region of 2x1μm is shown. The shape of the elevated regions does not change by heat treatment, however, the sizes change considerably. The average size of the elevated regions increases from 0.07μm at the as prepared state to 0.14μm after the last annealing step. Deep valleys between the elevated regions can be seen; the observed maximum depth is 250Å for the sample in the as prepared state and 500Å at 210°C.

In addition to the structural characterisation we have also performed measurements of the PLQY and the lifetime of long living photoexcitations at different wavelength[6] after each heat treatment step – the results are summarized in Fig. 4.

Discussion

Slow deposited films with an average thickness of 1500Å show strong structured surfaces with height differences up to several thousand Å[2]. Taking the sample thickness of 7000Å into consideration, the roughness of our hexaphenyl film surface is quite small. The roughness of the surface increases by heat treatment; the valleys between the elevated regions become considerably deeper. Similar observations were made on hexathiophene thin films. Rapid thermal annealing close to the melting point leads to major discontinuities in the film, cracks and holes develop[4,5].

We conclude that only a small part of amorphous hexaphenyl material crystallises due to heat treatment. Due to decreasing FWHM a better formation of the crystalline state is observed. The decreasing FWHM upon thermal treatment is in correlation with the PLQY measurements. This confirms the assumption that increased crystallinity leads to lower PLQY due to a specific intermolecular interaction between the molecules in the ordered regions[11]. This interaction obviously enhances the nonradiative recombinations. The fact that both lifetimes of the excited states absorbing at 1.8 and 2.2eV decrease monotonically with more elevated temperatures upon heat treatment clearly shows that the photoexcited states are stabilized by defects. Furthermore the concentration of these stabilizing defects can be decreased by moderate temperatures which implies that they are of structural and not chemical nature. The fact that order can influence the observed photoexcitations was recently shown for films of oligothiophene[12].

Acknowledgement

We thank the Austrian Fonds zur Förderung der wissenschaftlichen Forschung under the project number 9747-CHE.

References

1. S.Tasch, C.Brandstätter, F.Meghdadi, G.Leising, G.Froyer, L.Athouël, Adv. Mater., **9**, (1997) 33
2. R.Resel, N.Koch, F.Meghdadi, G.Leising, W.Unzog, K.Reichmann, Thin Solid Films, (1997) in press
3. E.M.Han, L.M.Do, N.Yamamoto, M.Fujihira, Mol. Cryst. Liq. Cryst., **267**, (1995) 411
4. L.Torsi, A.Dodabalapur, A.J.Lovinger, H.E.Katz, R.Ruel, D.D.Davis, K.W.Baldwin, Chem. Mater., **7**, (1995) 2247
5. A.J.Lovinger, D.D.Davis, R.Ruel, L.Torsi, A.Dodabalapur, H.E.Katz, J. Mater. Res., **10**, (1995) 2958
6. http://www.cis.tu-graz.ac.at/if/homepage/index.html
7. K.Fukao, H.Kawamoto, T.Horiuchi, K.Matsushige, Thin Solid Films, **197**, (1991) 157
8. M.Yamanaka, K.Sato, K.Inaoka, K.Yase, Jpn.J.Appl.Phys., **31**, (1992) L1632
9. K.N.Baker, A.V.Fratini, T.Resch, H.C.Knachel, W.W.Adams, E.P.Socci, B.L.Farmer, Polymer, **34**, (1993) 1571
10. A.Kawaguchi, M.Tsuji, S.Moriguchi, A.Uemura, S.Isoda, M.Ohara, J.Petermann, K.Katayama, Bull.Inst.Chem.Res., Kyoto Univ., **64** (1986) 54
11. J.Cornil, A.J.Heeger, J.L.Bredas, Chem. Phys. Lett., **272**, (1997) 463
12. P.A.Lane, X.Wei, V.Vardeny, Phys. Rev. Lett., **77**, (1996) 1544

SURFACES, COATINGS AND THEIR RESIDUAL STRESSES - NEW METHODOLOGICAL APPROACH

S.J. SKRZYPEK
Faculty of Metallurgy and Materials Science
University of Mining and Metallurgy,
Al. Mickiewicza 30, 30-059 Kraków - Poland

The residual stress (RS) becomes more and more important (investigated) property of the surface because of their influence on surface's and coating's properties. This is for the reason of its superposition with loading and important influence of RS on physical, chemical structural and microstructural phenomena e.g. stress induced phase transformation, piezoelectricity, linear relation of resistivity versus RS and stress-induced self-disintegration etc.

Although most methods of measurements of residual stresses are still under development they become more and more accurate. The non-destructive diffraction methods of measurement of various kind of residual stresses are suitable for surface examinations. The most popular $\sin^2\psi$ diffraction method elaborated for Bragg-Brentano geometry has one important fault i.e. penetration depth changes vs. the tilt angle ψ. This is particularly important for RS of coatings and surface layers with big gradients of RS. The new version of the $\sin^2\psi$ method designated as g-$\sin^2\psi$ due to grazing angle scattering geometry was applied to study residual macrostresses of TiN coatings.

1.Appearance and evaluation of residual stresses (RS).

The surface is responsible for thin film nucleation and growth, for passivation layers and for adsorption and chemisorption. The surface and an adjacent region should be considered as a controlling factor in wear and fracture of materials, also in fatigue, creep and stress-corrosion cracking and also as a quality factor of machining. The importance of the surface is manifested by two groups of factors i.e. mechanical features of the surface, such as roughness, waviness, etc. and the presence of thin films, layers and coatings. These films were already investigated in 1934 by Roscoe [1] who found the influence of the oxide film on flow stress of cadmium and by Cottrell and Gibbons in 1948 [2]. Generally the strengthening and weakening effects are due to surface layers. Both are connected with surface energy and its influence on thermodynamical processes between the surface and environment. **Any additional energy of the surface, e.g. in the form of residual stress can amplify the non-equilibrium state since the surface becomes more and more the place of active interaction with environment and with external forces.**

The magnitude and distribution of residual stresses play a key role in welds, heat-treated and plastically deformed materials in composites and in thin films and coatings. The RSs affect the fatigue strength, stress corrosion resistance and other stress-induced phenomena as well e.g. stress induced diffusion, phase transformation, piezoelectric effect and stress influences on electrical resistance. Long-term experiments in natural environment where samples were exposed to the atmosphere showed that longitudinal cracks in non-loaded specimen appeared as a result of the existing surface residual stresses [3]. There are known events where machine components destroy themselves and they can even self-disintegrate to a powder [4]. The tensile RS reduces indentation loop energy in HfN films [5].

The total stress σ_t, carried by material is sum of load stress σ_l, residual macrostress σ_m, and microstress of particular grains σ_μ. Where, σ_μ consist of σ_d and σ_g

i.e. σ_d - stress field around crystal defects and σ_g - stress field due to misfit between grains.

$$\sigma_t = \sigma_m + \sigma_\mu + \sigma_l,$$

Residual stresses can be introduced in two levels, i.e. in the macro and microscale. Residual macrostresses, also called the "first- order" ones are caused by the elastic interaction between different areas of the sample in terms of macroscale e.g. a surface layer in respect to bulk of the sample. Microstresses regarded at the grain-size scale are also called "the second order" stresses. Some authors also distinguish "the third-order" stresses or submicrostresses which appear around crystal lattice defects. Most authors recognise only residual macro and micro-stresses, however we should be aware of their origin in terms of microstructure and crystal lattice. The grains are subjected to different elastic deformation caused by incompatibilities/misfitting in their boundary regions. Other sources of macro and microstrains, i.e. inhomogeneity of chemical composition and defect density and anisotropy of the elastic constants should be considered as well.

2.The residual stresses generation in PVD/CVD processes

Deposited surface layers usually have different chemical and phase composition with respect to the bulk of material. Due to different physical and chemical properties they create residual strains/stresses caused by a temperature change. The simplest stress describing of the surface is the biaxial stress state which concerns only surface and very thin films. In this case the residual stress can be calculated according to the following formula [6]:

$$\sigma_c = \Delta(\alpha_c - \alpha_s) * E_c * (T_t - T_m)$$

where;

α_c - the thermal expansion coefficient (TEC) of coating and,
α_s - the TEC of metal substrate
E_c - the Young modulus of coating
T_t - the temperature of the deposition process
T_m - „ of measurement

In the case of metal A deposited on thick metal B the TEC's of these metals must be taken into account in residual stress evaluation. The temperature-induced stresses on the A-B boundary are an important properties of the connection between them. The strain induced by a change in temperature from T1 to T2 can be calculated as;

$$\varepsilon_1 = \varepsilon_2 = \Delta\alpha\Delta T,$$

$$\varepsilon_3 = \frac{2\Delta\alpha * \Delta T * \nu}{\nu - 1}$$ (despite of its own expansion/contraction).

For the plain stress state we can get the stress from Hooke`s law;

$$\sigma_1 = \sigma_2 = \Delta\alpha\Delta T * E/(1-\nu), \quad \sigma_3 = 0$$

The difference in TEC for the layers, precipitations, inclusions and the lattice defects of the matrix are the main sources and relaxation mechanisms of RS. The wider consideration in this field were provided by Oettel and Wiedemann [7]. Toghether with above appraoch they should explain discrepancies between calculated and measured RS.

The PVD process is carried out at 400-500 °C and CVD at 900 °C. RS created by this temperature range can be evaluated using above equations (Fig.1).

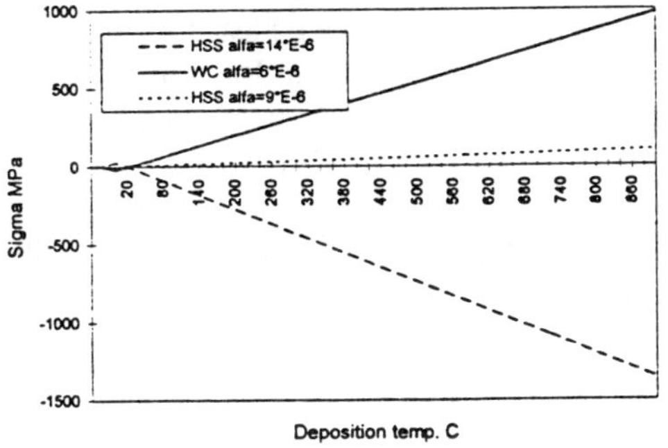

Fig. 1. Theoretically evaluated RS's for TiN coating deposited on high-speed steel (HSS) and on WC carbide.

3. Methods of residual stresses measurement

There are a number of methods in measuring the residual stresses, but the most important ones are based on diffraction of X-rays and neutrons.

The main method, called $\sin^2\psi$ [8], has been elaborated for Bragg-Brentano geometry and can be applied to "ω" and "ψ" goniometers. In this geometry the different penetration depth of measurement versus the tilt angle ψ appears (Fig.2). It implicates problems in measurement of RS in the case of its gradients. The importance of RS gradients in physically vapour-deposited coatings was recognized by Perry et al.[10] and Rickerby et al.[12].

Fig.2. Effective penetration depth for the standart $\sin^2\psi$ (a), and for g-$\sin^2\psi$ method (b), TiN, $\lambda CuK_{\alpha 1}$, Gx-the fraction of total diffracted intensity.

Unfortunately there are a lot of discrepancies in results obtained by X -ray diffraction methods and this is due to difficulties in interpreting of diffraction pattern in particular geomeries and methods are still under development. The diffraction methods are based on measurement of elastic strains of crystal lattice and they are applied as standard methods. On thc other hand, anisotropy of crystals and texture cause some difficulties for diffraction methods [15]. The most universal diffraction method „$\sin^2\psi$" enables measurement of directions and values of main stresses and anizotropic values of the mechanical compliances E_{hkl} and υ_{hkl} and the so-called diffraction elastic constants

(DEC) as well [9]. The diffraction methods have several important features such as a non-destructive character, a wide range of penetration depth (very small for X-ray and quite large for neutrons). They can be applied to isotropic characterisation and they are non-standard as well. Diffraction methods can measure any stress component in the space of the sample. The general principle of diffraction methods are used for stress measurement is the expression of strain through interatomic distances d_{hkl} measured by suitable diffraction technique.

There is a general assumption describing surface stresses as plain stresses which can be measured by the simple θ/2θ diffraction measurement according to the following formula which needs the standard d_0 value;

$$(\sigma_1 + \sigma_2) = -\nu / E * (d_{hkl} - d_0) / d_0$$

here; d_0 - the interplenar distance of (hkl) of the unstressed phase.

By linking Hook's equations for plain state of stresses with ellipsoid of strains in new co-ordinates L_i (Fig.3) we can derive the principal equation of $\sin^2\psi$ method;

$$\varepsilon_{\varphi_i \psi_j} = 1/2\, S_2\, \sigma_{\varphi_i} \sin^2 \psi_j + S_1 (\sigma_1 + \sigma_2) = (d_{\varphi_i, \psi_j} - d_0) / d_0$$

$1/2S_2 = (1+\nu)/E$ and $S_1 = -\nu/E$ are DECs (E -the Young modulus and ν-the Poisson ratio)

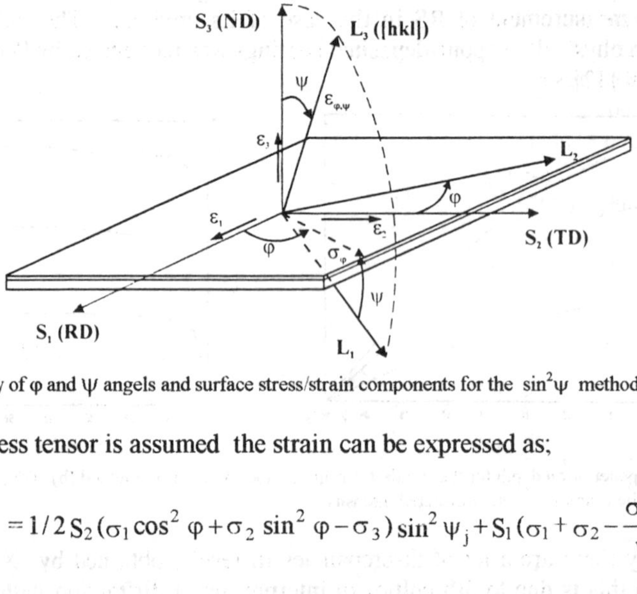

Fig.3. The geometry of φ and ψ angels and surface stress/strain components for the $\sin^2\psi$ method.

When space stress tensor is assumed the strain can be expressed as;

$$\varepsilon_{\varphi_i \psi_j} = 1/2\, S_2 (\sigma_1 \cos^2 \varphi + \sigma_2 \sin^2 \varphi - \sigma_3) \sin^2 \psi_j + S_1 (\sigma_1 + \sigma_2 - \frac{\sigma_3}{\nu})$$

Hence the measurement of $d_{\varphi,\psi}$ is necessary for calculating appropriate elastic strains of crystal lattice. The next step in calculation for the isotropic distribution of the crystallite orientations is an approximation of set of points ($d^{i}_{\varphi,\psi}$ - $\sin^2\psi(i)$) by a linear function using the least squares method. The stress can be calculated using a derivative of the main equation, since the value of d_0 can be eliminated, which makes this approach a non-standard method;

$$\sigma_{\varphi} = (\partial \varepsilon_{\varphi, \psi_i}) / \partial \sin^2 \psi_i * 1/(1/2 S_2)$$

For anisotropic solids the calculation procedure introduced by Baczmański et al. [15] needs the elastic constants for monocrystals and texture.

A new approach in measurement of the RS enables the grazing angle scattering (GAS) geometry of diffraction given by Quaeyhaegens [11]. Some additional elaboration, solutions and examples will be introduced. The main advantage of the application of this geometry to the RS measurement consist in almost constant penetration depth versus the tilt angle ψ (Fig.2). Hence distribution of measurement points due to RS gradients can be lowered. The calculation procedure is common for both geometries, although many features are quite different. Therefore for clear distinguishing both measurements , addition of „g" letter to traditional name i.e. „g-sin $^2\psi$" is proposed.

4. Material and experimental procedure

Two kind of substrates were aplied i.e. sintered high speed steel ASP60 and sintered WC. TiN coatings were deposited by PVD and CVD techniques [13,14]. In the case of PVD, ASP sinters were heat treated before, wheras PVD coated sinters were heat treated after deposition. Diffraction patterns were recorded with a XRD-7 diffractometer for 6 deg of the grazing angle what gave penetration depth very near to the coating thickness (Tab.1).

5.Examples of measurements

The test of this methodical approach was done by the RS measurement on TiN coatings deposited on sintered high speed steel and WC. The elaborated technology was done due to international research project [13] and is also described in paper [14]. Using above procedure and average elastic constant $1/2S_2 = 1.84*10^{-6}$ MPa^{-1} for TiN [11] the RS were measured (Tab.1).

Tab.1.Residual stresses in TiN coatings on various substrates. The error should not exceed 30% with significance level α=0.1 for g-sin$^2\psi$ and is larger for sin$^2\psi$ method. T15-High Speed Steel (HSS), 32-HSS ASP60, U30-WC carbide.

Sample no. phase	stress MPa (g-sin$^2\psi$)	lattice par.[nm]	deposition proc. thickness [um]	stress MPa (sin$^2\psi$)
T15,TiN(95)	1142		CVD/4.5-5.2	410
32,TiN (95)	-1040		PVD/3.7-4.7	-545
U30,TiN (95)	2809	0.4238	CVD/4-5.2	849
T15,Ti(CN) X	2558		CVD/3.5-5.5	
T15,Ti(CN) IX	3963		„	
HSS,(TiAl)N P1(97)	1374	0.4175	CVD	

4.Results and conclusions

1.Because of texture and gradients of the RS the standart $\sin^2\psi$ gives greater oscilations of the measured points giving larger errors.
2.Application of the g-$\sin^2\psi$ method for the RS measurement in coatings is profitable due to avoiding above problems.
3.Heat treatment of HSS (substrate) after CVD deposition complitly changes the RS in TiN.
4. Application of the grazing angle scattering geometry anables measurements at constant penetration depth in the wide range of the θ and ψ angles.

Acknowledgements

This work has been partialy founded by EU PECO Program (grant CIPA 92-3032) and by University of Mining and Metallurgy, Krakow, - Grant # 11 110 230. The author wishs thank Dr Carl Quaeyhaegens for the opportunity to perform some measurements.

References

1. R.Roscoe, *Nature*, London 133,(1934),912.
2. A.H.Cottrell, D.F.Gibbons, *Nature,* London 162,(1948), 488.
3. H.Haselmair, *Mat.Perform.* (1992), 60-63.
4. A.Wojtysiak, S.J.Skrzypek, *Archiwum Nauki o Materiałach*, (1990), 11,45.
5. R.Nowak,Y.Horino, Y.Ando, S.Maruno , *Appl.Phys.Lett.* 68 (26), (1996),3743-45
6. H.E.Evans, Inter. Mat.Rev. 40, (1995),1-40
7. H.Oettel, R.Wiedemann, *Surface and Coat.Technology,* 76/77, (1995), 265-273
8. E.Macherauch, H.Vohlfahrt, *HTM* 28,(1973),3, 201-211.
9. S.J.Skrzypek, J.Karp, K.Chruściel, *Przegląd Mechaniczny* z.20, (1989), 13-15.
10. A.J.Perry, J.A.Sue, P.J.Martin, *Surface and Coat.Technology,* 81,(1996), 17-28
11. C.Quaeyhaegens, G.Knuyt, L.M.M.Stals, *Surface and Coat. Technology*, 74/75, (1995),104-109.
12. Rickerby, A.M. Jones, B.A. Bellamy - *Surface and Coat. Technology*, 37, (1989), 111-137
13. FINAL REPORT of the Joint Research Project CIPA-CT92-30-23:" PVD and CVD Coated PM Cutting Tool Materials JRP", Project funded by the European Comission under PECO programme, Brussel 1997. Coordinator A.S.Wroński, Task 3-Coating Characterisation p.34-49.
14. S.J.Skrzypek, A.Ciaś, Makronaprężenia własne warstw TiN nakładanych na spiekane kształtki stali szybkotnacych. Międzynarodowe Seminarium „INZYNERIA POWIERZCHNI '97, Warszawa 12-14 maja,1997, s.75-80, Inst.Mech.Prec.
15. A.Baczmański, K.Wierzbanowski, S.J.Skrzypek, Residual Stresses in Textured Polycrysalline Material- Method for Experimental Determination. Proc. of the XVI Conf. Applied Crystallography, Cieszyn, Poland, 22-26 August 1994, p.111-115. World Scientific Singapore, N.Jersy, London, Editors; H.Morawiec, D.Stróż

DIFFRACTION CHARACTERISTICS OF SOME PHYSICAL TECHNOLOGIES FOR STEEL CUTTING

N. GANEV, G. GOSMANOVÁ, I. KRAUS
Faculty of Nuclear Sciences and Physical Engineering, Czech Technical University in Prague, Břehová 7, 115 19 Prague 1, Czech Republic

Residual stresses in a carbon and a tool steel after some unconventional technologies of cutting were investigated by means of X-ray diffraction. Stress measurements were performed in both the austenite and the ferrite phases simultaneously. Significant stress gradient was identified in surface layers of α-Fe.

1. Introduction

The assessment of surface characteristics of metals after various technologies of cutting is, as a rule, reduced only to their roughness description. Information about the residual stress state is very sporadic, especially in the case of unconventional machining [1]. At the same time even the residual stresses play an important role for service life of various tools and preparations (punching and stamping die edges etc.) produced by these technologies.

The type and magnitude of residual stresses generally depend on the material treated and on the parameters of the cutting technology applied. The character of stresses on the workpiece surface and beneath it affects its functionality. Compressive stresses increase fatigue limit and improve wear resistance, whereas tensile stresses decrease fatigue strength and make destruction of frictional surfaces easy.

2. Unconventional technologies of machining

Physical (unconventional) methods of machining are implied in the following chipless technologies: electro discharge machining, water jet and abrasive stream cutting, cutting by laser beam, electron beam machining, plasma beam machining, ion beam machining, electro-chemical machining, chemical machining and ultrasonic machining.

Considering that these methods are usually used as final technologies of machining, the precise determination of residual stresses can play an important role in influencing the performance of the cut surface.

Some specific features of unconventional technologies of machining:

- possibility of treating heat-resistant materials,
- possibility of using for production of complicated shearing and forming tools,
- using in production of sintered carbide cutting tools,
- manufacturing of clean-cut miniature details,
- for most of these technologies the machinability of materials does not depend on their mechanical properties, but it is related with their physical properties,
- heat loading of working pieces is minimal.

3. Investigated samples:

▸ Samples prismatical in shape $4\times10\times15$ mm^3.Cutting surfaces of steel ČSN 19 096 after electro discharge cutting using both the graphite and copper electrode: rough cut (096-CR, 096-CuR specimens); fine (clean) cut (096-CF, 096-CuF specimens). The cutting surfaces of steel ČSN 19662 were designated by analogy with the above ones.

▸ Cutting surfaces 096-W, 662-W of streels ČSN 19096, ČSN 19662, respectively, after water jet cutting.

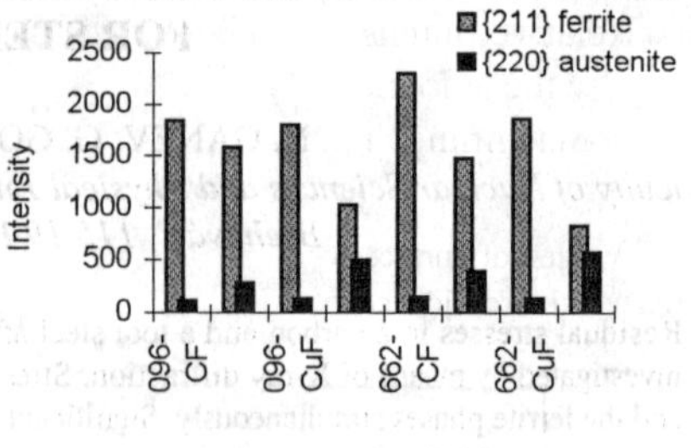

Figure 1 Comparative plot of relative intensities of the {211}-ferrite and {220}-austenite diffraction lines.

4. Diffraction techniques used:

• Diffractometer DRON UM1 with secondary monochromator and CuKα radiation was used for phase analysis.

• Stress analysis was performed in the longitudinal (σ_L) and transversal (σ_T) direction by means of an ω-goniometer SIEMENS with CrKα radiation. Calculation of surface stresses and their gradients within the penetration depth of applied radiation was based on the shift of the centroid of ferrite {211} and of austenite {220} diffraction lines and X-ray elastic constants $\frac{1}{2}s_2$ {211}= $5.76.10^{-6}$ MPa^{-1}, $\frac{1}{2}s_2$ {220}= $6.05.10^{-6}$ MPa^{-1}, $-s_1${211} = $1.25.10^{-6}$ MPa^{-1}, $-s_1${220} = $1.56.10^{-6}$ Mpa^{-1} [2].

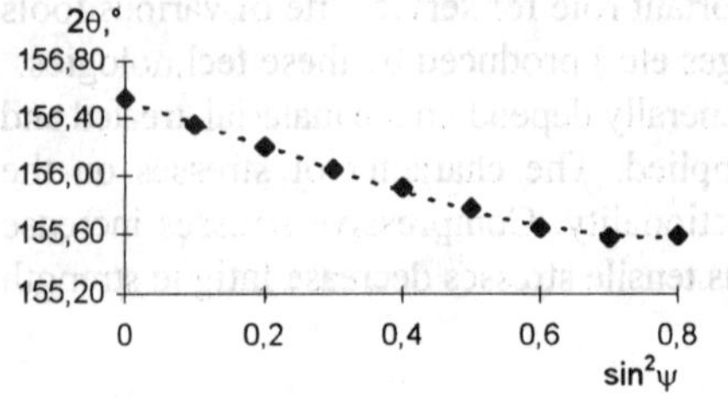

Figure 2 Distribution of 2θ vs. $\sin^2\psi$ obtained in ferritic phase of the 662-CR sample

5. Cutting technologies :

a) Electro discharge machining

Graphite electrode: roughing draught V_w=70mm^3.min^{-1},
fine cut draught V_w=2mm^3.min^{-1},

Copper electrode: roughing draught V_w=1.5mm^3.min^{-1},
fine cut draught being unmeasurable.

b) Abrasive water jet cutting

Water jet was 0.3mm in diameter, feed v_f=30mm.min^{-1}; water pressure p=275 MPa, the flow rate being 2.27l.min^{-1}; abrasive material - SiC beads (0.2-0.3)mm in diameter, flow rate Q_a=(100-250)g.min^{-1}.

6. Results of measurements:

Phase analysis:

Besides the ferritic phase, the austenitic phases Fe-C or Fe-Cu were identified on cuts investigated after electro discharge cutting with a C or Cu electrode. The FeN phase is also present, resulting from high temperature of the cutting process.

Residual stress analysis

a) Relative intensities of the ferrite {211} and austenite {220} diffraction lines characterise the amount of both the phases in diffracting volume of surface layers after electro discharge machining. They are illustrated in Fig.1.

Table 1 Values of surface residual stresses $\sigma(0)$ and gradients g_{11}, g_{33} in the ferritic phase of the samples investigated; the confidence interval is 90%

Sample	$\sigma(0)$, MPa	g_{11},MPa.μm^{-1}	g_{33},MPa.μm^{-1}
096-CF	327.7 ± 70.0	22.1 ± 22.0	1.9 ± 5.1
096-CR	191.4 ± 47.2	54.8 ± 14.9	0.7 ± 3.4
096-CuF	401.0 ± 49.4	28.3 ± 15.5	0.8 ± 3.5
096-CuR	23.4 ± 57.9	75.1 ± 18.2	0.2 ± 4.2
662-CF	500.1 ± 26.2	0.3 ± 8.3	0.5 ± 1.9
662-CR	103.6 ± 68.3	77.9 ± 21.5	2.3 ± 4.9
662-CuF	460.2 ± 55.6	19.5 ± 19.5	0 ± 4.0
662-CuR	-172.5 ± 57.1	100.1 ± 18.0	0.9 ± 4.2
096-V	-489.3 ± 18.3	7.7 ± 7.7	-0.3 ± 1.3
662-V	-572.0 ± 16.5	17.4 ± 5.2	-0.3 ± 1.2

b) Tables 1and 2 show the values of residual stresses calculated from interplanar spacings d_{211} (ferrite) and d_{220} (austenite).The results obtained for three-axial stress state in the surface layers of ferritic phase (surface stress $\sigma(0)$ and gradients g_{11}, g_{33} corresponding to $\sigma_{11} = \sigma_{22}$, σ_{33} respectively) are in Table I; the evaluation was based on the measurements for $\sin^2\psi$ from 0 to 0.8 (see Fig.2). Table 2 includes the results for biaxial stress state derived from experimental data $\sin^2\psi \in [0,0.5]$ for ferrite and $\sin^2\psi \in [0,0.6]$ for austenite. Values $\boldsymbol{\sigma}$ and sums of principle stresses $\boldsymbol{\sigma_1} + \boldsymbol{\sigma_2}$ present the average residual stress in a surface layer containing diffracting crystallites.

7. Conclusions:

Stress analysis of electro discharge cut samples:

• A three-axial stress state isotropic in a plane parallel to the surface was found in the ferritic phase; the gradient $g_{11} = g_{22}$ is statistically significant.
The stress of γ-Fe was biaxial isotropic in a plane parallel to the surface.

• All the surface (Table 1) and average (Table 2) residual macroscopic stresses in both the α-Fe and the γ-Fe are tensile except σ(0) of the sample 662-CuR.

• Statistical plausibility of average stresses in the austenitic phase corresponds with the detected intensities of {220}-diffraction lines (Table 2, Fig. 1)

• Gradients g_{11} of the surface stress component $\sigma_{11} = \sigma_{22}$ of the α-phase are steeper in the case of rough cutting than in the surface layers of the clean cuts.

Table 2 Average residual stresses σ and sums of principle stresses $\sigma_1 + \sigma_2$ in both the austenite and the ferrite phases of the samples investigated; the confidence interval is 90%

Sample	α-Fe		γ-Fe	
	σ_φ,MPa	$\sigma_1 + \sigma_2$,MPa	σ_φ,MPa	$\sigma_1 + \sigma_2$,MPa
096-CF	467.2±15.2	935.0±21.3	1014.1 ± 253.2	1892.3 ± 355.3
096-CR	471.1±23.9	942.9±33.2	587.0 ± 79.9	1118.8 ± 111.7
096-CuF	540.0±15.3	1080.5±21.3	817.7 ± 324.0	1539.0 ± 453.1
096CuR	415.8±26.1	830.9±36.5	383.4 ± 64.8	773.9 ± 90.9
662-CF	496.6±11.9	992.7±16.6	847.6 ± 191.6	1665.5 ± 267.9
662-CR	486.4±31.1	972.9±43.4	996.8 ± 135.4	1979.3 ± 189.2
662-CuF	562.6±18.7	1125.6±26.1	956.8 ± 181.8	2219.8 ± 258.8
662-CuR	342.2±40.9	684.5±56.9	746.6 ± 79.6	1555.3 ± 111.3
096-V	-445.8±7.1	-891.3±9.9	-	-
662-V	-447.4±10.5	-954.9±14.6	-	-

Stress analysis of water jet cut samples:

• A three-axial stress state isotropic in a plane parallel to the surface was found; the gradient $g_{11} = g_{22}$ is not statistically significant.

• Only compressive surface and average residual stresses were found.

References

1. M.A.Ayad, I.Kvasnička, Acta Polytechnica 34, 1/1994, 59
2. I.Kraus and N.Ganev: Diffraction Analysis of Mechanical Stresses, Czech TU in Prague, Prague 1995

This research is a part of the research project supported by the Grant Agency of the Czech Republic (Grant №101/96/1181)

INVESTIGATIONS OF MACROSTRESSES BY X-RAY INTEGRAL METHOD

D. SENCZYK
Poznań University of Technology, Institute of Materials Engineering, X-Ray Laboratory,
M. Skłodowskiej-Curie 5, 60-965 Poznań, Poland

The aim of this researches is non-destructive investigations the changes of residual macrostress tensor components in the technological surface layer of grinded constructional steel 0.55%C using integral method and x-ray diffractometer made by Rigaku-Japan. an influence of grinding parameters on stress tensor components in surface layer of polycrystalline materials has been showed (e. g. increasing depth of cut increases the residual normal macrostresses, cooling decreases the tensile residual macrostresses., principal stresses in grinding are tilted around an axis transverse to the grinding direction).

1 Introduction

The aim of this researches is non-destructive investigations the changes of stress tensor components in the technological surface layer of 0.55% carbon constructional steel.

The strain $\varepsilon_{\varphi\psi}$ measured in the direction, which forms the azimuth angle φ with the normal to the specimen surface and whose projection onto the specimen makes the angle ψ with the X axis of sample, is related to the strain tensor ε_{ij} by the equation:

$$\varepsilon_{\varphi\psi} = \varepsilon_{11}\cos^2\varphi\sin^2\psi + \varepsilon_{12}\sin 2\varphi\sin^2\psi + \varepsilon_{22}\sin^2\varphi\sin^2\psi +$$

$$+\varepsilon_{13}\cos\varphi\sin 2\psi + \varepsilon_{23}\sin\varphi\sin 2\psi + \varepsilon_{33}\cos^2\varphi. \quad (1)$$

Residual strain $\varepsilon_{\varphi\psi}$ can be determined from a change in interplanar spacing $\mathbf{d}_{\varphi\psi}$ of the diffracting planes, i. e.,

$$\varepsilon_{\varphi\psi} = \frac{d_{\varphi\psi} - d_o}{d_o}, \quad (2)$$

where d_o is the interplanar spacing of the unstrained material.

It is possible to determine the strain tensor ε_{ij} as a function of the azimuth angle φ for fixed values of angles ψ. This method is called the integral method.

Since x-ray penetrate the specimen surface, we measure the strain $\langle\varepsilon_{\varphi\psi}\rangle$ averaged over the penetration depth:

$$\langle\varepsilon_{\varphi\psi}\rangle = \langle\varepsilon_{11}\rangle\cos^2\varphi\sin^2\psi + \langle\varepsilon_{12}\rangle\sin 2\varphi\sin^2\psi + \langle\varepsilon_{22}\rangle\sin^2\varphi\sin^2\psi +$$

$$+\langle\varepsilon_{13}\rangle\cos\varphi\sin 2\psi + \langle\varepsilon_{23}\rangle\sin\varphi\sin 2\psi + \langle\varepsilon_{33}\rangle\cos^2\psi, \quad (3)$$

where $\langle \varepsilon_{ij} \rangle$ is given by:

$$\langle \varepsilon_{ij} \rangle = \frac{\int_0^z \varepsilon_{ij} \exp\left(-\frac{z}{\tau}\right) dz}{\int_0^z \exp\left(-\frac{z}{\tau}\right) dz}, \tag{4}$$

where z is depth and τ is absorption factor.

Absorption factor for Ω-diffractometer can be written as:

$$\tau = \frac{\sin^2 \Theta - \sin^2 \psi}{2\mu \sin\Theta \cos\psi}. \tag{5}$$

Factor μ is linear absorption factor and Θ is diffraction angle.

Absorption factor τ depends only the tilt angle ψ (when diffraction angle Θ and linear absorption factor μ is constant), it becomes advantageous to develop $\langle \varepsilon_{\varphi\psi} \rangle$ as a function of φ:

$$\langle \varepsilon_{\varphi\psi} \rangle = \frac{1}{2} A_o + A_1 \cos\varphi + A_2 \cos 2\varphi + B_1 \sin\varphi + B_2 \sin 2\varphi, \tag{6}$$

where:

$$A_o = \left[\langle \varepsilon_{11} \rangle + \langle \varepsilon_{22} \rangle\right] \sin^2 \psi + 2\langle \varepsilon_{33} \rangle \cos^2 \psi, \tag{7}$$

$$A_1 = \langle \varepsilon_{13} \rangle \sin 2\psi \tag{8}$$

$$A_2 = (1/2)\left[\langle \varepsilon_{11} \rangle - \langle \varepsilon_{22} \rangle\right] \sin^2 \psi, \tag{9}$$

$$B_1 = \langle \varepsilon_{23} \rangle \sin 2\psi, \tag{10}$$

$$B_2 = \langle \varepsilon_{12} \rangle \sin^2 \psi\ , \tag{11}$$

and A_n, B_n Fourier series coefficients can be determined by the relations:

$$A_n = \frac{1}{\pi} \int_0^{2\pi} \langle \varepsilon_{\varphi\psi} \rangle \cos n\varphi d\varphi, \tag{12}$$

$$B_n = \frac{1}{\pi} \int_0^{2\pi} \langle \varepsilon_{\varphi\psi} \rangle \sin n\varphi d\varphi. \tag{13}$$

2 Experimental

In this paper residual macrostress tensor measurements were made on a the surface layer of grinded constructional steel 0.55%C using integral method and x-ray diffractometer Strainflex PSF-1M made by Rigaku-Japan. The profiles of the (211) diffraction line were measured with filtered CrKα radiation in the step scanning mode using a scintillation counter. The data were taken as a function of the azimuth angle φ varying from 0° to 360° in steps of 15° at the fixed angles ψ varying between 0° and 45° in steps of 9°. The measurements of x-ray stress method have been done using our own computer programs „Acquisition" and „Stress tensor".

During grinding, the near-surface region of the workpiece is deformed plastically. Heating, due to the lack of cooling or high downfeed, produces tensile stresses because locally-heated regions are upset by the cooler surroundings. Thus, the stress pattern, is altered by many parameters, such as the depth of cut, cooling (or lack) and cutting direction.

3 Results

Some experimental results are shown in Table 1 and Table 2.

Table 1: Some experimental results of stress tensor measurement

INPUT DATA			
s_1 = -1,25E-06 MPa^{-1}	$1/2s_2$ =5,77E-06 MPa^{-1}		Θ_o =78,160°
Data in [°]			
$\Theta_{0,0}$ = 78,283	$\Theta_{0,+15}$ = 78,202	$\Theta_{0,+30}$ = 78,081	$\Theta_{0,+45}$ = 78,160
$\Theta_{0,0}$ = 78,283	$\Theta_{0,-15}$ = 78,301	$\Theta_{0,-30}$ = 78,251	$\Theta_{0,-45}$ = 78,148
$\Theta_{45,0}$ = 78,283	$\Theta_{45,+15}$ = 78,220	$\Theta_{45,+30}$ = 78,118	$\Theta_{45,+45}$ = 78,003
$\Theta_{45,0}$ = 78,283	$\Theta_{45,-15}$ = 78,289	$\Theta_{45,-30}$ = 78,237	$\Theta_{45,-45}$ = 78,140
$\Theta_{90,0}$ = 78,283	$\Theta_{90,+15}$ = 78,261	$\Theta_{90,+30}$ = 78,200	$\Theta_{90,+45}$ = 78,116
$\Theta_{90,0}$ = 78,283	$\Theta_{90,-15}$ = 78,260	$\Theta_{90,-30}$ = 78,198	$\Theta_{90,-45}$ = 78,114

COMPONENT OF STRESS TENSOR [MPa]		
σ_{11} = 404,162	σ_{12} = 13,561	σ_{13} = 62,654
σ_{21} = 13,656	σ_{22} = 305,565	σ_{-23} = -0,670
σ_{31} = 62,654	σ_{32} = -0,670	σ_{33} = 93,221

PRINCIPAL STRESSES [MPa] AND ITS ORIENTATIONS [°]		
σ_1 = 404,162	σ_2 = 303,859	σ_3 = 80,474
ϕ_x = 13,656	ϕ_y = 97,281	ϕ_z = 78,511
η_x = 11,275	η_y = -2,341	η_z = -78,478

Table 2: Influence of grinding parameters on stress tensor components in surface layer of polycrystalline materials

Specimen		1	2	3
Cooling		no cooling	water	water
Downfeed [μm]		5	5	10
Stress tensor [MPa]		408 −17 59 −17 416 6 59 6 108	390 14 63 14 306 −1 63 −1 93	534 −2 69 −2 468 −3 69 −3 95
Principal	σ_1	402,08	404,66	544,67
stresses	σ_2	433,37	304,21	467,94
[MPa]	σ_3	96,56	80,13	84,39

The results of our investigations have enabled us to formulate the following conclusions:

1° there is possibility to measure the changes of stress tensor in surface layer of polycrystalline materials by x-ray integral method,

2° research carried out of triaxial stress state have showed full reliability of the worked out methodology as well as the auxiliary computer program; this program accelerates time-consuming calculations associated with the measurements,

3° an influence of grinding parameters on stress tensor components in surface layer of polycrystalline materials has been showed (e. g. increasing depth of cut increases the residual normal macrostresses, cooling decreases the tensile residual macrostresses., principal stresses in grinding are tilted around an axis transverse to the grinding direction).

Acknowledgements

This research was supported by Poznań University of Technology under Grants No. 24-215/97 and 24-217/97.

References

1. Senczyk D., Computer programs for x-ray macrostress measurements, Proc. of European Conf. "Physics for industry - industry for physics", pp. 179-180, 17-19 Sept., 1991, Cracow - Poland.
2. Senczyk D., Metrological characterization of X-ray stress methods by means of cybernetics, Recent Advances in Experimental Mechanics, J.F. Silva Gomes et al., editors, v. 2, pp.763-766, A. A. Balkema, Rotterdam, Brookfield, 1994.

ROOM TEMPERATURE STUDIES ON OXYGEN ADSORPTION OF A TiNi ALLOY SURFACE UNDER LOW PRESSURE OF AIR OR OXYGEN

E. Rówiński, E. Łągiewka

Institute of Physics and Chemistry of Metals, University of Silesia, Bankowa12, 40-007 Katowice

Process of oxygen adsorption on a clean surface of the nearly equi-atomic TiNi alloy at the room temperature and under pressure of air of $3°10^{-9}$ and $1°10^{-6}$ hPa or under pressure of oxygen of $1^{*}10^{-6}$ hPa is described in this work. It was found that in the air on the studied surface chemisorption phenomenon resulting in creating the TiO oxide occurs as well as a weak physical adsorption of carbon, while in the oxygen atmosphere under $1^{8}10^{-6}$ hPa pressure the chemisorption of oxygen occurs which causes creation of the TiO_2 oxide. The oxygen chemisorption in the air atmosphere under pressure of $3*10^{-9}$ hPa follows the Langmuir model and under pressure of $3*10^{-9}$ hPa it occurs according to the cluster model and a mono-layer of the titanium oxide is formed. In the oxygen atmosphere under $1*10^{-6}$ hPa on the clean surface of the alloys there grow three mono-layers of the titanium dioxide and the kinetics of adsorption follows the cluster model.
No chemical reaction between nickel and oxygen and between titanium and carbon was found.

1 Introduction

The TiNi shape memory alloys show large chemical activity for oxidation and carbonisation which is attractive for some applications but is unfavourable for others [1-5]. Therefore studies on the mechanism and kinetics of interaction between the alloy surface, especially a clean surface, and a gas atmosphere in the early stages of the process are important not only because of the scientific reasons but also from the application point of view.

There are few works in the literature [6,7] which concern studies on layers formed on the TiNi alloy surface during the process of oxidation in the air, water steam, hydrogen peroxide solution at raised temperatures and different pressures. With the exception of [8] there is lack of studies concerning the mechanism and kinetics of the early stages of the interaction process between a gas and a clean alloy surface. These studies should elucidate the cause of the alloy activity in the interaction with gases and the structure of the forming layers.

In this work the process of interaction between a clean TiNi alloy surface with air of $3*10^{-9}$ and $1*10^{-6}$ hPa pressure and with oxygen of $1*10^{-6}$ hPa pressure. Yes studied known models of the adsorption kinetics were used for describing the changes of the degree of surface covering by an adsorbate.

2 Material and experimental

Two-component TiNi alloy containing 50.7 at.% Ni was the material of studies. It was prepared in the Krupp factory (Germany) to the final form of 3mm thick strips. The material was annealed at 700°C for 15 minutes at argon atmosphere and then cooled down in iced-water. In result the alloy was a one-phase material of the parent β phase structure [3].

The specimens were cut in form of squares of 12 x 12 mm sizes, ground on abrasive paper and polished on a felt soaked in water + Al_2O_3 till a shiny surface was obtained. Before inserting the specimens in the Auger spectrometer chamber in which the pressure was 1*10^{-9} they were washed with distilled water and dried in air. The further step of the surface cleaning was carried out in the chamber with the use of ion gun of ion beam energy 4 keV and density 0.72 μA/cm^2. The degree of cleaning was controlled taking the Auger spectra and observing the spectral lines of the alloy components and of oxygen and carbon. A surface from which only spectral lines of the alloy components and trace amounts of carbon were observed, was treated as a clean one. Then the etching process was stopped, the pressure in the chamber was established for 3*10^{-9}hPa or 1*10^{-6} hPa for air and 1*10^{-6} for introduced in the chamber technical oxygen and the spectra were taken after every 3 minutes. The total time of keeping the specimens in the chamber was about 180 minutes.

From the Auger spectrograms taken in a differential system the degree of the surface covering by oxygen and carbon versus time at certain pressure was evaluated by a method described in [9]. The method is based on the following formula:

$$\theta = \frac{I(X)\cdot I^{Hb}(Ni)\cdot c^{Hb}(X)}{I(Ni)\cdot I^{Hb}(X)}\cdot\left[1-\exp\left(\frac{-a}{\lambda_X \cos\varphi}\right)\right] \qquad 1)$$

where

I(X) - experimental intensity of the main Auger line of oxygen with energy E=509eV, or of carbon with energy E=272 eV.

I^{hb}(X) - catalogue intensity of the main Auger line of oxygen with energy E=509 eV, or of carbon with energy E=272 eV acc. to [10]

I(Ni) - experimental intensity of the main Auger line of nickel with energy E=846 eV

I^{Hb}(Ni) - cataloque intensities of Auger line of nickel with energy E=848 eV acc.to [10]

c^{Hb}(X) - oxygen or carbon concentration determined basing on intensities of spectral lines of oxygen, carbon and nickel acc. to the catalogue [10]

a - thickness of the mono-layer

λ_X - attenuation lenght of Auger electrons escape for oxygen or carbon.

In this work the following values were taken for determination of the degree of covering by oxygen and carbon: I^{Hb}(O) = 27.5, I^{Hb}(C) = 3.75, I^{Hb}(Ni)= 13.57, c^{Hb}(O) = 0.53, c^{Hb}(C) = 0.53, a = 1 ML, λ_O = 4.39a and according to [9], λ_C = 3.21a, cosφ = 0.7.

To determinate whether the surface of the TiNi alloy is covered by one or more mono-layers of oxygen a following criterion was applied: Assuming in the formula 1 that θ = 1, the theoretical value of $\frac{I_1(O)}{I(Ni)}$ at which the surface is totally covered by a mono-layer was calculated. It was calculated that $\frac{I_1(O)}{I(Ni)}$ = 0.69. Introducing this value to the formula [11]:

$$\frac{I_n(O)}{I(Ni)} = \left[1-\left(1-\frac{I_1(O)}{I(Ni)}\right)^n\right]$$

ranges of the $\frac{I_1(O)}{I(Ni)}$ values were calculated for which the surface is covered by 2, 3 mono-layers, respectively (Table 1). These values were used in calculation of the degree of covering when the experimental values of the ratios $\frac{I_1(O)}{I(Ni)}$ were in the adequate ranges.

Table 1. Theoretical values of the ratio $\frac{I_1(O)}{I(Ni)}$ calculated basing on the formula 1

number of the mono-layers	values of
1	0-0.69
2	0.69-0.90
3	0.90-0.97

3 Results and discussion

Analysis of the Auger spectrograms taken from the TiNi alloy surface kept in air at $3*10^{-9}$ and $1*10^{-6}$ hPa showed that the surface is covered by oxygen and carbon. The degree of covering increases with time. The maximal degree of coverage by carbon is 0.13 ML at the pressure of $3*10^{-9}$ hPa and 0.19 ML at the pressure of $1*10^{-6}$ hPa. Maximal degree of covering by oxygen tends to 0.8 ML (Fig. 1) independent of the pressure. The alloy surface kept in oxygen at $1*10^{-6}$ hPa is covering by oxygen. The maximal degree of the oxygen coverage was about 3 ML (Fig. 1).

Analysis of the main Auger line for titanium occurring on the spectra from surfaces obtained at $3*10^{-9}$ and $1*10^{-6}$ hPa air pressure showed that an additional line of energy of 412 eV occurred (Fig. 2). The line proves that a chemical reaction between titanium and oxygen took place and in result the TiO oxide was created. This line was observed in works [12] and it was interpreted as an Auger transition of the Ti(LM)O(V) type. It is also present in the catalogue [13] of the titanium spectral line in the TiO compound.

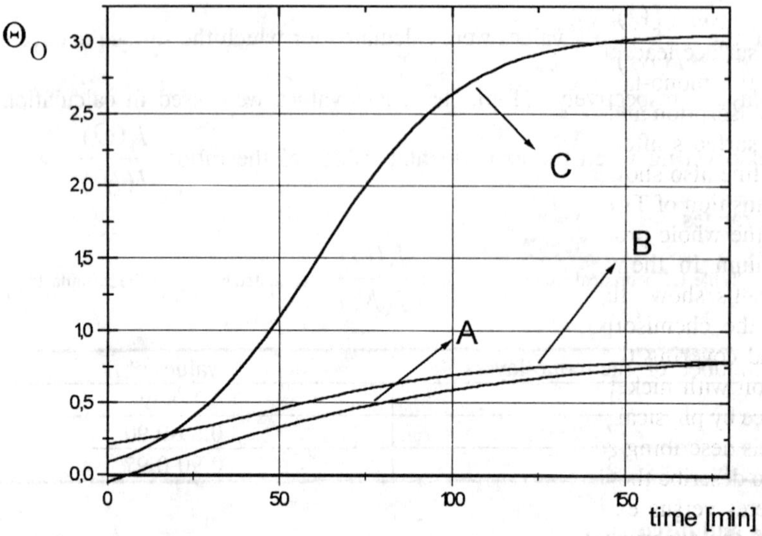

Fig. 1 Degree of covering of the TiNi alloy surface by oxygen
A curve - in air at $3*10^{-9}$ [hPa];
B curve - in air at $1*10^{-6}$ [hPa];
C curve - in oxygen at $1*10^{-6}$ [hPa]

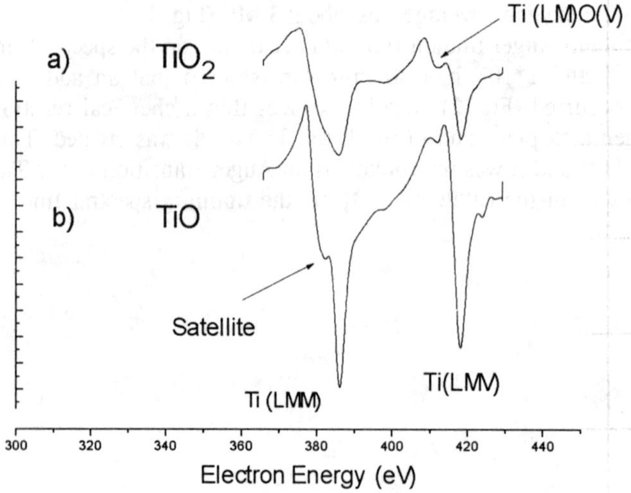

Fig. 2 Differential profiles of the titanium main line LMM and LMV in TiO and TiO_2
a) - in oxygen at $1*10^{-6}$ hPa for t = 180 min
b) - in air at $1*10^{-6}$ hPa for t = 180 min

The above results prove that in air atmosphere a chemisorption phenomenon occurs on the alloy surface leading to creation of the TiO compound. The surface is covered by a maximal 0.8 mono-layer. Thus coverage of the alloy surface by carbon is a result of physical adsorption and the covering value does not exceed 0.13 - 0.19 ML.

For surfaces after interaction with oxygen at $1*10^{-6}$ hPa the profile of the main titanium line also showed an additional line of 412 eV energy which was linked to the Auger transition of Ti(LM)O(V) (Fig. 2). The line, however, has a higher intensity and changes the whole profile of the titanium line comparing with the line in TiO and in pure titanium. In the works [8,12] it was stated that such a profile belongs to TiO_2. The above results show that on the TiNi alloy surfaces kept in oxygen of $1*10^{-6}$ hPa pressure the chemisorption phenomenon occurs which causes creation of the TiO_2 compound covering the surface by maximum 3 mono-layers. The reaction of oxygen and carbon with nickel does not occur and thus the covering of the surface by carbon takes place by physical adsorption phenomenon.

Models describing adsorption kinetics (the Langmuir and cluster models [12]) were applied to describe the degree of covering the TiNi alloy surface by oxygen in the given atmospheres versus exposure measured in langmuirs (1 langmuir = 1.33 10^{-6} hPa s). Fitting of the experimental curves to the theoretical formula describing the models are presented in Table 2. The correlation coefficient was a measure of fitting.

Table 2. The Langmuir and cluster models in kinetics of oxygen chemisorption on the TiNi alloy surface depending on the atmosphere and pressure

atmosphere pressure	Model	the obtained equation of oxygen adsorption kinetics	correlation coefficient
air $3*10^{-9}$ hPa	Langmuir $1/(1-\Theta) = K L$	$\Theta/(1-\Theta)=$ $0.24+0.69 L$	0.99721
air $3*10^{-9}$ hPa	cluster $-\ln(1-\Theta)=KL$	$-\ln(1-\Theta)=$ $0.199+0.31 L$	0.99275
air $1*10^{-6}$ hPa	Lamgmuir $1/(1-\Theta)=KL$	$\Theta/(1-\Theta)= -$ $0.038+0.00049 L$	0.9475
air $1*10^{-6}$ hPa	cluster $-\ln(1-\Theta)=KL$	$-\ln(1-\Theta)=$ $0.189+0.0002 L$	0.9750
oxygen $1*10^{-6}$ hPa	Langmuir $1/(1-\Theta)=KL$	$\Theta/(1-\Theta)=$ $-2.15+0.0033 L$	0.8900
oxygen $1*10^{-6}$ hPa	cluster $-\ln(1-\Theta)=KL$	$-\ln(1-\Theta)=$ $-0.49+0.0011 L$	0.9400

Analysis of the correlation coefficient showed that chemisorption of oxygen on the TiNi alloy surface in air pressure of $3*10^{-9}$ is close to the Langmuir model while chemisorption of oxygen in air and oxygen at $1*10^{-6}$ hPa corresponds to the cluster model.

With the use of these models and taking the determined K constant value a maximal number of chemisorption sites on the TiNi alloy surface was calculated from the equation: $N_{max} = 1.24*10^{16}*S_0/(K*T^{1/2})$, assuming that $S_0 = 1$. The N_{max} values are given in Table 3.

Table 3. Maximal number of oxygen chemisorption sites on the TiNi alloy surface kept in air and oxygen - versus pressure.

Atmosphere	pressure [hPa]	N_{max} [sites/cm^2] assuming $S_0 = 1$
air	$3 \cdot 10^{-9}$	$1.04 \cdot 10^{15}$
air	$1 \cdot 10^{-6}$	$3.4 \cdot 10^{18}$
oxygen	$1 \cdot 10^{-6}$	$6.5 \cdot 10^{17}$

From Table 3 it can be seen that the maximal number of oxygen chemisorption sites in air at the pressure of $3*10^{-9}$ hPa is of the order 10^{15}, while at the $1*10^{-6}$ hPa in air is 3 orders higher and in oxygen of the same pressure $1*10^{-6}$ hPa - 2 orders higher. The above results show that a real surface of the TiNi alloy at higher pressure has larger number of oxygen chemisorption sites for both air and oxygen atmospheres.

Basing on the carried out studies it can be stated that pure surface of the TiNi alloy after reaction with air at $3*10^{-9}$ hPa pressure and room temperature is reconstructed and as result of oxygen chemisorption a mono-layer of titanium oxide is created. The kinetics of this process is following the Langmuir model. In air at the $1*10^{-6}$ hPa the chemisorption of oxygen occurs according to the cluster model. In air at the pressure of $3*10^{-9}$ and $1*10^{-6}$ hPa a weak physical adsorption of carbon was found which resulted in covering the surface by carbon below 0.2 ML.

In air at the pressure of $1*10^{-6}$ hPa on a clean alloy surface three mono-layers of titanium dioxide are formed similarly as in [8], as a result of oxygen chemisorption phenomenon. Kinetics of this phenomenon can be described by the cluster model.

Conclusions:

1. A clean surface of the TiNi alloy in air at the pressure of $3*10^{-9}$ and $1*10^{-6}$ hPa at room temperature is covered by a mono-layer of titanium oxide and the maximal degree of the covering does not depend on pressure.

2. Kinetics of oxygen chemisorption on a clean surface of the TiNi alloy in air at the pressure of $3*10^{-9}$ hPa and at room temperature is following the Langmuir model while in air and oxygen at $1*10^{-6}$ hPa the cluster model can be applied.

3.No chemical reactions between nickel and oxygen or titanium and carbon were on these alloys surface. The covering of the TiNi alloy surface by carbon is a result of physical adsorption.

References

[1] Chu Youyi: Research and aplications of shape Memory Alloys in China, Proceedings of the MRS International Meeting on Advanced Materials, V.9, Shape Memory Materials , (31.05-3.07) 1988, Tokyo, Japan
[2] Shibi Lu M.D. Medical Aplications of NiTi Alloys in China , in Engineering Aspects of Shape Memory Alloys p.445-451, Butterworth- Heinemann,London Boston Singapore Sydney Toronto Wellington 1990
[3] E. Rówiński, G. Syska, E. Łągiewka: Procecdings of XV the Conference Applied Crystallography, Cieszyn, Poland , 9-12 August 1992, Singapore 1993, p 234
[4] E. Rówiński, Z. Lekston, E. Łągiewka: Procecdings of XVI the Conference Applied Crystallography, Cieszyn, Poland , 22-26 August 1994, World Scientific Co. Pte. Ltd. Singapore, p. 452-454
[5] CL Chu, SK Wu, YC Yen: Materials Science and Engineering A, Vol 216, Iss 1-2 , 1996 pp 193-200
[6] S.A. Shabalovskaya: J. de Physiqe III, Colloque C8, 5 (12) (1995), p 1199 -1204
[7] S.A. Shabalovskaya and J.W. Anderegg: J. Vac. Sci. Technol. A 13(5), Sep/Oct 1995 p 2624-2632
[8] P.H. Mc Breen, M. Polak: Surface Sci, vol 179, no 2-3 1987 p 483-497
[9] U. Otterbein and S. Hofmann: Surface and interface analysis, Vol.24 (1996) p 263-270
[10] P.W. Palmberg, G.E. Riach, R.E. Weber, N.C. MacDonald: Handbook of Auger Electron Spectroscopy, Physical Electronics Industries, Edina, Minnesota 1976
[11] D. Chattarji: The Theory of Auger Transitions, Academic Press 1976 p. 250
[12] A. Azoulay, N. Shamir, E. Fromm, M.H. Mintz: Surface Science 370 (1997) 1-16
[13] D.Biggs and M.P. Seah: Practical Surface Anatysis (Second Edition) John Wiley and Sons ltd 1990

Materials Structure

CRYSTAL STRUCTURE AND MICROWAVE DIELECTRIC PROPERTIES OF TUNGSTEN BRONZE-TYPE $Ba_{6-3x}R_{8+2x}Ti_{18}O_{54}$ (R=La, Pr, Nd and Sm) SOLID SOLUTIONS

H. OHSATO, M. MIZUTA and T. OKUDA
Materials Science and Engineering, Nagoya Institute of Technology, Gokiso-cho, Showaku, Nagoya 466, Japan

S. NISHIGAKI
Daiken Chemical Co., Ltd., 2-7-9 Hanaten-Nishi, Joto-ku, Osaka 573, Japan

The dielectric properties of the $Ba_{6-3x}R_{8+2x}Ti_{18}O_{54}$ (R=rare earth) solid solutions with La, Pr, Nd and Sm are discussed base on the crystal structure. The structural formula is derived as $(R_{8+2x}Ba_{2-3x})(Ba_4)Ti_{18}O_{54}$ in the range of $0 \leq x \leq 2/3$ and $(R_{9.33+2(x-2/3)})(Ba_{4-3(x-2/3)})Ti_{18}O_{54}$ in the range of $2/3 \leq x \leq 1$. The quality factor $Q \cdot f$ changed non linearly according to both functions of composition x and ionic radius of R ions. Inner stress generated by the distribution of large cations R and Ba, and by decreasing of the difference of ionic radius between R and Ba ions reduces the $Q \cdot f$ vales. High quality dielectric properties such as ε_r=89, $Q \cdot f$=8018 GHz, τ_f=75 ppm/ °C for Nd-system and ε_r=80, $Q \cdot f$=9302 GHz, τ_f=-9.8 ppm/ °C for Sm-system are obtained.

1 Introduction

Microwave dielectric ceramics have received attention due to the rapid progress in microwave telecommunication such as car telephone system, potable telephone and satellite broadcasting.[1-2] The desirable properties in microwave dielectric resonators are a high dielectric constant (ε_r), high quality factor ($Q \cdot f$) and low temperature coefficient of the resonant frequency (τ_f). The ε_r reduces the dimensions of resonators, because the wavelength (λ) in dielectrics is inversely proportional to $\sqrt{\varepsilon_r}$ of the wavelength (λ_0) in vacuum ($\lambda = \lambda_0/\sqrt{\varepsilon_r}$). The Q, which is the inverse of the dielectric loss (tanδ), is required to achieve high frequency selectivity and stability in microwave transmitter components. Since the Q value varies inversely with the frequency in the microwave region, $Q \cdot f$ is used to evaluate the quality instead of the Q value. And the τ_f is required to be as close to 0 ppm/ °C as possible in the temperature range required for natural conditions.

The microwave dielectric ceramics with high quality in the BaO-R_2O_3-TiO_2 (R = rare earth) system were investigated by many researchers such as Bolton[3], Kolar *et al.*[4], Wakino *et al.*[5], Kawashima *et al.*[6], Nishigaki *et al.*[7] and so on. The last three groups in Japanese company made the resonators develop for commercial products. The $Ba_{6-3x}R_{8+2x}Ti_{18}O_{54}$ solid solutions[8-9] were reported as ternary compounds existing on the tie line between $BaTiO_3$ and $R_2Ti_2O_9$ composition in the BaO-R_2O_3-TiO_2 ternary system as shown in Fig. 1. The $BaO \cdot R_2O_3 \cdot 4TiO_2$ ternary compound [10-14] known as microwave resonator materials corresponds exactly to x=0.5, and $Ba_{3.75}Pr_{9.5}TI_{18}O_{54}$ reported by Matveeva *et al.* corresponds to x=0.75 in the $Ba_{6-3x}R_{8+2x}Ti_{18}O_{54}$ solid solution formula. These solid solutions contain two kinds of large cations which are different in valence, such as divalent Ba^{2+} and trivalent R^{3+}. To maintain electrostatic stability, three Ba^{2+} ions should be replaced by two R^{3+} ions and one vacancy. Therefore, there exists very char-

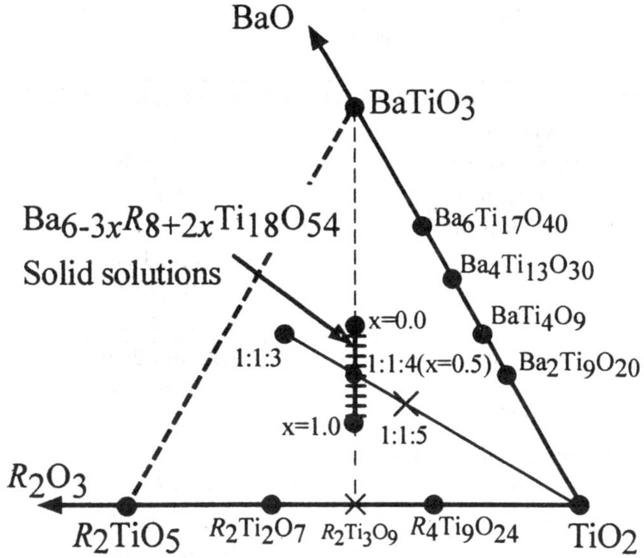

Figure 1: BaO-R_2O_3-TiO_2 ternary system.

acteristic behavior along the tie line in which the substitution of 2 R^{3+} for 3 Ba^{2+} occurs. The structure of the solid solutions will be described in the next section.

We described first the dielectric properties of $Ba_{6-3x}R_{8+2x}Ti_{18}O_{54}$ solid solutions (Sm-system) as a function of composition[15] and Fukuda *et al.*[16] also described those of Pr-system. We also report those with 0 ppm/ °C in the $Ba_{6-3x}(Sm,R)_{8+2x}$ $Ti_{18}O_{54}$ solid solutions mixed Sm-system with negative temperature coefficient and Nd- and La-systems with positive coefficient.[17] Moreover, we reported the dielectric properties of the solid solutions with La, Pr, Nd and Sm as a function of the composition in the previous paper.[18] In this study, after describing the crystal structure, we discuss the dielectric properties of the $Ba_{6-3x}R_{8+2x}Ti_{18}O_{54}$ (R=La, Pr, Nd and Sm) solid solutions as a function of ionic radius of rare earth elements.

2 Crystal structure and structural formula

The crystal data of the $Ba_{6-3x}Sm_{8+2x}Ti_{18}O_{54}$ solid solutions with x=0.71 are shown in Table 1, which structure was refined by single crystal X-ray structure analysis (Ohsato[19-20]). This compound has a superlattice with two times the fundamental lattice interplanar spacing along the c-axis, which was found out in an oscillation photograph with c-axis rotation.[21] The space groups were determined by Weissenberg and precession photographs of the single crystals: either $Pbn2_1$ (No.33) or *Pbnm* (No.62) for superlattice and either *Pba*2 (No.32) or *Pbam* (No.55) for fundamental lattice.[21-22] The crystal structure of the solid solutions[19-20,23] is a new tungsten bronze-type with 2x2 perovskite blocks (A_1-site) and pentagonal columns (A_2-site) as shown in Fig. 2. It is three-dimensional framework of $[TiO_6]^{8-}$ octa-

Table 1: Crystal data of a single crystal with x=0.71.Z: number of the compositional formula in the unit cell.

Composition of the single crystal	$Ba_{3.78}Sm_{9.42}Ti_{18}O_{54}$($x$=0.71)
Crystal system	orthorhombic
Laue groupe	*mmm*
Fundamental lattice parameter a	12.131(3)Å
b	22.271(5)Å
c	3.819(3)Å
Space group	*Pba*2(No.32) or *Pbam*(No.55)
Z	1
Superlattice parameter c	7.639(5)Å
Space group	*Pbn*2_1(No.33) *Pbnm*(No.62)
Z	2

hedra connected with each other at all apices. The solid solutions are located on a tie line with Ti:O=1:3 ratio including perovskite $BaTiO_3$ as described in the introduction section. R and Ba cations with large ionic radius are mainly located in the A_1- and A_2-sites, respectively. The terminal composition with x=0 of the solid solutions is $Ba_6R_8Ti_{18}O_{54}$ which occupy all A_1- and A_2-sites having 10 and 4 spaces, respectively. Then, the basic structural formula is $[A_1]_{10}[A_2]_4B_{18}X_{54}$. The solid solutions formula is formed from the terminal composition according to the substitution mode $3\,Ba^{2+} \rightarrow 2\,R^{3+}+\square$. At the terminal composition, A_1-sites are occupied by eight R and two Ba ions and A_2-sites are by four only Ba ions. Here, as ionic radius of Ba ions in A_1-sites is greater than that of other R ions, inner stress arises around the Ba ions as discussed later. The structural formula of the terminal composition is $(R_8Ba_2)(Ba_4)Ti_{18}O_{54}$. That of the solid solutions is $(R_{8+2x}Ba_{2-3x})(Ba_4)Ti_{18}O_{54}$ in the range of $0 \leq x \leq 2/3$. Here, when the number of Ba ions in A_1-sites becomes zero, x=2/3. Moreover, continuing the substitution in the range of $2/3 \leq x \leq 1$, the formula is $(R_{9+1/3+2(x-2/3)})(Ba_{4-3(x-2/3)})Ti_{18}O_{54}$, indicating that the R ions occupy the A_1-sites preferentially and that vacancies are created in the A_2-sites. The occupation by the two kinds of large cations of these A_1- and A_2-sites in the range of $2/3 \leq x \leq 1$ has not yet been determined exactly.

3 Experimental

The microwave dielectric ceramics resonators with the desired composition for characterization were sintered as disks in 10 mm diameter and 5 mm height as described in the previous papers.[18] Rare earth elements in the $Ba_{6-3x}R_{8+2x}Ti_{18}O_{54}$ solid

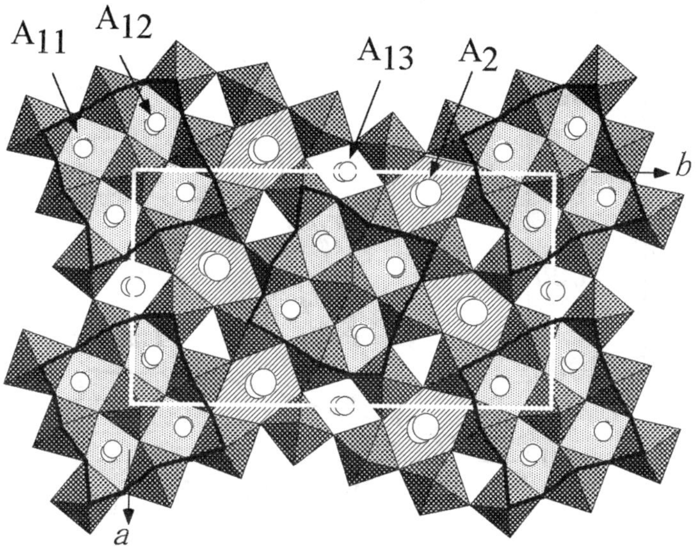

Figure 2: Crystal structure of the solid solution with super lattice.

solutions are varied as La, Pr, Nd and Sm. Dielectric properties of the disks with the ratio of diameter and height: 1:2 were measured between a pair of parallel conducting Ag plates in the TE_{011} mode using the method of Hakki and Colleman.[24] We selected the samples with maximum $Q \cdot f$ values for results and discussion. Accurate lattice parameters of the solid solutions were determined to refine the powder diffraction patterns obtained by the step scanning measurement technique using Si (99.99%) as an internal standard. A program used for the refinement of the lattice parameters is the whole powder pattern decomposition (WPPD) program made by Toraya.[25]

4 Results and discussion

The dielectric properties of the $Ba_{6-3x}R_{8+2x}Ti_{18}O_{54}$ solid solutions with La, Pr, Nd and Sm were measured as a function of composition as reported in a previous paper.[18] The characteristic phenomenon obtained is that $Q \cdot f$ values varied non linearly as a function of composition as shown in Fig. 3. Other properties ε_r and τ_f are varied linearly with proportion each other. The reason that the $Q \cdot f$ values steeply increased in the vicinity of x=0.5 is explained by inner stress due to large cations distribution. In the early stage of the solid solutions $0 \leq x \leq 2/3$, Ba ions with larger ionic radius are existing in perovskite blocks mainly occupied by rare earth with a bit small ionic radius as seen in the structural formula $(R_{8+2x}Ba_{2-3x})(Ba_4)Ti_{18}O_{54}$. As cuboctahedron containing a Ba ion becomes large volume in comparison with other cuboctahedra with R ions in the perovskite block, inner stress might be generated around the Ba ion. The inner stress might bring lower $Q \cdot f$ values as seen in Fig. 3.

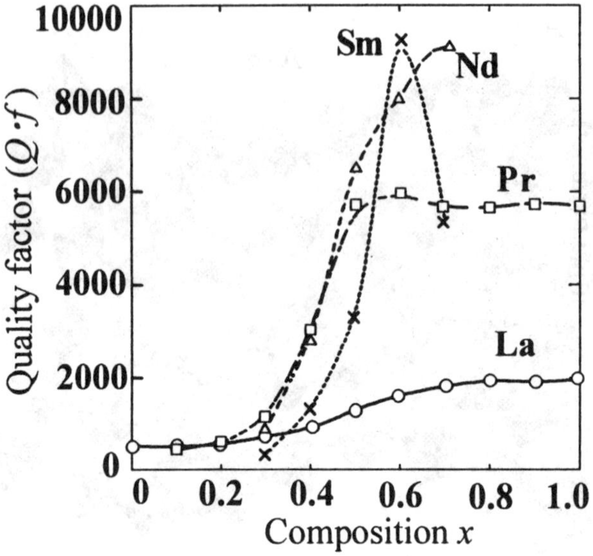

Figure 3: Quality factor of $Ba_{6-3x}R_{8+2x}Ti_{18}O_{54}$ (R=La, Pr, Nd and Sm) solid solutions as function of composition.

According to the substitution R for Ba, $Q\cdot f$ increases around x=0.5 composition. When Ba atoms in the perovskite blocks become zero at x=2/3, Ba and R ions locate separately in pentagonal columns and perovskite blocks, respectively. The large cations distribution arranged with same size ions brings no inner strain to show high $Q\cdot f$.

The lattice parameters and the dielectric properties of the solid solutions are shown as a function of ionic radius of R elements in Fig. 4 and 5, respectively. The x=0.6 composition of each R-system is selected, because the quality factor is best in the solid solutions composition. As seen in Fig. 4, the lattice parameters are decreased linearly according to the lanthanide shrinkage of R elements. Dielectric constant (ε_r) and temperature coefficient of the resonant frequency (τ_f) also decrease linearly as a function of ionic radius. By the way as seen in Fig. 5, $Q\cdot f$ values change non linearly as a function of the ionic radius. Those of the three cases: Pr-, Nd- and Sm-system, show linear relationship but that of La-system deviates from the linear line. If the change of the $Q\cdot f$ are affected only with ionic radius, the line should be linear. The reason of the deviation might be considered being inner stress which is created by distribution of cations. There are two cation sites: perovskite blocks and pentagonal columns with different size as described before, which are occupied by different size cations. As the difference of the ionic radius between Ba and La is not large in comparison with other R ions, the inner stress always exists as an intrinsic quality in the case of La-system. So, the inner stress might make the $Q\cdot f$ value decrease from the expected line.

By means of rewriting the horizontal axis shown ionic radius with the lattice parameter, we could see the properties as function of the lattice parameter. The

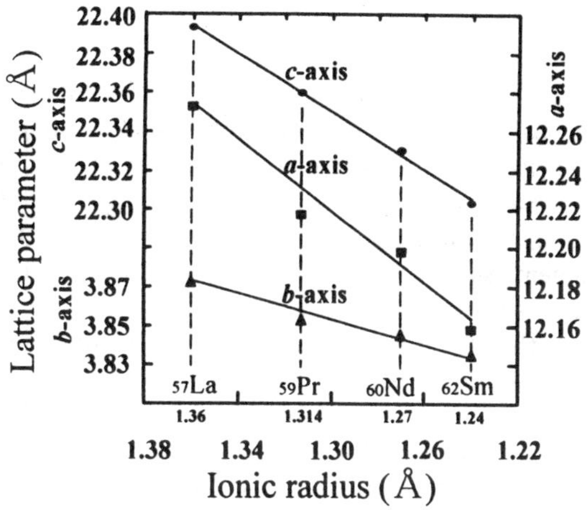

Figure 4: Lattice parameters of tungsten bronze-type solid solutions as a function of ionic radii.

dielectric constant decreases linearly according to the lattice parameter which show the volume of TiO_6 octahedron, because the polarization of this compound arises with displacement of Ti ion in the octahedron. In this case, the inner stress does not effect the dielectric constant. The range of ε_r on the solid solutions is located in 80 to 140, which value is largest group in the ceramics resonators. The $Q \cdot f$ values are also obtained as high quality more than 5,000 GHz, especially that of Sm-system is 9,302 GHz. We have to notify that the τ_f on Sm-system is -9.8 ppm/ °C. The τ_f values with zero were derived on the solid solutions between Sm-system and Nd- or La-system as reported in a previous paper.[16] The reason why the τ_f is in proportion to ε_r, has not been cleared yet.

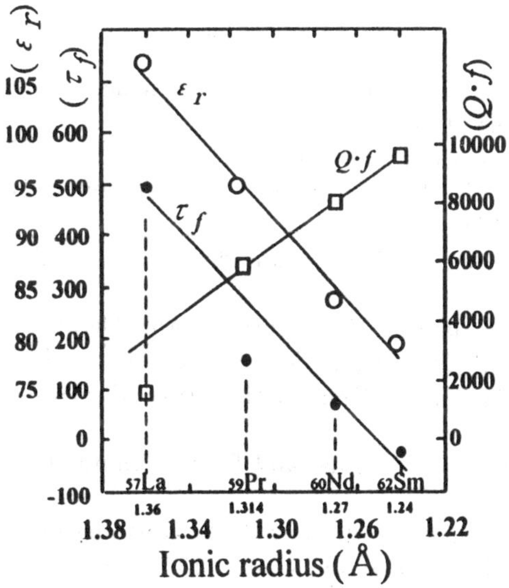

Figure 5: Dielectric properties of tungsten bronze-type solid solutions as a function of ionic radii.

5 Conclusion

The dielectric properties of the $Ba_{6-3x}R_{8+2x}Ti_{18}O_{54}$ solid solutions with La, Pr, Nd and Sm are discussed base on the crystal structure which is tungsten bronze-type with 2x2 perovskite blocks and pentagonal columns. The structural formula is derived as $(R_{8+2x}Ba_{2-3x})(Ba_4)Ti_{18}O_{54}$ in the range of $0 \leq x \leq 2/3$ and $(R_{9+1/3+2(x-2/3)})(Ba_{4-3(x-2/3)})Ti_{18}O_{54}$ in the range of $2/3 \leq x \leq 1$. When $x=2/3$, both sites are occupied with single element, that is, R ions for A_1-sites and Ba ions for A_2-sites. In the Ba rich region, Ba ions are also included in A_1-sites, which produce inner stress. The stress might reduce the $Q \cdot f$. We explained the reason that the $Q \cdot f$ value varied non linearly as a function of composition. We also showed the dielectric properties of the solid solutions as a function of the ionic radius of R ions. The dielectric constant and temperature coefficient of the resonant frequency are changed linearly, but the $Q \cdot f$ of La-system was deviated from the linear line according to other Pr, Nd and Sm. The inner stress due to the smaller difference of the radius between La ions occupied A_1-sites and Ba ions in A_2-sites might brings the deviation as intrinsic qualities. We have to detect the inner stress by means of X-ray diffraction analysis in our following research.

Acknowledgements

The authors thank Daiken Chemical Co., Ltd. for their assistance with the measurement of the dielectric properties, Murata Manufacturing Co., Ltd. for supplying the standard sample for the measurement of the dielectric properties, Mr. R. D. Saxena for discussion of technical expressions. This work was supported in part by the Japan Sheet Glass Foundation for Materials Science and Engineering.

References

1. S. Nishigaki: FC Rep. **5** (1987) 413. Translation: FC Annual Report for Overseas Readers (1988) 32.
2. S. Nishigaki: New Ceramics. **No.9** (1996) 25.
3. R. L. Bolton: PhD. Thesis, Ceramic Engineering, University of Illinois, Urbana, Illinois, 1968.
4. D. Kolar, Z. Stadler, S. Gaberscek and D. Suvorov: Ber. Dtsh. Keram. Ges. **55** (1978) 346.
5. K. Wakino, K. Minai and H. Tamura: J. Am. Ceram. Soc. **67** (1984) 278.
6. S. Kawashima, M. Nishida, I. Ueda and H. Ouchi: Ext. Abstr. 87th Annu. Meet. Am. Ceram. Soc., Cincinnati, (1985) (Electronics Division Paper No. 15-E-85).
7. S. Nishigaki, H. Kato, S. Yano and R. Kamimura: Ceram. Bull. **66** (1987) 1405.
8. M. B. Varfolomeev, A. S. Mironov, V. S. Kostomarov, L. A. Golubtsova and T. A. Zolotova: Russ. J. Inorg. Chem. **33** (1988) 607.
9. H. Ohsato, T. Ohhashi, S. Nishigaki, T. Okuda, K. Sumiya and S. Suzuki: Jpn. J. Appl. Phys. **32** (1993) 4323.

10. E. S. Razgon, A. M. Gens, M. B. Varfolomeev, S. S. Korovin and V. S. Kostomarov: Zh. Neorg. Khim. **25** (1980) 1701. Translation: Russ. J. Inorg. Chem. **25** (1980) 945.
11. E. S. Razgon, A. M. Gens, M. B. Varfolomeev, S. S. Korovin and V. S. Kostomarov: Zh. Neorg. Khim. **25** (1980) 2298. Translation: Russ. J. Inorg. Chem. **25** (1980) 1274.
12. A. M. Gens, M. B. Vafolomeev, V. S. Kostomarov and S. S. Korovin: Zh. Neorg. Khim. **26** (1981) 896. Translation: Russ. J. Inorg. Chem. **26** (1981) 482.
13. J. Takahashi, T. Ikegami and K. Kageyama: J. Am. Ceram. Soc. **74** (1991) 1868.
14. J. Takahashi, T. Ikegami and K. Kageyama: J. Am. Ceram. Soc. **74** (1991) 1873.
15. H. Ohsato, T. Ohashi, H. Kato, S. Nishigaki and T. Okuda: Jpn. J. Appl. Phys. **32** (1995) 187.
16. K. Fukuda and R. Kitoh: J. Mater. Res. **10**, No. 2 (1995) 312.
17. H. Ohsato, H. Kato, M. Mizuta, S. Nishigaki and T. Okuda: Jpn. J. Appl. Phys. **32** (1995) 5413.
18. H. Ohsato, M. Mizuta, T. Ikoma, Y. Onogi, S. Nishigaki and T. Okuda: J. Ceram. Soc. Japan. (to be published).
19. H. Ohsato, T. Ohhashi and T. Okuda: Ext. Abstr. AsCA 93 Conf., Singapore, November, **14U-50**.
20. H. Ohsato, H. Kato,M. Mizuta and T. Okuda: Ext. Abstr. AsCA 95 Conf., Bangkok, November, **1P40** (1995).
21. H. Ohsato, S. Nishigaki, T. Okuda: Jpn. J. Appl. Phys. **31** (1992) 3136.
22. D. Kolar, S. Gaberscek and B. Volavsek: J. Solid State Chem. **38** (1981) 158.
23. R. G. Matveeva, M. B. Varfolomeev and L. S. Il'yushchenko: Zh. Neorg. Khim. **29** (1984) 31. Translation: Russ. J. Inorg. Chem. **29** (1984) 17.
24. B. W. Hakki and P. D. Coleman: IRE Trans. Microwave Theory & Tech. **MTT-8** (1960) 402.
25. H. Toraya: J. Appl. Cryst. **19** (1986) 440.

X - RAY DIFFRACTION STUDIES OF SOME COMPOSITES BELONGING TO THE Al_2O_3 - SiO_2 AND Al_2O_3 - ZrO_2 SYSTEMS

L. GÓRSKI
Institute of Atomic Energy, 05-400 Świerk Otwock, Poland

In this work composites containing two groups of materials based on the Al_2O_3 - SiO_2 and Al_2O_3 - ZrO_2 systems are studied. These composites are used in high temperature resistant protective coatings deposited mainly by plasma spraying technology. X - ray diffraction methods are applied in order to determine all changes in phase composition due to coating deposition and subsequent thermal treatment.

In the first group of composites (containing silica) the stability of mullite type of crystal lattice have been observed. Small lattice distortions leading to slight differences in some peaks intensities may be connected with some kind of polytypism. In the second group of composites complex phase transitons concerning both oxides causing formation of different crystal modifications have been determined.

1. Introduction

The presented studies are connected with works on protective coatings resistant to different forms of degradation in the temperature range 1000 - 1500 K especially thermal barrier coatings being products of advanced technology. This type of coatings are applied in diffrerent branches of industry and technology among others in gas turbines and Diesel engines [1-3]. The main component of the above coatings are ceramic layers based on Al_2O_3 and ZrO_2 . Phase transitions occuring in these coatings are caused by two groups of factors. The first group is connected with existence of both oxides in different crystal modifications. The second group is connected with the process of coating deposition (mainly by plasma spraying technology) and further coatings applications in the above mentioned temperature region. Therefore X -ray diffraction measurements are necessary and very important part of works on high temperature resistant protective coatings.

In the result of earlier studies on ceramic materials based on ZrO_2 with the additions of Y_2O_3 and rare earth oxides and materials based on Al_2O_3 with the additions of NiO and TiO_2 some coatings compositions resistant among others to 200 thermal shocks in the temperature up to 1500 K have been obtained [4 - 7].

In this work materials and coatings belonging to Al_2O_3 - SiO_2 and Al_2O_3 - ZrO_2 systems are studied. The purpose of this work has a twofold character. First research side - to determine phase transitions and thermal stability for the above materials and coatings. The second practical side is to choice the best coatings compositions of increasing resistance for thermal fatigue processes.

2. Experimental

Several materials belonging to above mentioned Al_2O_3 - SiO_2 and Al_2O_3 - ZrO_2 systems are studied. Coatings from these materials are sprayed by Ar - H_2 plasma arc using Polish plasmatrones PN -120. The details of applied thermal spraying process and coatings preparation are described elsewhere [5,7]. It is however worthwhile to remark that this process are characterized by short time of powder particles flight (range of ms) by the temperature region about 10 -15 kK and resulting high cooling rate which is in the range of mega K/s. As can be seen, there are favourite conditions to different phase transitions.

Powders and coatings are annealed in furnace in oxidising atmosphere by 30 and 50 hours in the temperature range about 1200 K. Some specimens after annealing are cooled in liquid nitrogen. All powders and coatings before and after any thermal treatment are studied by x - rays (characteristic Cu and Co radiation) in order to determnine all changes in phase composition and possible phase transitions caused by the applied processes.

3. Results and discussion

Obtained results illustrated in listed below X - ray patterns differ in considerable degree for two studied groups of materials.. Materials belonging to Al_2O_3 - SiO_2 system show crystal lattice existence of mullite type. Mulite with the formula $3Al_2O_3$ - 2 SiO_2 is a material well known nd widely spread in high temperature applications [8,9]. But protective coatings containing mullite are, as yet, rarely encountered . Mullite crystallize in orthorhombic system with space group Pban [10]. It shows thermal stability in the conditions of performed experiments. Neither effects of mullite dissociation to component oxides nor new phases appearance have been observed. One visible effect on x - ray patterns from fig. 1 is a difference in some peaks intensites between powder pattern (fig. 1a) and coating pattern before and after thermal treatment (fig. 1b and 1c respectively). Three reasons of this effect may be taking into consideration.

1) solid solutions forming beyond stechiometry with other silicates and component oxides

2) recrystallisation with order change

3) polytypism

First reason is rather improbable because there is no displacement of diffraction peaks positions. But recrystallisation with change of crystallite sizes and disorder due to plasma spraying conditions and slow ordering due to subsequent coatings annealing is an effect encountered in earlier works [7,11]. Polytypism is also possible especially with doubling of c - axis and probable as a reason of the peaks intensities change. It is worthwhile to remark that intensities in annealed coating diffraction pattern (fig. 1c) are the closest to those listed in standard pattern from JCPDS Database [10]. It means that the conditions after specimens annealing become near to thermodynamical equilibrium. Further experiments and connected works on interpretation of the above effects are still in progress.

Materials belonging to Al_2O_3 -ZrO_2 system (both experimental and commercial) contain about 40 ZrO_2 weight %. In all cases we observe the presence of two separate phases: Al_2O_3 in the crystal form of α - corundum and ZrO_2 in the monoclinic form (fig. 2a). Diffraction patterns of the coatings as sprayed and after annealing (fig. 2b and 2c respectively) show appearance of phase transitions concerning both oxides. Alumina transforms from thermodynamically stable α - form to metastable γ - type form . This effect is characteristic for plasma sprayed coatings [5,7]. But more interesting effect is partial transition of zirconia monoclinic form to higher symmetry forms tetragonal and cubic existing in the room temperature region only in stabilised materials. After coatings annealing we can observe further increasing of higher symmetry peaks intensities which became comparable with monoclinic form peaks intensities (fig. 2c). In Al_2O_3 reverse transformation take place from metastable γ - type form into stable α - form (not entirely in used experimental conditions). Additional effect visible in both coatings diffraction patterns is the background gain connected with partial amorphisation, which may be caused by plasma spraying conditions.

On the first sight of view above materials might be treated as unsuitable to further works, but after described effects analysis it is worhwhile, at least in my opinion, to direct it to further thermal experiment especially to thermal shocks realised on special arrangement described earlier. The thermal shocks experiments on this arrangement may be treated as exploitation test of thermal barrier coatings [7,11]. These experiments are still in progress

3. Conclusions

1 Materials and coatings from the Al_2O_3 - SiO_2 system exist in the mullite crystal form. This form show stablity in the conditions of coatings forming and annealing.
2. Materials from the Al_2O_3 - ZrO_2 system exist in the form of separate oxides undergoing phase transitions in the course of performed experiments. Transitions in ZrO_2 are interested from the practical point of view.
3. Full determination of the above materials usefulness in high temperature resistant coatings and particularly in thermal barrier coatings will be performed after finishing of thermal shocks experiments.

References

1. Nicoll A.H., Gruner H., Surf.Eng., **1**,11 (1985).
2. Stecura S., Thin Solid Films, **159**,15 (1987).
3. Brandon A.R., Taylor S., Surf.Coat. Technology, **50**,285 (1992).
4. Górski L., Wolski T., in Euro-Ceramics II, Vol. II, Structural Ceramics and Composites, DKG, Koln 1993, p.1739.
5. Górski L., Nucleonics, **T39**, nr 3,127 (1994).
6. Górski L., in Applied Crystallography, ed. H. Morawiec, D. Stróż, World Scientific, Singapore - London 1995, p.379.
7. Górski L., Wolski T., Gostyński D., Changes Caused by Thermal Shocks in Plasma Sprayed Coatings from Materials Based on Al_2O_3,,, , Research Report IAE - 19/A (1996).
8. Ahay A., Dublis D.M., J.Am>Cer.Soc., **74**, 2341 (1991).
9. Sudreau F., Clagnan C., et al. in Euro-Ceramics II, Vol.II, Structural Ceramics and Composites, DKG, Koln 1993, p.1249.
10. Powder Diffraction System Ver 2.2, DHN, 1993.
11. 11.Górski L., Phase Transitions in Complex Ceramic Oxides Systems Occuring at the Processes with Short Time High Temperature Action, Research Report IAE (in press).

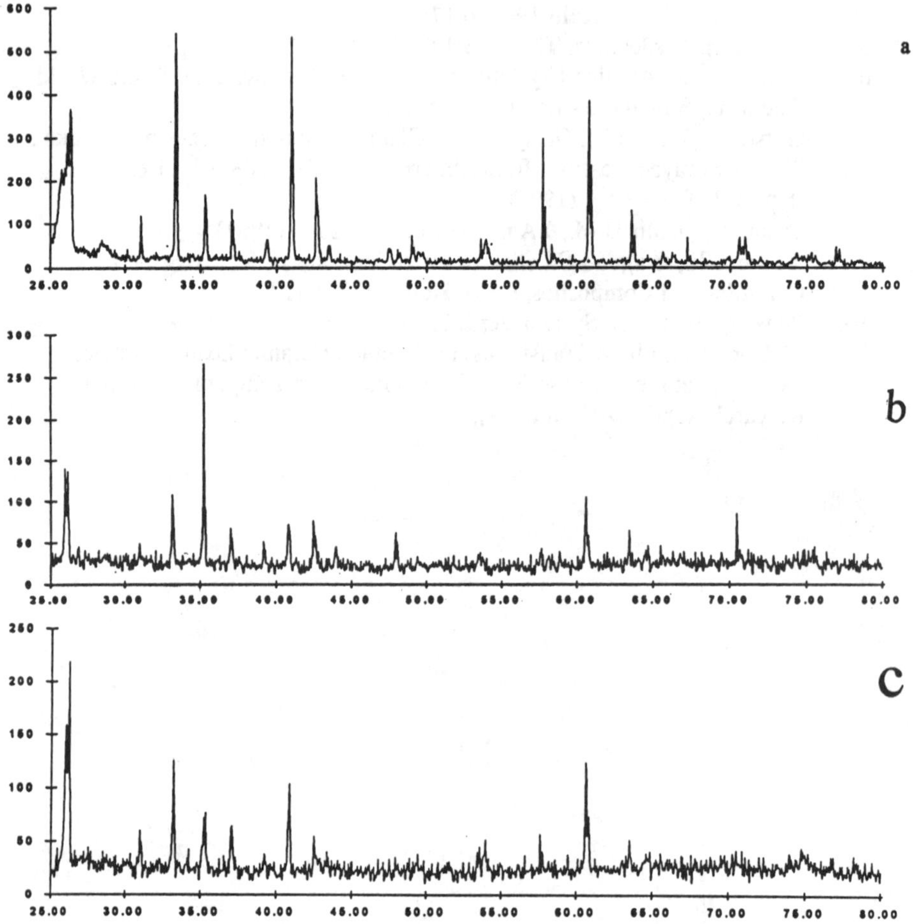

Figure 1 Diffraction patterns of mullite
a- powder before spraying, b- coating as sprayed, c- coating after 50 h annealing in 1200 K

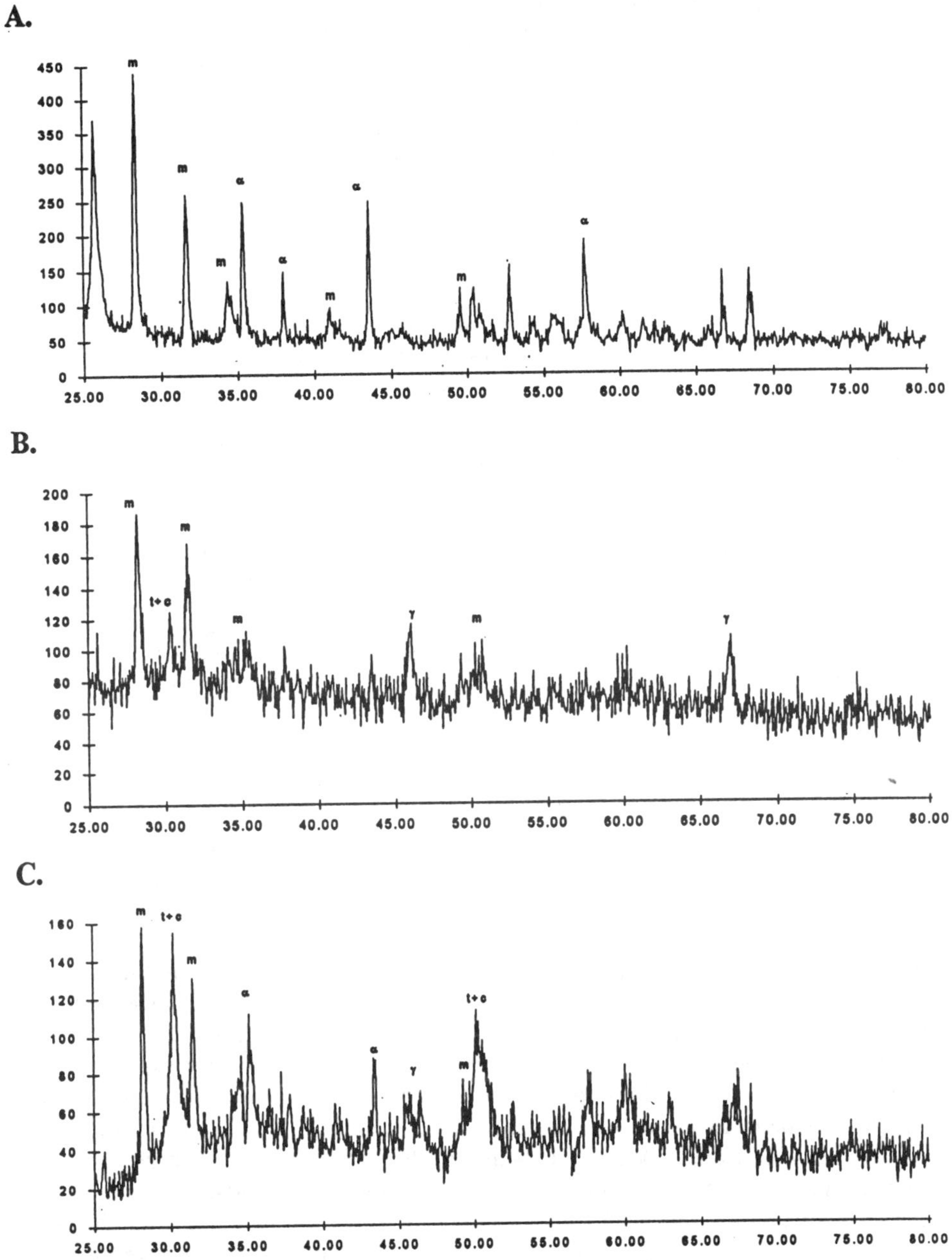

Figure 2 Diffraction patterns of Al_2O_3 - ZrO_2 system
A- powder before spraying, B - coating as sprayed, C - coating after annealing 50 h in 1200 K
c - cubic, t - tetragonal, m - monoclinic ZrO_2, α, γ - polymorphs of Al_2O_3

INFLUENCE OF FREE CHARGE AND ELASTIC PROPERTIES ON NEGATIVE THERMAL EXPANSION OF SEMICONDUCTORS

M. LESZCZYNSKI,
High Pressure Research Center UNIPRESS, Polish Academy of Sciences, Sokolowska 29/37, 01-142 Warsaw, Poland.
e-mail: mike@iris.unipress.waw.pl

V. B. PLUZHNIKOV
B. Verkin Institute of Low Temperature Physics and Engineering, 47 Lenin Ave. Ukrainian Academy of Sciences, Kharkov 310164, Ukraine

A. CZOPNIK, A. PIETRASZKO
Institute of Low Temperatures and Structural Research
Polish Academy of Sciences, ul. Okólna 2, Wroclaw, Poland

J. BAK-MISIUK
Institute of Physics Polish Academy of Sciences, Al. Lotnikow 32/46 Warsaw, Poland

The paper reports on thermal expansion of GaAs, GaN and ZnTe in the temperature range of 10-300 K. These semiconductors were chosen because of their different elastic properties. A special attantion was paid to the lowest temperature range for which a phenomenon of the negative thermal expansion occurs for many semiconductors. The negative thermal expansion was more pronounced for the GaAs sample with a high free-electron concentration. No correlation with elastic properties of the examined materials was found.

Introduction

Thermal expansion of semiconductors is a topic of some technological as well as scientific interest since it plays an important role in determining the residual stresses in the layer structures used for constructing electronic and optoelectronic devices. Recently, we gave an experimental evidence[1,2] for the doped GaAs, AlGaAs and GaN that thermal expansion coefficient (TEC) increases proportionally to the free electron concentration at temperatures 300-750 K. In this article we explore the low-temperature range, which is particularly interesting because of the negative thermal expansion (NTE) which was described by a number of authors (for example, Ref. 3-5) and examined experimentally (for example, Ref.6,7). The NTE phenomenon is attributed to the negative Grüneisen parameters (phonon frequency increases as crystal volume increases) of the transverse acoustic phonon near the Brillouin-zone boundary. In particular, the NTE depends on the difference in the relative strength of the central forces and angular forces. Experimentally, the NTE was found for silicon[6], AlGaAs[7], but not for diamond[3], suggesting that the NTE may easier occur for softer materials. This makes the problem particularly interesting as free electrons soften the lattice of semiconductors simultaneously increasing the thermal expansion at higher temperatures.

This paper reports on the measurements for GaN, GaAs and ZnTe, three basic compound semiconductors. The study was performed in order to answer the question up to what extent the NTE depends on elastic properties of the semiconductor and on the free electron concentration.

Experimental

Measurements of the thermal expansion were performed using X-ray diffractometry and dilatometry at variable temperature. The former method consists in lattice-parameter measurement using a high-resolution diffractometer equipped with a four-reflection monochromator of the primary beam and a two-reflection analyser of the reflected beam. The lattice parameters were determined with an error $\Delta a/a$ of about 10^{-5} with respect to a perfect silicon crystal. In the latter method, the dilatometry, the capacitance was measured using an ultra-precise capacitance bridge. For a sample of a length, l, of about 10 mm, the measurement precision $\Delta l/l$ was of about 10^{-8}. The instrument was callibrated against the thermal expansion of spectrally pure copper. Elastic properties of the GaAs and ZnTe samples were examined using ultrasound method.

Gallium Nitride

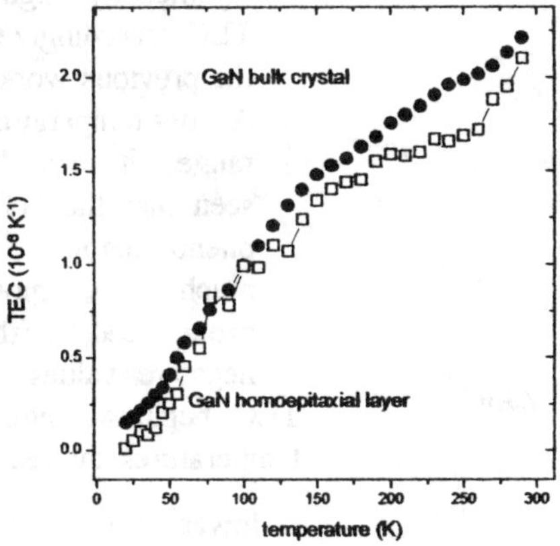

Fig.1
Thermal expansion coefficients for GaN bulk crystal (high free-electron concentration and for GaN homoepitaxial layer (low free-electron concentration). X-ray diffraction data

The examined sample consisted of a bulk crystal grown at high pressure of nitrogen (1.2 GPa, 2000 K) with a homo-epitaxial layer grown on top of it using metal-organic chemical vapour deposi-tion (MOCVD). The layer had a free- electron concentration of about 10^{17} cm^{-3}, whereas for the bulk crystal this concentration was of about $5x10^{19}$ cm^{-3} caused most probably by a high concen-tration of oxygen. Such a high free-electron concentration caused a significant lattice expansion by about 0.01% and allowed us to examine simultaneously the substrate and the layer using X-ray diffractometry as the Bragg peaks were separated.

Figure 1 shows the values of TEC for the GaN layer and the substrate. It can be seen that the GaN bulk crystal expands more with temperature, accordingly to expectations. However, within the experimental accuracy (the dilatometry could not be used because the sample was too small), no NTE was observed.

Gallium Arsenide

Two samples of the bulk crystals of GaAs were examined. Both were grown using the Czochralski method, but one of the samples was heavily doped with Te to about $1.1x10^{19}$ cm^{-3}. Similarly to GaN, the doped sample had a lattice parameter larger by about 0.1% with respect to the undoped sample.

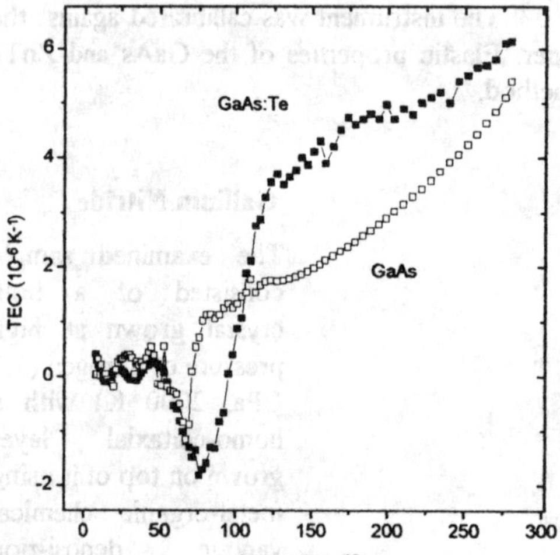

Fig.2
Thermal expansion coefficients for GaAs samples. X-ray diffraction and dilatometry data.

Figure 2 shows values of the thermal expansion coefficient for GaAs samples. At higher tempera-tures, the doped sample has a significantly higher TEC accordingly to the previous work.[1] At low-temperature range, it can be seen that the NTE pheno-menon is much more pronounced (the negative values of TEC begin at higher temperatures and reach lower values for the doped sample).

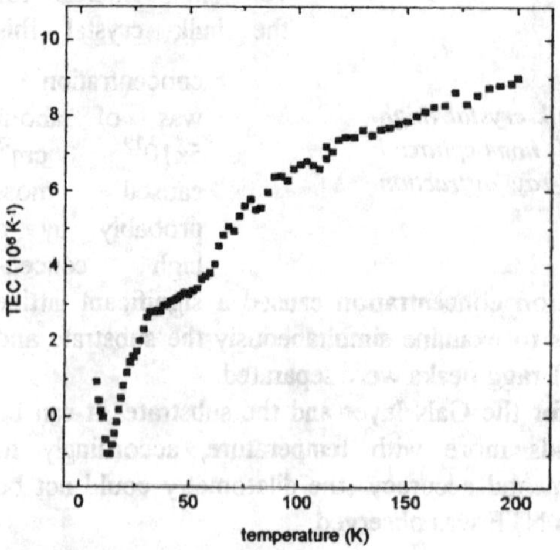

Fig.3
Thermal expansion coefficient for ZnTe. X-ray and dilatometry data.

Zink Telluride

The bulk crystals of ZnTe were grown using vapour-phase technique. Figure 3 shows the values of thermal expansion coffi-cient for this sample. It can be seen that the NTE phenomenon is evident, but occurs for a very small temperature range, only between 15 and 20 K.

Elastic properties

Table I shows elastic properties of the examined semiconductors.

sample	c_{11} (10^8Nm^{-2})	c_{12} (10^8Nm^{-2})	c_{44} (10^8Nm^{-2})	B (GPa)
GaN[a]	370	150	100	200
GaAs[c]	118	58	58	78
GaAs:Te[c]	119	52	59	74
ZnTe[b]	713	407	312	52

[a] - Reference 8
[b] - Reference 9
[c] - this work

It can be seen that there are large differences in elastic parameters of the chosen semiconductors. Accordingly to expectations, the free-electron concentration in Te-doped GaAs makes the lattice softer.

Conclusions

The experimental data give a base to draw the following conclusions:

a) the free-electrons expand the lattice, they cause an increase of thermal expansion at higher temperatures and make the phenomenon of the negative thermal expansion more pronounced (the data were available for GaAs and GaN, not for ZnTe),

b) no correlation was found between the negative thermal expansion and the elastic properties for the examined three different compound semiconductors, GaN, GaAs and ZnTe. However, the sample of GaAs doped with Te, softer than the undoped sample, exhibits the NTE phenomenon to be much more pronounced.

Acknowledgements

The work was sponsored by the grant of the Polish Committee for Scientific Research (KBN) no 175/P03/96/10.

References

1. J. Bak-Misiuk, M. Leszczynski, J.Domagala, Z. Zytkiewicz, *J.Appl. Phys.* **78**, 6994, (1995).
2. M. Leszczynski, T. Suski, H. Teisseyre, P. Perlin, I. Grzegory, S. Porowski, J. Appl. Phys. **76**, 4909 (1996).
3. C. H. Xu, C. Z. Wang, C. T. Chan and K. M. Ho, *Phys. Rev.* **B 43**, 5024 (1991).
4. S. Biernacki and M. Scheffler, *Phys. Rev. Lett.* **63**, 290 (1989).
5. T. Soma, J. Satoh and H. Matsuo, *Solid State Commun.* **42**, 889 (1982).
6. H. Ibach, *Phys. Stat. Solidi* **31**, 625 (1969).
7. G. Clec'h, G. Calvarin, P. Auvray and M. Baudet, *J. Appl. Cryst.* **22**, 372 (1989).
8. R. B. Schwarz, K. Khaachaturyan and E. R. Weber, *Appl. Phys. Lett.* **70,** 1122 (1997).
9. J.G. Collins, G. K. White, J. A. Birch and T. F. Smith, *J. Phys. C* **13**, 1649 (1980).

CRYSTALLIZATION OF Ga-Mg-Zn ALLOY

P.JELEŃ, J.GWÓŹDŹ AND M. SUROWIEC
Institute of Physics and Chemistry of Metals, University of Silesia, Bankowa 12,
40007 Katowice,Poland.

Abstract.
This paper reports on technological trials to obtain single quasicrystals in Ga-Mg-Zn system. Ivestigations have been curried out with the Laue method, X-ray diffractometer, X-ray microanalizer, metalographical microscope and differential scanning calorimeter (DSC). Till now no quasicrystal phase in Ga-Mg-Zn alloy has not been observed. However, the important characteristic of the obtained material was the occurence of many shrinkage cavites in which polyhedral crystal forms have been observed.

1. Introduction.

The interest in Ga-Mg-Zn alloy ensues from the fact that the literature data [1,2,3] are indicative of the existence of this alloy in quasicrystal phase. The icosahedral phase of composition $Ga_{1.0}$-$Mg_{2.1}$-$Zn_{3.0}$ has been obtained by Ohashi and Speapen [3]. The purpose of our studies was realization of technological trials to obtain the quasicrystal alloy.

2. Experimental.

The technological process has been executed in the constructed technological system consisting of the heating system and the pressure-chamber affording possibilities for crystalization of the alloy under an atmosphere composed of hydrogen and helium.

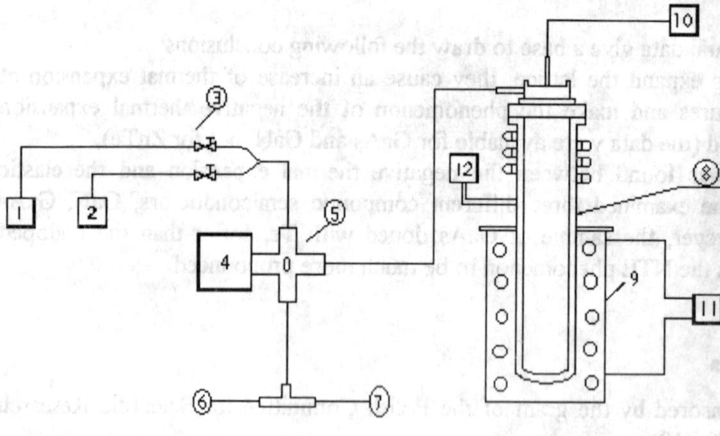

1. helium
2. hydrogen
3. reducer
4. vacum system
5. control system of gas flow
6. rubber gas container
8. autoclave
9. double zone
10. temperature control sysyem and thermoelement
11. feed heating system
12. cooling systemconduit gas outle

Fig.1. Technological systems for crystalization of the alloy under an composed atmosphere.

Investigations have been curried out using the Laue method, X-ray diffractometer for polycrystalline samples, X-ray microanalizer, metalographical microscope and the differential scanning calorimeter (DSC).

3.Results.

The important characteristic of the obtained material was the occurrence of many shrinkage cavites in which polyhedral crystal forms have been observed.

On the ground of realized studies with the Laue method we noted that the examined polyhedral single crystals grown in shrinkage cavites showed the typical symmetry for the Laue diffraction classof 6/mmm. The single crystals exhibited flat faces and heksagonal cross section.

The X-ray powder diffraction showed homogenity of the specimen in respect of the phase content all over its volume.

Fig.2. Polyhedral crystal forms grown in shrinkage cavites of Ga-Mg-Zn alloy.

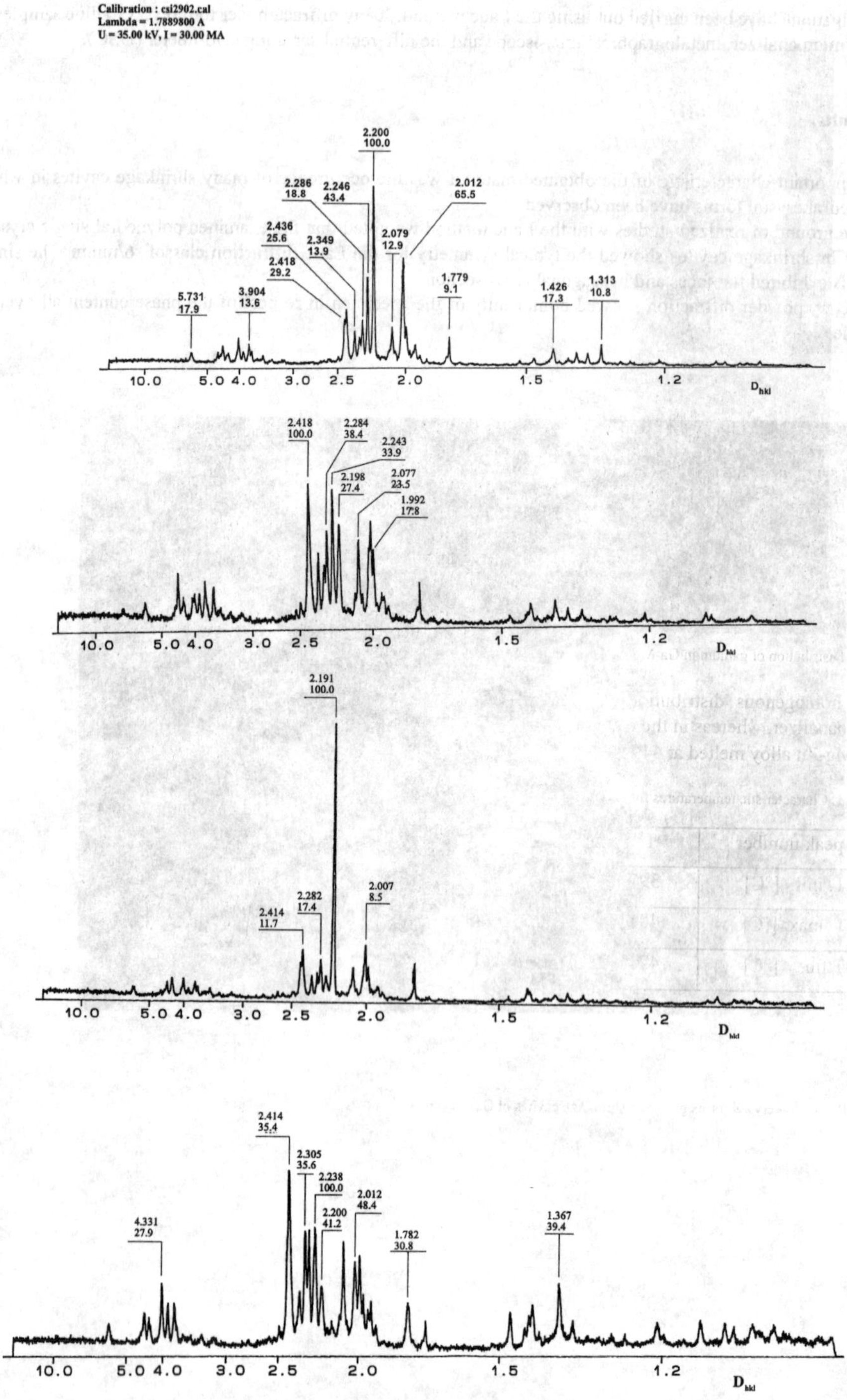

Fig.3. A set of four powder diffraction patterns taken from diffrent regions of the sample

The alloy contained at least two phases which are not yet fully identified.

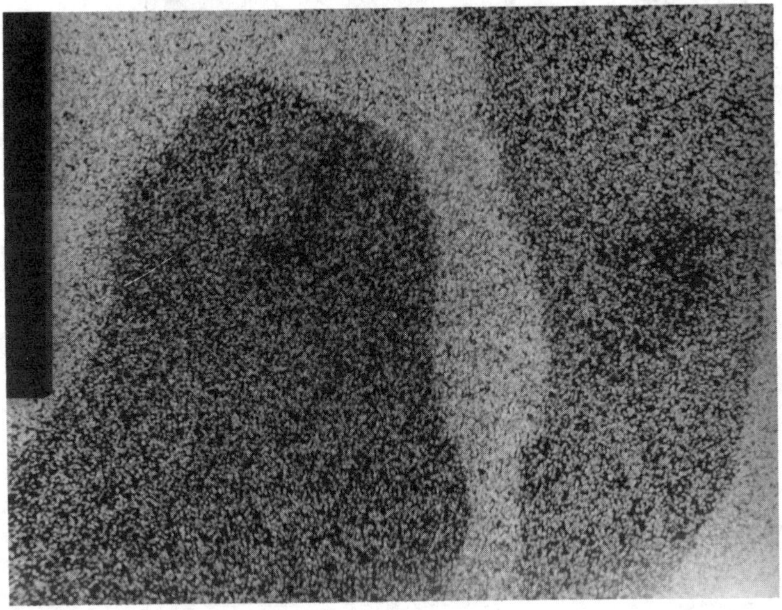

Fig.4. Distribution of gallium in Ga-Mg-Zn alloy. Two phase region is clearly visible.

The homogenous distribution of elements such as magnesium and zinc has been confirmed by X-ray microanalizer, whereas in the case of gallium there are heterogenous regions. The DSC method showed that the Ga-Mg-Zn alloy melted at 449 °C.

Table.1. Characteristic temperatures for melting process.

peak number	I	II	III	IV
T. init. [°C]	397	445	462	486
T. max. [°C]	401	449	463,5	489
T.fin. [°C]	421	452	466	498

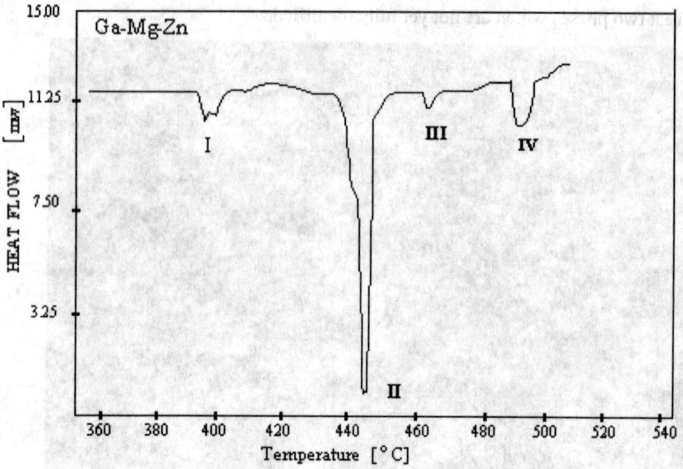

Fig.5. Heat flow during melting process of Ga-Mg-Zn alloy. (DSC curve).

4. Conclusions.

Till now our studies has not certified to existence of a quasicrystal phase in the Ga-Mg-Zn alloy. However, some obtained results are very inviting (for example : polyhedral single crystals grown in the shrinkage cavites). All phases of obtained alloy are not yet fully identified but our studies are still going on and each day can bring the succesful results.

References.

[1] H.S.Chen and A.Inoue, Scripta Met. 21 (1987) 527.
[2] C.Beeli and H.U.Nissen, Philos.Mag. B, 68 (1993) 487.
[3] W. Oshashi and F.Speapen, Nature, 330 (1987) 555.

Defect Structure

INFLUENCE OF LATTICE ROTATIONS ON THE INTENSITY DISTRIBUTION OF X-RAY SCATTERING

R.I.BARABASH

Metal Physics Department, National Technical University of Ukraine, pr.Pobedy,37,Kiev 252056 Ukraine

P.KLIMANEK

Institut fur Metallkunde, Freiberg University of Mining and Technology, Bergacademie, Gustav-Zeuner-Str.5, D-09 596

Dislocation structure consisting of different kind dislocation walls results in the appearance of misorientations and rotations in the crystal structure. We used the model when the boundary is formed by two sets of parallel edge dislocations inclined towards another by an angle b. In this case the intensity distribution is close to Gaussian in radial direction and to Lorents in azimuth direction. Expressions for scattering intensity are true, when the averaging over a large statistical ensemble of defects takes place. If this condition is not valid then intensity distribution in azimuth plane will consist of several spikes. The mean difference between these spikes in the reciprocal lattice space is connected with misorientations between the walls. At larger stages of deformation there is often formed the excess of dislocation walls resulting in non-zero rotations in some comparatively large part of the sample (for example, connected with uniform bending or torsion). The main effects caused by such rotation fields in crystals will mostly change the intensity distribution in azimuth plane. Such intensity distributions can be measured for example by rocking curve technique or by restoring the whole reciprocal lattice space around different reciprocal lattice points. In this case the intensity distribution in the azimuth direction will have so called "table-like" shape.

1 Introduction

Different dislocation structures result in the formation of complex intensity distribution in the reciprocal lattice space. Around each reciprocal lattice site with Miller indices (hkl) corresponding to certain reciprocal lattice vector G_{hkl} the appearing intensity distribution changes, if the diffraction vector $Q=k_2-k_1$, (k_1,k_2-wave vectors of initial and scattered waves) is deviating from G_{hkl}. For the deviation vector $q=Q-G_{hkl}$ the most important are two directions: radial ($q_\parallel$ parallel to G_{hkl}, which in the case of Debye diffractogramm usually is defined as $q_D \equiv q_\parallel$) and azimuth ($q_\perp$ perpendicular to G_{hkl}). In these two directions intensity distribution often has quite different character, is described by different functions and correlates with different parameters of dislocation structure. At early stages of deformation a uniform dislocation distribution is often formed. It corresponds to smooth broadened X-ray intensity distributions in all directions around reciprocal lattice point. Such profiles were observed for example in classic paper of M.Wilkens [1], Ungar etc. [2]. They have observed the dependence of broadening on the extent of deformation.

2 Evidence of Lattice Rotation in X-Ray Rocking Curves

During plastic deformation there is often a dislocation structure formed consisting of dislocation walls of different kind. They result in the appearance of misorientations and rotations in the crystal structure. For example during experimental investigation of bended Be blades in [3] there were obtained complex intensity distributions in azimuth direction by rocking curve technique. They show the appearance of disorientation in crystal structure (Fig.1).

Interesting results were obtained during experimental investigation of pseudosingle crystal tubes obtained by electron beam zone remelting. Experimental intensity distributions were obtained both in radial and azimuth directions [4]. Another example of the formation of rotational fields in the crystal structure was observed during creep investigation by rocking curve technique. The intensity distribution in azimuth direction in this case often consisted of several spikes [5].

From these examples it can be seen that different dislocation structures sometimes result in almost close intensity distribution along radial direction $q_{|}$, and it is necessary to analyse azimuth intensity distribution $q_{\perp}$ to distinguish them and to have more information about dislocation structure characteristics.

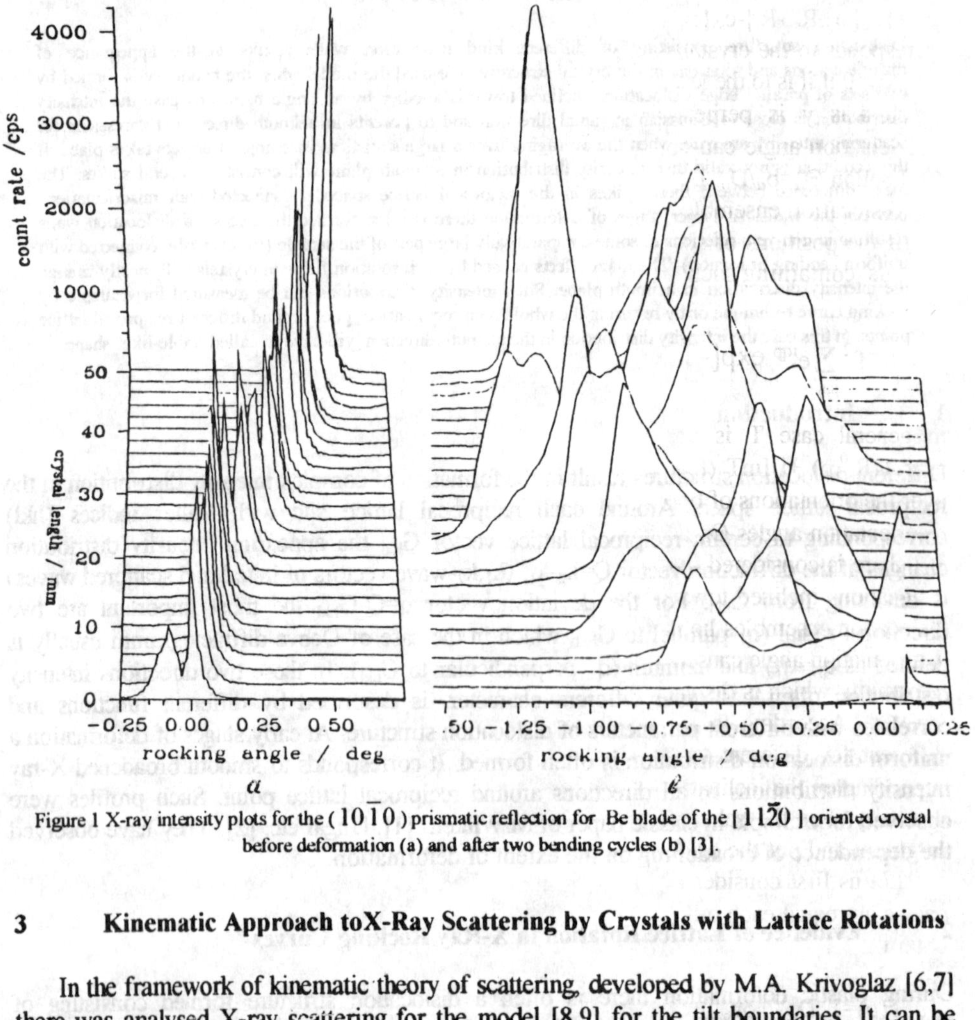

Figure 1 X-ray intensity plots for the ($10\bar{1}0$) prismatic reflection for Be blade of the [$11\bar{2}0$] oriented crystal before deformation (a) and after two bending cycles (b) [3].

3 Kinematic Approach toX-Ray Scattering by Crystals with Lattice Rotations

In the framework of kinematic theory of scattering, developed by M.A. Krivoglaz [6,7] there was analysed X-ray scattering for the model [8,9] for the tilt boundaries. It can be described by a system α of edge dislocations, the unit vector parallel to the rotation axis ω_α coincides with unit vector along the dislocation lines τ_α. For twist boundaries formed, for example, by two sets of orthogonal screw dislocations, it coincides with the unit vector perpendicular to the plane containing dislocation lines n_α. For more complex dislocations ensemble generally ω_α is at an acute angle to n_α. Let vector R_r extends from the rotation axis to the r-th plane. If dislocations are regularly arranged inside dislocation ensemble forming the boundary with orientation α and occuping the position r we may regard it as a single defect, producing certain rotation field $\omega_{sr\alpha}$. For such case the stresses caused by such defect decrease exponentially with the increase of the distance from the defect so, that such defects mainly produce rotation field in the crystal. We shall consider the case, when dislocation ensemble results only in rotations and there are no strains between them connected with translation part of the displacement field caused by each such defect $u_{sr\alpha}$. Let us for simplicity assume, that certain tilt boundary rotates only one part of the crystal, for instance the half to the right side of it. The difference of displacements due to the rotation produced by such small angle tilt boundary in two points s and s' can be written as following [8]: $u_{ss'r\alpha}=u_{sr\alpha}-u_{s'r\alpha}=\{c_{sr}[\omega_\alpha,R_s^0-R_r]-c_{s'r}[\omega_\alpha,R_{s'}^0-R_r]\}\Theta_\alpha$, if the condition $R_{sr},R_{s'r}\rangle\rangle D_0\rangle\rangle b$ is fulfilled. Here R_s^0 corresponds to the crystal without any defects; $c_{sr}=1$ if the point s is to the right side of the wall, $c_{sr}=0$ if it is to the left side of the wall; R_{sr} - the distance from the point s to the plane r in which axis y is perpendicular to the dislocation lines, forming the boundary, Θ_α - disorientation angle caused by single α-type dislocation wall. For a simple case of wedge disclination $\Theta_\alpha=b_\alpha/D_0$. D_0 is equal to the mean distance between neibouring dislocations in this dislocation ensemble forming, the boundary.

If the positions of such defects *r* are random then the intensity of X-ray scattering for crystals, containing such defects can be written as following:

$$I=f^2\sum_{s,s'}e^{iq\rho}\exp[-T(R^0_s,\rho)];T=\sum_\alpha c_\alpha\sum_r[1-\exp(iQu_{s's r\alpha})] \qquad (1)$$

in general case T is complex and depends on both R_s^0 and ρ, so that $T(R_s^0,\rho)=$ $=\mathrm{Re}\,T(R_s^0,\rho)+i\,\mathrm{Im}T(R_s^0,\rho)$, where $\rho=R_s^0-R_{s'}^0$. Here the summation over α is performed over the orientations of boundaries formed by dislocation walls (over n_α), over the values of disorientation angles Θ_α and over the distances A_δ between the nearest α-type walls. In [8-10] there is considered the case when the boundary is formed by two sets of parallel edge dislocations inclined towards another by an angle β. (Fig.4). The dislocation lines in both dislocation ensembles lie in the same plane. In this case distribution function for a resulting disorientation angle caused by simultaneous influence of these two dislocation ensembles has maximum, which is displaced from zero and its values lie in the interval x=0,4-1.0, where $x=\Theta/\Theta^m$. According to [8-10] the distribution function $f(\Theta_\alpha)$ for disorientation angles is different for different types of the positions of these two initial dislocation ensembles depending on the inclination angle β between the sets of dislocation ensembles, forming this boundary. For the sake of definiteness let us fix the front left bottom of certain crystal fragment and put the origin of coordinates at its edge.

Let us first consider the case, when probabilities for boundaries with opposite rotation angles are equal each other. Then $\mathrm{Im}T(\rho)=0$ and after the summation of (1) over r we shall obtain:

$$T(\rho)=\Sigma_\alpha D_\alpha^{-1}|\rho n_\alpha|+\Sigma_\alpha f(\Theta_\alpha)D_\alpha^{-1}|R_s^0 n_\alpha|\{1-\cos(Q[\omega_\alpha\rho]\Theta_\alpha)\} \qquad (2)$$

The last equation is true, if the condition $\rho\rangle\rangle 1/Q\Theta$ is valid. Here D_α is the mean distance between neighbouring α-type boundaries of any sign. The first term in (2) depends on the terms in the sum over r in (1), for which the plane r is between the points s and s' and the second term depends on the terms for which both points s and s' are situated at the same side of the plane r.

The last expression for $T(\rho)$ results in the sharp asymmetry of X-ray scattering intensity distribution in the vicinity of the reciprocal lattice space. If $\rho \| Q$ the second term in (2) vanishes. That's why in radial direction, or in Debye diffractogram, which is taken for $q \| G$, and hence for $\rho \| Q$ the last equation will look like: $T(\rho)=\Sigma_\alpha D_\alpha^{-1} |\rho n_\alpha|$ and such expression for T results in the Lorenz's intensity distribution in radial direction $I_D(q_D)=I_{Di}q_{Di}/(\pi^2 q_D^2+q_D^2)$, where $q_{Di}=\pi/\overline{D}$, $\overline{D}=\Sigma_\alpha D_\alpha/|mn_\alpha|$, or in the case, when D_α do not differ essentially for different types of boundaries, it can be replaced by its average value D, so that $q_{Di}=\pi\varsigma/D$, $\varsigma=\Sigma_\alpha |mn_\alpha|$, where $m=\mathbf{Q}/Q$ and D is an averaged value of the distancies between neighbouring boundaries of all types. In angle units an integral line width for last case is equal $2\delta\theta_i=\varsigma\lambda\sec\theta/2D$. The factor ς determines the dependence of line width on the boundary orientation and on the diffraction vector Q. Let us analyse the intensity distribution in the azimuth direction analogous to the method used in [6,11,12].

In the azimuth direction the intensity distribution is mainly influenced by the second term in (2). As the coefficient $R_s^0D^{-1}\sim LD^{-1}$ is rather high, then if disorientation angle is not very large, then we can retain only quadratic terms in (2) in the expansion of the second term in (2) in powers of ρ:

$$T(\rho)-\Sigma_\alpha D_\alpha^{-1}|\rho n_\alpha|=R_s^0Q^2B_{\xi\eta}\rho_\xi\rho_\eta,$$
$$2B_{\xi\eta}=\Sigma_\alpha\Theta_\alpha^2 f(\Theta_\alpha)|R_s^0n_\alpha|[m\omega_\alpha]_\xi[m\omega_\alpha]_\eta(R_s^0D_\alpha)^{-1} \qquad (3)$$

Here co-ordinate axis ξ and η are perpendicular to vector G_{hkl}. As was shown in papers [8,9], distribution function $f(\Theta_\alpha)$ has maximum, which slightly depends on the inclination angle β between the two sets of of dislocation ensembles, which form the boundary. Let us for the simplicity assume, that $\beta=90^0$. Under such β value distribution function $f(\Theta)$ is the most simple $\Theta_\alpha=\Theta_\alpha\sigma^{-2}\exp(-\Theta_\alpha^2/2\sigma^2)$. For each value R_s^0 matrix (3) can be diagonalized by proper choice of ξ and η axis. Substituting (3) into (1) we shall obtain for the intensity distribution (for the case, when ξ and η are the main axis and when the distance between boundaries of different types D_α is almost equal its mean value D) :

$$I(q)=2\pi^2 f^2 v^{-2}Q^{-2}q_{Di}/(\pi^2q_D^2+q_{Di}^2)\bullet$$
$$\int dR_s^0D(R_s^0)^{-1}(A_{\xi\xi}A_{\eta\eta})^{-1/2}\exp[-0{,}25D(R_s^0)^{-1}Q^{-2}(q_\xi^2/A_{\xi\xi}+q_\eta^2/A_{\eta\eta})] \qquad (4)$$

where $2A_{\xi\eta}=\Sigma_\alpha\Theta_\alpha^2 f(\Theta_\alpha)|R_s^0n_\alpha|[m\omega_\alpha]_\xi[m\omega_\alpha]_\eta(R_s^0)^{-1}$. Then the section of the intensity distribution in the vicinity of the reciprocal lattice point with Miller indices (hkl) by the plane $q_\perp \perp G_{hkl}$ in the direction of the axis ξ can be written as following:

$$I(q_\xi)=2\pi(f^2D/v^2Q^2\varsigma)\int dR_s^0(D/R_s^0\sqrt{A_{\xi\xi}A_{\eta\eta}})\exp(-q_\xi^2D/4R_s^0Q^2A_{\xi\xi}) \qquad (5)$$

This distribution (5) is close to Gaussian. Being measured by rocking curve method for the case, when rotation axis coincides with axis ξ, if all D_α are practically equal the mean value D, halfwidth of intensity distribution (5) is equal $\delta q_\perp = =2(\pi A_0L/D)^{1/2}Q$, where $(A_0)^{-1/2}=V^{-1}\int dR_s^0(L/A_{\xi\xi}R_s^0)^{1/2}$. Here $A_{\xi\xi}$ depends on the distribution function $f(\Theta_\alpha)$ and Θ_α.

When there exist several types of the walls with essentially different values D_α these expressions will look like $\delta q_\perp = 2(\pi B_0L)^{1/2}Q$, $(B_0)^{-1/2}=V^{-1}\int dR_s^0(L/B_{\xi\xi}R_s^0)^{1/2}$.

Expression (4) is true, when the averaging over a large statistical ensemble of defects takes place. As the crystal contains L/D boundaries formed, then such averaging will take place, if the condition $L \gg D^3Q^2\Theta^2$ is fulfilled. Here Θ-is the mean value of disorientation angle.

In some cases this condition is not valid. Then intensity distribution in azimuth plane will consist of several spikes. The mean difference between these spikes in the reciprocal lattice space is equal $\delta q_{\perp} \sim (D/L)^{-1/2}\Theta Q$.

At larger stages of deformation there is often formed the excess of dislocation walls, resulting in non-zero rotations in some comparatively large part of the sample (for example, if a band structure is formed). Then opposite to the above case the imaginary part of T will not disappear $\mathrm{Im}\, T(R_s^0,\rho) = -\Sigma_\alpha c_\alpha \Sigma_t\, \rho\partial(Qu_{st\alpha})/\partial R_s^0$. So that in a whole T can be written as a following sum: $T=T_1+T_2'+T_2''$: $T_1=i\Sigma_{t\alpha}c_\alpha(\rho\nabla)\bullet(Qu_{st\alpha})$; $T_2'=-\Sigma_{t\alpha}c_\alpha\{1-\cos[(\rho\nabla)(Qu_{st\alpha})]\}$; $T_2''=0{,}5i\Sigma_{t\alpha}c_\alpha\cos[(\rho\nabla)(Qu_{st\alpha})](\rho\nabla)^2(Qu_{st\alpha})$. Here it was convinient to write separately linear part of the imaginary part of T. This linear contribution to the value T_1 is connected with rotational fields of uniform bending or torsion. When we consider such rotational fields caused by excess dislocation ensembles, we can speak about the corresponding density of excess dislocations $\Delta n_\alpha=(D_{0\alpha})^{-1}[(D_\alpha^+)^{-1}-(D_\alpha^-)^{-1}]$ connected with these excess disclination of α type, providing such rotational fields. Here $D_{0\alpha}$ is the distance between dislocations in the wall, forming the boundary, and D_α^+, D_α^- are correspondingly the distancies between α-type boundaries, having positive or negative direction of rotation axis correspondingly. In the framework of continuum dislocation theory analogous to [12-15] corresponding mean distortion field W_{mn} is determined by the averaged tensor ρ_{ik} of dislocation density by following equation $\varepsilon_{klm}\partial W_{mn}/\partial x_l=-\rho_{kn}$, where ε_{klm} is unit antisymmetric tensor. When the crystal is not in the external stress field so, that on its boundary the condition $\sigma_{kl}n_l|_S=0$ is fulfilled, the solutions of the last equation do not depend on the shape of the crystal surface S. Then T_1 for such case can be written as following:

$$T_1(R_s^0,\rho) = -i\, R_s^0\rho\Sigma_\alpha\Theta_\alpha f(\Theta_\alpha)[(D_\alpha^+)^{-1}-(D_\alpha^-)^{-1}]Q\, B_\alpha\,, \qquad (6)$$

where vector B_α depends on the type of dislocation wall forming the boundary and on the boundary conditions at the crystal surface.

In the crystals with free boundary for one system α of dislocation walls with $\tau_\alpha \| 0Z$, formed by the set of edge dislocations, which lines are situated in the plane with the normal n_α parallel to axis 0X only one component of excess dislocation density tensor ρ_{ik} will not vanish: $\rho_{zx}=\Theta_\alpha f(\Theta_\alpha)[(D_\alpha^+)^{-1}-(D_\alpha^-)^{-1}]$ and distortion tensor will have only the components: $W_{xy} = -W_{yx} = x\Theta_\alpha f(\Theta_\alpha)[(D_\alpha^+)^{-1}-(D_\alpha^-)^{-1}]$

In this case vector B_α in (6) will look like: $B_\alpha=(e_s n_\alpha)[\tau_\alpha m]$, where $e_s=R_s^0/R_s^0$.

For crystals with free boundaries containing twist boundaries formed by sets of screw dislocations which dislocation lines are parallel to axis OZ, only one component of dislocation density tensor does not vanish: $\rho_{zz}=\Theta_\alpha f(\Theta_\alpha)[(D_\alpha^+)^{-1}-(D_\alpha^-)^{-1}]$. For each set of α-type screw dislocations we shall choose its own co-ordinate system in which axis OZ will be parallel to the direction of positive α-type screw dislocations set. In that case rotation field will be just a torsion field and will be described by antisymmetric distortion tensor $W_{xy}=0{,}5z\Theta_\alpha f(\Theta_\alpha)[(D_\alpha^+)^{-1}-(D_\alpha^-)^{-1}]$; $W_{xz}=0{,}5y\Theta_\alpha f(\Theta_\alpha)[(D_\alpha^+)^{-1}-(D_\alpha^-)^{-1}]$; $W_{mn}= -W_{nm}$; $W_{yz}= -0{,}5x\Theta_\alpha f(\Theta_\alpha)[(D_\alpha^+)^{-1}-(D_\alpha^-)^{-1}]$..

For the case of elastically isotropic crystals with free boundaries and with torsion fields vector B_α will look as following $B_\alpha= 0{,}5\{[e_s m]-2(\tau_\alpha e_s)[\tau_\alpha m]\}$.

For general case, when the crystal contains both bending and torsion fields the summation over α in (6) must include all types of dislocation ensembles. From the above expressions it is essential, that vector B_α is always perpendicular to Q. It means, that the main effects caused by rotational fields in crystals will mostly change the intensity distribution in the plane ⊥Q (in azimuth plane). Such intensity distributions can be measured for example by rocking curve technique or by restoring the whole reciprocal lattice space around different reciprocal lattice points. For such type of dislocation substructure intensity distribution in the azimuth direction will have so called "table-like" shape, and will differ essentially from previous case.

4 Conclusions

- Structures with compensated rotation fields result in Gaussian X-ray intensity distribution. Broadening of rocking curve depends on the distribution function and on the extent of local misorientation.
- When the size of scettering fragments is of the order of irradiated area intensity distribution in azimuth plane consists of several spikes.
- Structures with uncompensated rotation fields result in 'table-shape' intensity distribution in azimuth plane. Broadening of rocking curve being determined by the excess of dislocation walls and uncompensated rotations.

References

1. M.Wilkens,*Appl.Cryst.*,**-12,** 119 (1979)
2. T.Ungar, H.Mughrabi, M.Wilkens,*Acta Metall*, **30,** 1861 (1982)
3. C.May,P.Klimanek,A.Magerl, *Nuclear Instruments & Methods in Physics Research,* **A** 357, 511 (1995)
4. Barabash O.M., Barabash R.I., Masimov M.V., Karasevskaya O.P., Zaderii B.A.*Metal Physics and modern technology*. 16, № 5, 33 (1994).
5. R.I. Barabash, V.A. Kononenko, O.P. Karasevskaya. *Metallofizika*, v.10, N3, 106 (1988)
6. M.A.Krivoglaz, *X-ray and Neutron Diffraction in Nonideal Crystals, Springer, Berlin-NewYork,* (1996)
7. M.A.Krivoglaz, K.P.Ryaboshapka, R.I.Barabash: *Fiz. Metal. Metalloved.* 3**0**, 1134 (1970) [English transl.*Phys.Metals Metallog*.**30**, (1970)]
8. W.Pantleon:*Scripta Metall*. **35**, 4, 511 (1996)
9. D.Franke, W.Pantleon, P.Klimanek, *Computational Materials Science*, **5**, 111 (1995)
10. D.Franke, F.Zimmerman, W.Pantleon and P.Klimanek, *Proc. of the 4th European Conference on Advanced Materials and Processes in Padua/ Venice, Italy,* Vol.3, 233, Associazione Italiana de Metallurgia, Milano (1995)
11. R.I.Barabash, O.P.Karasevskaya, V.A.Kononenko, K.P.Ryaboshapka: *in Ser. Metallofizika*, ed. By V.N. Svechnikov, Vol. 70 (Naukova Dumka, Kiev 1977)
12. K.P.Ryaboshapka, N.A. Stelmashenko, O.N. Grigorijev: *Materials Sci.&Eng.,* A127, 65 (1990)
13. R.I.Barabash, M.A.Krivoglaz, K.P.Ryaboshapka: *Fiz. Metal. Metalloved.* **41**, 33 (1976) [English transl.:*Phys. Metals Metallog*. **41** (1976)]
14. R.I.Barabash, K.P.Ryaboshapka: *Fiz. Metal. Metalloved.* **41**, 264 (1976) [English transl.:*Phys. Metals Metallog*.**41**, (1976)]
15. O.V.Martynenko, K.P.Ryaboshapka: *Fiz. Metal. Metalloved.* **46**, N2, 231 (1978) [English transl.:*Phys. Metals Metallog*.**46**, (1978)]

HRTEM IMAGE SIMULATION IN THE STUDY OF PLANAR DEFECTS IN GaAs: Te CRYSTALS

J. BORYSIUK, J.A. KOZUBOWSKI
Faculty of Materials Science and Engineering, Warsaw University of Technology, Narbutta 85, 02-524 Warsaw, Poland

T. SŁUPIŃSKI
Institute of Experimental Physics, Warsaw University, Hoża 69, 00-681 Warsaw, Poland

The structure of planar defects (i.e., extrinsic stacking faults and microtwins) in heavily doped GaAs:Te (n ~$1\times10^{19}cm^{-3}$) was analyzed by high-resolution transmission electron microscopy (HRTEM) and multislice image-simulation technique. It is proposed that these planar defects are formed as a Ga-Te precipitate plates in consequence of change in chemical composition of supersaturated solid solution. The computer simulated images are used to show differences in contrast of images of defects created by native atoms and defects which contains impurity atoms.

1 Introduction

In n-type GaAs crystal heavily doped with Te a correlation between planar defects creation and free electron concentration changes caused by annealing was observed. [1]
The annealing at low temperature decreases free electron concentration and generates extended type defects, such as dislocation loops formed on {111} planes and containing extrinsic stacking fault. According to Hutchinson [2] the nature of faulted loops was interpreted as a presence of the interstitial Ga and As atoms in the matrix. Verner [3] proposed that existence of the faults was related to planar Ga-Te precipitates.
In the present studies the computer simulated images were used to find the optimal HRTEM imaging conditions for the planar defects. These conditions correspond to the largest differences between contrasts of the defects with impurity atoms and defects composed of native atoms. The computer simulated images are essential to get information about chemical composition of observed defects when chemical microanalyses (EDX) did not show clear differences in composition.

2 Experimental Conditions

Investigations were performed on the highly Te-doped GaAs samples (estimated free electron concentration $n=1.4\times10^{19}cm^{-3}$). Specimens were annealed at two temperatures: T_1=1150 °C for 2h (homogenization) and T_2=700 °C for 74 h, with quenching in water. Thin foils of [110] orientation were examined in JOEL JEM 3010 electron microscope operated at 300 kV in the high-resolution imaging mode. Simulated high-resolution TEM images were calculated using the EMS multislice program described by Stadelmann. [4]
The following conditions were assumed: accelerating voltage U= 300 kV; spherical aberration coefficient C_s= 1.06mm; beam semi-convergence S_c= 0.8mrad; objective aperture diameter $12nm^{-1}$; Gaussian spread of focus d_s= 10nm.

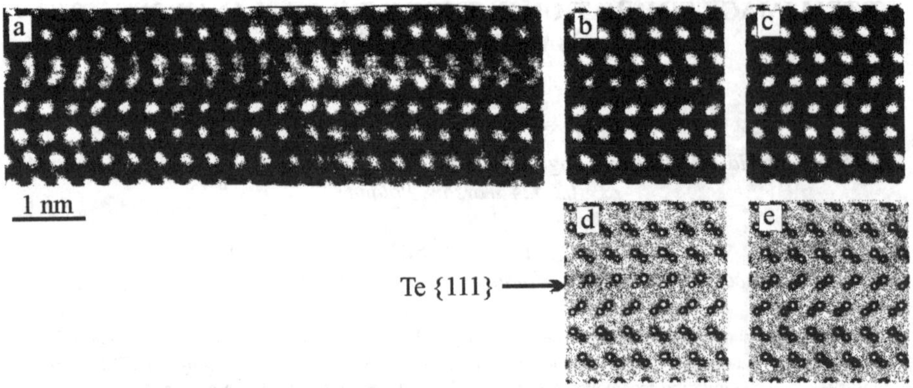

Figure 1: HRTEM experimental image of an extrinsic stacking fault in a <110> oriented GaAs: Te (a), simulated images of stacking fault with Te atoms (b) and ordinary stacking fault (c) for sample thickness t = 12 nm and defocus Δf = - 60 nm, images of wave function (d,e) for defects (b,c); inserted {111}Te plane is indicated (arrow).

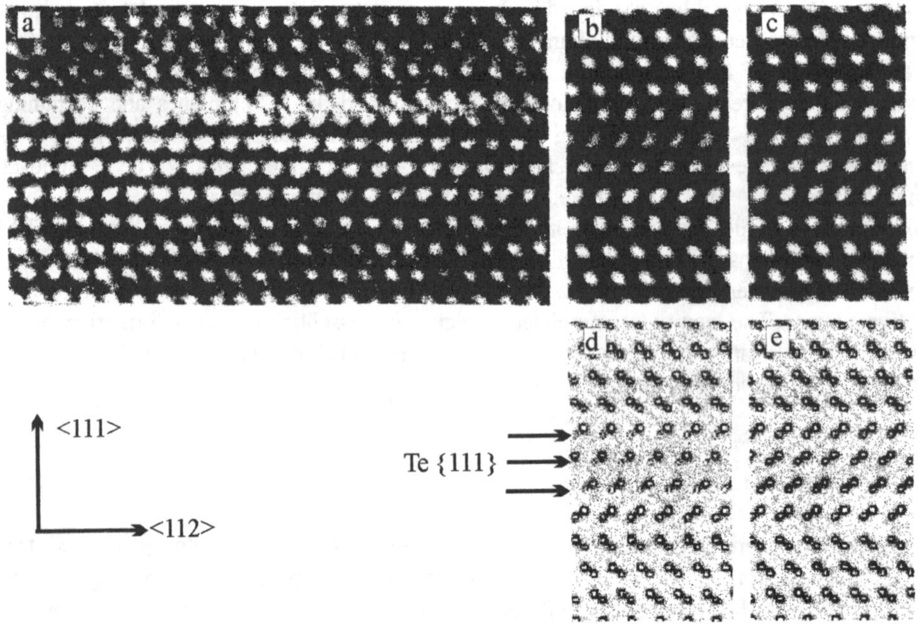

Figure 2: HRTEM experimental image of an micro-twin (a), simulated images of micro-twin with Te atoms (b) and ordinary micro-twin in GaAs (c) for sample thickness t =12 nm and defocus Δf = - 60 nm, images of wave function (d,e) for defects (b,c); inserted {111}Te planes are indicated (arrows).

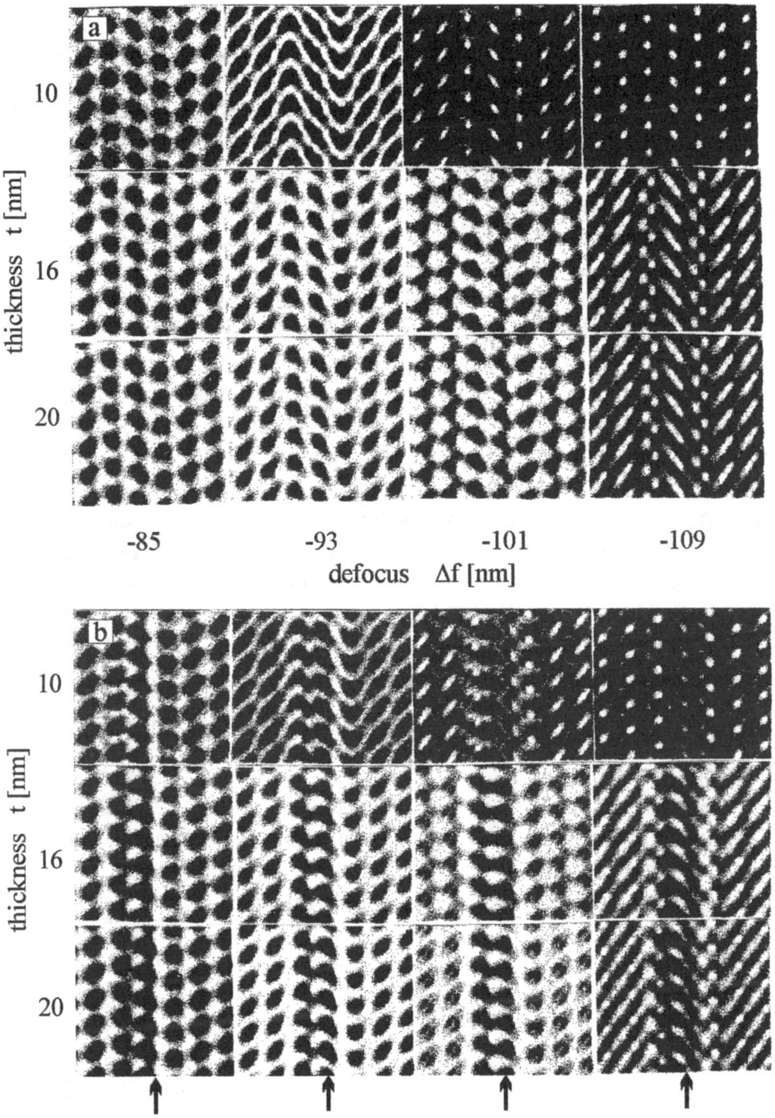

Figure 3: A series of simulated images as a function of defocus Δf and crystal thickness t which demonstrates the changes in image contrast of ordinary extrinsic stacking fault (a) and stacking fault with impurity atoms (b); arrowed rows contains {111}Te planes.

3 Results and Discussion

The annealing at T= 700°C caused a drop in free electron concentration to $n=4\times10^{18}cm^{-3}$ and creation of dislocation loops (Burgers vector b=(1/3)a<111>) containing extrinsic stacking fault on {111} planes. An example of HRTEM images show the extrinsic stacking fault - Fig.1(a) and the micro-twin - Fig.2(a).

The possibility of creation of Ga-Te precipitate plates on {111}GaAs planes was assumed for calculations of the defects with impurity atoms. These precipitates are a consequence of a segregation of Te and Ga atoms on the {111} GaAs planes. The plates consist of two kinds of atom planes; one of them (Ga_{Ga} {111}) contains Ga atoms only and the other (Te_{As}{111}) - Te atoms. The stacking fault and micro-twin contrasts are asymmetric and lattice planes bending at the edge of the defects is observed. The fitting of Ga-Te planes in the GaAs matrix may cause a local change of the lattice parameter, which was not included in calculations. The simulated images corresponding to HRTEM images are shown in Fig.1(b) - stacking fault with Te atoms, Fig.1(c) - ordinary stacking fault, Fig.2(b) - micro-twin with Te atoms, Fig.2(c) - ordinary micro-twin. Every bright spot in simulated images corresponds to pair of columns (Ga-As or Ga-Te), the microscope resolution is insufficient to resolve the columns. The images of wave functions of defects are presented in Figs.1(d,e) and 2(d,e); inserted {111}Te planes are indicated (arrows). The simulated images are close to the experimental HRTEM images if sample thickness is between 8 and 14 nm and observations are performed near the Scherzer defocus $\Delta f \sim -60$ nm. The considerable differences of the electron scattering at impurity atoms and native atoms was not noticed. These differences were revealed at defocus values $\Delta f = -85 \div -109$ nm and sample thickness $t \geq 10$ nm, as it shown in Fig.3. The conditions of the optimal contrast are the same for the stacking fault and the micro-twin.

4 Conclusions

The optimal imaging conditions of the defects of dissimilar chemical composition were found. Distinct differences of the contrast of simulated images can be observed at defocus values $\Delta f = -85 \div -109$ nm and sample thickness $t \geq 10$nm.

Acknowledgments

JEOL 3010 electron microscope used in the study was founded by the Foundation for Polish Science. The permission of Prof. P.A. Stadelmann for using his EMS program for simulation of images is greatly acknowledged.

References

1. J. Borysiuk, J.A. Kozubowski, T. Słupiński, *Acta Phys.Polon.* A **90**, 739 (1996).
2. P.W. Hutchinson, P.S. Dobson, *J. Mater. Sci.* **10**, 1636 (1975).
3. V.D. Verner, S.K. Maksimov, D.K. Nichugowski, *Phys. Stat. Sol.* (a) **33**, 755 (1976).
4. P.A. Stadelmann, *Ultramicroscopy* **21**, 131 (1987).

QUANTITATIVE TEM CHARACTERIZATION OF THE DISLOCATION SUBSTRUCTURE IN INCOLOY MA956

A. WASILKOWSKA, B. DUBIEL, A. CZYRSKA-FILEMONOWICZ
University of Mining and Metallurgy, Faculty of Metallurgy and Materials Science
Al. Mickiewicza 30, PL-30059 Kraków, Poland

The paper is devoted to TEM quantitative characterization of the dislocation substructure formed during high temperature steady state deformation of the oxide dispersion strengthened (ODS) alloy INCOLOY MA956. Practical methods for dislocation density and Burgers vector analyses are briefly described. The results provide support for the theoretical understanding of deformation behaviour of ferritic ODS alloys.

1 Introduction

The quantitative characterization of dislocation substructure is becoming an essential parameter in materials research, in particular for achieving a better understanding of deformation mechanism and in modelling the plastic behaviour of materials.

In the present investigations the analyses of dislocation density, dislocation substructure and active slip systems in an oxide dispersion strengthened (ODS) alloy INCOLOY MA956 deformed at high temperatures has been performed using transmission electron microscopy (TEM).

2 Experimental details

INCOLOY MA956 supplied as hot extruded and recrystallized bar was deformed at 950 and 1050°C under constant load (creep tests) or constant strain rate (tensile tests). Detailed description of the as-received material is given in reference [1]. TEM microstructural analyses has been performed by JEM2010ARP electron microscope[a] using thin foils prepared far from the specimen necking area.

The practical aspects of defect analyses using TEM are described by Head et al [2]. For quantitative dislocation analyses (Burgers vector, dislocation density) it is necessary to obtain an image under two-beam dynamical conditions: - on the diffraction pattern only the incident beam spot and one strong diffracted spot are intense; - the bright Kikuchi line is lying close outside to the strong diffraction spot. The procedure involves taking for the same dislocation a series of electron micrographs and corresponding selected area diffraction patterns for a number of different beam directions and diffracting conditions. To ensure that sufficient information is obtained, at least three non-complaner beam directions should be used. Orientation maps identical with a portion of a standard stereographic projection are particularly useful for these experiments, because the sense and approximate value of tilt can be directly determined (Figure 1).

For Burgers vector determination using the "invisibility criterion" it is necessary to compare the bright field and selected area diffraction patterns for the same dislocation. The "invisibility criterion" has to be fulfilled for at least two different diffraction conditions: $\mathbf{g_1}\cdot\mathbf{b} = 0$ and $\mathbf{g_2}\cdot\mathbf{b} = 0$. A Burgers vector is then parallel to the result of

[a]JEM 2010ARP electron microscope was funded by the Foundation for Polish Science under project SEZAM'94

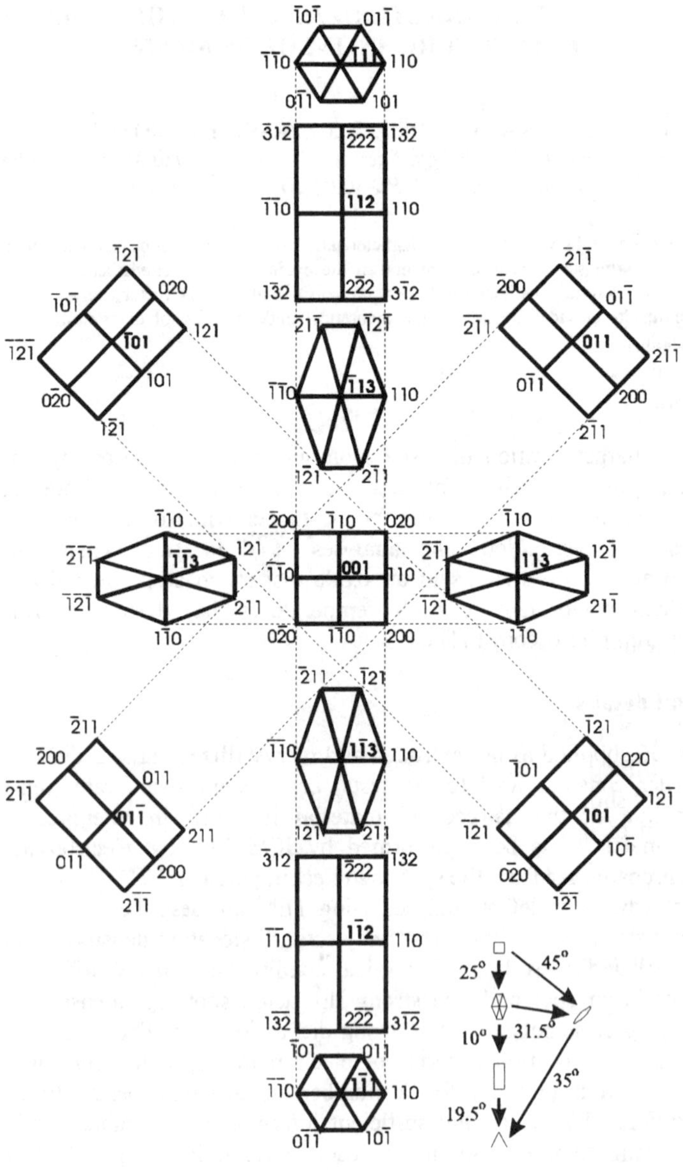

Figure 1 : Orientation map for A2 crystall structure (cutting plane 001)

$\mathbf{g}_1 \times \mathbf{g}_2$. By measuring the angle between the direction of $\mathbf{g}_{hkl}$ and of dislocation projection $\mathbf{u}_i$, on micrographs taken at least in three beam directions, the true dislocation direction **u** can be determined using a stereographic net [2]. An example is shown in Figure 2.

a)

beam	0 0 1	0 0 1	$1\ \bar{1}\ 3$	1 1 3
g_{hkl}	2 0 0	$\bar{1}\ \bar{1}\ 0$	$2\ \bar{1}\ \bar{1}$	$\bar{1}\ \bar{1}\ 0$
$-\ \bar{g}\ \bar{u}_i$	+46°	×	+18°	90°

$$b = 1/2\, a[\ \bar{1}\ 1\ 1]$$
$$u = [\ 0\ 0\ 1]$$

b)

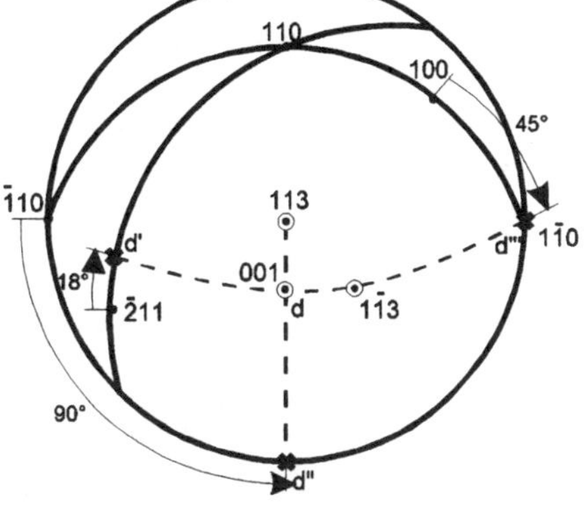

Figure 2: An example of determination the true dislocation direction **u**. a) angles between $\mathbf{g}_{hkl}$ and dislocation projection $\mathbf{u}_i$ in three beam directions and b) solution by stereographic net

The determination of dislocation density was performed by counting the intersection points of dislocations with grid lines placed on the micrograph [3]:

$$\rho = x \cdot \frac{n_1/l_1 + n_2/l_2}{t}\ [m^{-2}]$$

where n_i - number of intersection points with horizontal (vertical) grid lines
l_i - total length of horizontal (vertical) grid lines, t - foil thickness

Quantitative TEM measurements are possible for the range of dislocation densities from 10^{12} m^{-2} to 10^{15} m^{-2}. The accurate foil thickness measurements were done by converged beam electron diffraction (CBED) described in reference [4]. The value of calculated dislocation density should be increased by the appropriate factor x, which compensates for the proportion of invisible dislocation under the conditions of electron diffraction (Table 1).

Table 1: Correction factors used in the imaging of <111> dislocation in A2 lattice.

Imaging vector g_{hkl}	Percentage of invisible dislocations %	Correction factor x
{110}	50	2
{112}	25	4
{123}	25	4

3 Results and discussion

Figure 3 shows the dislocation substructure formed during high temperature creep deformation at 1050°C. The results in the medium strain rate regime give a good correlation with the creep model developed by Rösler and Arzt, which predicts the main role of thermally activated detachment of dislocations from dispersoids as a rate controlling mechanism. At all strain rates examined, dislocations pinned at the dispersoids were observed. Moving dislocations had an a/2 <111> Burgers vector and mixed orientation. The dominant slip plane determined as **b** × **u** was $\{1\ 1\ \bar{2}\}$. In specimen deformed at 1050°C the arrangement of dislocations in a substructure was not found even at very high strain rates up to $1 \cdot 10^{-3}\ s^{-1}$. At the lower temperature of 950°C the role of solid solution strengthening and dislocation climb becomes dominant and fragments of sub-boundaries were detected. Similar results were reported for the microstructure of ferritic ODS alloy PM2000 after steady state creep deformation [5].

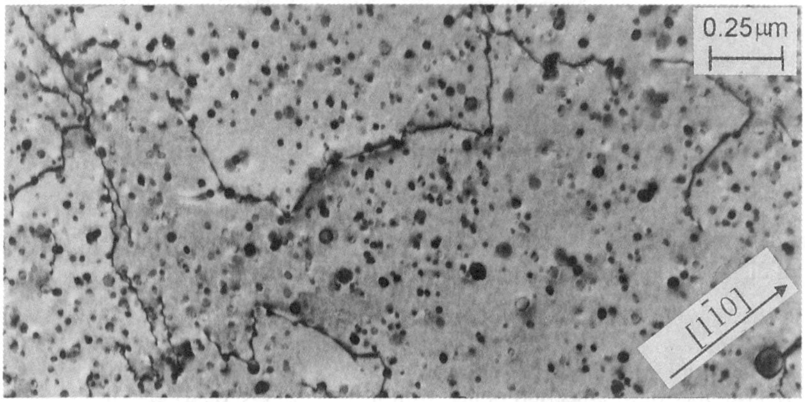

Figure 3 : The dislocation substructure formed during high temperature creep deformation at 1050°C.

4 Summary

Quantitative microstructural analyses of INCOLOY MA956 after steady state creep deformation have been carried out by TEM. The established parameters of the dislocation substructure enable the modelling of high temperature deformation behaviour. The analyses presented here are relatively easy procedures and are of practical importance for quantitative characterization of a microstructure of deformed materials.

Acknowledgements

The authors appreciate the financial support of the University of Mining and Metallurgy (grant nr 10.110.31). The valuable contribution of P. J. Ennis (Research Centre Jülich) to joint studies of the properties of ODS alloys is gratefully acknowledged.

References

1. A. Wasilkowska, W. Ratuszek, A. Czyrska-Filemonowicz, *Determination of Grain Orientations in recrystallized ODS alloy MA956,* this volume
2. A.K. Head, P. Humble, L.M. Clarebrough, A.L. Morton, C.T. Forwood in *Defects in Crystalline Solids*, eds. S. Amelinckx, R. Gevers, J. Nihoul, **vol. 7**, p. 39, (North-Holland Publishing Company, Amsterdam, 1973).
3. U. Martin, U. Muehle, H. Oettel, *Prakt. Metallogr.* **32**, 467 (1995).
4. A.Czyrska-Filemonowicz, B. Dubiel, K. Wiencek, *Acta Stereologica*, in press
5. A. Wasilkowska, R. Herzog, A. Czyrska-Filemonowicz, *Archives of Metallurgy* **42**, 89 (1997).

OXYGEN DIFFUSION AND STRUCTURE OF OXYGEN-RELATED DEFECTS IN SILICON: MOLECULAR QUANTUM CHEMICAL MODELS AND RECIPROCAL LATTICE MAPPING

J. DZELME, G. RATNIEKS, B. ZAPOL
Institute of Chemical Physics, University of Latvia
Raina b. 19, Riga, LV-1586, Latvia

A. MISIUK
Institute of Electron Technology, Al. Lotników 32/46,
02-668 Warszawa, Poland

J. BAK-MISIUK
Institute of Physics, PAS, Al. Lotników 32/46,
02-668 Warszawa, Poland

Ab initio quantum chemical simulation of oxygen-related defects O_i , $V\text{-}O_2$, and $V\text{-}O_4$ in oxygen-containing silicon was performed. The defect geometry and electronic structure were studied. The molecular cluster used to simulate $V\text{-}O_4$ defect with C_{2v} symmetry was shown to have only slight deviation from D_{2d} at atmospheric pressure, a strong deviation at 5.0 per cent lattice compression, and two energetically near (by 0.15 eV) geometries at lattice compression 0.37 per cent, the less distorted geometry being lower in energy. The cluster used for $V\text{-}O_2$ defect (C_{2v} symmetry again) was shown to have only slight deviation from D_{2d} for all studied pressure values.The migration activation energy for O_i defect was estimated as 2.73 eV at the atmospheric pressure, and 2.70, 2.68, and 1.92 eV for the lattice compressed by 0.25, 0.37, or 5.0 per cent, respectively. X-ray reciprocal lattice mapping was performed on Cz-Si samples treated at 1000K for 5h under atmospheric or 10^9Pa pressure. They showed significantly smaller size and higher concentration of defects in case of high-pressure treatment. Both theoretical and experimental results support our suggestion that high pressure increases the effective radius of O_i interaction with nucleation centre and yields a lower preexponential factor of the diffusion coefficient due to O capture by nucleation centres.

1 Introduction

The aim of our investigations of oxygen-related defects (OD) in silicon is to search direct and indirect links between the structure and properties of Czochralski-grown silicon (Cz-Si) crystals, on one hand, and modifications of OD, their migration and characteristics, on the other hand [1-8]. Cz-Si is the most used material in microelectronics. The properties of Cz-Si and their changes due to defect diffusion and reactions are very important for applications: creation of new technologies and evaluation of reliability of Cz-Si based integrated circuits. Characteristics of oxygen point defects can be used to study Cz-Si crystal local properties such as 2D and 3D defects and strain. New applications can appear with the discovery of room temperature visible photoluminescence from pressure-annealed Cz-Si [1].

Models for localized OD in Si were considered in [2-6]. Possibilities to investigate phase transitions and related phenomena in crystals were discussed in [7]. In the present work, we performed a quantum chemical simulation of three types of OD in Si: $V\text{-}O_2$, $V\text{-}O_4$ (Si vacancy and 2 or 4 O atoms), which were studied in [3], and O_i (interstitial O atom) studied in [9], as well as the X-ray reciprocal lattice mapping on Cz-Si samples with very high initial interstitial oxygen concentration, $c_o = 1\times10^{18}cm^{-3}$, annealed in argon at 1000K for 5h under either atmospheric pressure (AP) or 10^9Pa. These theoretical and experimental results were applied to analysis of O diffusion and its precipitation in Si.

2 Structure of Oxygen-Related Defects and Oxygen Diffusion

The *ab initio* SCF method was used to calculate electronic structure of defects. The reference ideal crystal was represented by a molecular cluster consisting of 17 or 41 silicon atoms, with hydrogen pseudoatoms at dangling bonds. The value of 0.23514 nm was adopted for Si-Si bond length in the perfect crystal at AP and room temperature. Pressure effect was simulated by putting a smaller value for the bond length, i. e. effective reduction of the lattice constant. The geometry optimization was performed, using (at a given pressure) the constant volume model, i.e. keeping positions of the pseudoatoms and the adjacent outer Si atoms fixed. Hydrogen pseudoatom basis function parameter and Si-H bond length were chosen so as to minimise charges at Si atoms of the cluster (these charges did not exceed 0.1 of electron charge). All defects were considered in electrically neutral state. Si vacancy was placed at the central position in the cluster. The band structure of the ideal crystal was assumed to be represented by occupied and unoccupied molecular orbitals (MO) in case of the valence and the conduction band, respectively.

Regarding Chadi results [3] corresponding to case of AP, we used the cluster with C_{2v} symmetry for V-O_4 defect for all studied pressures. For AP, the geometry showed only a slight deviation from D_{2d} symmetry, but for the case of high pressure (HP) causing 5.0 per cent lattice compression the deviation came out to be significant. In case of 0.37 per cent compression, a defect geometry with significant deviation from D_{2d} was found with energy higher by 0.15 eV than that for another geometry where the deviation is slight. This could indicate changes in structure of the defect at HP. Such changes could have interesting applications. For example, it was shown [10] that HP induced changes of defect structure in LiF and related changes in thermostimulated luminescence spectra and lioluminescence kinetics can be used to perform measurements of HP.

The Chadi model [3] was used also for the V-O_2 defect. The used symmetry of cluster with defect was C_{2v} , and deviation from D_{2d} was found to be small for different pressures. The distance between the oxygen atoms is 0.247 nm for AP, and they are not chemically bound with each other but only with the two nearest silicons, saturating their dangling bonds. In case of very high pressure (5.0 per cent reduction of the lattice constant), the distance between O atoms is 0.228 nm, and their chemical interaction is still weak. The defect energy level was found to be under and close to the conduction band bottom (0.754 eV for AP, 1.039 eV for HP) and unoccupied in the ground state of the defect. The width of the gap increased with pressure.

The geometry and diffusion mechanism of O_i defect were studied using cluster model with 17 Si atoms and one O atom near central Si. It is known that diffusion coefficient D depends on activation energy E_a and preexponential factor D_0 as follows: $D=D_0\exp(-E_a/kT)$. E_a values were calculated with correlation correction accounted by Møller-Plesset perturbation theory. E_a was found to equal 2.73 eV in case of AP. Under a uniform compression causing 0.25, 0.37, or 5.0 per cent reduction of the lattice constant, E_a was obtained 2.70, 2.68, and 1.92 eV, respectively, with obvious trend to decrease with pressure. This agrees qualitatively with experiments by Antonova et al. [7].

According to results by Blöchl et al. [11, 12], at high temperature (300 - 1500K) the

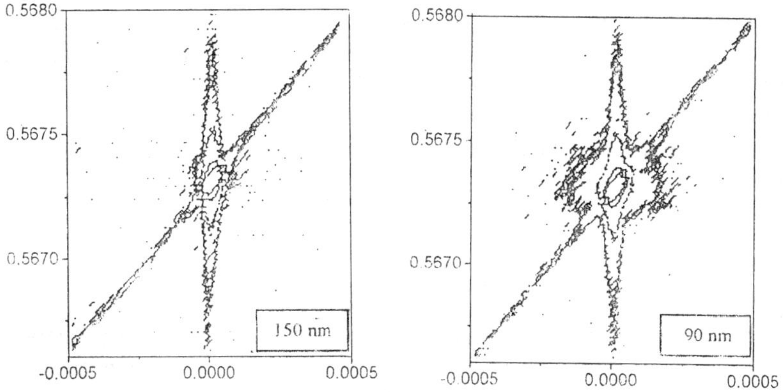

Fig. 1. Two-dimensional reciprocal space maps recorded near the 400 reciprocal lattice point for [001] oriented p-type Cz-Si samples with initial $c_O = 1.1\times10^{18}cm^{-3}$, preannealed at 720K - 10^5Pa for 20h ($c_O = 9.1\times10^{17}cm^{-3}$) and afterwards annealed additionally at 1000K - 10^5Pa (A) and at 1000K - 10^9Pa (B) for 5h. $CuK_{\alpha 1}$ radiation was used. The resulting O concentration was $c_O = 8.32\times10^{17}cm^{-3}$ (A) and $c_O = 5.56\times10^{17}cm^{-3}$ (B). The mean defect dimensions (determined from Huang scattering data) are shown in the boxes.

preexponential is insensitive and varies only by a factor of 5 [11]. Supposing this remains valid in case of OD in Si, the pressure would increase the diffusion rate due to decrease in activation energy.

The observed decrease of the precipitate dissolution rate [7, 8] and more effective small defect formation at HP could be explained by: 1) interaction of O with Si interstitials and Si_i-O_{ix} complex formation with lower diffusion rate than that of O_i [7]; 2) creation of attractive potentials (nucleation centres) [8] in regions neighbouring with O interstitials; 3) cancellation of fast diffusion ways along 1- and 2-dimensional defects such as dislocations.

Though the first mechanism looks reasonable, it seems not to be the most important, because Si_i-O_{ix} complex formation is controlled by Si_i concentration and the rate of its reaction with oxygen that is not specific for HP only.

A mechanism of nucleation centre creation in Cz-Si at HP conditions has been suggested by Misiuk [8]. The HP-induced nucleation centres could catch the migrating O_i and thus effectively reduce its migration speed. The presence of nucleation centres in Si follows from interpretation of our data on reciprocal lattice mapping and Huang scattering analysis [13] given below. We have investigated Cz-Si samples preannealed during 20h at 720K. As it is seen at Fig. 1, annealing at 1000K for 5h under 10^9Pa pressure yields significantly higher concentration of smaller size defects as compared to the normal pressure case that shows the increase in number of nucleation centres. This supports our suggestion that HP makes O_i capture by nucleation centre more effective, reducing its diffusion length. The increase of effective radius of O_i interaction with nucleation centre yields a smaller preexponential factor of the diffusion coefficient due to O capture by nucleation centre, with lower probability of diffusion outwards. A similar conclusion that HP augments the effectivity of the nucleation centre interaction with O_i

follows from X-ray topography [14]. Significantly lower O_i concentration after HP treatment at high temperature (1620K) in Si samples with initially present large oxygen precipitates also evidences the higher effectivity of O_i capture by nucleation centre. This shows that HP suppresses dissolution of oxygen-related defects and lowers the diffusivity of oxygen by suppressing the diffusion of oxygen outwards from the oxygen containing centres.

3 Conclusions

The quantum chemical models used in our work allow to compare local characteristics of the defects such as energy barrier for defect migration with changes of the crystal properties such as lattice structure distortions. Joint effect of high pressure induced enhancements of migration speed and of interaction of migrating oxygen with nucleation centres reduces the effective diffusion rate and creates more numerous but smaller OD. The models of defect structure and chemical properties allow to use the results of study of OD defects creation at HP as a source of information about properties of crystals and in elaborating methods for their investigation.

Acknowledgements

This work was supported partially by the Polish Committee for Scientific Research (grant No 8T11B 024 11), the Foundation for Polish-German Collaboration (project No 1184/LN/94), the Latvian Council of Science (project Nr. 93.0270), and the Commission of the European Communities (contract ERB CIPDCT 940008, amendment to ERB CHRX CT 930134, the European network SSASS). The high resolution X-ray MRD Philips difractometer used for reciprocal lattice mapping was bought from the funds of Polish Science Foundation (the SEZAM'94 programme).

References

1. G.P. Karwasz, A. Misiuk et al., *Appl. Phys. Lett.* **69**, 2900 (1996).
2. J. Michel, L.C.Kimerling, *Semiconductors and semimetals*, **42**, 251 (1994).
3. D.J. Chadi, *Phys. Rev.* B **41**, 10595 (1990).
4. V.V. Voronkov, *Semicond. Sci. Technol.* **8**, 2037 (1993).
5. J.J. Qian et al., *J. Appl. Phys.* **68**, 954 (1990).
6. N. Meilwes et al., *Semicond. Sci. Technol.* **9**, 1623 (1994).
7. I.V. Antonova, A. Misiuk et al., *Physica* B **225**, 251 (1996).
8. A. Misiuk, *Proc. XII Conf. Solid State Crystals* (Zakopane, 1996), in press.
9. M. Ramamoorthy and S.T. Pantelides, *Phys. Rev. Lett.* **76**, 267 (1996).
10. L.T. Bugaenko, D.P. Erts, J.R. Dzelme, K.P. Burdina, J.J. Tiliks, J.A. Kalashnikov. Method for Measurement of High Pressure. *SU Inventions* Nr. 1315838, Nr. 1399648.
11. P. Blöchl et al., *Phys. Rev. Letters* **64**, 1401 (1990).
12. P. Blöchl et al., *Phys. Rev. Letters* **70**, 2435 (1993).
13. J.R. Patel, J. Appl. Cryst. **8**, 186 (1975).
14. L. Datsenko et al., *Acta Phys. Polonica* **91**, 929 (1997).

RECIPROCAL LATTICE MAPPING OF PRESSURE - ANNEALED CZOCHRALSKI GROWN SILICON

A.MISIUK

Institute of Electron Technology, Al.Lotników 46, 02-668 Warsaw

J.BĄK - MISIUK

Institute of Physics, PAS, Al. Lotników 46, 02-668 Warsaw

J.JUN

High Pressure Research Center, PAS, Sokołowska 29, 01-142 Warsaw, Poland

An effect of annealing of Czochralski grown silicon at 720 - 1230K under enhanced (up to 1.5GPa) pressure was investigated by reciprocal lattice mapping, diffuse X-ray Huang scattering and related measurements. Increased pressure during annealing resulted in massive oxygen precipitation. Stress - induced increase (as compared to effect of annealing at 10^5 Pa) of concentration of oxygen - related defects and diminishing of their sizes were detected.

1. Introduction

Czochralski grown single crystalline silicon, Cz-Si, typically in the form of about 0.6mm thick wafers, is the basic semiconductor used in microelectronics and so carefully investigated. It contains oxygen atoms in the interstitial positions, O_i, with concentration, c_o, up to above 10^{18} cm^{-3}. Some features of Cz-Si annealed at enhanced temperatures, HT, under stress (e.g. at enhanced hydrostatic pressure, HP, of ambient) are still not well understood, especially as it concerns creation and transformation of oxygen - related defects.

An effect of HP on features (size, concentration) of oxygen - containing defects created by annealing of Cz-Si at 720 - 1020 (1230)K has been investigated in this work.

Structural properties of Cz-Si with distributed oxygen - related defects created in effect of its annealing at HP were measured by X-ray reciprocal space mapping for the Bragg case, XRSP [1] and other defect - sensitive methods. The strain fields related to presence of defects (mostly oxygen atom clusters and precipitates) were estimated from the diffusion scattering in the Huang region (Huang scattering, HS [2]).

The fundamental theory of diffuse scattering from the crystals with defects has been presented e.g. in [3]. In general it is possible to distinguish between two regions of diffuse scattering. In one of them the deviation from the Bragg position, **q**, is small (**q** is defined as **q** = **K** - **h**, where **K** is the scattering vector and **h** - the reciprocal space vector). In this region, known as the Huang scattering region, the scattering function $S(\mathbf{K}) \sim \mathbf{q}^{-2}$. The contribution to this region of scattering at small **q** in the reciprocal space comes in real space from the region distant from the defect core. Scattering at larger **q** values origins from the region placed close to the defect core and it is proportional to $\mathbf{q}^{-4}$. For the region of small **q**, where $q < R_o^{-1}$ (R_o means the mean defect radius), the symmetrical part of diffuse scattering intensity (for the minimum value of **q**, $\mathbf{q}_o$, where $|\mathbf{q}_o| = |\mathbf{h}|\Delta\theta \cos\theta_{\mathbf{B}}$), $I^s(q_o)$ can be expressed as:

$$I^s(q_o) \sim A \ln |e^{1/2}/q_o R_o| \qquad (1)$$

The $\Delta\Theta$ means deviation from the Bragg angle, Θ_B. Term A involves the size and number of defects (oxygen atom clusters) as well as the wavelength of X-ray radiation.

Thus a plot of $I^s(q_o)$ vs ln q_o should be linear and intercept the I^s axis at $I^s(q_o) = 0$ for $q_o = e^{1/2}/R_o$. So it is possible to estimate the mean cluster radius R_o from the angular dependence of diffuse scattering.

2. Experimental

Cz-Si samples of {001} orientation, with c_o up to $1.1x10^{18}cm^{-3}$ were subjected to the HP-HT treatment at 720 - 1230K at argon pressure up to 1.5GPa. Some samples were preannealed at (720-1020)K-$10^5/10^9$ Pa, applied to create „initial" oxygen - related defects, nucleation centers for oxygen precipitation at higher temperatures. At such conditions the oxygen interstitials are removed from the crystal lattice creating oxygen clusters, precipitates and other defects.

X-ray diffraction measurements were performed using Philips High Resolution Diffractometer MRD. Determination of the diffuse scattering intensity around the 004 reciprocal lattice point was carried out using $CuK\alpha_1$ radiation. Size of defects was calculated using eq. (1). The symmetrical part of diffuse scattering intensity was estimated from relation:

$$I^s = [I(+q_o) + I(-q_o)]/2 \qquad (2)$$

The c_o value was determined by Fourier Transform Infrared Spectroscopy, FTIR [4]. The kind and density of created oxygen - related defects were determined also by ellipsometry [5], X - ray topography and selective chemical etching. The thermal donors, TDs, concentration (TDs creation is related to oxygen clustering) was determined from electrical resistivity and CV measurements.

3. Results and Discussion

Conditions of sample preparation and some sample features (for the p - type Cz-Si samples with initial $c_o \approx 1x10^{18}cm^{-3}$) are listed in Table 1.

During preannealing at T≤ 1020K, performed sometimes also at enhanced pressure (samples 9 and 11), small oxygen - related defects were created (some with thermal donor activity), serving as the nucleation centers for oxygen precipitation at higher temperatures. In the case of annealing at 720 - 1000K, pronounced HP - induced creation of thermal donors was observed [6]. For example, the no. 3 sample indicated concentration of holes, $Np = 1.48x10^{15}cm^{-3}$, much lower than that ($1.93x10^{15}cm^{-3}$) for the same sample annealed at the same conditions but at 10^5Pa. Such diminishing of Np is related to effect of hole compensation by HP - stimulated creation of TDs, representing in fact small oxygen atom clusters.

In the case of samples no. 1 - 7 an effect of HP applied at comparatively low temperatures, HT ≤ 1000K, on the samples with previously created nucleation centers [7] was investigated. The defect structure of the samples preannealed at ≤ 1020K - 10^5Pa (typical conditions for creation of the mentioned nucleation centers) was determined.

In the case of samples 8 - 11 an effect of ambient pressure - annealing at 1230K on creation of oxygen - related defects in the samples with "HP - modified" nucleation centers was investigated. The massive oxygen precipitation occurs at such conditions on the before - created nucleation centers.

Intensity of diffuse scattering of X-rays was much enhanced in the case of HT - HP

- treated Cz-Si samples. Typical reciprocal space maps of the investigated samples are presented in Fig. 1. Sample designations in this figure correspond to that in Table 1.

Similar XRSP investigations were performed on other Cz-Si samples subjected to HP - treatments at temperatures up to 1230K and the similar results were obtained.

Table 1: Preannealing / treatment conditions (pressure, temperature, time) and values of c_o of R_o (as determined from HS measurements and ellipsometry [5], the last values with asterisk*) and of concentration of defects, N_d (determined by ellipsometry and by chemical selective etching, the last values with two asterisks concern Cz-Si samples with initial $c_o = 8x10^{17}cm^{-3}$) in processed Cz-Si samples.

Sample	Preannealing Pa, K, hrs	Treatment Pa, K, hrs	c_ox10^{17}, cm^{-3}	R_o, nm	N_dx10^6, cm^{-3}
1 (A).	10^5/720/10	3x10^6/870/10	9.5	115/85*	1.5
2 (B).	10^5/720/10	10^9/870/10	8.1	60/15*	2.6
3.	-	10^9/1000/5	6.9	75/35*	2.9
4.	10^5/720/20	10^5/1000/5	8.3	150	0.27**
5.	10^5/720/20	10^9/1000/5	5.6	90	1.41**
6.	10^5/1020/8	10^5/1000/5	7.6	95/25*	2.7
7.	10^5/1020/8	10^9/1000/5	8.1	100/15*	2.5
8 (C).	10^5/1000/5	10^5/1230/5	3.1	140	2.3**
9 (D).	10^9/1000/5	10^5/1230/5	2.7	85	2.8**
10.	10^5/1020/8 +10^5/1000/5	10^5/1230/5	7.1	190/28*	2.1
11.	10^5/1020/8 +10^9/1000/5	10^5/1230/5	6.3	62/13*	2.1

X-ray topographs of the samples HP - treated at 720 - 1020K did not exhibit pronounced dependence on HP. Contrary to that, such dependence on HP was distinct as it concerned defect sizes (estimated from HS and other measurements) and their concentrations, the first diminishing and the second - increasing with HP. It must be mentioned that defect size estimation from the HS data gives rather too high values in comparison to that obtained by other methods. It is, however, not surprising since the diffuse scattering measurements are much more sensitive to the strain fields that surround an oxygen precipitate, rather than to the precipitate dimension [8].

The following effects were most characteristic for the HP - treated at 720 - 1020K Cz-Si samples:

- massive HP - induced oxygen precipitation [4] (Table 1),

- HP - stimulated generation of small defects (also detected by XRSP, Fig.1), and

- rise of defects concentration (especially as it concerns the TD activity) with diminishing of their dimensions (Table 1).

The observed phenomena can be attributed to the HP - stimulated oxygen precipitate nucleation with subsequent enhanced creation of oxygen - related defects causing HP - induced disturbance of surrounding silicon matrix [7].

Our investigations also confirmed usefulness of XRSP for revealing structural characteristics of Cz-Si with small oxygen - related defects. Between others, we have demostrated a possibility to estimate the size and strain fields of small defects in pressure- annealed Cz - Si from X-ray diffuse scattering data.

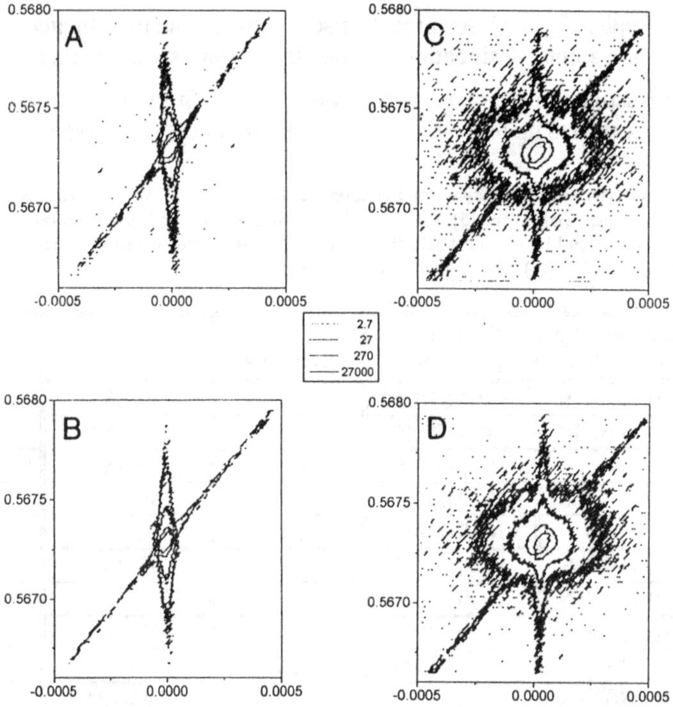

Fig. 1: Two - dimensional reciprocal space maps recorded near 400 reciprocal lattice point for (001) oriented A, B, C and D samples (see Table 1). Reciprocal space coordinates are given in $\lambda/2d$ units. The horizontal axis is perpendicular and the vertical one - parallel to the diffraction vector.

Acknowledgements

This work was supported partially by the Polish Commmittee for Scientific Research (grant no. 8T11B 024 11) and the Foundation for Polish - German Collaboration (project no. 1184/LN/94). The high resolution X-ray MRD Philips diffractometer was bought from the funds of Polish Science Foundation (the SEZAM'94 programme).

The authors are grateful to M.sc. A.Kamińska, M.sc. B.Surma and Mr M.Rozental for performing ellipsometrical and FTIR measurement as well as for sample preparation.

References

1. G.Bauer, J.H.Li. V.Holy, *Acta Phys.Polon.* **A 89,** 115 (1996).
2. J.R.Patel, *J.Appl.Cryst.* **8**, 186 (1975).
3. B.C.Larson, *ibid,* **8**, 150 (1975).
4. A.Misiuk, B.Surma, J.Hartwig, *Mater. Sci. Eng.* **B36**, 30 (1996).
5. A.M.Kamińska, *Electron Techn.* **26**, 109 (1993).
6. A.Misiuk, W.Jung, *phys.stat.sol. (b)* **198,** 565 (1996).
7. A.Misiuk, *Proceed. XII Conf. Solid State Crystals, Zakopane 1996,* in press.
8. R. Matyi, M..Melloch, K.Zhang, D.Miller, *Phys.D: Appl.Phys.***28,** A139 (1995).

TEM INVESTIGATIONS OF LATTICE DISLOCATIONS INTERACTION WITH INTERPHASE BOUNDARIES

T. AKSMAN, R. OBRUBKO, W. ZIELIŃSKI, W.A. ŚWIĄTNICKI
Faculty of Materials Science and Engineering, Warsaw University of Technology
Narbutta 85, 02-524 Warsaw, Poland

Interaction between the lattice dislocations and the interphase boundaries [IBs] during tensile deformation of austeintic-ferritic steel has been investigated by transmission electron microscopy [TEM]. *In situ* TEM deformation experiment delivered details of the mechanism of the slip transfer through IBs. It has been found that the character of slip transmission through IBs depends on the grain orientation relationship. The crystallography of the slip systems in austenite and ferrite was examined by the slip traces method and confirmed by a computer simulation. Results are discussed in terms of the relation between the crystallography of IBs and the mechanical properties of the material.

1 Introduction

It has already been known that stresses induced during plastic deformation as a result of the elastic and plastic incompatibilities between neighbouring grains, may be enhanced or relaxed by the interfaces [1]. In the case of grain boundaries the stresses may be accommodated by dislocation slip transmission [2], dislocation generation from a boundary [1,3] and the accommodation of dislocations into the boundary structure [4,5]. The mechanisms and structural conditions of these effects are well known [1,2,4,5]. There is however a lack of knowledge regarding the stress accommodation at the IBs and its influence on the mechanical properties of multiphase material.

The main goal of this work is to find the correlation between the mechanism of slip transmission from one phase to another and the crystallography of IBs and the geometry of the operating slip systems.

2 Experiment

The material used in the study was austenitic-ferritic steel of the composition: 0.03%C, 0.24%Mn, 0.21%Si, 0.005%P, 0.014%S, 26.5%Cr, 6.0%Ni and Fe balance. Annealing at 1260^0 C for 30 min in vacuum and then oil quenching produced a ferritic microstructure with a mean grain size of about 1 mm. In order to obtain an austenitic-ferritic microstructure in two different states (A and B), two treatments were applied. State A - annealing at 960^0 C for 1 h and oil quench resulted in the austenite phase being precipitated in the form of elongated grains of 7 μm long and 0.5 μm wide in the ferritic matrix. State B - cold rolling with a 50% reduction, then annealing at 960^0 C for 1h and oil quench gave fully recrystallized duplex microstructure of equiaxial austenite and ferrite grains with a mean size of 2 μm.

Observations of the lattice dislocations reaction with IBs were performed by *in situ* and *ex post* TEM. The *in situ* TEM experiment was conducted on standard 3 mm diameter A and B state samples. The *ex post* TEM observations were made on samples deformed up to $\varepsilon = 0.5\%$ prior to the observations.

The crystallography of IBs and of the slip systems as well as the dislocation character were determined from the Kikuchi lines pattern, the **g . b** = 0 vanishing condition and a computer simulation of the dislocation contrast using the commercial TWODIS software [7].

3 Results and discussion

The crystallographic analysis of the IBs in state A has shown that most of the austenitic grains possess the Kurdiumov-Sachs (K-S) or Nishiyama-Wassermann (N-W) special orientation relationship (OR) with respect to the ferritic matrix. In these orientations one of the easy slip planes {111} in austenite is parallel to one of the easy slip planes{110} in ferrite. In the state B, the ORs were deviated from the exact K-S and N-W relations. This means, that the main difference between A and B states is due to the character of the IBs orientation relationship and may be considered to be special in the case of state A and random in the case of state B.

Figure 1 shows the transmission of dislocations through an austenite/ferrite IB observed *ex post* in the sample strained to 0.5%. Dislocations, generated by a source located at the opposite IB, went through a low angle grain boundary in austenite and piled-up against IB. The stress induced by the dislocation pile-up was partially accommodated by generation of several dislocations to the adjacent grain of ferrite.

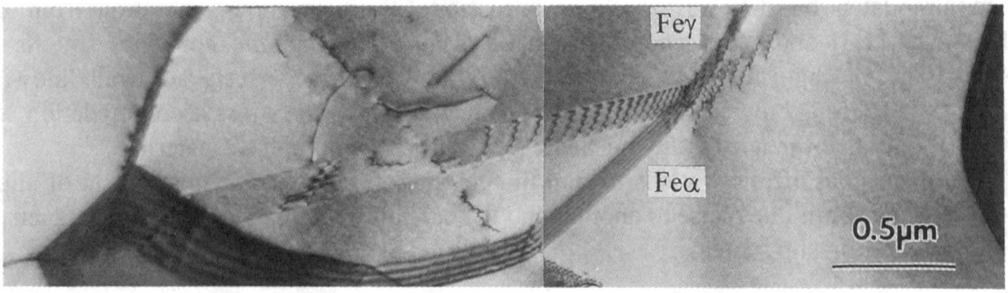

Figure 1. Dislocation slip transmission through an austenite/ferrite IB.

An analysis of the electron diffraction patterns indicates that the orientation relationship of austenite and ferrite is deviated by 30^0 from the Kurdiumow-Sachs orientation. Based on the two satisfied vanishing conditions for each set of dislocations, it was found that the Burgers vector of dislocations in austenite is $1/2[0\bar{1}1]$ whereas in ferrite it is $\frac{1}{2}[\bar{1}\bar{1}\bar{1}]$. In order to determine the slip planes that operates in austenite and ferrite, we used a stereographic projection, which allowed us to find the slip plane indices: $(\bar{1}\bar{1}\bar{1})$ in austenite and $(\bar{1}\bar{3}2)$ in ferrite. The angle between the Burgers vectors of dislocations in austenite and ferrite was estimated at 33^0, whereas the angle between the slip planes - at 8^0.

In order to confirm the indices of the Burgers vector, some theoretical computations of the dislocation diffraction contrast were performed. The TEM image of dislocation near the IB in austenite is shown in Figure 2a. Using several estimated crystallographic parameters such as the dislocation line, foil normal, beam direction and diffraction vector it was found that, with the Burgers vector $1/2[\bar{1}\bar{1}1]$ the simulated image was most similar to the real one, see Figure 2b. To confirm the choice of the Burgers vector, the computations were repeated for the same dislocation at the orientation that satisfied the vanishing condition. The results are shown in Figure 3.

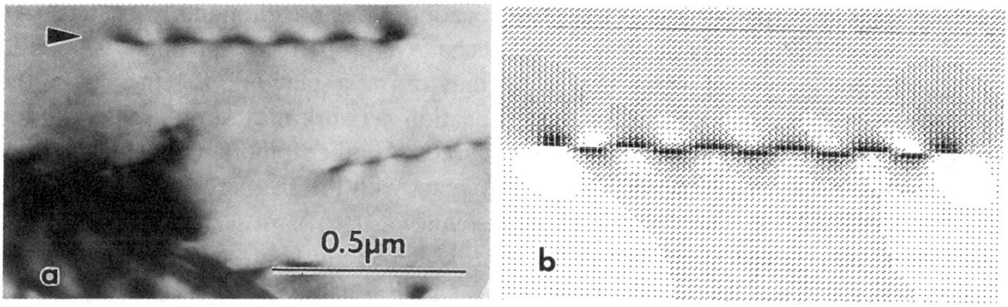

Figure 2. Analysis of the Burgers vector. a) TEM image of a dislocation. b) Simulated image.

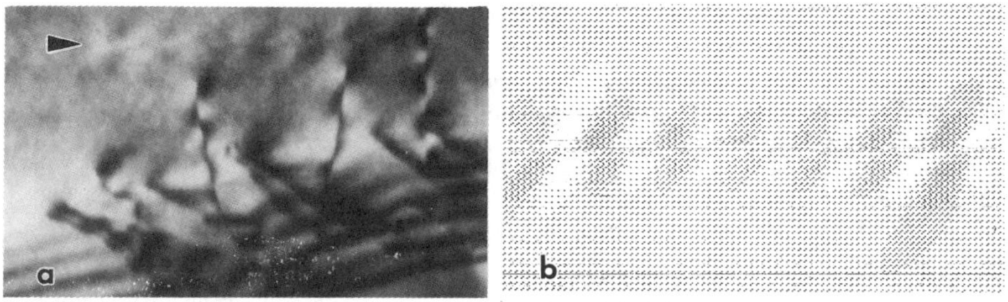

Figure 3. Analysis of the Burgers vector. a) TEM image of a vanish dislocation. b) Simulated image.

An *in situ* TEM straining experiment revealed that the mechanism of the dislocation slip transfer through the IB in the state A differs from that in state B. In both the cases, in the region where in austenite the dislocation pile-up against IB occurs, dislocation activity in ferrite was observed.

Figure 4 shows two stages of the dislocation slip transfer through an IB observed in the state A, where IBs with the K-S and N-W orientation relationships are present. The slip lines left by the moving dislocations in ferrite are quite straight and they follow the direction of the dislocation pile-up in austenite.

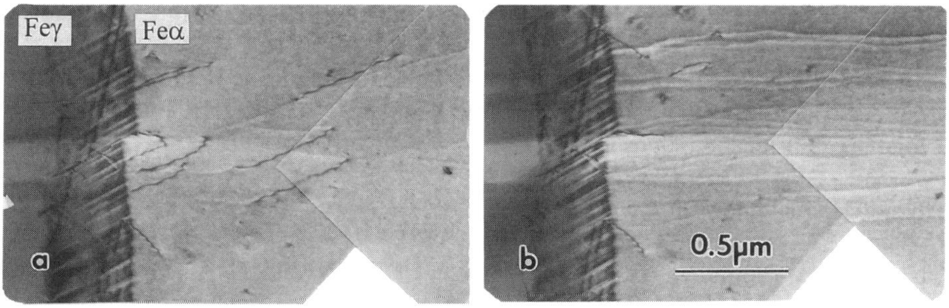

Figure 4. Two stages of the dislocation slip transmission through an IB with special orientation. a) ε_1, b) $\varepsilon_1 + \varepsilon_2$

Figure 5 shows two stages of the dislocation slip transfer through an IB observed in the state B, where randomly oriented IBs are present. The slip lines in ferrite do not follow the dislocation pile-up direction. Moreover, many dislocation loops were found to appear near the boundary in ferrite. The dislocation loops took active part in the slip transfer process. The loops were stable up to a size of 20 nm. Under the stress applied they expanded very fast. When tracing the slip lines, we could see, that mostly one part of the dislocation loop moved into the grain interior whereas the other - into the IB.

During the slip transfer, the process of dislocation loops creation and evolution may be significant from the point of view of the mechanical properties of the materials. First, the dislocations produced from the loops may form a network together with the dislocations emitted from the IB. Such a dislocation network may strongly hamper the further emission of dislocations from the IB sources. Another important meaning is, that the dislocations from the expanded loops may react with the IB thereby decreasing the stress induced by the dislocation pile-up in austenite.

This phenomenon was also observed during the tensile test. It was revealed that the yield stress for state A (with special IBs) was smaller than the flow stress for state B (with random IBs) [6]. The differences in the mechanical behaviour of steel in the two states examined can be attributed to the IBs crystallographic orientation and the slip transfer mechanism.

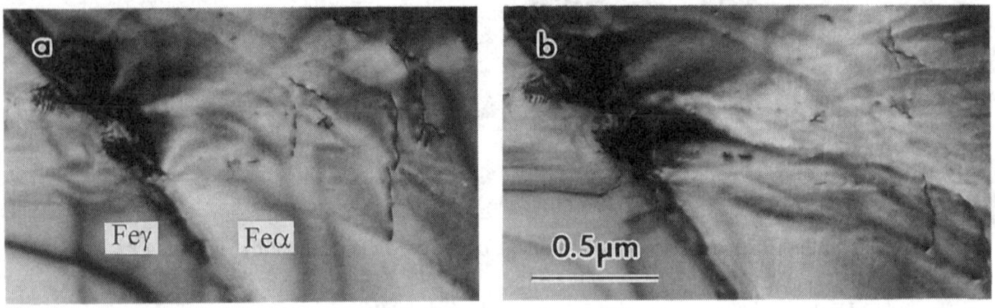

Figure 5. Two stages of dislocation slip transmission through an IB with random orientation.

a) ε_1, b) $\varepsilon_1 + \varepsilon_2$

4 Conclusions

1. The microscopic mechanism of stress relaxation by interphase boundary depends on the orientation relationship between phases
2. In the case of the special orientation relationship, the plastic deformation transfer from austenite to ferrite requires a relatively small stress to occur; this is due to the compatibility of the slip systems.
3. In the case of the random orientation relationship the plastic deformation transfer from austenite to ferrite requires a relatively high stress because of the formation and evolution of dislocation loops in ferrite.

Acknowledgements

This work was supported by European Contract COPERNICUS No CIPA CT 93 0112.

References

1. M. W.Grabski, *J.Physique* **46**, C4-567 (1985).
2. T.C. Lee, I.M. Robertson, H.K. Birnbaum, *Phil. Mag. A* **62**, 131 (1990).
3. K.J. Kurzydłowski, R.A. Varin, W. Zieliński, *Acta Metall.* **32**, 71 (1984).
4. M.W. Grabski, *J.Physique* **49**, C5-497 (1988).
5. L. Priester, *Interface Science* **4**, 205 (1997).
6. W. Swiatnicki, M.W. Grabski, *Materials Science forum,* **207-209**, 769 (1996).
7. A.K. Head et al., *Computed Electron Micrographs and Defects Identification,* ed. North-Holland Publishing Company (1973).